鸿蒙HarmonyOS应用开发从入门到精通

第2版

柳伟卫◎著

北京大学出版社
PEKING UNIVERSITY PRESS

内 容 简 介

华为自主研发的HarmonyOS（鸿蒙系统）是一款面向未来、面向全场景（移动办公、运动健康、社交通信、媒体娱乐等）的分布式操作系统。本书采用HarmonyOS最新版本作为基石，详细介绍如何基于HarmonyOS进行应用的开发，包括HarmonyOS架构、DevEco Studio、应用结构、Ability、安全管理、公共事件、通知、ArkTS、ArkUI、Stage模型、设备管理、数据管理、线程管理、视频、图像、网络管理等多个主题。本书辅以大量的实战案例，图文并茂，使读者易于理解和掌握。同时，本书的案例选型偏重于解决实际问题，具有很强的前瞻性、应用性和趣味性。加入HarmonyOS生态，让我们一起构建万物互联的新时代！

本书主要面向的是对HarmonyOS应用开发感兴趣的学生、开发人员、架构师。

图书在版编目(CIP)数据

鸿蒙HarmonyOS应用开发从入门到精通 / 柳伟卫著. 2版. -- 北京：北京大学出版社，2025. 1. -- ISBN 978-7-301-35546-6

Ⅰ. TN929.53

中国国家版本馆CIP数据核字第202492C1H9号

书　　名	鸿蒙HarmonyOS应用开发从入门到精通（第2版） HONGMENG HarmonyOS YINGYONG KAIFA CONG RUMEN DAO JINGTONG（DI-ER BAN）
著作责任者	柳伟卫　著
责任编辑	王继伟　吴秀川
标准书号	ISBN 978-7-301-35546-6
出版发行	北京大学出版社
地　　址	北京市海淀区成府路205号　100871
网　　址	http://www.pup.cn　　新浪微博：@ 北京大学出版社
电子邮箱	编辑部 pup7@pup.cn　总编室 zpup@pup.cn
电　　话	邮购部 010-62752015　发行部 010-62750672　编辑部 010-62570390
印 刷 者	北京溢漾印刷有限公司
经 销 者	新华书店
	787毫米×1092毫米　16开本　38印张　915千字
	2025年1月第1版　2025年1月第1次印刷
印　　数	1-4000册
定　　价	129.00元

再版序

INTRODUCTION

时光荏苒，岁月匆匆，距离《鸿蒙HarmonyOS应用开发从入门到精通》2022年4月第1版面世已逾两载。热心的读者对于本书也投以极大的关注，伴随着本书的成长，提出了很多中肯的建议。对于这些意见，不管褒贬，一并全收，于是才有了第2版的可能。

对于技术型的书籍创作，笔者更加倾向于采用当今软件开发主流的方式——敏捷。敏捷写作打通了编写、校稿、出版、发行的整个流程，让知识可以第一时间呈现给读者。读者在阅读本书之后，也可以及时对书中的内容进行反馈，从而帮助作者完善书中内容，最终形成良好的反馈闭环。所以，第2版所更新的内容，应该正是读者所期待的。

近些年HarmonyOS版本迭代较快，发展迅猛，特别是HarmonyOS 3版本引入了ArkTS语言，产生了新的编程模式。因此，本书第2版修改篇幅较大，各章节都做了大幅度更新。完整的修改内容，可以参阅本书附录部分。

柳伟卫

本书献给我的父母，

愿他们健康长寿！

前言

INTRODUCTION

写作背景

中国信息产业一直是“缺芯少魂”，其中的“芯”指的是芯片，而“魂”则是指操作系统。自2019年5月16日起，美国陆续把包括华为在内的中国高科技企业列入其所谓的“实体清单”，标志着科技再次成为中美博弈的核心领域。

随着谷歌暂停与华为的部分合作，包括软件和技术服务的转让，华为在国外市场面临升级Android版本、搭载谷歌服务等方面的困境。在这种背景下，华为顺势推出HarmonyOS，以求在操作系统领域不受制于人。

HarmonyOS是一款全新的面向未来、面向全场景（移动办公、运动健康、社交通信、媒体娱乐等）的分布式操作系统。作为操作系统领域的新成员，HarmonyOS势必会面临Bug多、学习资源缺乏等众多困难。为此，笔者在开源社区以开源方式推出了免费系列学习教程《跟老卫学HarmonyOS开发》，以帮助HarmonyOS爱好者入门。同时，为了让更多的人了解并使用HarmonyOS，笔者将自身工作、学习中遇到的问题、难题进行了总结，形成了本书，以填补市场空白。

内容介绍

全书大致分为以下三部分。

- 入门（1～4章）：介绍HarmonyOS的背景、开发环境搭建，并创建一个简单的HarmonyOS应用。
- 进阶（5～15章）：介绍HarmonyOS核心功能的开发，内容包括Ability、安全管理、公共事件与通知、ArkTS、ArkUI、Stage模型、设备管理、数据管理、线程管理、视频、图像、网络管理等。
- 实战（16～19章）：演示 HarmonyOS 在各类场景下的综合实战案例，包括智能穿戴、智慧屏和手机应用。

本书采用的技术及相关版本

技术的版本非常重要，因为不同版本之间存在兼容性问题，而且不同版本的软件对应的功能也不同。本书列出的技术在版本上相对较新，都经过了笔者的大量测试。因此，读者在自行编写代码时可以参考本书列出的版本，从而避免因版本兼容性产生的问题。建议读者将相关开发环境设置得与本书一致，或者不低于本书所列配置。详细的技术和版本配置参考如下。

- 操作系统：Windows10 64位。
- 内存：8GB及以上。
- 硬盘：100GB及以上。
- 分辨率：1280×800像素及以上。
- DevEco Studio 3.1.1 Release（Buid Version: 3.1.0.501）。
- HarmonyOS SDK 3.1.0（API 9）。
- OpenHarmony SDK 3.2.13.5（API 9）。

勘误和交流

本书如有勘误，会在以下网址发布：https://github.com/waylau/harmonyos-tutorial/issues。

由于笔者能力有限、时间仓促，书中难免有疏漏之处，欢迎读者批评指正。读者可以通过以下方式与笔者联系。

- 博客：https://waylau.com。
- 邮箱：waylau521@gmail.com。
- 微博：http://weibo.com/waylau521。
- GitHub：https://github.com/waylau。

学习资源

本书提供的素材和源代码可从以下网址下载：https://github.com/waylau/harmonyos-tutorial。

读者也可以扫描右侧二维码关注“博雅读书社”微信公众号，输入本书77页的资源下载码，即可获得本书的配套学习资源。

致谢

感谢北京大学出版社的各位工作人员为本书的出版所做的努力。

感谢家人对我的理解和支持。

感谢关心和支持我的朋友、读者、网友。

特别感谢华为技术有限公司的李毅、欧建深、谢炎等技术专家对于本书内容方面的指导。

柳伟卫

目录
CONTENTS

第3章 牛刀小试——开发第一个HarmonyOS应用

第4章 应用初探——探索HarmonyOS应用

第5章 Ability基础知识

第6章 安全管理

第7章 Ability公共事件与通知

第8章 用ArkUI开发UI

第12章 线程管理

第13章 视频

第14章 图像

第15章 网络管理

第1章

鸿蒙缘起——HarmonyOS简介

鸿蒙是中国神话传说中的远古时代，传说在开天辟地之前，世界是一团混沌的元气，这种自然的元气叫作鸿蒙。本章介绍 HarmonyOS（鸿蒙系统）产生的历史背景、特点及技术架构。

1.1 HarmonyOS产生的背景

2024年4月17日，第21届华为分析师大会在深圳开幕。华为副董事长、轮值董事长徐直军在大会上透露，华为希望通过2024年一年的时间，先在中国市场把智能手机上使用超过99%时间的5000个应用全面迁移到鸿蒙原生操作系统上，真正实现操作系统和应用生态的统一。“当把这5000个应用以及其他成千上万的应用都从安卓生态迁移到鸿蒙操作系统上时，我们的鸿蒙操作系统就真正完成了打造，并真正成为除了苹果iOS和谷歌安卓外的全球第三个移动操作系统。”

那么，为什么需要HarmonyOS？到底什么是HarmonyOS？

1.1.1 为什么需要HarmonyOS

2019年5月16日，美国商务部宣布，把华为及其70家关联企业列入其所谓的“实体清单”（Entity List）。这意味着，此后如果没有美国政府的批准，华为将无法向美国企业购买元器件和技术。“实体清单”是美国为维护其所谓的“国家安全利益”而设立的出口管制条例。在未得到许可证前，美国各出口商不得帮助这些名单上的企业获取受本条例管辖的任何物项。简单地说，“实体清单”就是一份“黑名单”，相关企业一旦进入此榜单，实际上是被剥夺了在美国的贸易机会。

随着中国国力的增强，美国的“实体清单”不断扩容，体现了美国对中国高科技企业的限制升级，科技再次成为中美博弈的核心领域。

作为中国科技领域的头部企业，华为首当其冲。华为虽然早就建立了自己的芯片企业——海思，但海思生产的芯片还不能完全覆盖自己的产品线，华为依然需要直接采购美国芯片厂商的产品。受到“实体清单”的影响，美国全面封锁华为在全球的芯片采购，直接导致了华为忍痛出售旗下手机品牌——荣耀。

不光是芯片等硬件产品，在“实体清单”的限制下，软件等技术同样受到限制。谷歌已暂停与华为的部分合作，包括软件和技术服务的转让。华为在国外市场面临着升级Android版本、搭载谷歌服务等方面的困境。华为智能手机业务在海外市场占有率明显下降。

早在1999年，中国科技部原部长徐冠华曾说，“中国信息产业缺芯少魂”。其中的“芯”指的是芯片，而“魂”则是指操作系统。当时，中国曾大力扶持国产芯片和操作系统，也曾诞生过一些亮眼的产品，比如红旗Linux、龙芯等。然而，20多年过去了，中国信息产业依然是“缺芯少魂”，这次美国对华为的封杀，第一个禁的是芯片，第二个禁的就是操作系统。

为了避免被人“卡脖子”，华为展开了自救和反击。2019年5月17日凌晨2点，华为海思总裁何庭波发表致员工的一封信，信中称：“公司做出了极限生存的假设，预计有一天，所有美国的先进芯片和技术将不可获得……为了这个以为永远不会发生的假设，数千海思儿女，走上了科技史上最为悲壮的长征，为公司的生存打造‘备胎’。”“今天，命运的年轮转到这个极限而黑暗的时刻，超级大国毫不留情地中断全球合作的技术与产业体系，做出了最疯狂的决定，在毫无依据的条件下，把华为公司放入了实体名单。”“今后的路，不会再有另一个十年来打造备胎然后再换胎了，缓冲区已经消失，每一个新产品一出生，将必须同步‘科技自立’的方案。”

因此，在这个背景下，除了加大海思的研发投入之外，华为推出了自己的操作系统——HarmonyOS。正如其中文“鸿蒙”的寓意，这意味着HarmonyOS将会开启一个开天辟地的时代。

1.1.2 HarmonyOS概述

HarmonyOS在2019年8月9日华为开发者大会上首次公开亮相，余承东进行主题演讲并正式公开了HarmonyOS。

HarmonyOS也称为鸿蒙系统，或者鸿蒙OS，是一个全新的面向全场景的分布式操作系统。HarmonyOS以人为中心，将人、设备、场景有机地联系在一起，尤其是面向IoT（Internet of Things，物联网）领域，将多种智能设备的体验进行系统级融合，使人、设备、场景不再是孤立的存在，为用户适应不同场景带来最佳体验。

HarmonyOS是一款面向未来、面向全场景（移动办公、运动健康、社交通信、媒体娱乐等）的分布式操作系统。在传统的单设备系统能力的基础上，HarmonyOS提出了基于同一套系统能力、适配多种终端形态的分布式理念，能够支持手机、平板、智能穿戴（Wearable）、智慧屏（TV）、车机（Car）等多种终端设备。

对消费者而言，HarmonyOS用一个统一的软件系统，从根本上解决了消费者使用大量终端体验割裂的问题。HarmonyOS能够将生活场景中的各类终端进行能力整合，可以实现不同的终端设备之间的快速连接、能力互助、资源共享，匹配合适的设备，为消费者提供统一、便利、安全、智慧化的全场景体验。

对应用开发者而言，HarmonyOS采用了多种分布式技术，整合各种终端硬件能力，形成一个虚拟的“超级终端”。开发者可以基于“超级终端”进行应用开发，使应用程序的开发实现与不同终端设备的形态差异无关。这能够让开发者聚焦上层业务逻辑，而无须关注硬件差异，更加便捷、高效地开发应用。

对设备开发者而言，HarmonyOS采用了组件化的设计方案，可以按需调用“超级终端”能力，带来“超级终端”的创新体验。HarmonyOS根据设备的资源能力和业务特征进行灵活裁剪，满足不同形态的终端设备对于操作系统的要求。

举例来说，当用户走进厨房，用HarmonyOS手机一接触微波炉，就能实现设备极速联网；用HarmonyOS手机接触一下豆浆机，立刻就能实现无屏变有屏。

HarmonyOS能够把手机的内核级安全能力扩展到其他终端，进而提升全场景设备的安全性，通过设备能力互助，共同抵御攻击，保障智能家居网络安全；HarmonyOS通过定义数据和设备的安全级别，对数据和设备进行分类分级保护，确保数据流通安全可信。

1.1.3 OpenHarmony、HarmonyOS、HarmonyOS NEXT与鸿蒙的联系与区别

1. OpenHarmony

OpenHarmony是由开放原子开源基金会（OpenAtom Foundation）孵化及运营的开源项目，目标是面向全场景、全连接、全智能时代，基于开源的方式，搭建一个智能终端设备操作系统的框架和平台，促进万物互联产业的繁荣发展。开放原子开源基金会由华为、阿里、腾讯、百度、浪潮、招商银行、360等企业共同发起组建。

OpenHarmony暂时还没有中文名字，业界一般俗称为“开源鸿蒙”。

OpenHarmony开源项目主要包括两部分：一是华为捐献的“鸿蒙操作系统”的基础能力；二是其他参与者的贡献。因此，OpenHarmony是“鸿蒙操作系统”的核心底座。

2. HarmonyOS

HarmonyOS就是“鸿蒙操作系统”，或者简称为“鸿蒙OS”，是基于OpenHarmony、AOSP等开源项目的商用版本。

这里需要注意以下两点。

- 一是，HarmonyOS不是开源项目，而是商用版本。
- 二是，HarmonyOS手机和平板之所以能运行Android应用，是因为HarmonyOS实现了现有Android生态应用（AOSP）的运行。

3. HarmonyOS NEXT

2023年8月4日，在华为开发者大会上，华为发布HarmonyOS NEXT开发者预览版本。据介绍，HarmonyOS NEXT系统底座全线自研，去掉了传统的AOSP代码，不再兼容安卓开源应用，仅支持鸿蒙内核和系统的应用，因此也被称为“纯血鸿蒙”。

HarmonyOS NEXT可以理解为HarmonyOS面向未来的、自研程度更高的下一代鸿蒙系统。

4. 鸿蒙生态

鸿蒙生态包括OpenHarmony、HarmonyOS和HarmonyOS NEXT，当然还包括开发工具以及周边的一些开发库。当我们在说“鸿蒙”的时候，也许就是指鸿蒙生态。

1.1.4 HarmonyOS应用开发

为了进一步扩大HarmonyOS的生态圈，面对广大的硬件设备厂商，HarmonyOS通过SDK（Software Development Kit，软件开发工具包）、源代码、开发板/模组和HUAWEI DevEco Studio等装备共同构成了完备的开发平台与工具链，让HarmonyOS设备开发易如反掌。

应用创新是一款操作系统发展的关键，应用开发体验更是如此。在一个完整的应用开发生态中，应用框架、编译器、IDE（Integrated Development Environment，集成开发环境）、API（Application

Programming Interface，应用程序接口）、SDK都是必不可少的。为了赋能开发者，HarmonyOS提供了一系列构建全场景应用的完整平台工具链与生态体系，让应用能力可分可合可流转，轻松构筑全场景创新体验。

本书就是介绍如何针对HarmonyOS进行应用的开发。可以预见的是，HarmonyOS必将是近些年的热门话题。对于能在早期投身于HarmonyOS开发的技术人员而言，其意义不亚于当年Android的开发，HarmonyOS必将带给开发者广阔的前景。同时，基于HarmonyOS提供的完善的平台工具链与生态体系，笔者相信广大读者一定也能轻松入门HarmonyOS。

5G网络准备就绪，物联网产业链也已经渐趋成熟，在物联网爆发前夜，正亟须一套专为物联网准备的操作系统，而华为的HarmonyOS正逢其时。Windows成就了微软，Android成就了谷歌，HarmonyOS是否能成就华为，让我们拭目以待。

1.2 技术理念

在万物智联时代的重要机遇期，HarmonyOS结合移动生态发展的趋势，提出了三大核心技术理念（如图1-1所示）：一次开发，多端部署；可分可合，自由流转；统一生态，原生智能。

图1-1　三大核心技术理念

1.2.1　一次开发，多端部署

“一次开发，多端部署”指的是一个工程，一次开发上架，多端按需部署。目的是支撑开发者高效地开发多种终端设备上的应用。为了实现这一目的，HarmonyOS提供了几个核心能力，包括多端开发环境、多端开发能力及多端分发机制（如图1-2所示）。

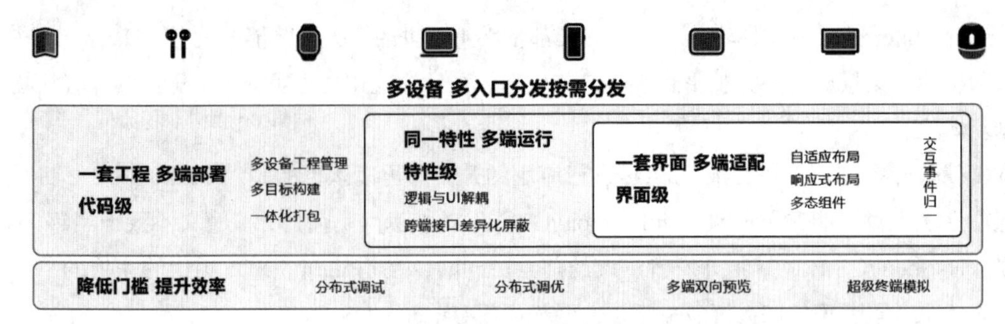

图1-2 “一次开发，多端部署”示意图

1. 多端开发环境

HUAWEI DevEco Studio是面向全场景多设备构建的一站式开发平台，支持多端双向预览、分布式调优、分布式调试、超级终端模拟、低代码可视化开发等能力，帮助开发者降低成本、提升效率、提高质量。

HUAWEI DevEco Studio提供的核心能力如图1-3所示。

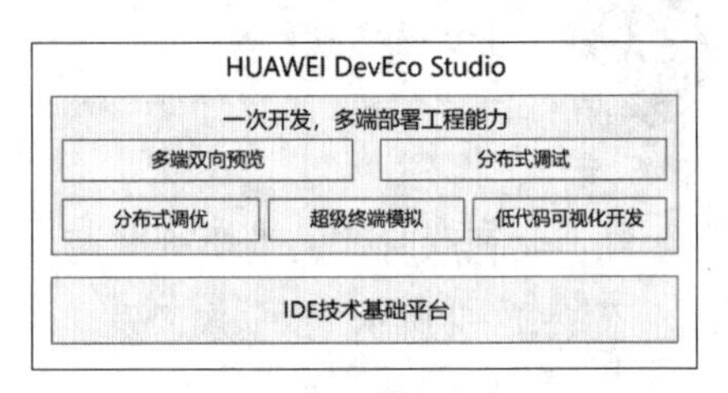

图1-3 HUAWEI DevEco Studio提供的核心能力

- 多端双向预览：在HarmonyOS应用的开发阶段，因不同设备的屏幕分辨率、形状、大小等差异，开发者需要在不同设备上查看UI界面显示，确保实现效果与设计目标一致。在传统的开发模式下，开发者需要获取大量不同的真机设备用于测试验证。HUAWEI DevEco Studio提供了多种设备的双向预览能力，支持同时查看UI代码在多个设备上的预览效果，并支持UI代码和预览效果的双向定位修改。
- 分布式调试：HarmonyOS应用具有天然的分布式特征，体现在同一个应用在多个设备之间会有大量的交互。在开发过程中，对这些交互进行调试时，需要对每个设备分别建立调试会话，并且需要在多个设备之间来回切换，容易造成调试不连续、操作烦琐等问题。为了提升开发效率，HUAWEI DevEco Studio提供了分布式调试功能，支持跨设备调试，通过代码断点和调试堆栈可以方便地跟踪不同设备之间的交互，用于定位多设备互动场景下的代码缺陷。
- 分布式调优：分布式应用的运行性能至关重要。在跨端迁移场景中，需要应用在目标设备上快速启动，以实现和原设备之间的无缝衔接；在多端协同场景中，需要应用在算力和资源不同的多个设备上都能高效运行，以获得整体的流畅体验。以往开发者在分析分布式应用的性能问题时，需要单独查看每个设备的性能数据，并手动关联分析这些数据，操作烦琐，复杂度高。HUAWEI DevEco Studio提供了分布式调优功能，支持多设备分布式调用链跟踪、跨设备调用堆栈缝合，同时采集多设备性能数据并进行联合分析。
- 超级终端模拟：移动应用开发时需要使用本地模拟器来进行应用调试，实现快速开发的目的。HarmonyOS应用需要运行在多种不同类型的设备上，为此HUAWEI DevEco Studio提供了不同类型的终端模拟，支持开发者在多个模拟终端上进行开发调试，降低门槛、节约成本。同时，多个模拟终端、真机设备也可以自由地组成超级终端，进一步降低开发者获取分布式调测环境的难度。

- 低代码可视化开发：低代码开发提供UI可视化开发能力，支持自由拖曳组件和可视化数据绑定，可快速预览效果，所见即所得。通过拖曳式编排、可视化配置的方式，帮助开发者减少重复性的代码编写，快速地构建多端应用程序。低代码开发的产物，如组件、模板等可以被其他模块的代码引用，并且能通过跨工程复用，支持开发团队协同完成复杂应用的开发。

2. 多端开发能力

应用如需在多个设备上运行，需要适配不同的屏幕尺寸和分辨率、不同的交互方式（如触摸和键盘等）、不同的硬件能力（如内存差异和器件差异等），开发成本较高。因此，多端开发能力的核心目标是降低多设备应用的开发成本。为了实现该目标，HarmonyOS提供了以下几个核心能力：支持多端UI适配、交互事件归一、设备能力抽象。这可以帮助开发者降低开发与维护成本，提升代码复用度。

- 多端UI适配：不同设备屏幕尺寸、分辨率等存在差异，HarmonyOS将对屏幕进行逻辑抽象，包括尺寸和物理像素，并提供丰富的自适应/响应式的布局和视觉能力，方便开发者进行不同屏幕的界面适配。
- 交互事件归一：不同设备间的交互方式等存在差异，如触摸、键盘、鼠标、语音、手写笔等，HarmonyOS将不同设备的输入映射成归一交互事件，从而简化开发者适配逻辑。
- 设备能力抽象：不同设备间的软、硬件能力等存在差异，如设备是否具备定位能力、是否具备摄像头、是否具备蓝牙功能等，HarmonyOS要对设备能力进行逻辑抽象，并提供接口来查询设备是否支持某一能力，方便开发者进行不同软、硬件能力的功能适配。在HarmonyOS中，使用SystemCapability（简写为 SysCap）定义每个部件对应用开发者提供的系统软硬件能力。应用开发者可基于统一的方式访问不同设备的能力。

3. 多端分发机制

如果需要开发多设备上运行的应用，一般会针对不同类型的设备多次开发并独立上架。开发和维护的成本高，为了解决这个问题，HarmonyOS提供了“一次开发，多端部署”的能力，开发者开发多设备应用，只需要一套工程，一次打包出多个HAP（Harmony Ability Package），统一上架，即可根据设备类型按需进行分发。

除了可以开发传统的应用，开发者还可以开发元服务。元服务是一种面向未来的服务提供方式，是具有独立入口、免安装、可为用户提供一个或多个便捷服务的应用程序形态。HarmonyOS为元服务提供了更多的分发入口，方便用户获取，同时也增加了元服务露出的机会。

1.2.2 可分可合，自由流转

元服务是HarmonyOS提供的一种全新的应用形态，具有独立入口，用户可通过点击、碰一碰、扫一扫等方式直接触发，无须显式安装，由程序框架后台静默安装后即可使用，可为用户提供便捷服务。

传统移动生态下，开发者通常需要开发一个原生应用版本，如果提供小程序给用户，往往需要开发若干个独立的小程序。在鸿蒙生态下，鸿蒙原生支持元服务开发，开发者无须维护多套版本，通过业务解耦将应用分解为若干元服务独立开发，根据场景按需组合成复杂应用。

元服务基于HarmonyOS API开发，支持运行在1+8+N设备上，供用户在合适的场景、合适的设备上便捷使用。元服务是支持“可分可合，自由流转”的轻量化程序实体，帮助开发者让服务更快触达用户。其具备如下特点。

- 触手可及：元服务可以在服务中心发现并使用，同时也可以基于合适场景被主动推荐给用户使用，例如用户可在服务中心和“小艺建议”中发现系统推荐的服务。
- 服务直达：元服务无须下载安装，“秒开体验”，即点即用，即用即走。
- 万能卡片：支持用户无须打开元服务便可获取服务内重要信息的展示和动态变化，如天气、关键事务备忘、热点新闻列表。
- 自由流转：元服务支持运行在多设备上并按需跨端迁移，或者多个设备协同起来给用户提供最优的体验。例如，手机上未完成的邮件，迁移到平板上继续编辑；用手机进行文档翻页和批注，配合智慧屏完成分布式办公；在分布式游戏场景中，手机可作为手柄，与智慧屏配合玩游戏，获得新奇游戏体验。

1.2.3 统一生态，原生智能

1. 统一生态

移动操作系统和桌面操作系统的跨平台应用开发框架不尽相同，从渲染方式的角度可以归纳为WebView渲染、原生渲染和自渲染这三类，HarmonyOS对应提供系统WebView、ArkUI框架和XComponent能力来支撑三种类型的跨平台框架的接入。

主流跨平台开发框架已有版本正在适配HarmonyOS，基于这些框架开发的应用可以以较低成本迁移到HarmonyOS。

2. 原生智能

HarmonyOS内置强大的AI能力，面向鸿蒙生态应用的开发，通过不同层次的AI能力开放，满足开发者的不同开发场景下的诉求，降低应用的开发门槛，帮助开发者快速实现应用智能化。

1.3 架构介绍

HarmonyOS整体遵从分层架构设计，从下向上依次为：内核层、系统服务层、框架层和应用层。系统功能按照“系统 > 子系统 > 组件”逐级展开，在多设备部署场景下，支持根据实际需求裁剪某些非必要的组件。HarmonyOS技术架构如图1-4所示。

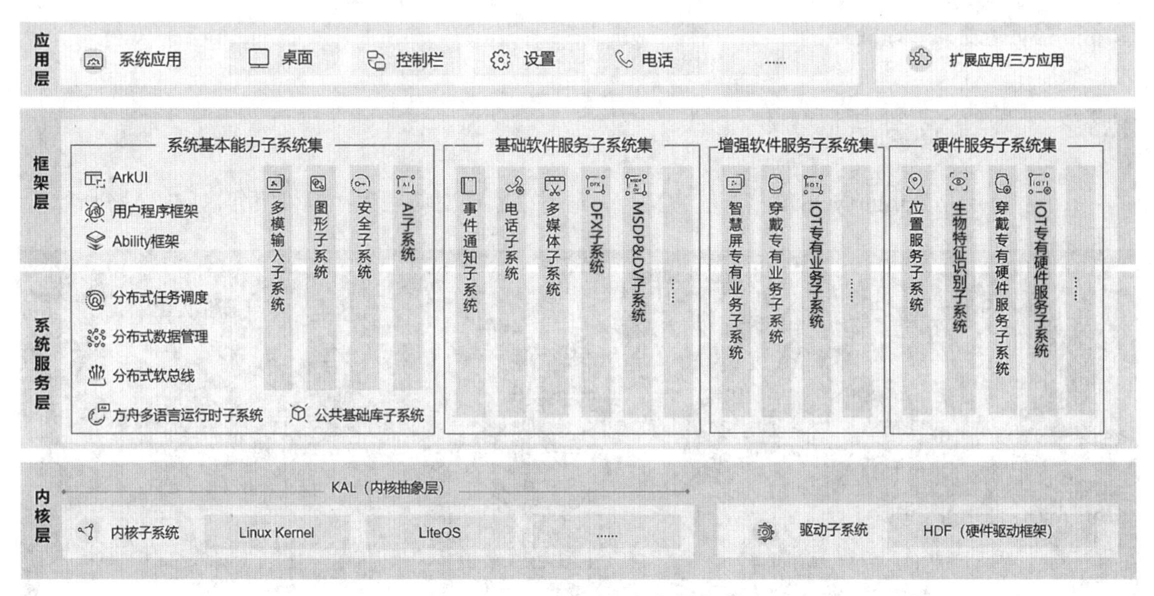

图1-4　HarmonyOS技术架构图

1.3.1　内核层

内核层主要分为以下两部分。

- 内核子系统：HarmonyOS采用多内核设计，支持针对不同资源受限设备选用适合的OS内核。内核抽象层（KAL，Kernel Abstract Layer）通过屏蔽多内核差异，对上层提供基础的内核能力，包括进程/线程管理、内存管理、文件系统、网络管理和外设管理等，如图1-5所示。

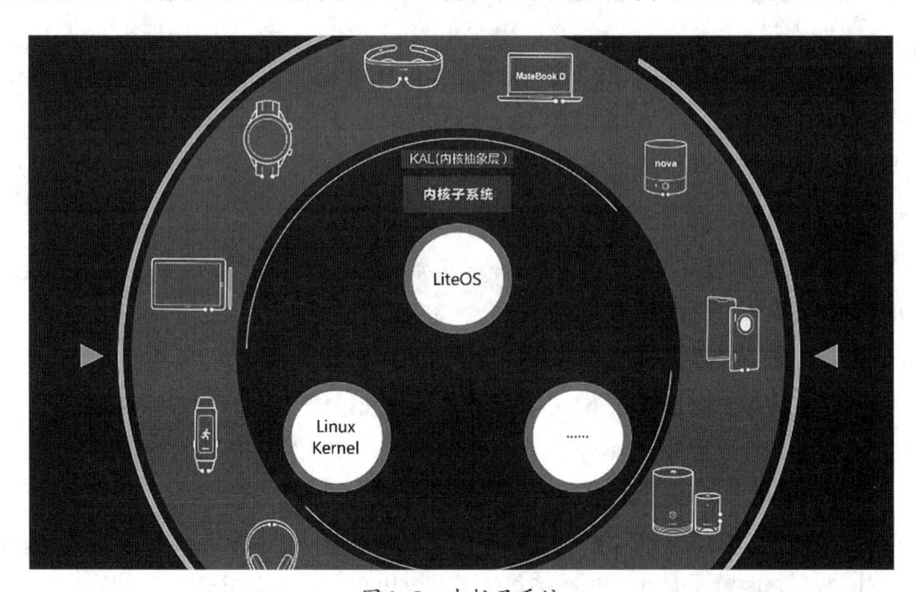

图1-5　内核子系统

- 驱动子系统：硬件驱动框架（HDF）是HarmonyOS硬件生态开放的基础，提供统一外设访问能力和驱动开发、管理框架。

1.3.2 系统服务层

系统服务层是HarmonyOS的核心能力集合，通过框架层对应用程序提供服务，如图1-6所示。

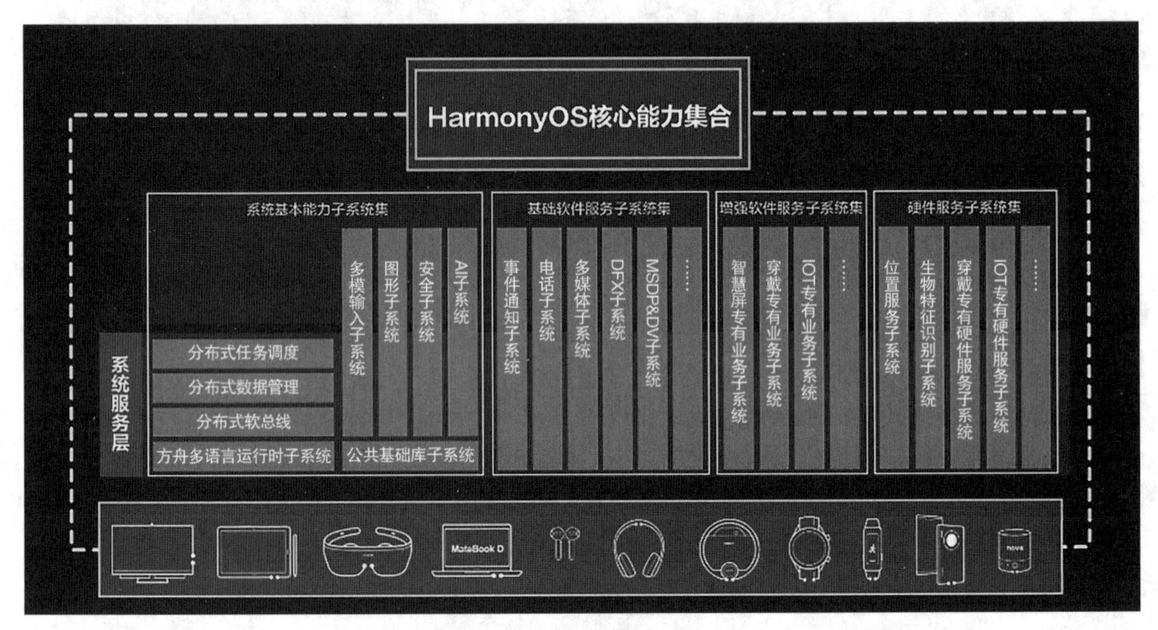

图1-6 系统服务层

图1-6展示了系统服务层的能力集合。该层包含以下几个部分。

- 系统基本能力子系统集：为分布式应用在HarmonyOS多设备上的运行、调度、迁移等操作提供了基础能力，由分布式软总线、分布式数据管理、分布式任务调度，以及方舟多语言运行时、公共基础库、多模输入、图形、安全、AI等子系统组成。其中，方舟多语言运行时子系统提供了C/C++/JS/ArkTS多语言运行时和基础的系统类库，也为使用方舟编译器静态化的Java程序（应用程序或框架层中使用Java语言开发的部分）提供运行时。
- 基础软件服务子系统集：为HarmonyOS提供公共的、通用的软件服务，由事件通知、电话、多媒体、DFX（Design For X）、MSDP&DV等子系统组成。
- 增强软件服务子系统集：为HarmonyOS提供针对不同设备的、差异化的能力增强型软件服务，由智慧屏专有业务、穿戴专有业务、IoT专有业务等子系统组成。
- 硬件服务子系统集：为HarmonyOS提供硬件服务，由位置服务、生物特征识别、穿戴专有硬件服务、IoT专有硬件服务等子系统组成。

根据不同设备形态的部署环境，基础软件服务子系统集、增强软件服务子系统集、硬件服务子系统集内部可以按子系统粒度裁剪，每个子系统内部又可以按功能粒度裁剪。

1.3.3 框架层

框架层为HarmonyOS应用开发提供了Java/C/C++/JS等多语言的用户程序框架和Ability框架，如适用于Java语言的Java UI框架，适用于JS语言的ArkUI框架，以及各种软硬件服务对外开放的多语言框架API。根据系统的组件化裁剪程度，采用HarmonyOS的设备支持的API也会有所不同。

图1-7展示了框架层所涵盖的功能。

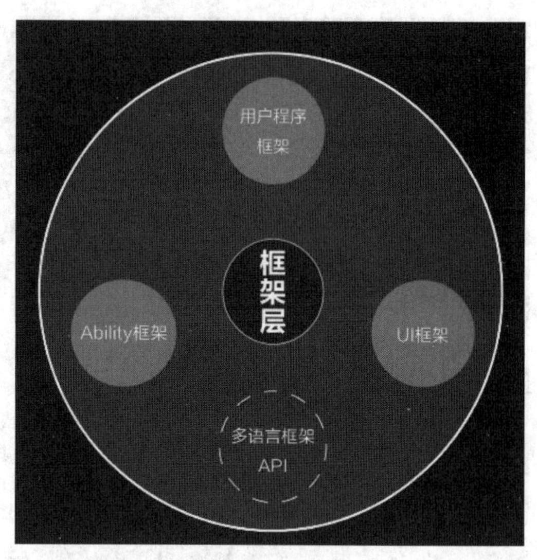

图1-7 框架层

1.3.4 应用层

应用层包括系统应用和第三方非系统应用。HarmonyOS的应用由一个或多个FA（Feature Ability）或PA（Particle Ability）组成。其中，FA有UI界面，提供与用户交互的能力；而PA无UI界面，提供后台运行任务的能力及统一的数据访问抽象。基于FA/PA开发的应用，能够实现特定的业务功能，支持跨设备调度与分发，为用户提供一致、高效的应用体验。

图1-8展示的是一个视频通话应用的组成情况。

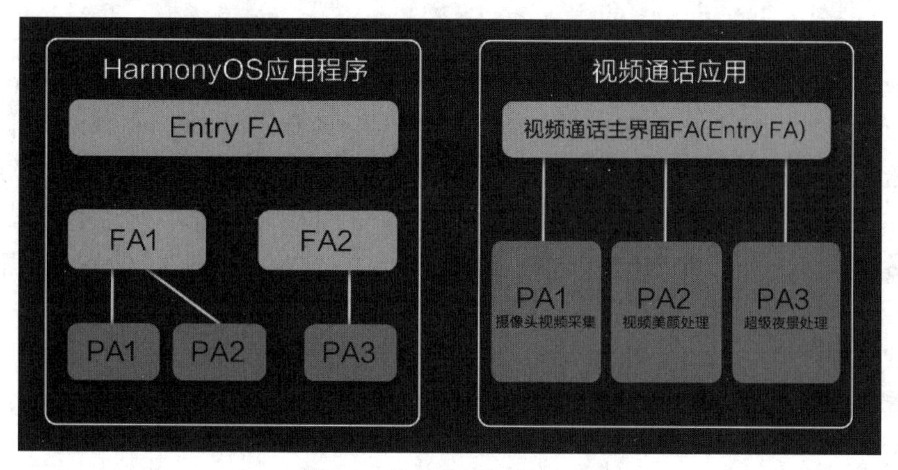

图1-8 视频通话应用的组成

在一个视频通话应用中，往往会有一个作为视频通话的主界面的FA和若干个PA。FA提供UI界面，用于与用户进行交互，PA1用于摄像头视频采集，PA2用于视频美颜处理，PA3用于超级夜景处理。这些FA、PA可以按需下载、加载和运行。

图1-9展示了不同设备下载相同应用时的不同表现。当手机下载该应用时，将同时拥有FA主界面、PA1摄像头视频采集、PA2视频美颜处理、PA3超级夜景处理；而当智慧屏下载该应用时，如果智慧屏不支持视频美颜处理、超级夜景处理功能，则只会下载FA主界面、PA1摄像头视频采集。

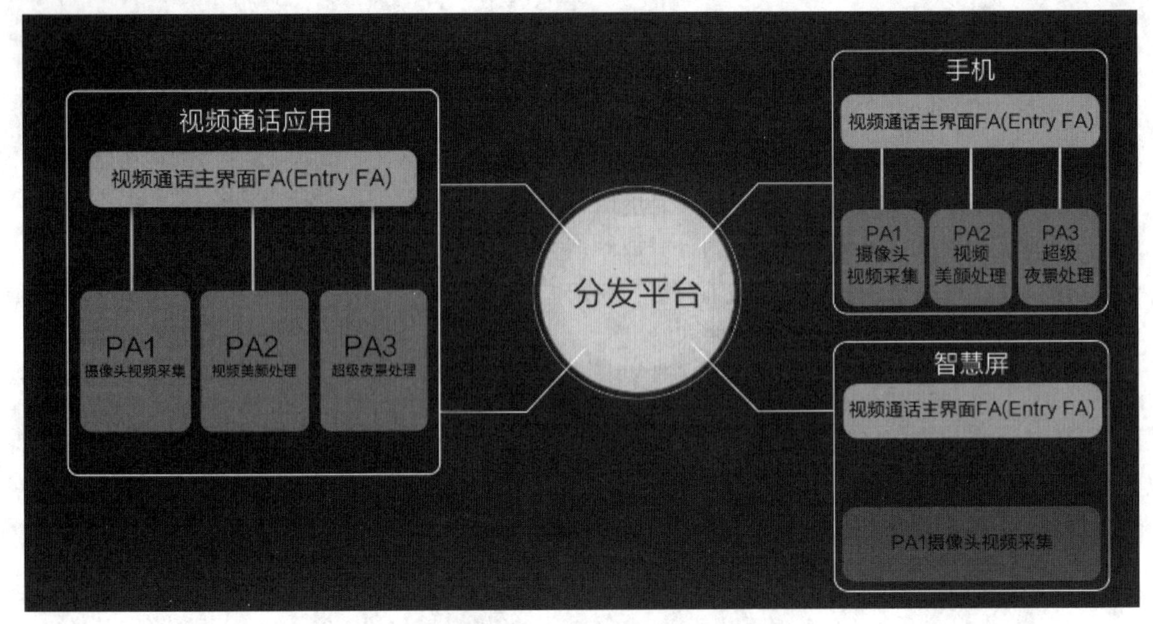

图 1-9　不同设备下载相同应用时的不同表现

HarmonyOS 2新特性概述

自HarmonyOS诞生以来，经过一年多的发展，人们终于迎来了HarmonyOS 2，HarmonyOS 2也带来了更多惊喜，最为重要的是该版本开始支持手机应用场景。HarmonyOS 2给应用带来更多流量入口，给设备带来更好的互联体验。该系统实现硬件互助、资源共享；同时该系统也能够实现“一次开发，多端部署”及“统一OS，弹性部署”，并且在安全和隐私方面也有较大的提升。

HarmonyOS 2在分布式能力上进行了全面提升，升级后的分布式软总线、分布式数据管理和分布式安全，为开发者和消费者都带来了不少新鲜感。

分布式软总线让多设备融合为“一个设备”，带来设备内和设备间高吞吐、低时延、高可靠的流畅连接体验。分布式数据管理让跨设备数据访问如同访问本地，大大提升了跨设备数据远程读写和检索性能等。

分布式安全确保正确的人用正确的设备来正确使用数据。当用户进行解锁、付款、登录等行为时，系统会主动拉出认证请求，并通过分布式技术可信互联能力，协同身份认证确保“正确的人”；通过把手机的内核级安全能力扩展到其他终端，从而提升全场景设备的安全性；通过定义数据和设备的安全级别，进行分类分级保护，从而确保数据流通安全可信。

HarmonyOS不是手机系统的一个简单替代，它是面向未来全场景融合的操作系统，其核心底座就是分布式技术，一方面其分布式技术有了本质提升；另一方面除了支持华为自身的设备之外，也开始支持第三方设备。

HarmonyOS 2提供了用户程序框架、Ability框架及UI框架，支持应用开发过程中多终端的业务逻辑和界面逻辑进行复用，能够实现应用的一次开发、多端部署，提升了跨设备应用的开发效率。其中，UI框架支持Java和JS两种开发语言，并提供了丰富的多态控件，可以在手机、平板、智能穿戴、智慧屏、车机上显示不同的UI效果。采用业界主流设计方式，提供多种响应式布局方案，支持栅格化布局，满足不同屏幕的界面适配能力。

HarmonyOS 2支持万能卡片功能。相比于现在大家使用的App图标，万能卡片功能可省去点击操作和等待时间，而且卡片功能信息实时更新，只需一瞥即可获得想要的信息。只需要上滑App图标就能生成万能卡片，还能根据自己的喜好随心定制，放在桌面的任意位置。更方便的是，卡片还能在手机、手表、平板各设备之间分享，一点即开，无须下载。

1.4.1 Harmony 2.0 Canary

当前版本在HarmonyOS 1.1.0的基础上，增加标准系统版本，具备的主要功能如下。

- 新增22个子系统，支持全面的OS能力，支持内存大于128M的带屏设备开发等。
- 提供系统三大应用：桌面、设置和SystemUI。
- 提供全新的Open Harmony应用框架能力、Ability Cross-platform Engine能力。
- 提供JS应用开发能力。
- 提供媒体框架，支持音视频功能开发。
- 提供图形框架能力，支持窗口管理和合成，支持GPU能力。

1.4.2 HarmonyOS v2.2 Beta2

标准系统新增特性功能如下。

- 新增分布式远程拉起能力端到端的构建。
- 新增系统基础应用的拖曳能力和新增若干Sample应用。
- 新增媒体三大服务能力，提供更好的媒体系统功能。

轻量和小型系统新增特性功能如下。

- 新增小型系统Linux版本构建能力。
- 新增轻量级内核能力增强，包括文件系统增强、内核调试工具增强支持、内核模块支持可配置、三方芯片适配支持、支持ARM9架构等。
- 轻量级图形能力增强支持，包括支持多语言字体对齐、支持显示控件轮廓、支持点阵字体、供统一多后端框架支持多芯片平台等。
- DFX能力增强支持，包括HiLog功能增强、HiEvent功能增强，提供轻量级系统信息dump工具，提供重启维测框架等。

- AI能力增强支持，包括新增Linux内核适配支持、AI引擎支持基于共享内存的数据传输。

1.5 HarmonyOS 3新特性概述

HarmonyOS 3支持更多设备加入超级终端，并提升了鸿蒙智联、万能卡片、流畅性能、隐私安全、信息无障碍等方面性能。此外，搭载该系统的手表还支持运动时收听音乐，其卡片样式更加多样，可一键升级智能桌面布局。

标准系统新增特性功能如下。

- 用户程序框架支持服务能力和线程模型。
- 支持文件安全访问，即文件转成URI和解析URI打开文件的能力。
- 支持设备管理PIN码认证的基本能力。
- 支持关系型数据库、分布式数据管理基础能力。
- 支持方舟JS编译工具链和运行时，支持OpenHarmony JS UI框架应用开发和运行。
- 支持远程绑定ServiceAbility、FA跨设备迁移能力。
- 支持应用通知订阅与应用通知消息跳转能力。
- 支持输入法框架及支持输入基础英文字母、符号和数字。
- 相机应用支持预览、拍照和录像基础能力。
- 支持CS（Circuit Switched，电路交换）基础通话、GSM（Global System for Mobile Communications，全球移动通信系统）短信能力。
- 支持定时器能力，提供定时时区管理能力。
- 在标准设备间的分布式组网下，提供应用跨设备访问对端资源或能力时的权限校验功能。

轻量和小型系统新增特性功能如下。

- 轻量级分布式能力增强，支持从轻量级系统启动标准系统上的Ability。
- 软总线能力增强支持，提供认证通道传输能力，用于设备绑定。
- 轻量级全球化能力增强支持，新增31种语言支持。
- 轻量级系统上新增权限属性字段及其写入接口，上层应用可通过该字段实现相关业务。

1.5.1 HarmonyOS 3.1 Release

标准系统基础能力增强，表现如下。

- 本地基础音视频播放能力，视频硬编码、硬解码，相机基础预览、拍摄能力。
- RenderService新渲染框架、2D/3D绘制能力、新的动画和显示效果框架。
- 窗口管理新框架，提供更加灵活的窗口框架能力，支持全屏、分屏、窗口化形态，支持跨窗口拖曳能力。

- display管理能力，支持分屏并增强亮灭屏管理能力；支持窗口动画和效果。
- 鼠标、键盘、触摸板的基本功能支持，常见传感器加速度、陀螺仪、霍尔、马达振动等的基本能力支持。
- 语言区域选择、新增/增强国际化特性、系统资源、rawfile资源。
- 支持时间时区同步，新增对剪贴板、锁屏服务、静态壁纸、下载服务管理能力的支持。
- 新增系统服务状态监控、跨设备oncall能力（跨设备启动FA能力）、长时、短时及延迟任务调度能力。
- 增强了内存管理、电源管理、进程调度等特性。
- 本地账号、域账号与本地账号绑定功能，分布式账号ID派生和状态管理功能，本地应用权限管理、分布式权限的管理能力。
- 提供Wi-Fi STA、AP、P2P相关基本能力及JS API；新增蓝牙BR、SPP、BLE相关能力及JS API；新增位置服务子系统，提供位置服务框架能力。

标准系统分布式能力增强，表现如下。

- 分布式软总线网络切换组网、P2P连接、流传输、蓝牙等能力。
- 支持硬件互助、资源共享，支持镜像和扩展投屏等。
- 设备上下线和PIN码认证等增强功能，存储管理、应用沙箱隔离、公共数据沙箱隔离等能力，支持分布式数据库、分布式数据对象，支持本地数据库访问和跨应用数据库访问等能力。

标准系统应用程序框架能力增强，表现如下。

- ArkUI自定义绘制能力和Lottie动画能力，键盘、鼠标交互操作能力。
- 声明式Web组件、XComponent组件能力。
- 卡片能力，提供卡片添加、删除、刷新等基础能力。
- 多用户能力，提供适应多端的基础JS工具链及运行时能力，对多HAP安装、隐式查询、多用户、权限管理的基本功能支持，支持分布式通知和通知模板功能。

标准系统应用能力增强，表现如下。

系统应用构建，包含系统桌面、SystemUI、系统设置、相机、图库、通话、联系人、信息、备忘录、文件选择器、输入法等应用支持。

1.5.2 HarmonyOS 3.2 Release

HarmonyOS 3.2版本标准系统能力进一步完善，支持采用ArkTS语言+Stage应用模型进行大型应用、原子化服务开发；对ArkCompiler进行了优化，其中Taskpool机制可以提升应用的运行性能；ArkUI组件能力增强，强化图形渲染能力和系统安全能力，丰富分布式业务开发；提供API Level 9稳定接口，在HarmonyOS 3.1 Release版本的基础上，进一步提升系统的整体性能。

1.6 HarmonyOS 4新特性概述

HarmonyOS 4开发套件同步升级到API 10，相较于3.2 Release版本，新增4000多个API，应用开发能力更强；HDF新增200多个HDI接口，硬件适配更加便捷；持续优化图形框架和方舟编译器（ArkCompiler），用户交互体验得到进一步提升；ArkUI组件定制化能力和组件动效能力也得到进一步增强；分布式硬件支持的范围扩大到音频和输入领域；分布式数据为开发者数据分享带来了全新的统一数据管理框架。另外，该版本在媒体、安全和隐私保护等方面也得到了进一步增强。

> **注意：** HarmonyOS 4目前仅在部分手机上得到了安装，其开发工具、SDK并未向开发者全面开发，因此本书所编写的示例，主要是基于HarmonyOS SDK 3构建。

1.7 Java与ArkTS如何抉择

本节详细分析基于HarmonyOS，使用Java与ArkTS进行应用开发的区别，力求解答读者在选择开发语言时的一些困惑。

1.7.1 HarmonyOS编程语言的发展

HarmonyOS的编程语言的发展，是随着HarmonyOS的发展而演进的。

早期的HarmonyOS支持的开发语言包括JS（JavaScript）、C/C++。其中，JS主要用于应用开发，而C/C++主要用于设备开发。因为，早期的HarmonyOS只支持手表等智能穿戴设备，所以用JS是能够胜任的。

从HarmonyOS 2开始，HarmonyOS开始兼容Android，引入了Java开发语言，能够支持手机、平板、智能穿戴、智慧屏、车机、PC、智能音箱、耳机、AR/VR眼镜等多种终端设备，提供全场景（移动办公、运动健康、社交通信、媒体娱乐等）业务能力。此时的HarmonyOS才算真正意义上的“鸿蒙操作系统”，因为已经具备“鸿蒙操作系统”的三大特征：

- 硬件互助，资源共享。
- 一次开发，多端部署。
- 统一OS，弹性部署。

HarmonyOS 3开始引入华为自研的ArkTS开发语言（前身是eTS）、方舟编译器等。ArkTS基于TypeScript（简称TS）语言扩展而来，是TS的超集。其最大的亮点是，ArkTS在TS基础上主要扩展了声明式UI能力（ArkUI），让开发者以更简洁、更自然的方式开发高性能应用。

但比较遗憾的是，直到HarmonyOS 4，ArkTS开发语言能实现的功能还是比较有限，甚至无法兑现"鸿蒙操作系统"三大特征的承诺。具体表现为：

- 模拟器支持的设备有限，只支持Phone和Tablet；
- 没有相机等功能的开发；
- 没有平行视界；
- 没有AI功能；
- 没有流转、跨端迁移和多端协同；

……

总之，ArkTS现阶段能实现的功能还不如Java强大。

1.7.2 HarmonyOS编程语言的选择

在了解上述HarmonyOS编程语言的区别之后，相信各位对于HarmonyOS编程语言的选择有了自己的判断。如果选择Java来开发应用则需要选择API 7及以下版本；如果选择ArkTS来开发应用则需要选择API 8及以上版本。

选择时可以综合考虑以下几个方面。

1. 个人的偏好

开发者对于编程语言是有一定偏好的，选ArkTS或是Java也可以尊重个人的喜好。

如果你熟悉Java或Android则可以选Java；如果你熟悉JS或TS，就可以选ArkTS。

2. 职业的需要

如果想要实现一个强大、完整的可商用的HarmonyOS应用功能，那么现阶段只有Java能做到，ArkTS还处在初级阶段。

从项目的稳定性和用户广度而言，越早的API，用的人越多，稳定性相对越好。目前市面上很多HarmonyOS 4的手机（比如华为畅享70），只支持API 6。

如果你是纯粹想学习一门新语言，想不断跟随语言的演进，那么ArkTS也是不错的选择。从未来发展上看，HarmonyOS NEXT有可能会将ArkTS作为主力开发语言。

3. 本书的选择

本书对于Java和ArkTS的内容都会涉及。每个章节的前半部分采用Java编写示例，后半部分采用ArkTS编写示例，力求面面俱到。当然，部分章节的知识点，如果ArkTS还不具备相关能力，则仅用Java编写示例。

第2章

先利其器——开发环境搭建

本章介绍 HarmonyOS 开发环境搭建。HUAWEI DevEco Studio（以下简称 DevEco Studio）是基于 IntelliJ IDEA Community 开源版本打造，为运行在 HarmonyOS 和 OpenHarmony 系统上的应用和服务提供一站式的开发平台。本章会详细介绍 DevEco Studio 的安装和配置方法，以及使用技巧。

2.1 注册华为开发者联盟账号

要进行HarmonyOS应用的开发，开发者首先要有华为开发者联盟账号。华为开发者联盟开放诸多功能和服务，助力联盟成员打造优质应用。开发者需要注册华为开发者联盟账号，并且实名认证才能享受联盟开放的各类能力和服务。

账号注册完成后，可选择认证成为企业开发者或个人开发者。本节主要介绍如何认证成为个人开发者。

2.1.1 开发者享受的权益

以下是个人开发者和企业开发者的权益列表。从表2-1中可以看到，企业开发者比个人开发者享受的服务更多。

表2-1 个人开发者和企业开发者的权益

开发者类型	享受的服务/权益
个人开发者	应用市场、主题、商品管理、账号、PUSH、新游预约、互动评论、社交、HUAWEI HiAI、手表应用市场等
企业开发者	应用市场、主题、首发、支付、游戏礼包、应用市场推广、商品管理、游戏、账号、PUSH、新游预约、互动评论、社交、HUAWEI HiAI、手表应用市场、运动健康、云测、智能家居等

2.1.2 注册、认证准备的资料

对于个人开发者而言，注册、认证联盟账号需要准备的资料如下。

- 注册：可接收验证码的手机号码或电子邮箱地址。
- 人工审核：身份证原件正反面扫描件或照片；手持身份证正面照片。
- 个人银行卡认证：个人银行卡号。

华为开发者联盟将在1～2个工作日内完成审核。审核完成后，将向提交的认证信息中的联系人邮箱发送审核结果。

2.1.3 注册账号

要注册华为开发者联盟账号，请按以下步骤操作。

1. 进入注册页面

打开华为开发者联盟官网，点击“注册”进入注册页面。

2. 进行注册

可以通过电子邮箱或手机号码注册华为开发者联盟账号。

如果用电子邮箱注册，输入正确的电子邮箱地址和验证码并设置密码后，单击“注册”，如图 2-1 所示。

如果使用手机号码注册，输入正确的手机号码和验证码并设置密码后，单击“注册”，如图 2-2 所示。

图 2-1　用电子邮箱注册页面

图 2-2　用手机号码注册页面

2.1.4　登录账号

使用华为商城账号、华为云账号和花粉论坛账号均可登录华为开发者联盟官网，单击官网右上角“登录”，则看到图 2-3 所示的登录界面。输入账号密码，单击“登录”即可。也可使用华为移动服务 App 扫一扫登录华为开发者联盟官网。

图 2-3　登录页面

2.1.5　实名认证

账号注册完后，需要完成实名认证才能享受联盟开放的各类功能和服务。

1. 提交实名认证申请

打开华为开发者联盟官网，登录账号，单击账号名下拉框中的“去认证”，可进入“实名认证”页面，如图 2-4 所示。

在开发者实名认证页面，单击图2-5所示的“个人开发者”图标或“下一步”，进入个人认证方式选择页面。

图2-4 “实名认证”入口页面

图2-5 开发者实名认证页面

个人开发者实名认证分为个人银行卡认证、身份证人工审核认证、华为云授权认证、人脸识别认证4种方式，个人银行卡认证、身份证人工审核认证、人脸识别认证总体流程，如图2-6所示。

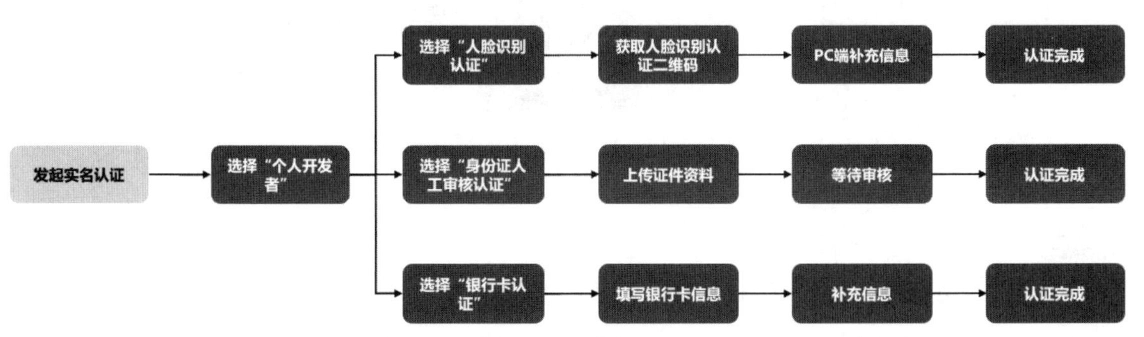

图2-6 个人银行卡认证、身份证人工审核认证、人脸识别认证总体流程

华为云授权认证总体流程，如图2-7所示。

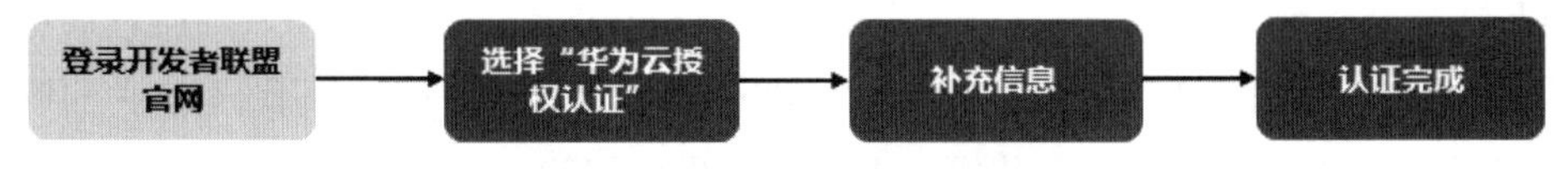

图2-7 华为云授权认证总体流程

最后完善个人信息，签署《华为开发者联盟与隐私的声明》和《华为开发者服务协议》，单击“提交”，完成认证。

2. 个人资料人工审核认证

选择身份证人工审核认证，进入人工审核认证页面，如图2-8所示。

图2-8 人工审核认证页面

完善个人信息，签署《华为开发者联盟与隐私的声明》和《华为开发者服务协议》，单击“下一步”，等待审核，如图2-9所示。审核结果会

在1～2个工作日发送至联系人邮箱。

图2-9　完善个人信息

2.2 DevEco Studio下载安装

作为一款开发工具，除了具有基本的代码开发、编译构建及调测等功能外，DevEco Studio还具有如下特点。

- 高效智能代码编辑：支持Java、XML、ArkTS、JS、C/C++等语言的代码高亮、代码智能补齐、代码错误检查、代码自动跳转、代码格式化、代码查找等功能，提升代码编写效率。
- 低代码可视化开发：丰富的UI界面编辑能力，支持自由拖曳组件和可视化数据绑定，可快速预览效果，所见即所得；同时支持卡片的零代码开发，降低开发门槛和提升界面开发效率。
- 多端双向实时预览：支持UI界面代码的双向预览、实时预览、动态预览、组件预览及多端设备预览，便于快速查看代码运行效果。
- 多端设备模拟仿真：提供HarmonyOS本地模拟器，支持手机等设备的模拟仿真，便捷获取调试环境。

DevEco Studio支持Windows和Mac系统，在两个系统的安装步骤类似，因此，下面只针对Windows操作系统的软件安装方式进行介绍。

2.2.1 运行环境要求

为保证DevEco Studio正常运行，建议计算机配置满足如下要求。

- 操作系统：Windows10 64位、Windows11 64位。
- 内存：8GB及以上。
- 硬盘：100GB及以上。
- 分辨率：1280×800像素及以上。

2.2.2 下载和安装DevEco Studio

1. 下载DevEco Studio

进入DevEco Studio产品页，如图2-10所示。

图2-10 进入DevEco Studio产品页

找到DevEco Studio下载页面，如图2-11所示，根据自己的操作系统，下载对应的DevEco Studio安装包。

DevEco Studio 3.1.1 Release

platform	DevEco Studio Package	Size	SHA-256 checksum	Download
Windows(64-bit)	devecostudio-windows-3.1.0.501.zip	843.6M	fbe79d92017d642ee91b2471b36c3e22ff3c186a0df36f3ae683129cfd445d9c	
Mac(X86)	devecostudio-mac-3.1.0.501.zip	942.9M	1a380b8b4a172b0f00af476b3bdcd83ee2dab24937c00b72d20d9121db99f5b7	
Mac(ARM)	devecostudio-mac-arm-3.1.0.501.zip	934.8M	f3e77ba60e596c9e49cd5fc3ab67f3f944efd235f2fa7b298b501abcfc668f04	

图2-11 DevEco Studio下载页面

2. 安装DevEco Studio

下载完成后，双击下载的“deveco-studio-××××.exe”，进入DevEco Studio安装向导。如果之前有安装过老版本，则会提示先卸载老版本。

在图2-12所示的界面选择安装路径，默认安装于“C:\Program Files”路径下，也可以单击“Browse...”指定其他安装路径，然后单击“Next”按钮。

在图2-13所示的安装选项界面勾选“DevEco Studio”，便于后续创建快捷方式用，而后单击“Next”按钮，直至安装完成。

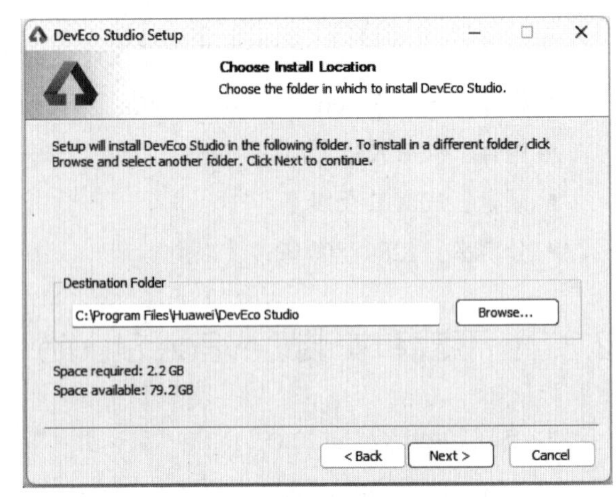

图2-12　界面选择安装路径

安装完成后，先不要勾选“Run DevEco Studio”选项，接下来请根据实际情况，检查和配置开发环境，如图2-14所示。

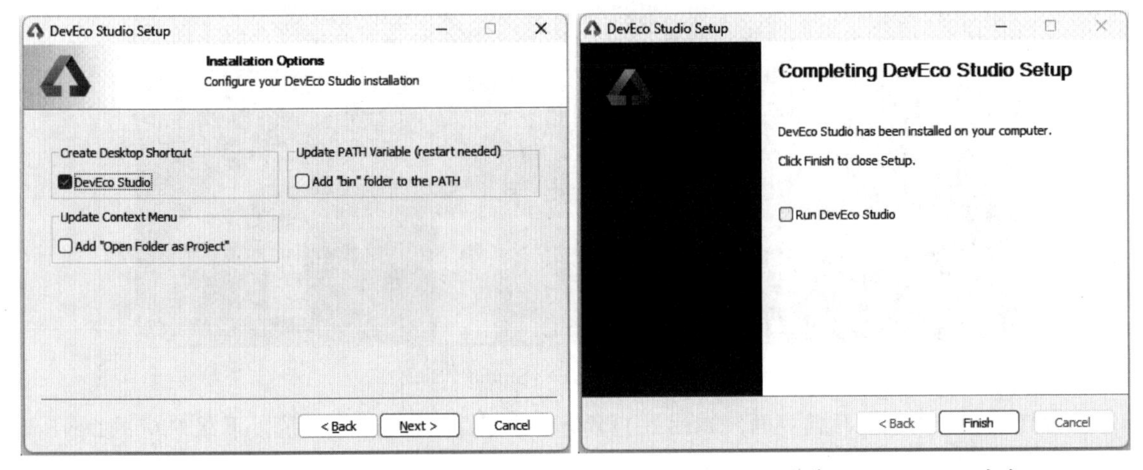

图2-13　安装选项界面勾选“DevEco Studio”　　图2-14　安装DevEco Studio完成

2.3 设置DevEco Studio

设置DevEco Studio可以分为基本设置和详细设置。

2.3.1　基本设置

双击DevEco Studio桌面快捷方式以启动DevEco Studio。

首次使用DevEco Studio时会有图2-15所示的提示信息，单击“Agree”继续执行下一步。

如果你之前使用过DevEco Studio并保存过DevEco Studio的配置，则可以导入DevEco Studio的配置，否则，选择“Do not import settings”执行下一步，如图2-16所示。

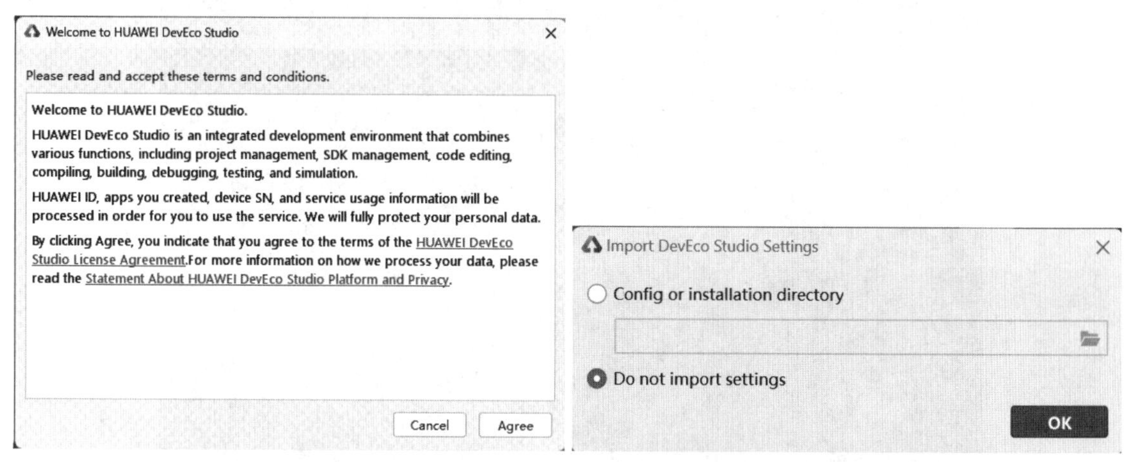

图2-15　单击“Agree”　　　　图2-16　选择“Do not import settings”

安装Node.js与ohpm。单击“Local”，可以指定本地已安装的Node.js或ohpm路径位置。如果本地没有合适的版本，可以选择“Install”按钮，选择下载源和存储路径后进行在线下载，单击“Next”进入下一步，如图2-17所示。

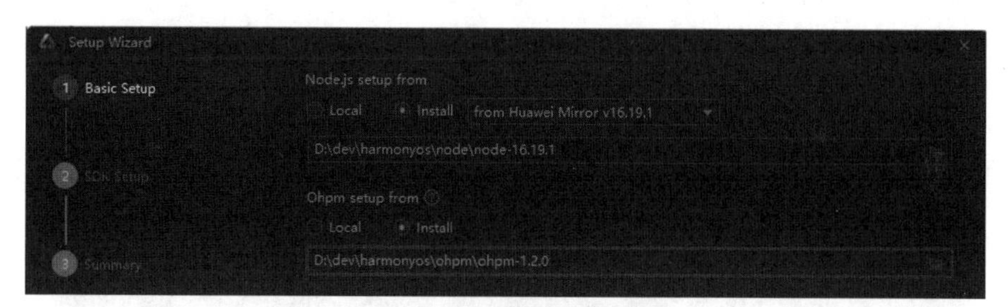

图2-17　安装Node.js与ohpm

在SDK Setup界面，设置HarmonyOS SDK存储路径，单击“Next”进入下一步，如图2-18所示。

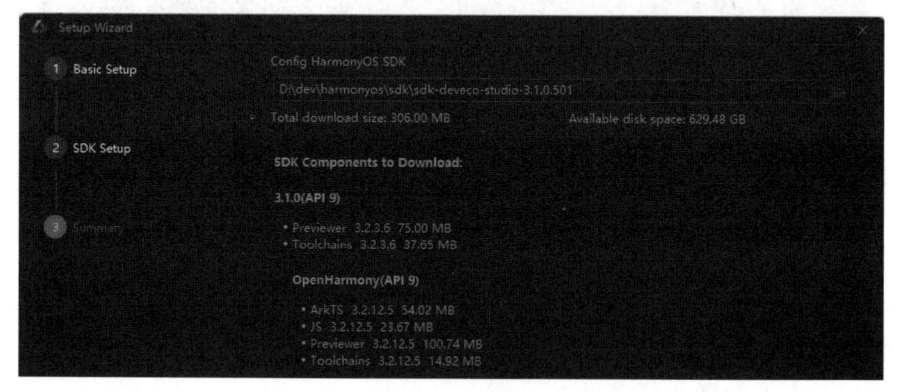

图2-18　设置SDK

在弹出的SDK下载信息页面，单击“Next”，并在弹出的License Agreement窗口，阅读License协议，同意License协议后，单击“Next”进入下一步，如图2-19所示。

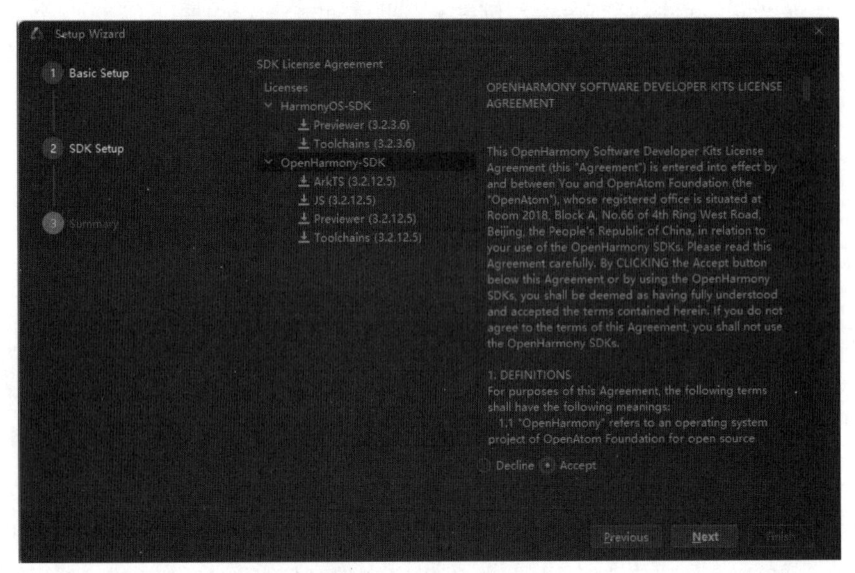

图2-19　下载SDK

确认设置项的信息，单击“Next”开始安装。等待Node.js、ohpm和SDK下载完成后，单击“Finish”，界面会进入DevEco Studio欢迎页，如图2-20所示。

图2-20　DevEco Studio欢迎页

2.3.2　详细设置

单击页面左下角“Configure”，可以看到进一步的设置，单击“Settings”，如图2-21所示进入详细设置页面。可以设置主题，比如，将颜色设置为更加护眼的“IntelliJ Light”，如图2-22所示，也可以设置字体和字号。

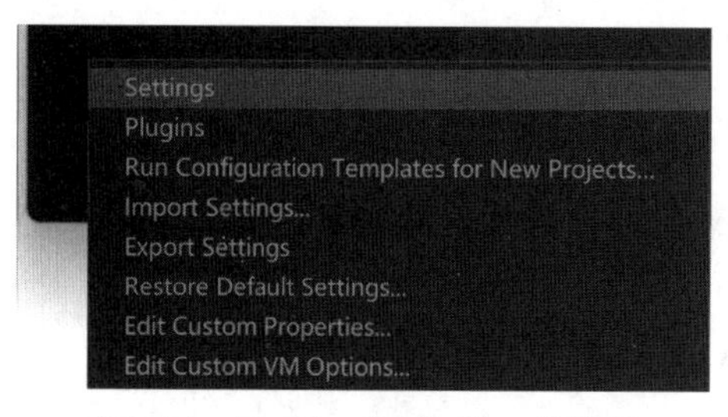

图2-21 DevEco Studio进一步设置

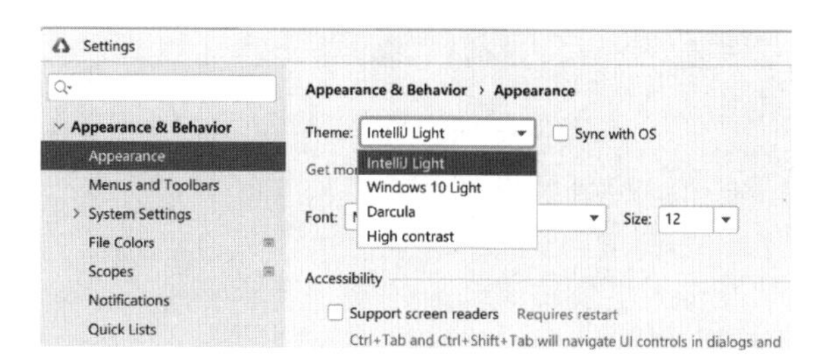

图2-22 DevEco Studio设置主题

在SDK一栏，可以看到DevEco Studio所预装的SDK版本，如图2-23所示。

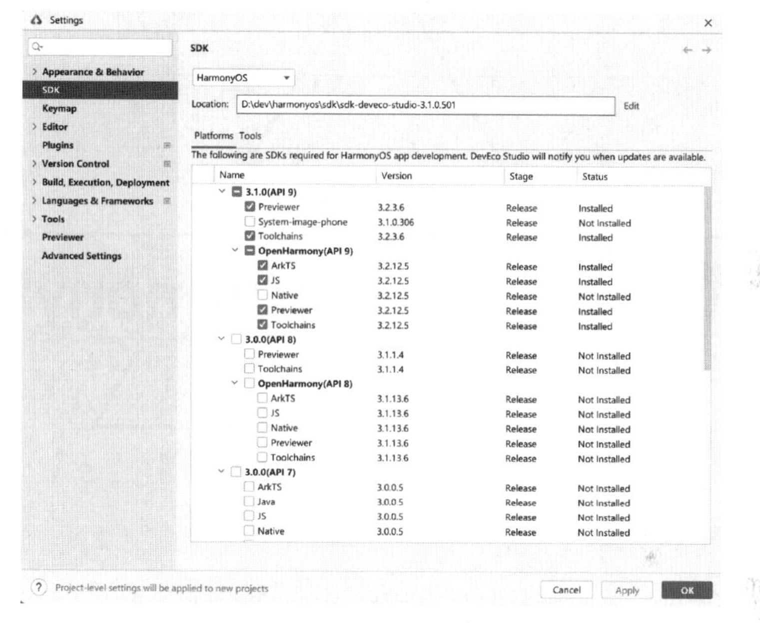

图2-23 DevEco Studio预装的SDK

可以按照自己的项目需求去安装相应版本的SDK。比如，如果想要用Java来开发HarmonyOS应用，就要把相关的SDK都勾选下载，如图2-24所示。

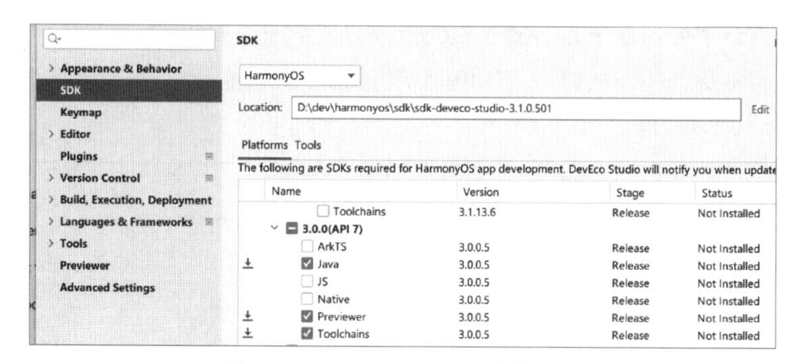

图2-24 DevEco Studio设置SDK

同理，如果你试图打开一个老版本的HarmonyOS应用，则DevEco Studio会提示你安装相应版本的SDK，或者自动帮你在后台安装。

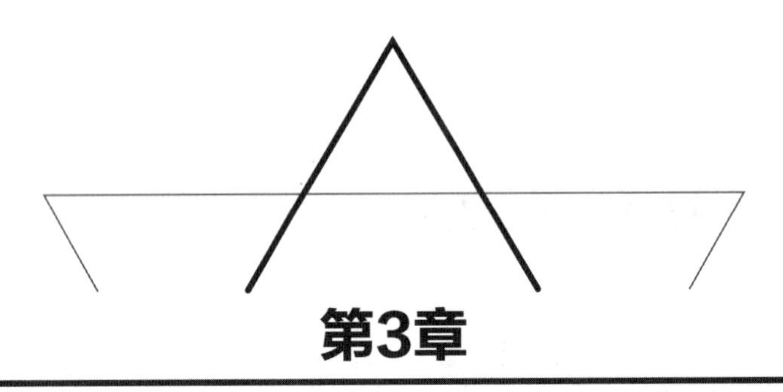

第3章

牛刀小试——开发第一个HarmonyOS应用

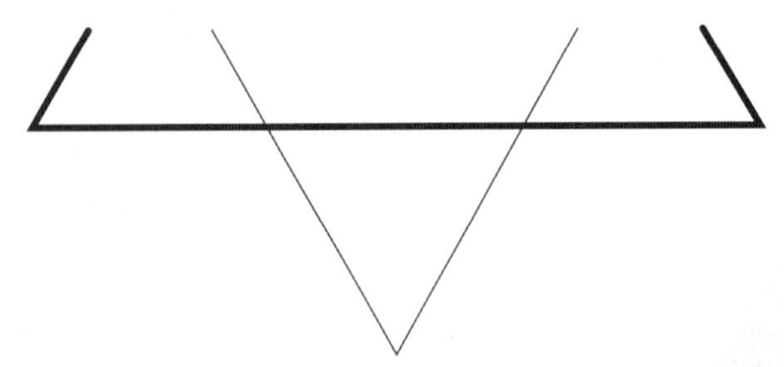

本章演示了如何基于 DevEco Studio 来开发第一个 HarmonyOS 应用。在本章，我们将不费“一枪一弹”（不用编写一行代码），就能开发出一个能够直接运行的 HarmonyOS 应用。

3.1 使用Java创建一个新应用

根据上一章的学习，我们已经安装好了DevEco Studio，终于可以进入激动人心的开发环节了。我们将演示如何基于DevEco Studio来开发第一个HarmonyOS应用。按照编程惯例，第一个应用称为“Hello World”应用。按照如下步骤来选择创建工程。

3.1.1 选择创建新工程

打开DevEco Studio后，我们单击“Create Project”来创建一个新工程。

3.1.2 选择设备应用类型的模板

此时，可以看到如图3-1所示的界面，这个界面让你选择不同类型的模板。在本例，选择一个空模版“Empty Ability”。有关Ability的概念，我们后续再介绍。这里可简单认为Ability就是你应用的一个功能。换言之，我们将要创建的是一个没有功能的应用。单击“Next”进行下一步。

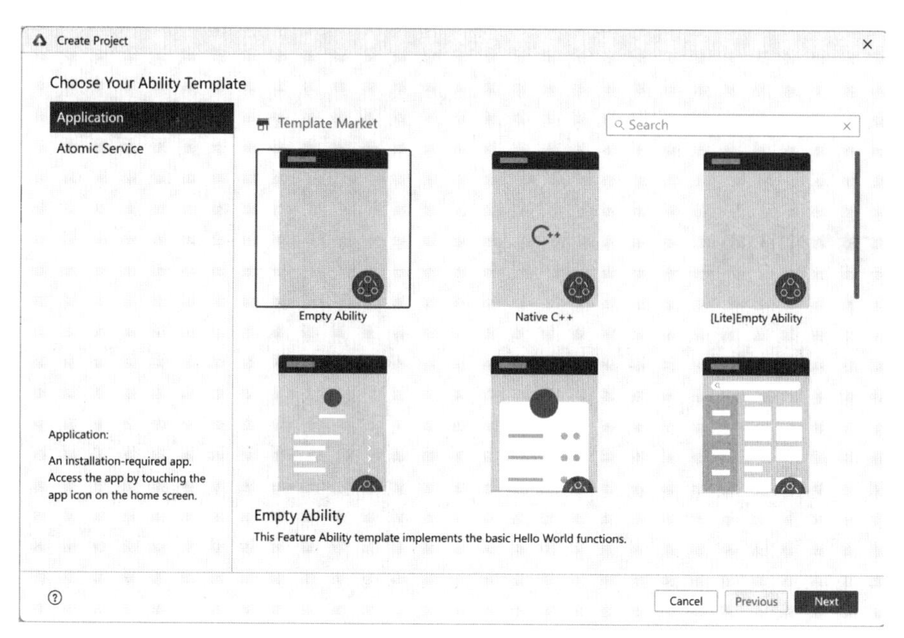

图3-1　选择不同设备应用类型的模板

3.1.3 配置项目的信息

下一步是配置项目的信息，比如项目名称、包名、位置、SDK版本等，按照个人实际情况填即可。

如图3-2所示，上述示例创建的“HelloWorld”应用，选用的SDK版本为3.0.0(API 7)，编程语言是Java，设备类型是Phone。

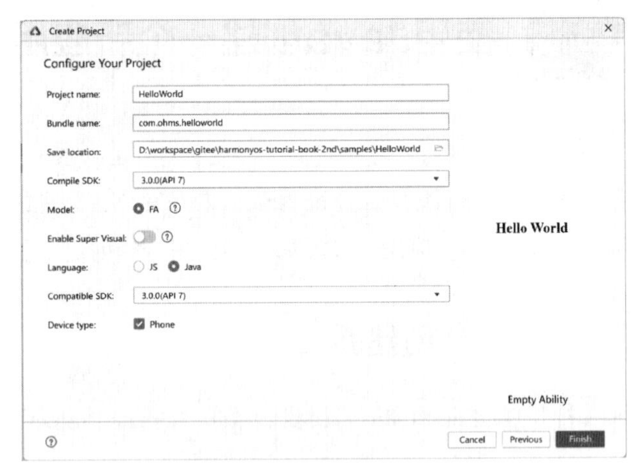

图3-2　配置项目的信息

3.1.4　自动生成工程代码

单击“Finish”之后，DevEco Studio就会创建好整个应用，并且自动生成工程代码。由于Java语言开发的HarmonyOS应用是采用Gradle构建的，因此，可以在控制台看到会自动下载Gradle安装包。Gradle下载完成之后，就会对工程进行配置，因此可以看到如图3-3所示的控制台配置成功的提示信息。

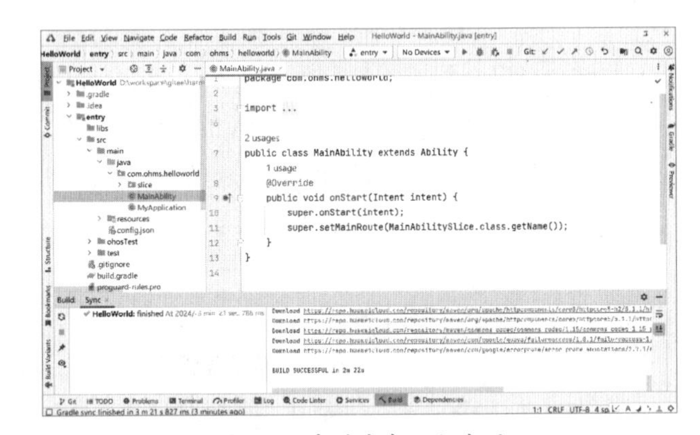

图3-3　自动生成工程代码

3.2　运行项目

构建好应用之后，如何来运行应用呢？ HarmonyOS支持本地模拟器、远程模拟器、本地真机、远程真机等多种方式来运行项目。

上述方式各有利弊，比如，本地模拟器不需要通过华为开发者联盟账号登录使用，但需要在本地安装模拟器，占用计算机的资源；远程模拟器需要通过华为开发者联盟账号登录，如果用的人多的话，在可用性上也有限制；本地真机需要用户自己准备具备有HarmonyOS系统的设备；远程真机也需要使用华为开发者联盟账号，是部署在云端的真机设备资源，但使用过程中需要给应用签名，同时还需要登录AppGallery Connect创建项目和应用，因此过程上比较烦琐。本书推荐采用本地安装模拟器和远程模拟器结合的方式来运行项目。

3.2.1　启动远程模拟器

打开“View>Device Manager”进入设备管理界面。在该界面选择“Remote Emulator”进入远程模拟器，如图3-4所示。

此时需要使用华为开发者联盟账号进行登录，并根据提示对设备进行授权，如图3-5所示。

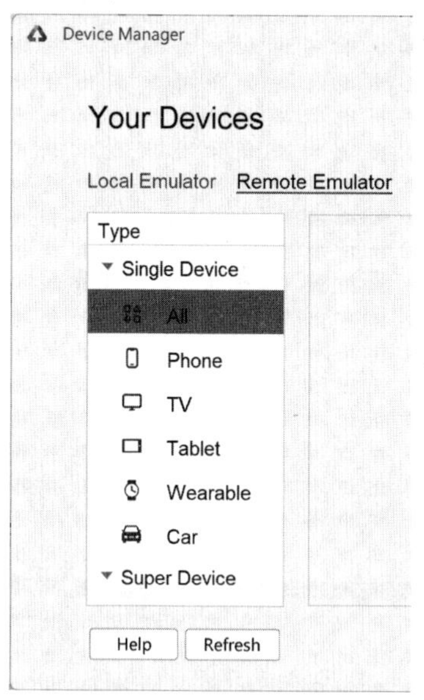

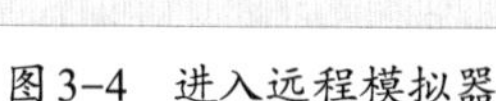
图3-4　进入远程模拟器

图3-5　对设备进行授权

单击“允许”进行下一步操作。授权完成之后，再次返回DevEco Studio，此时，看到如图3-6所示的各种类型的设备模拟器列表。可以单击启动任意API不小7的Phone模拟器。

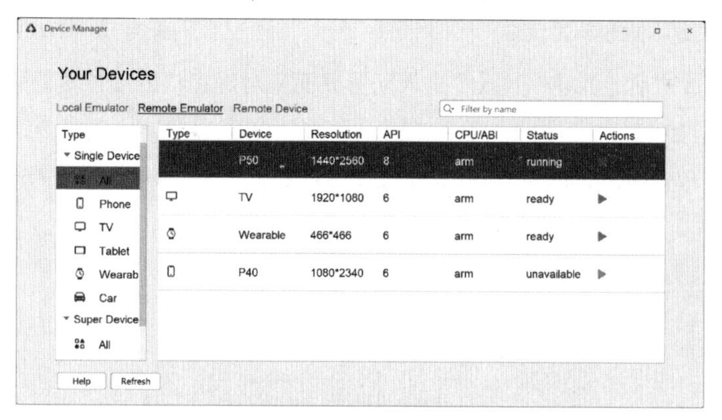

图3-6　启动远程模拟器

3.2.2　启动本地模拟器

打开“View>Device Manager”进入设备管理界面。在该界面选择“Local Emulator”进入本地模拟器，如图3-7所示。

单击“Install”下载本地模拟器。本地模拟器下载完成之后，可以看到如图3-8所示的配置界面，单击“New Emulator”创建本地模拟器设备。

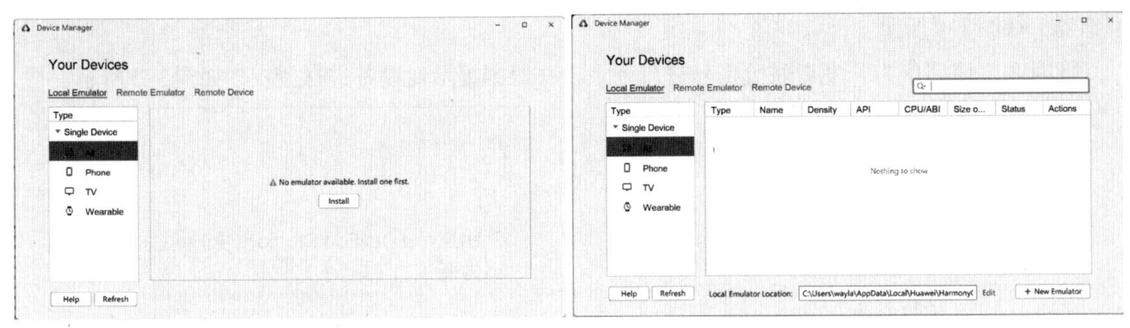

图 3-7　进入本地模拟器　　　　　图 3-8　本地模拟器配置界面

选择“Phone”类型的设备，单击“Next”如图3-9所示。

单击下载API 9的手机镜像，如图3-10所示。

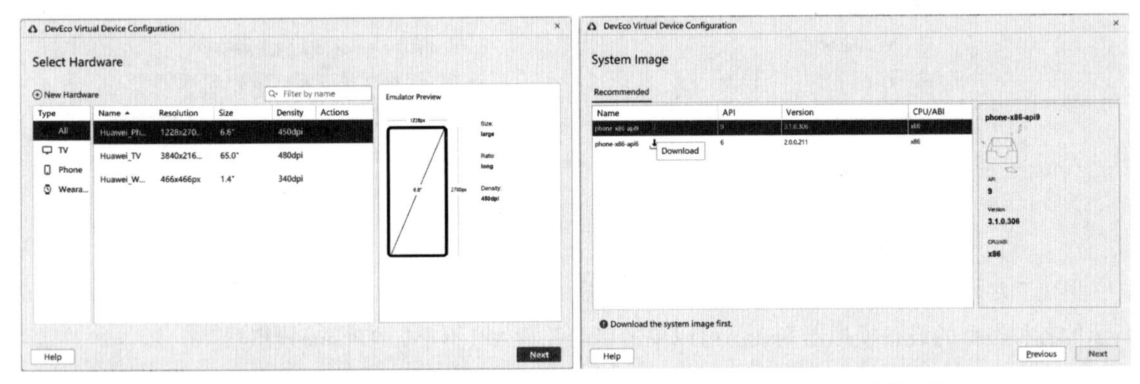

图 3-9　创建本地模拟器设备　　　　　图 3-10　下载镜像

手机镜像下载完成之后，单击“Next”进入配置界面来对虚拟机的硬件资源进行分配，如图3-11所示。

配置完成之后，单击“Finish”回到本地模拟器界面，此时就能看到可用的本地模拟器设备列表了，如图3-12所示。

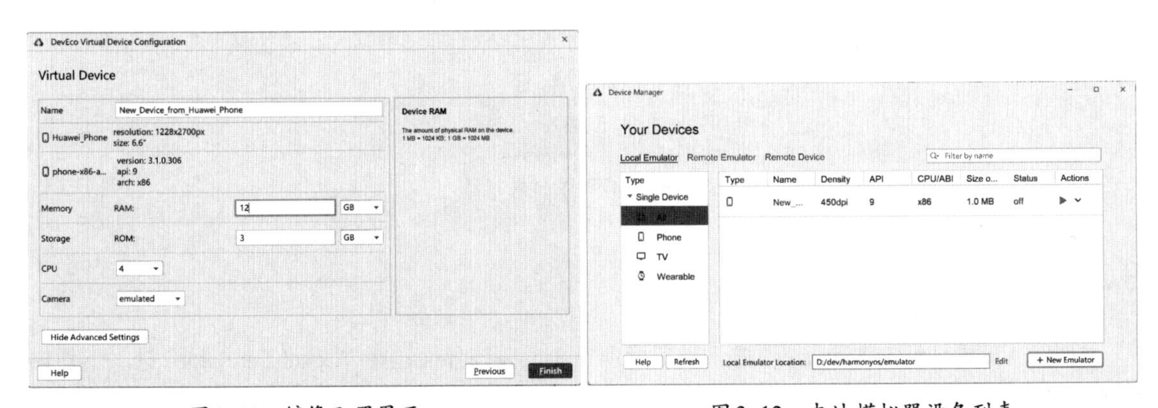

图 3-11　镜像配置界面　　　　　图 3-12　本地模拟器设备列表

可以单击启动上述列表里面的模拟器设备。

3.2.3　在模拟器里面运行项目

在本地安装模拟器或远程模拟器启动设备之后，就能看到可选的设备列表，如图3-13所示。

选中设备列表后，单击三角形按钮以启动项目，如图3-14所示。

项目在本地模拟器里面的运行效果如图3-15所示。

图3-13　可选的设备列表

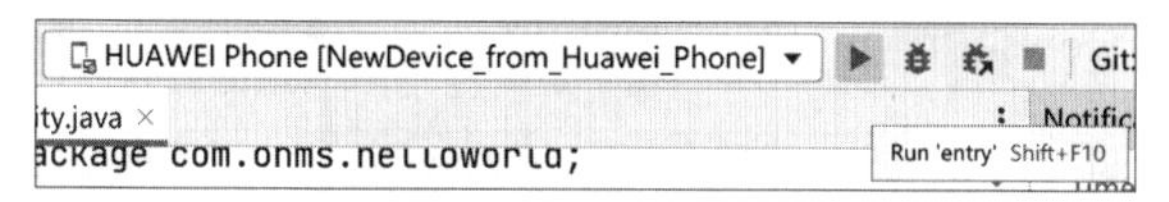

图3-14　启动项目

图3-15　项目运行效果

以上就是运行项目的完整过程。

3.3 使用DevEco Studio预览器

在前面的章节中，我们已经认识到如何来创建一个最为简单的“HelloWorld”应用，并且可以通过设备模拟器或真机运行应用。

但是使用设备模拟器或真机运行应用有一个缺点，那就是启动相对来说比较慢。如果只是调试一个简单的界面，却要等待非常久的时间，那可能就消磨人的耐心了。此时，推荐的方式是使用预览器。

3.3.1　如何安装预览器

在使用预览器查看应用界面的UI效果前，需要确保SDK中已下载安装了Previewer，如图3-16所示。

同时，启用Java Previewer，如图3-17所示。

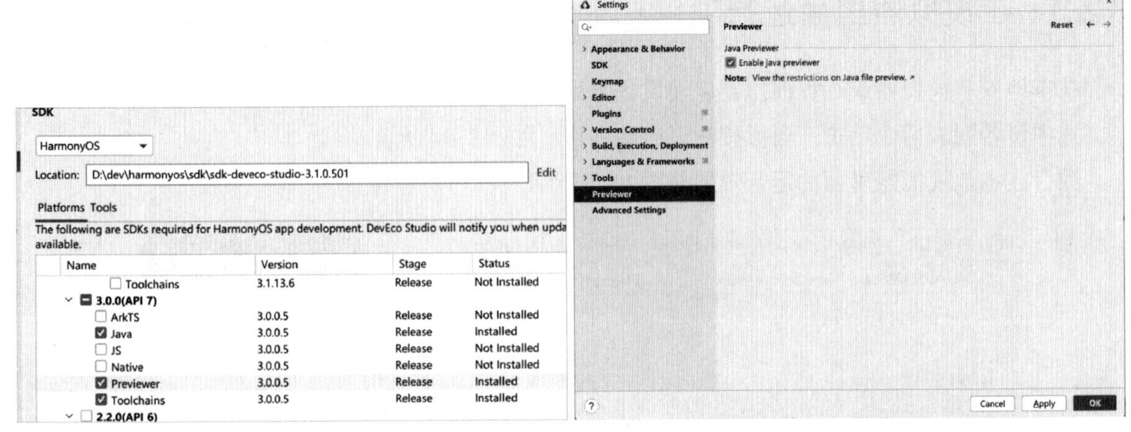

图 3-16　安装预览器

图 3-17　启用 Java Previewer

3.3.2　如何使用预览器

打开预览器有以下两种方式。

- 通过菜单栏单击“View>Tool Windows>Previewer”，打开预览器。
- 在编辑窗口右上角的侧边工具栏单击“Previewer”，打开预览器。

显示效果如图 3-18 所示。

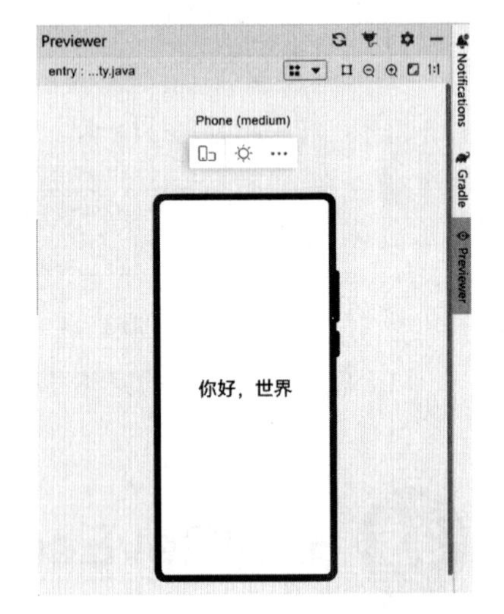

图 3-18　使用预览器

3.4 使用ArkTS创建一个新应用

在前面的章节中，已经演示了如何基于DevEco Studio来开发第一个HarmonyOS应用。第一个HarmonyOS应用是采用Java语言编写的。本节演示通过ArkTS来创建新应用。

3.4.1　选择创建新工程

打开DevEco Studio后，我们单击“Create Project”来创建一个新工程。

3.4.2　选择设备应用类型的模板

选择一个空模版“Empty Ability”，单击“Next”进行下一步，如图3-19所示。

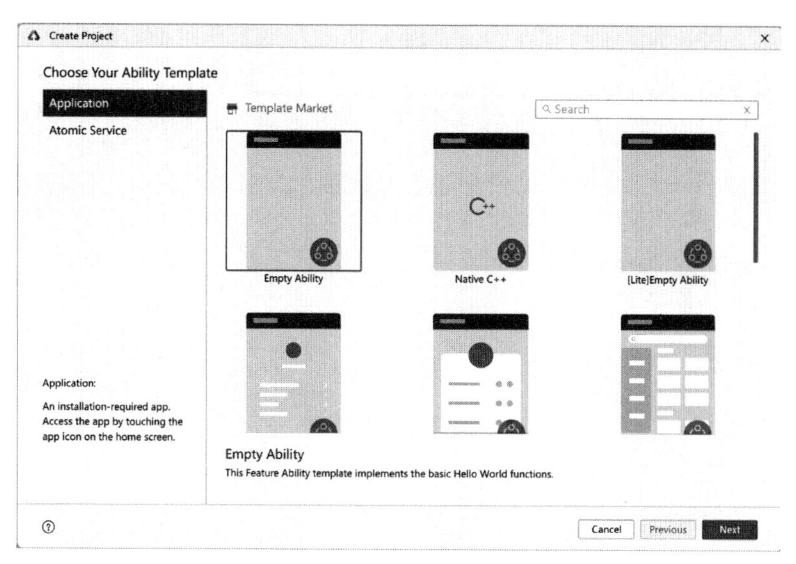

图3-19　选择一个空模版“Empty Ability”

3.4.3　配置项目的信息

下一步是配置项目的信息，比如项目名称、包名、位置、SDK版本等，按照个人实际情况填即可。

如图3-20所示，上述示例创建的ArkTSHelloWorld应用，选用的SDK版本为3.1.0(API 9)，编程语言是ArkTS，设备类型是Phone。

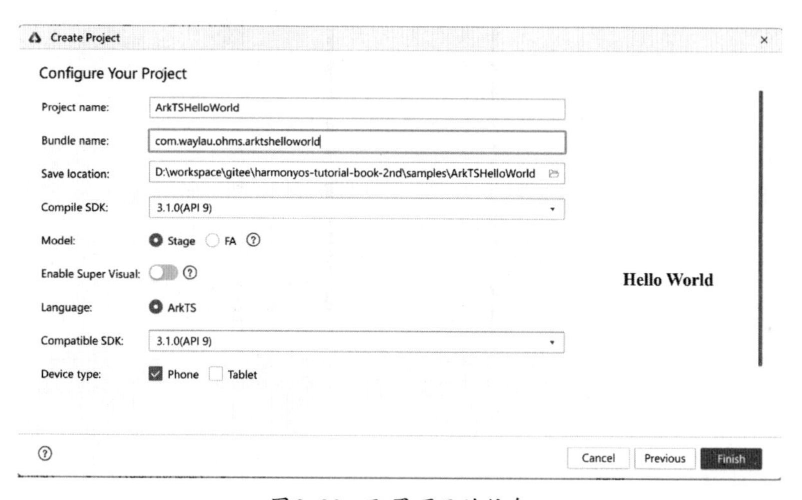

图3-20　配置项目的信息

3.4.4　自动生成工程代码

单击“Finish”之后，DevEco Studio就会创建好整个应用，并且自动生成工程代码，如图3-21

所示。

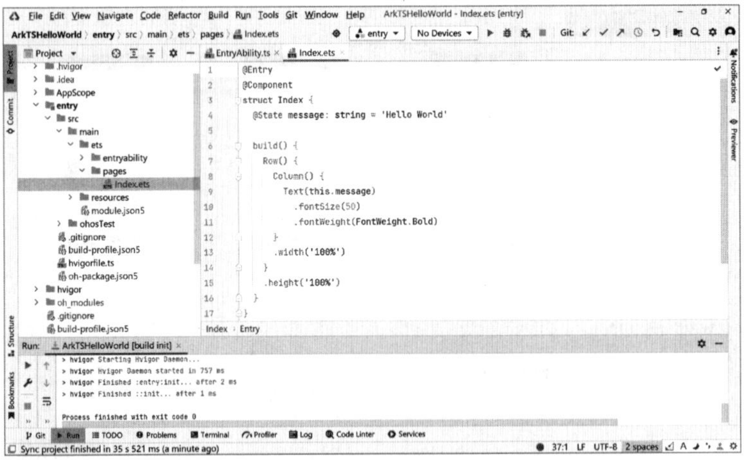

图 3-21　自动生成工程代码

3.4.5　查看应用运行效果

可以通过模拟器或预览器的方式来运行应用，查看实际的运行效果，如图 3-22 所示。

图 3-22　应用在预览器里面的运行效果

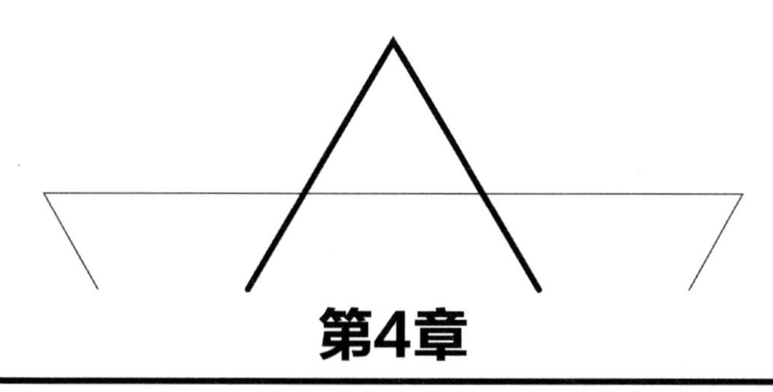

第4章

应用初探——探索HarmonyOS应用

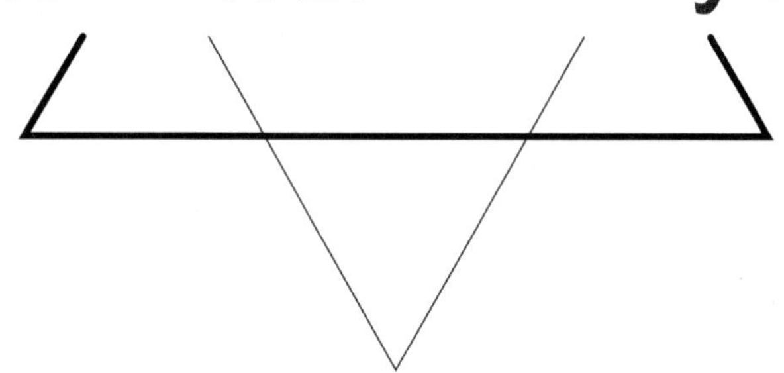

在上一章，我们没有敲一行代码，轻松创建了一个能够运行的 HarmonyOS 应用。那么这个 HarmonyOS 应用里面到底包含哪些内容呢？它为什么可以直接在设备上运行呢？本章将进行深入探索。

4.1 App

在上一章DevEco Studio为我们自动创建了如图4-1所示的目录结构的工程源码。那么这些目录结构到底是什么含义？每个文件的作用是什么？这些都是接下来我们要探索的内容。

在本节，我们将探索App这个话题。

4.1.1 什么是App

现在，我们认识下什么是App。

App就是应用。比如，在Android手机上安装一个软件，这个软件，我们就称之为App。

Android平台是以APK（Android application package，Android应用程序包）形式来发布的，当我们要在Android手机上安装一个App时，首先要找到这个App所对应的APK安装包，执行该APK安装包就能安装App了。

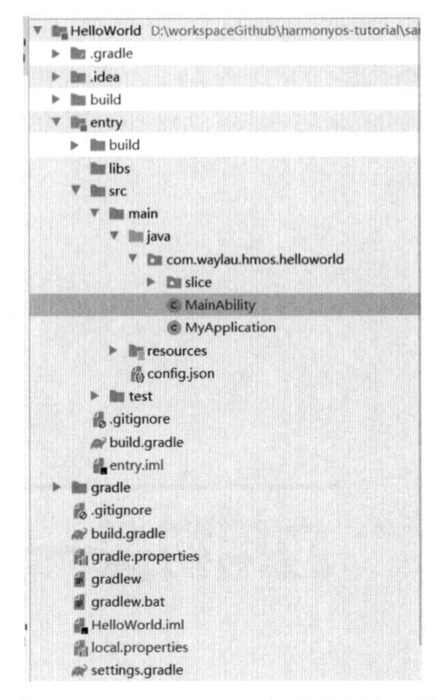

图4-1　DevEco Studio自动创建的工程

同理，HarmonyOS的应用软件包以App Pack（Application Package）形式发布，它是由一个或多个HAP（HarmonyOS Ability Package）及描述每个HAP属性的pack.info组成。HAP是Ability的部署包，HarmonyOS应用代码围绕Ability组件展开。

一个HAP是由代码、资源、第三方库及应用配置文件组成的模块包，可分为Entry和Feature两种模块类型，如图4-2所示。

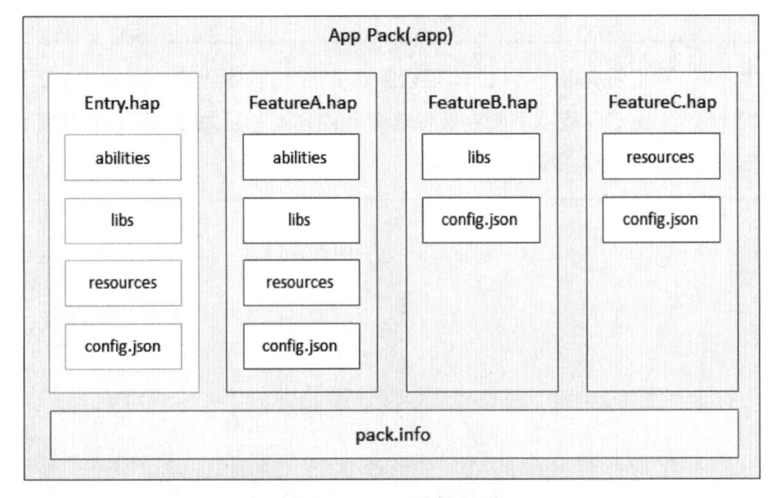

图4-2　App逻辑视图

- Entry：应用的主模块。一个App中，对于同一设备类型必须有且只有一个Entry类型的HAP，可独立安装运行。
- Feature：应用的动态特性模块。一个App可以包含一个或多个Feature类型的HAP，也可以不包含。只有包含Ability的HAP才能够独立运行。

我们在应用的build目录下，找一个名为“entry-debug-unsigned.hap”的文件，如图4-3所示，该文件就是HarmonyOS的应用软件包。

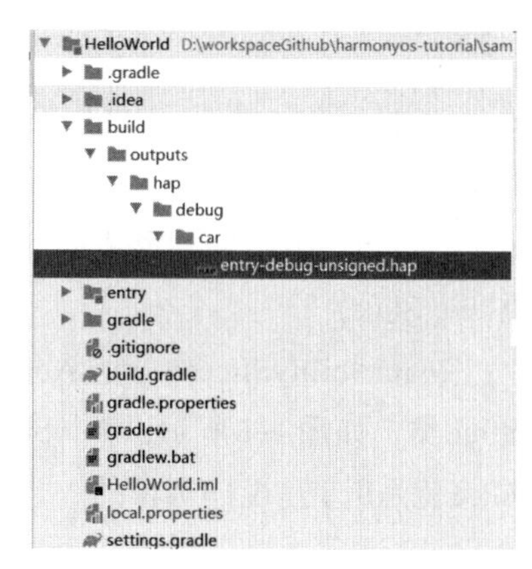

图4-3　HarmonyOS应用软件包

4.1.2　代码层次的应用

在代码层次，我们可以看到图4-4所示的MyApplication就是整个应用的入口。

从代码可以看到，MyApplication继承自AbilityPackage。AbilityPackage是用来初始化每个HAP的基类。

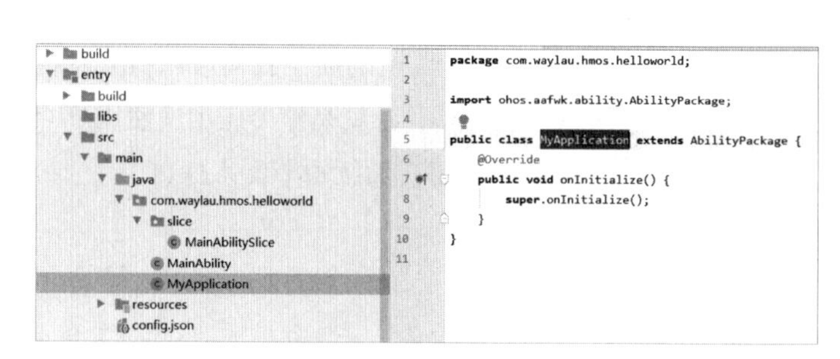

图4-4　App逻辑视图

4.2 Ability

Ability是应用所具备的能力的抽象，一个应用可以包含一个或多个Ability。

对于FA模型而言，Ability分为以下两种类型。

- FA（Feature Ability）。
- PA（Particle Ability）。

4.2.1　Ability类

FA/PA是应用的基本组成单元，能够实现特定的业务功能。两者主要的区别是FA有UI界面，而PA无UI界面。如图4-5所示，MainAbility就是一个FA。

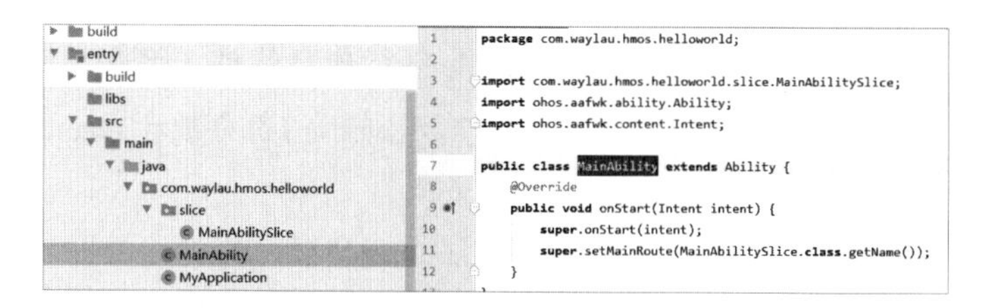

图4-5　MainAbility

4.2.2 AbilitySlice类

MainAbility继承自Ability类。同时，从代码可以看出，MainAbility设置了一个路由，可以路由到MainAbilitySlice。

MainAbilitySlice继承自AbilitySlice类，如图4-6所示，而AbilitySlice就是用于呈现UI界面的。

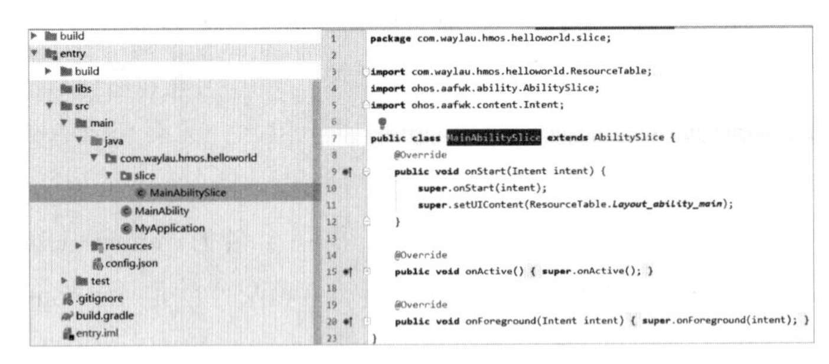

图4-6　AbilitySlice

4.2.3 UI界面

那么UI界面又是定义在哪里的呢？我们可以打开resources的base目录，如图4-7所示，该base目录就是整个应用所使用UI界面元素。

有关Ability的内容，还将在后续章节深入探讨。

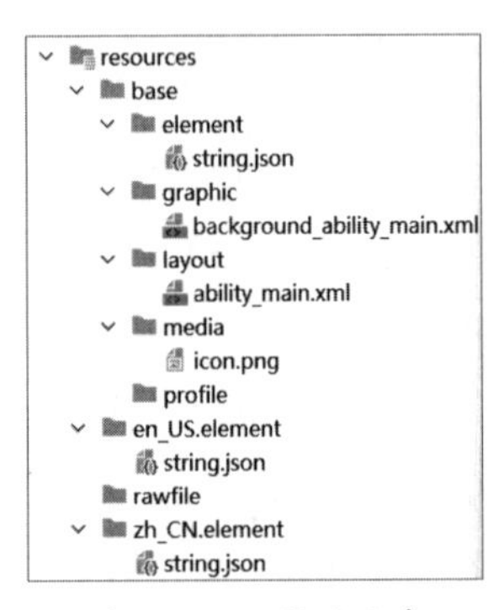

图4-7　UI界面元素

4.3 库文件

如果你有过Java开发或Android开发的经验，那么对于库文件就不会陌生。

库文件是应用依赖的第三方代码（如so、jar、bin、har等二进制文件），都存放在libs目录。libs目录位置如图4-8所示。

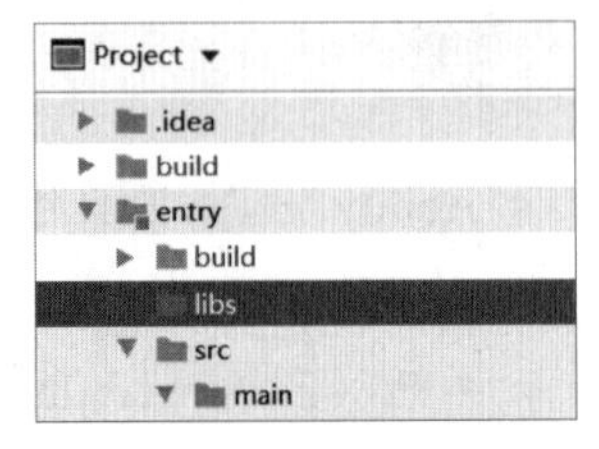

图4-8　libs目录位置

4.4 资源文件

HarmonyOS应用采用Gradle进行项目管理，因此与Maven类似，应用的资源文件（字符串、图片、音频等）都存放于resources目录下，便于开发者使用和维护。

图4-9所示的就是应用的resources目录。

resources目录包括两大类目录，一类为base目录与限定词目录，另一类为rawfile目录，如表4-1所示。

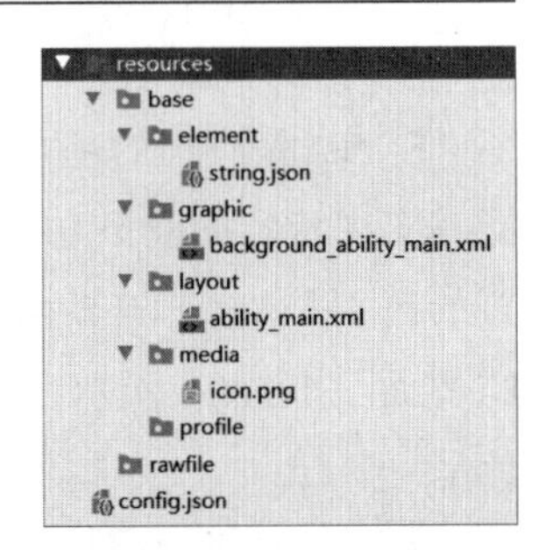

图4-9　resources目录

表4-1　resources的两大类目录

分类	base目录与限定词目录	rawfile目录
组织形式	按照两级目录形式来组织，目录命名必须符合规范，以便根据设备状态匹配相应目录下的资源文件。 一级子目录为base目录和限定词目录。base目录是默认存在的目录。当应用的resources资源目录中没有与设备状态匹配的限定词目录时，会自动引用该目录中的资源文件。限定词目录需要开发者自行创建。目录名称由一个或多个表征应用场景或设备特征的限定词组合而成。 二级子目录为资源目录，用于存放字符串、颜色、布尔值等基础元素，以及媒体、动画、布局等资源文件	支持创建多层子目录，目录名称可以自定义，文件夹内可以自由放置各类资源文件。rawfile目录的文件不会根据设备状态匹配不同的资源
编译方式	目录中的资源文件会被编译成二进制文件，并赋予资源文件ID	目录中的资源文件会被直接打包进应用，不经过编译，也不会被赋予资源文件ID
引用方式	通过指定资源类型（type）和资源名称（name）来引用	通过指定文件路径和文件名来引用

4.4.1　限定词目录

限定词目录可以由一个或多个表征应用场景或设备特征的限定词组合而成，包括语言、文字、国家或地区、横竖屏、设备类型和屏幕密度，共6个维度，限定词之间通过下划线（_）或中划线（-）连接。开发者在创建限定词目录时，需要掌握限定词目录的命名要求，以及限定词目录与设备状态的匹配规则。

1. 限定词目录的命名要求

限定词的组合顺序：语言_文字_国家或地区-横竖屏-设备类型-屏幕密度。开发者可以根据应用的使用场景和设备特征，选择其中的一类或几类限定词组成目录名称。 限定词的连接方式：语言、文字、国家或地区之间采用下划线（_）连接，除此之外的其他限定词之间均采用中划线（-）连接。例如：zh_Hant_CN、zh_CN-car-ldpi。

限定词的取值范围：每类限定词的取值必须符合表4-2中的条件，否则将无法匹配目录中的资源文件。

表4-2　限定词取值要求

限定词类型	含义与取值说明
语言	表示设备使用的语言类型，由两个小写字母组成。例如：zh表示中文，en表示英语。详细取值范围，参见ISO 639-1（ISO制定的语言编码标准）
文字	表示设备使用的文字类型，由一个大写字母（首字母）和三个小写字母组成。例如：Hans表示简体中文，Hant表示繁体中文。详细取值范围，参见ISO 15924（ISO制定的文字编码标准）

续表

限定词类型	含义与取值说明
国家或地区	表示用户所在的国家或地区，由2～3个大写字母或三个数字组成。例如：CN表示中国，GB表示英国。详细取值范围，参见ISO 3166-1（ISO制定的国家和地区编码标准）
横竖屏	表示设备的屏幕方向，取值包括vertical（竖屏）、horizontal（横屏）
设备类型	表示设备的类型，取值包括Phone（手机）、Tablet（平台）、Car（车机）、TV（智慧屏）、Wearable（智能穿戴）等
屏幕密度	表示设备的屏幕密度（单位为dpi）。sdpi表示小规模的屏幕密度（Small-scale Dots Per Inch），适用于dpi取值为(0, 120]的设备。mdpi表示中规模的屏幕密度（Medium-scale Dots Per Inch），适用于dpi取值为(120, 160]的设备。ldpi表示大规模的屏幕密度（Large-scale Dots Per Inch），适用于dpi取值为(160, 240]的设备。xldpi表示特大规模的屏幕密度（Extra Large-scale Dots Per Inch），适用于dpi取值为(240, 320]的设备。xxldpi表示超大规模的屏幕密度（Extra Extra Large-scale Dots Per Inch），适用于dpi取值为(320, 480]的设备。xxxldpi表示超特大规模的屏幕密度（Extra Extra Extra Large-scale Dots Per Inch），适用于dpi取值为(480, 640]的设备

2. 限定词目录与设备状态的匹配规则

在为设备匹配对应的资源文件时，限定词目录匹配的优先级从高到低依次为：

```
区域（语言_文字_国家或地区）> 横竖屏 > 设备类型 > 屏幕密度。
```

如果限定词目录中包含语言、文字、横竖屏、设备类型限定词，则对应限定词的取值必须与当前的设备状态完全一致，该目录才能够参与设备的资源匹配。例如，限定词目录zh_CN-car-ldpi不能参与en_US设备的资源匹配。

4.4.2 资源组目录

base目录与限定词目录下面可以创建资源组目录（包括element、media、animation、layout、graphic、profile），用于存放特定类型的资源文件。

1. element

element表示元素资源，以下每一类数据都采用相应的JSON文件来表征。

- boolean：布尔型。
- color：颜色。
- float：浮点型。
- intarray：整型数组。
- integer：整型。
- pattern：样式。
- plural：复数形式。

- strarray：字符串数组。
- string：字符串。

element目录中的文件名称建议与下面的文件名保持一致。每个文件中只能包含同一类型的数据。

- boolean.json。
- color.json。
- float.json。
- intarray.json。
- integer.json。
- pattern.json。
- plural.json。
- strarray.json。
- string.json。

2. media

media表示媒体资源，包括图片、音频、视频等非文本格式的文件。

文件名可自定义，例如：icon.png。

3. animation

animation表示动画资源，采用XML文件格式。

文件名可自定义，例如：zoom_in.xml。

4. layout

layout表示布局资源，采用XML文件格式。

文件名可自定义，例如：home_layout.xml。

5. graphic

graphic表示可绘制资源，采用XML文件格式。

文件名可自定义，例如：notifications_dark.xml。

6. profile

profile表示其他类型文件，以原始文件形式保存。

文件名可自定义。

4.5 配置文件

HarmonyOS应用的每个HAP的根目录下都存在一个配置文件，如果是用Java编写的应用配置文件，称为config.json；而如果是用ArkTS编写的应用配置文件，则分为app.json5和module.json5

两部分。图4-10所示的是用Java编写的应用配置文件config.json所在位置。

配置文件内容主要涵盖以下三个方面。

- 应用的全局配置信息，包含应用的包名、生产厂商、版本号等基本信息。
- 应用在具体设备上的配置信息，包含应用的备份恢复、网络安全等能力。
- HAP包的配置信息，包含每个Ability必须定义的基本属性（如包名、类名、类型及Ability提供的能力），以及应用访问系统或其他应用受保护部分所需的权限等。

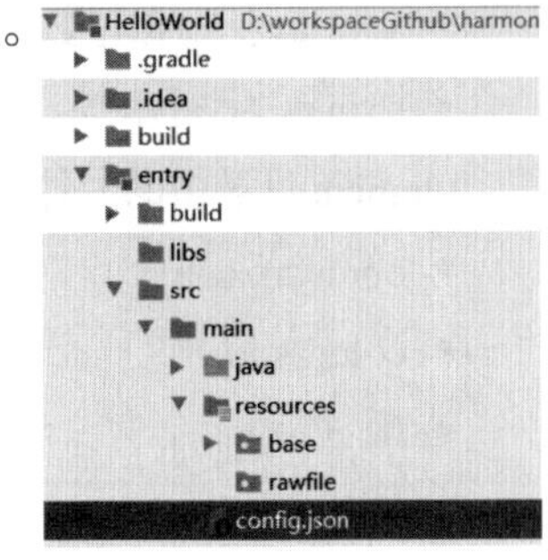

图4-10　配置文件

4.5.1　config.json配置文件的组成

配置文件config.json采用JSON文件格式，其中包含一系列配置项，每个配置项由属性和值两部分构成。

- 属性：属性出现顺序不分先后，且每个属性最多只允许出现一次。
- 值：每个属性的值为JSON的基本数据类型（数值、字符串、布尔值、数组、对象或null类型）。

应用的配置文件config.json中由app、deviceConfig和module三个部分组成，缺一不可。配置文件的内部结构说明参见表4-3。

表4-3　配置文件的内部结构说明

属性名称	含义	数据类型	是否可缺省
app	表示应用的全局配置信息。同一个应用的不同HAP包的App配置必须保持一致	对象	否
deviceConfig	表示应用在具体设备上的配置信息	对象	否
module	表示HAP包的配置信息。该标签下的配置只对当前HAP包生效	对象	否

以下是HelloWorld应用的配置文件：

```
{
  "app": {
    "bundleName": "com.ohms.helloworld",
    "vendor": "example",
    "version": {
      "code": 1000000,
      "name": "1.0.0"
    }
  },
  "deviceConfig": {
```

```
  },
  "module": {
    "package": "com.ohms.helloworld",
    "name": ".MyApplication",
    "mainAbility": "com.ohms.helloworld.MainAbility",
    "deviceType": [
      "phone"
    ],
    "distro": {
      "deliveryWithInstall": true,
      "moduleName": "entry",
      "moduleType": "entry",
      "installationFree": false
    },
    "abilities": [
      {
        "skills": [
          {
            "entities": [
              "entity.system.home"
            ],
            "actions": [
              "action.system.home"
            ]
          }
        ],
        "name": "com.ohms.helloworld.MainAbility",
        "description": "$string:mainability_description",
        "icon": "$media:icon",
        "label": "$string:entry_MainAbility",
        "launchType": "standard",
        "orientation": "unspecified",
        "visible": true,
        "type": "page"
      }
    ]
  }
}
```

接下来，我们详解介绍上述配置的含义。

4.5.2 app.json5配置文件的组成

app.json5 主要包含以下内容。

- 应用的全局配置信息，包含应用的包名、开发厂商、版本号等基本信息。

- 特定设备类型的配置信息。

以下是ArkTSHelloWorld应用的配置文件：

```
{
  "app": {
    "bundleName": "com.waylau.ohms.arktshelloworld",
    "vendor": "example",
    "versionCode": 1000000,
    "versionName": "1.0.0",
    "icon": "$media:app_icon",
    "label": "$string:app_name"
  }
}
```

4.5.3 module.json5配置文件的组成

module.json5主要包含以下内容。

- Module的基本配置信息，例如Module的名称、类型、描述、支持的设备类型等基本信息。
- 应用组件信息，包含UIAbility组件和ExtensionAbility组件的描述信息。
- 应用运行过程中所需的权限信息。

以下是ArkTSHelloWorld应用的配置文件：

```
{
  "module": {
    "name": "entry",
    "type": "entry",
    "description": "$string:module_desc",
    "mainElement": "EntryAbility",
    "deviceTypes": [
      "phone"
    ],
    "deliveryWithInstall": true,
    "installationFree": false,
    "pages": "$profile:main_pages",
    "abilities": [
      {
        "name": "EntryAbility",
        "srcEntry": "./ets/entryability/EntryAbility.ts",
        "description": "$string:EntryAbility_desc",
        "icon": "$media:icon",
        "label": "$string:EntryAbility_label",
        "startWindowIcon": "$media:icon",
        "startWindowBackground": "$color:start_window_background",
```

```
          "exported": true,
          "skills": [
            {
              "entities": [
                "entity.system.home"
              ],
              "actions": [
                "action.system.home"
              ]
            }
          ]
        }
      ]
    }
}
```

更多配置项的含义，请参阅官方文档。

4.6 pack.info

pack.info用于描述应用软件包中每个HAP的属性，由IDE编译生成，应用市场根据该文件进行拆包和HAP的分类存储。HAP的具体属性包括以下几种。

- delivery-with-install: 表示该HAP是否支持随应用安装。true表示支持随应用安装；false表示不支持随应用安装。
- name：HAP文件名。
- module-type：模块类型，Entry或Feature。
- device-type：表示支持该HAP运行的设备类型。

第5章 Ability基础知识

本章介绍 Ability 的基础知识。Ability 是 HarmonyOS 应用所具备能力的抽象，也是 HarmonyOS 应用程序的核心组成部分。

5.1 Ability概述

在HarmonyOS应用中，有一个非常核心的概念，那就是Ability。正如其字面含义，Ability可以理解为HarmonyOS应用所具备能力的抽象。一个HarmonyOS应用可以具备多少种能力，也就会包含多少个Ability。HarmonyOS支持应用以Ability为单位进行部署。

5.1.1 Ability分类

如果是用Java编写的应用，Ability主要分为以下两种类型。

- FA（Feature Ability）。
- PA（Particle Ability）。

如果是用ArkTS编写的应用，Ability主要分为以下两种类型。

- UIAbility。
- ExtensionAbility。

前一种应用模型称为FA模型，而后一种应用模型称为Stage模型。Stage模型是HarmonyOS 3.1 Developer Preview版本开始新增的模型，是目前主推且会长期演进的模型。在该模型中，由于提供了AbilityStage、WindowStage等类作为应用组件和Window窗口的“舞台”，因此称这种应用模型为Stage模型。

5.1.2 FA

FA支持Page Ability。就目前而言，Page模板是FA唯一支持的模板，用于提供与用户交互的能力。一个Page实例可以包含一组相关页面，每个页面用一个AbilitySlice实例表示。简言之，FA就是承担前端与用户交互的。

5.1.3 PA

PA支持Service Ability和Data Ability两种模板。

- Service模板：用于提供后台运行任务的能力。
- Data模板：用于对外部提供统一的数据访问抽象。

5.1.4 UIAbility

UIAbility组件是一种包含UI界面的应用组件，主要用于和用户交互。

UIAbility组件是系统调度的基本单元，为应用提供绘制界面的窗口。一个UIAbility组件中可以通过多个页面来实现一个功能模块。每一个UIAbility组件实例，都对应一个最近任务列表中的任务。

5.1.5 ExtensionAbility

ExtensionAbility组件是基于特定场景提供的应用组件，以便满足更多的使用场景。

每一个具体场景对应一个ExtensionAbilityType，各类型的ExtensionAbility组件均由相应的系统服务统一管理，例如，InputMethodExtensionAbility组件由输入法管理服务统一管理。当前支持的ExtensionAbility类型有以下两种。

- FormExtensionAbility：FORM类型的ExtensionAbility组件，用于提供服务卡片场景相关能力。
- WorkSchedulerExtensionAbility：WORK_SCHEDULER类型的ExtensionAbility组件，用于提供延迟任务注册、取消、查询的能力。

5.2 Stage模型介绍

Stage模型是HarmonyOS 3.1版本开始新增的模型，也是目前HarmonyOS主推且会长期演进的模型。如果采用ArkTS语言开发HarmonyOS应用，则建议始终采用Stage模型作为开发方式。

5.2.1 Stage模型的设计思想

Stage模型之所以成为主推模型，源于其设计思想。Stage模型的设计基于如下3个出发点。

1. 为复杂应用而设计

Stage模型可以简化应用复杂度。

- 多个应用组件共享同一个ArkTS引擎（运行ArkTS语言的虚拟机）实例，应用组件之间可以方便地共享对象和状态，同时减少复杂应用运行对内存的占用。
- 采用面向对象的开发方式，使复杂应用代码可读性高、易维护性好、可扩展性强。

2. 支持多设备和多窗口形态

应用组件管理和窗口管理在架构层面解耦。

- 便于系统对应用组件进行裁剪（无屏设备可裁剪窗口）。
- 便于系统扩展窗口形态。
- 在多设备（如桌面设备和移动设备）上应用组件可使用同一套生命周期。

3. 平衡应用能力和系统管控成本

Stage模型重新定义应用能力的边界，平衡应用能力和系统管控成本。

- 提供特定场景（如卡片、输入法）的应用组件，以便满足更多的使用场景。

- 规范化后台进程管理：为保障用户体验，Stage模型对后台应用进程进行了有序治理，应用程序不能随意驻留在后台，同时应用后台行为受到严格管理，防止恶意应用行为。

5.2.2 Stage模型的Ability生命周期

在Ability的使用过程中，会有多种生命周期状态。掌握Ability的生命周期，对于应用的开发非常重要。

为了实现多设备形态上的裁剪和多窗口的可扩展性，系统对组件管理和窗口管理进行了解耦。Ability的生命周期包括Create、Foreground、Background、Destroy四个状态，如图5-1所示。WindowStageCreate和WindowStageDestroy为窗口管理器（WindowStage）在Ability中管理UI界面功能的两个生命周期回调，从而实现Ability与窗口之间的弱耦合，如图5-2所示。

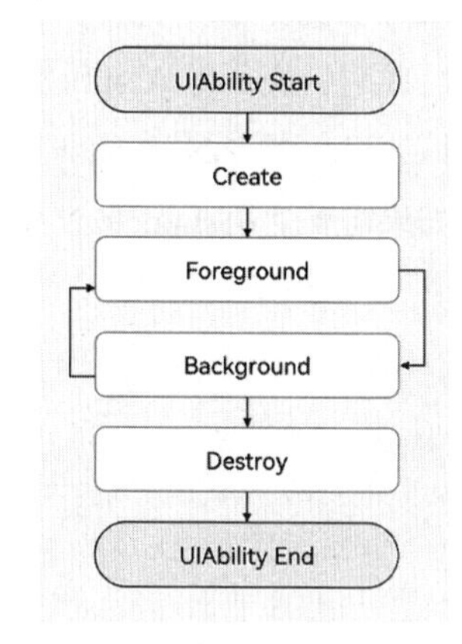

图5-1 Stage模型的Ability生命周期

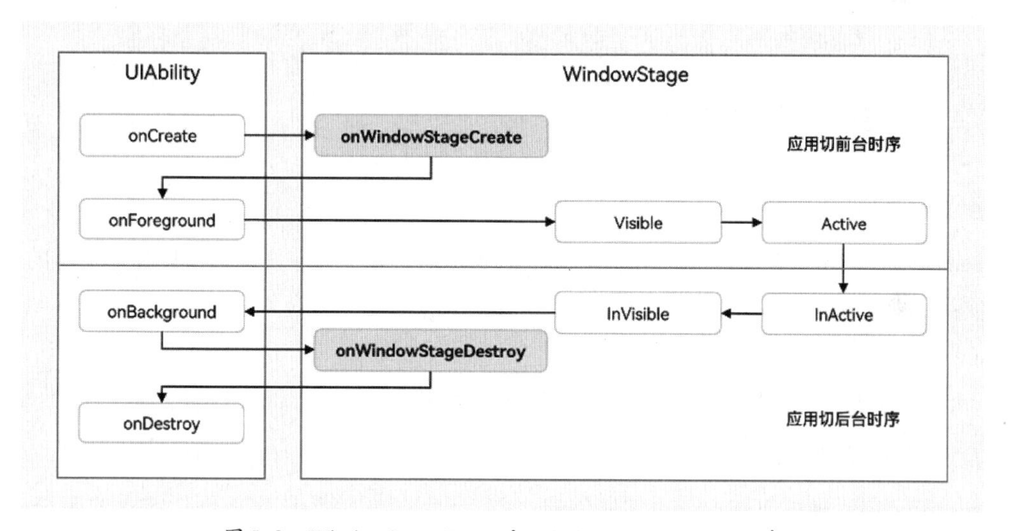

图5-2 WindowStageCreate和WindowStageDestroy状态

5.2.3 Stage模型的Ability启动模式

Ability的启动模式是指Ability实例在启动时的不同呈现状态。针对不同的业务场景，系统提供了以下三种启动模式。

- singleton（单实例模式）。
- standard（标准实例模式）。
- specified（指定实例模式）。

1. singleton 启动模式

singleton是默认情况下的启动模式。

每次调用startAbility()方法时，如果应用进程中该类型的Ability实例已经存在，则复用系统中的Ability实例。系统中只存在唯一该Ability实例，即在最近任务列表中只存在一个该类型的Ability实例。此时，应用的Ability实例已创建，当再次调用startAbility()方法启动该Ability实例时，只会进入该Ability的onNewWant()回调，不会进入其onCreate()和onWindowStageCreate()生命周期回调。

如果需要使用singleton启动模式，将module.json5配置文件中的launchType字段配置为singleton即可。

```
{
  "module": {
    // ...
    "abilities": [
      {
        "launchType": "singleton",
        // ...
      }
    ]
  }
}
```

2. standard 启动模式

在standard启动模式下，每次调用startAbility()方法时，都会在应用进程中创建一个新的该类型Ability实例，即在最近任务列表中可以看到多个该类型的Ability实例。这种情况下可以将Ability配置为standard。

standard启动模式的开发使用，需将module.json5配置文件中的launchType字段配置为standard。

3. specified 启动模式

在specified启动模式下，Ability实例创建之前，允许开发者为该实例创建唯一的字符串Key，创建的Ability实例绑定Key之后，后续每次调用startAbility()方法时，都会询问应用使用哪个Key对应的Ability实例来响应startAbility请求。运行时由Ability内部业务决定是否创建多实例，如果匹配有该Ability实例的Key，则直接拉起与之绑定的Ability实例，否则创建一个新的Ability实例。

例如，用户在应用中重复打开同一个文档时，启动的均是最近任务列表中的同一个任务；在应用中重复新建文档时，启动的均是最近任务列表中的新的任务。这种情况下可以将Ability配置为specified。当再次调用startAbility()方法启动该Ability实例，且AbilityStage的onAcceptWant()回调匹配到一个已创建的Ability实例，此时再次启动该Ability，只会进入该Ability的onNewWant()回调，不会进入其onCreate()和onWindowStageCreate()生命周期回调。

如果要使用specified启动模式，需将module.json5配置文件的launchType字段配置为specified。

5.3 Page Ability

正如前文所讲，Page Ability代表应用表示层的功能，用于提供与用户交互的能力。

5.3.1 Page Ability基本概念

Page模板（以下简称Page）是FA中目前唯一支持的模板。

一个Page可以由一个或多个AbilitySlice构成，AbilitySlice是指应用的单个页面及其控制逻辑的总和。

在HelloWorld应用中，MainAbility类就是一个Page。代码如下：

```
public class MainAbility extends Ability {
    @Override
    public void onStart(Intent intent) {
        super.onStart(intent);
        super.setMainRoute(MainAbilitySlice.class.getName());
    }
}
```

在HelloWorld应用中，MainAbility这个Page是由一个AbilitySlice构成的，这个AbilitySlice就是MainAbilitySlice类。代码如下：

```
public class MainAbilitySlice extends AbilitySlice {
    @Override
    public void onStart(Intent intent) {
        super.onStart(intent);
        super.setUIContent(ResourceTable.Layout_ability_main);
    }

    @Override
    public void onActive() {
        super.onActive();
    }

    @Override
    public void onForeground(Intent intent) {
        super.onForeground(intent);
    }
}
```

5.3.2 多个AbilitySlice构成一个Page

当一个Page由多个AbilitySlice共同构成时，这些AbilitySlice页面提供的业务能力应具有高度相关性。例如，新闻浏览功能可以通过一个Page来实现，其中包含两个AbilitySlice：一个AbilitySlice用于展示新闻列表，另一个AbilitySlice用于展示新闻详情。Page和AbilitySlice的关系如图5-3所示。

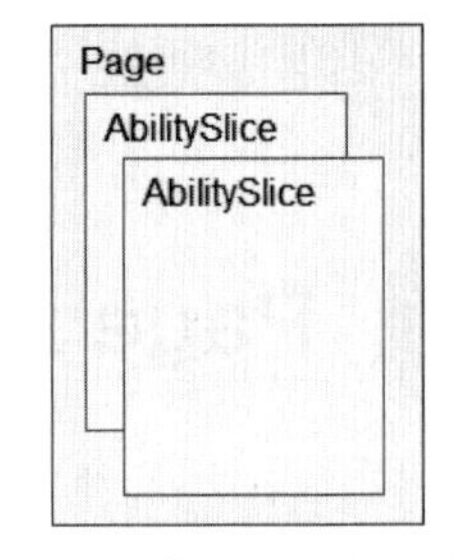

图5-3 Page和AbilitySlice的关系

相比于桌面场景，移动场景下应用之间的交互更为频繁。通常，单个应用专注于某个方面的能力开发，当它需要其他能力辅助时，会调用其他应用提供的能力。例如，快递应用提供了联系快递员的业务功能入口，当用户在使用该功能时，会跳转到通话应用的拨号页面。与此类似，HarmonyOS支持不同Page之间的跳转，并可以指定跳转到目标Page中某个具体的AbilitySlice。

5.3.3 AbilitySlice路由配置

虽然一个Page可以包含多个AbilitySlice，但是Page进入前台时界面默认只能展示一个AbilitySlice。默认展示的AbilitySlice是通过setMainRoute()方法来指定的。

比如，在HelloWorld应用中，MainAbility类就指定了MainAbilitySlice类作为默认展示的AbilitySlice。代码如下：

```
public class MainAbility extends Ability {
    @Override
    public void onStart(Intent intent) {
        super.onStart(intent);

        // 指定默认展示的 AbilitySlice
        super.setMainRoute(MainAbilitySlice.class.getName());
    }
}
```

如果需要更改默认展示的AbilitySlice，可以通过addActionRoute()方法为此AbilitySlice配置一条路由规则。setMainRoute()方法与addActionRoute()方法的使用示例如下：

```
public class MainAbility extends Ability {
    @Override
    public void onStart(Intent intent) {
        super.onStart(intent);

        // 指定默认展示的 AbilitySlice
        super.setMainRoute(MainAbilitySlice.class.getName());

        // 配置路由规则
```

```
        addActionRoute("action.pay", PayAbilitySlice.class.getName());
    }
}
```

下一节，我们将完整展示如何实现AbilitySlice的路由和导航。

5.3.4 不同Page间导航

不同Page中的AbilitySlice相互不可见，因此无法通过present()或presentForResult()方法直接导航到其他Page的AbilitySlice。

AbilitySlice作为Page的内部单元，以Action的形式对外暴露，因此可以通过配置Intent的Action导航到目标AbilitySlice。Page间的导航可以使用startAbility()或startAbilityForResult()方法，获得返回结果的回调为onAbilityResult()。在Ability中调用setResult()可以设置返回结果。

实战：多个AbilitySlice间的路由和导航

本节主要演示在一个Page包含多个AbilitySlice时，这些AbilitySlice之间是如何路由和导航的。

5.4.1 创建应用

创建一个名为“JavaAbilitySlice Navigation”的应用，如图5-4所示。该应用主要用于测试AbilitySlice之间的路由和导航。

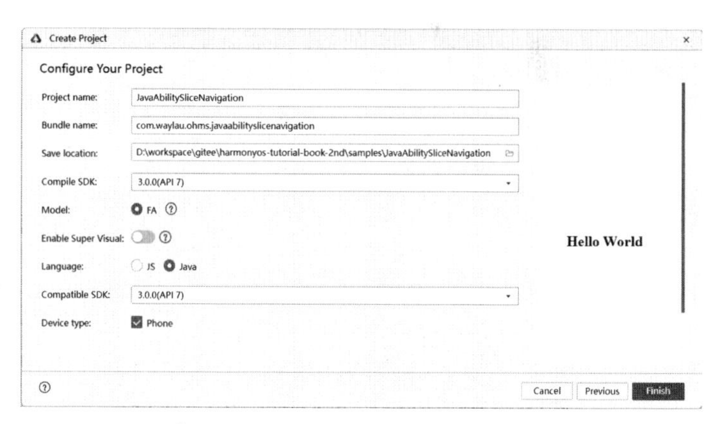

图5-4 JavaAbilitySliceNavigation应用

5.4.2 新建Page Ability

在初始化应用时，JavaAbilitySliceNavigation应用已经包含一个主Ability（MainAbility.java）和主AbilitySlice（MainAbilitySlice.java），因此，还需要新增一个Page Ability。

右击应用，通过菜单选项来创建一个名为“PayAbility”的Page Ability，如图5-5所示。

设置PayAbility的属性，如图5-6所示。

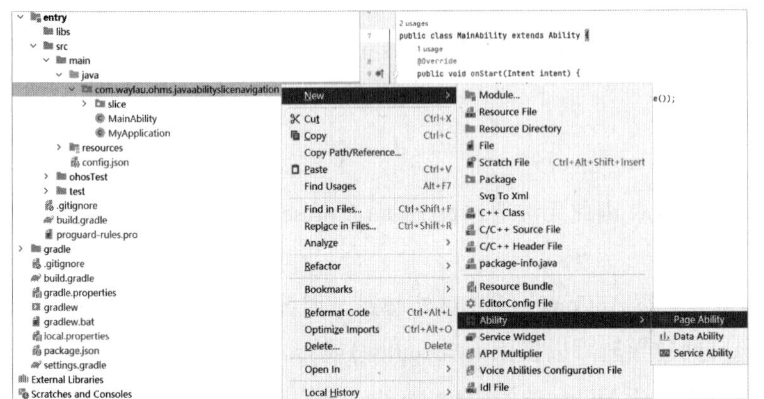

图5-5 通过菜单选项来创建Page Ability

New Page Ability

Configure Your Ability

Ability name: PayAbility

Language: Java

Layout name: ability_pay

Package name: com.waylau.ohms.javaabilityslicenavigation

Launcher ability:

图5-6 设置PayAbility的属性

Page Ability创建完成之后，会自动在应用目录下创建以下4个文件（如图5-7所示）。

- PayAbility。
- PayAbilitySlice。
- ability_pay.xml：表示布局资源。
- background_ability_pay.xml：表示可绘制资源。

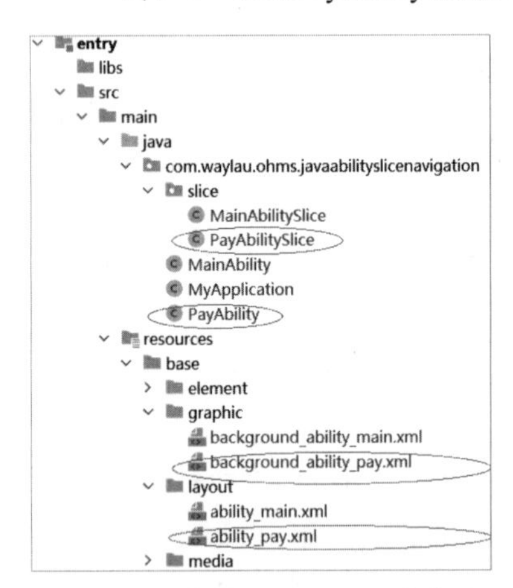

图5-7 新建Page Ability时生成的4个文件

上述4个文件基本与MainAbility、MainAbilitySlice.java及资源文件ability_main.xml、background_ability_main.xml内容相似。

同时，在config.json文件中，会自动添加PayAbility相关配置内容。代码如下：

```
{
  "app": {
    "bundleName": "com.waylau.ohms.javaabilityslicenavigation",
    "vendor": "example",
    "version": {
      "code": 1000000,
      "name": "1.0.0"
    }
  },
  "deviceConfig": {
  },
  "module": {
    "package": "com.waylau.ohms.javaabilityslicenavigation",
    "name": ".MyApplication",
```

```
    "mainAbility": "com.waylau.ohms.javaabilityslicenavigation.MainAbility",
    "deviceType": [
      "phone"
    ],
    "distro": {
      "deliveryWithInstall": true,
      "moduleName": "entry",
      "moduleType": "entry",
      "installationFree": false
    },
    "abilities": [
      // 为节约篇幅，省略部分代码
      // 自动添加 PayAbility 相关的配置内容
      {
        "name": "com.waylau.ohms.javaabilityslicenavigation.PayAbility",
        "description": "$string:payability_description",
        "icon": "$media:icon",
        "label": "$string:entry_PayAbility",
        "launchType": "standard",
        "orientation": "unspecified",
        "type": "page"
      }
    ]
  }
}
```

5.4.3 修改PayAbilitySlice样式布局

为了体现MainAbilitySlice和PayAbilitySlice的不同，我们需要在“面子”上“整容”一下。打开ability_pay.xml文件，可以看到文件内容如下：

```
<?xml version="1.0" encoding="utf-8"?>
<DirectionalLayout
    xmlns:ohos="http://schemas.huawei.com/res/ohos"
    ohos:height="match_parent"
    ohos:width="match_parent"
    ohos:alignment="center"
    ohos:orientation="vertical">

    <Text
        ohos:id="$+id:text_helloworld"
        ohos:height="match_content"
        ohos:width="match_content"
        ohos:background_element="$graphic:background_ability_pay"
```

```
        ohos:layout_alignment="horizontal_center"
        ohos:text="$string:payability_HelloWorld"
        ohos:text_size="40vp"
        />

</DirectionalLayout>
```

上述文件使用了Text组件，其Text组件的id由于与ability_main.xml里面相同，因此将id改为$+id:text_pay文件。该组件所要显示的文本内容引用了元素资源payability_HelloWorld。元素资源位于图5-8所示的3个目录的string.json文件中。

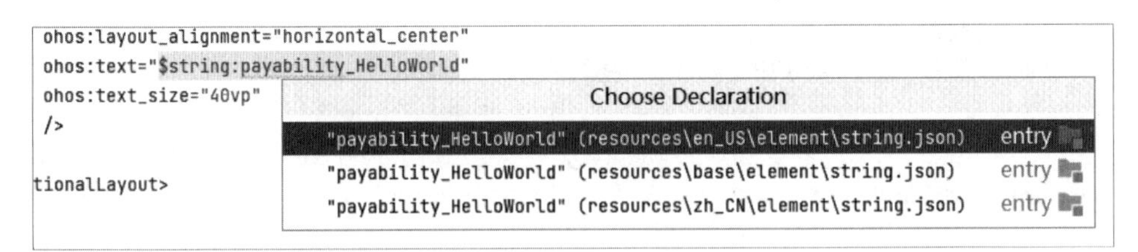

图5-8　元素资源

根据当前的语言环境，实际生效的是resources_CN.json。该string.json文件内容如下：

```
{
  "string": [
    {
      "name": "entry_MainAbility",
      "value": "entry_MainAbility"
    },
    {
      "name": "mainability_description",
      "value": "Java_Empty Ability"
    },
    {
      "name": "mainability_HelloWorld",
      "value": "你好，世界"
    },
    {
      "name": "payability_description",
      "value": "Java_Empty Ability"
    },
    {
      "name": "payability_HelloWorld",
      "value": "你好，世界"
    },
    {
      "name": "entry_PayAbility",
      "value": "entry_PayAbility"
```

```
    }
  ]
}
```

因此，payability_HelloWorld目前所对应的值是“你好，世界”。将值改为“你好，支付”，以便与MainAbilitySlice显示内容区分开。

5.4.4 实现AbilitySlice之间的路由和导航

下面演示如何实现AbilitySlice之间的路由和导航，请参考如下步骤。

1. 设置路由

在MainAbility中，通过addActionRoute方法来将PayAbilitySlice添加到路由。代码如下：

```
package com.waylau.ohms.javaabilityslicenavigation;

import com.waylau.ohms.javaabilityslicenavigation.slice.MainAbilitySlice;
import com.waylau.ohms.javaabilityslicenavigation.slice.PayAbilitySlice;
import ohos.aafwk.ability.Ability;
import ohos.aafwk.content.Intent;

public class MainAbility extends Ability {
    @Override
    public void onStart(Intent intent) {
        super.onStart(intent);
        // 指定默认显示的 AbilitySlice
        super.setMainRoute(MainAbilitySlice.class.getName());

        // 使用 addActionRounte 方法添加路由
        addActionRoute("action.pay", PayAbilitySlice.class.getName());
    }
}
```

其中，上述“action.pay”是指定路由动作的名称。这个名称还需要在config.json的actions数组中进行添加，配置如下：

```
"abilities": [
  {
    "skills": [
      {
        "entities": [
          "entity.system.home"
        ],
        "actions": [
          "action.system.home",
```

```
            "action.pay"  // 指定路由动作的名称
          ]
          }
      ],
      "name": "com.waylau.ohms.javaabilityslicenavigation.MainAbility",
      "description": "$string:mainability_description",
      "icon": "$media:icon",
      "label": "$string:entry_MainAbility",
      "launchType": "standard",
      "orientation": "unspecified",
      "visible": true,
      "type": "page"
    }
  ]
```

2. 设置点击事件触发导航

在MainAbilitySlice中，为文本设置点击事件，以便能够触发导航到PayAbilitySlice。代码如下：

```
package com.waylau.ohms.javaabilityslicenavigation.slice;

import com.waylau.ohms.javaabilityslicenavigation.ResourceTable;
import ohos.aafwk.ability.AbilitySlice;
import ohos.aafwk.content.Intent;
import ohos.agp.components.Text;

public class MainAbilitySlice extends AbilitySlice {
    @Override
    public void onStart(Intent intent) {
        super.onStart(intent);
        // 指定 ability_main.xml 定义的 UI
        super.setUIContent(ResourceTable.Layout_ability_main);

        // 添加点击事件来触发导航
        Text text = (Text) findComponentById(ResourceTable.Id_text_
helloworld);
        text.setClickedListener(listener ->
                present(new PayAbilitySlice(), new Intent()));;
    }

    @Override
    public void onActive() {
        super.onActive();
    }

    @Override
```

```
    public void onForeground(Intent intent) {
        super.onForeground(intent);
    }
}
```

在上述代码中，当发起导航的AbilitySlice和导航目标的AbilitySlice处于同一个Page时，可以通过present()方法实现导航。

同理，在PayAbilitySlice中，为文本设置点击事件，以触发导航到MainAbilitySlice。代码如下：

```
package com.waylau.ohms.javaabilityslicenavigation.slice;

import com.waylau.ohms.javaabilityslicenavigation.ResourceTable;
import ohos.aafwk.ability.AbilitySlice;
import ohos.aafwk.content.Intent;
import ohos.agp.components.Text;

public class PayAbilitySlice extends AbilitySlice {
    @Override
    public void onStart(Intent intent) {
        super.onStart(intent);

        // 指定 ability_pay.xml 定义的 UI
        super.setUIContent(ResourceTable.Layout_ability_pay);

        // 添加点击事件来触发导航
        Text text = (Text) findComponentById(ResourceTable.Id_text_pay);
        text.setClickedListener(listener ->
                present(new MainAbilitySlice(), new Intent()));
    }

    @Override
    public void onActive() {
        super.onActive();
    }

    @Override
    public void onForeground(Intent intent) {
        super.onForeground(intent);
    }
}
```

5.4.5 运行

运行应用，选中MainAbility类，再单击预览器，可以看到如图5-9所示的界面效果。

点击文本“你好，世界”后，可以切换到“你好，支付”，界面效果如图5-10所示。

再点击文本“你好，支付”，可以切换到“你好，世界”。至此实现了同一个Page下多个AbilitySlice之间的路由和导航。

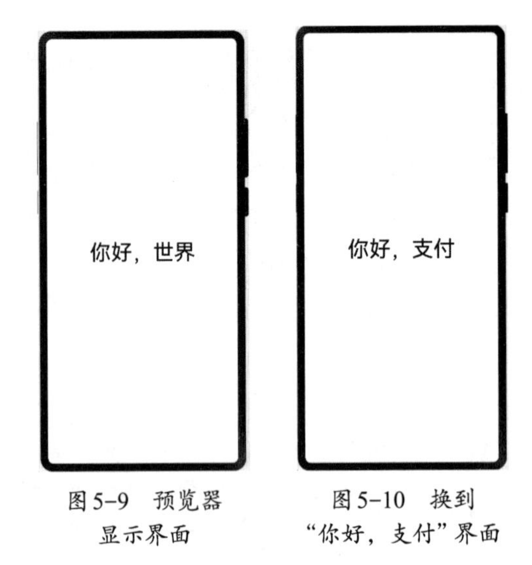

图5-9　预览器显示界面

图5-10　换到“你好，支付”界面

5.5 Page与AbilitySlice生命周期

本节介绍Page与AbilitySlice的生命周期。系统管理或用户操作等行为均会引起Page实例在其生命周期的不同状态之间进行转换。ability类提供的回调机制能够让Page及时感知外界变化，从而正确地应对状态变化（比如释放资源），这有助于提升应用的性能和稳健性。

5.5.1 Page生命周期

Page生命周期的不同状态转换及其对应的回调，如图5-11所示。

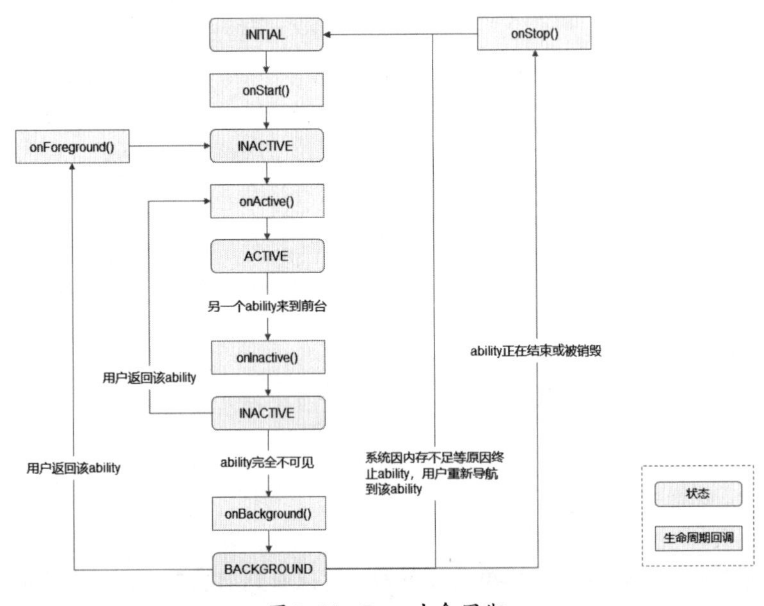

图5-11　Page生命周期

Page生命周期包含以下回调方法。

1. onStart()

当系统首次创建Page实例时，触发该回调。对于一个Page实例，该回调在其生命周期过程中仅触发一次，Page在该逻辑执行后将进入INACTIVE状态。开发者必须重写该方法，并在此配置默认展示的AbilitySlice。

比如，在“HelloWorld”应用中，经常会在onStart()回调方法中设置默认显示的AbilitySlice。代码如下：

```
@Override
public void onStart(Intent intent) {
    super.onStart(intent);

    // 指定默认显示的 AbilitySlice
    super.setMainRoute(MainAbilitySlice.class.getName());
}
```

2. onActive()

Page会在进入INACTIVE状态后来到前台，然后系统调用此回调。Page在此之后进入ACTIVE状态，该状态是应用与用户交互的状态。Page将保持在此状态，除非某类事件发生导致Page失去焦点，比如用户点击返回键或导航到其他Page。当此类事件发生时，会触发Page回到INACTIVE状态，系统将调用onInactive()回调。此后，Page可能重新回到ACTIVE状态，系统将再次调用onActive()回调。因此，开发者通常需要成对实现onActive()和onInactive()，并在onActive()中获取在onInactive()中被释放的资源。

3. onInactive()

当Page失去焦点时，系统将调用此回调，此后Page进入INACTIVE状态。开发者可以在此回调中实现Page失去焦点时应表现的恰当行为。

4. onBackground()

如果Page不再对用户可见，系统将调用此回调，通知开发者用户进行相应的资源释放，此后Page进入BACKGROUND状态。开发者应该在此回调中释放Page不可见时无用的资源，或在此回调中执行较为耗时的状态保存操作。

5. onForeground()

处于BACKGROUND状态的Page仍然驻留在内存中，当重新回到前台时（比如用户重新导航到此Page），系统将先调用onForeground()回调通知开发者，而后Page的生命周期状态回到INACTIVE状态。开发者应当在此回调中重新申请在onBackground()中释放的资源，最后Page的生命周期状态进一步回到ACTIVE状态，系统将通过onActive()回调通知开发者用户。

6. onStop()

系统要销毁Page时，将会触发此回调函数，通知用户进行系统资源的释放。销毁Page的可能原因包括以下几个方面：

- 用户通过系统管理能力关闭指定Page，例如使用任务管理器关闭Page。
- 用户行为触发Page的terminateAbility()方法调用，例如使用应用的退出功能。
- 配置变更导致系统暂时销毁Page并重建。
- 系统出于资源管理目的，自动触发销毁处于BACKGROUND状态的Page。

5.5.2 AbilitySlice生命周期

AbilitySlice作为Page的组成单元，其生命周期是依托于其所属Page生命周期的。AbilitySlice和Page具有相同的生命周期状态和同名的回调，当Page生命周期发生变化时，它的AbilitySlice也会发生相同的生命周期变化。此外，AbilitySlice还具有独立于Page的生命周期变化，在同一Page中的AbilitySlice之间进行导航时，Page的生命周期状态不会改变。

AbilitySlice生命周期回调与Page的相应回调类似，因此不再赘述。由于AbilitySlice承载具体的页面，开发者必须重写AbilitySlice的onStart()回调，并在此方法中通过setUIContent()方法设置页面。

比如，在JavaAbilitySliceNavigation应用中，经常会在onStart()回调方法中通过super.setUIContent()来指定新增的样式布局。代码如下：

```
@Override
public void onStart(Intent intent) {
        super.onStart(intent);

    // 指定 ability_pay.xml 定义的 UI
    super.setUIContent(ResourceTable.Layout_ability_pay);
}
```

AbilitySlice实例的创建和管理通常由应用负责，系统仅在特定情况下会创建AbilitySlice实例。例如，通过导航启动某个AbilitySlice时，是由系统负责实例化；但是在同一个Page中不同的AbilitySlice间导航时，则由应用负责实例化。

5.5.3 Page与AbilitySlice生命周期的关联

当AbilitySlice处于前台且具有焦点时，其生命周期状态随着所属Page的生命周期状态的变化而变化。当一个Page拥有多个AbilitySlice时，例如在JavaAbilitySliceNavigation应用中，MainAbility下有MainAbilitySlice和PayAbilitySlice，当前MainAbilitySlice处于前台并获得焦点，且即将导航到PayAbilitySlice，在此期间的生命周期状态变化顺序如下。

- MainAbilitySlice从ACTIVE状态变为INACTIVE状态。
- PayAbilitySlice从INITIAL状态首先变为INACTIVE状态，然后变为ACTIVE状态（假定此前PayAbilitySlice未曾启动）。
- MainAbilitySlice从INACTIVE状态变为BACKGROUND状态。

对应两个Slice的生命周期方法回调顺序为：

```
MainAbilitySlice.onInactive() --> PayAbilitySlice.onStart() -->
PayAbilitySlice.onActive() --> MainAbilitySlice.onBackground()
```

在整个流程中，MainAbility始终处于ACTIVE状态。但是，当Page被系统销毁时，其所有已实例化的AbilitySlice将联动销毁，而不仅是处于前台的AbilitySlice。

5.6 实战：Page与AbilitySlice生命周期的例子

为了更好理解Page与AbilitySlice生命周期，在JavaAbilitySliceNavigation应用中增加对生命周期回调方法的处理。

5.6.1 修改MainAbilitySlice

对MainAbilitySlice进行修改，代码如下：

```
package com.waylau.ohms.javaabilityslicenavigation.slice;

import com.waylau.ohms.javaabilityslicenavigation.ResourceTable;
import ohos.aafwk.ability.AbilitySlice;
import ohos.aafwk.content.Intent;
import ohos.agp.components.Text;
import ohos.hiviewdfx.HiLog;
import ohos.hiviewdfx.HiLogLabel;

public class MainAbilitySlice extends AbilitySlice {
    static final HiLogLabel logLabel =
            new HiLogLabel(HiLog.LOG_APP, 0x00001, "MainAbilitySlice");

    @Override
    public void onStart(Intent intent) {
        super.onStart(intent);
        super.setUIContent(ResourceTable.Layout_ability_main);

        // 添加点击事件来触发导航
```

```
        Text text = (Text) findComponentById(ResourceTable.Id_text_
helloworld);
        text.setClickedListener(listener ->
                present(new PayAbilitySlice(), new Intent()));

        HiLog.info(logLabel, "onStart");
    }

    @Override
    public void onActive() {
        super.onActive();
        HiLog.info(logLabel, "onActive");
    }

    @Override
    public void onForeground(Intent intent) {
        super.onForeground(intent);
        HiLog.info(logLabel, "onForeground");
    }

    @Override
    public  void onInactive() {
        HiLog.info(logLabel, "onInactive");
    }

    @Override
    public  void onBackground() {
        HiLog.info(logLabel, "onBackground");
    }

    @Override
    public  void onStop() {
        HiLog.info(logLabel, "onStop");
    }
}
```

在MainAbilitySlice类中：

- MainAbilitySlice重写了AbilitySlice的生命周期回调方法；
- 在每个生命周期回调方法中，通过HiLog日志工具将状态进行了输出，以便能够在控制台看到实时的状态。

5.6.2 修改PayAbilitySlice

修改PayAbilitySlice代码如下：

```java
package com.waylau.ohms.javaabilityslicenavigation.slice;

import com.waylau.ohms.javaabilityslicenavigation.ResourceTable;
import ohos.aafwk.ability.AbilitySlice;
import ohos.aafwk.content.Intent;
import ohos.agp.components.Text;
import ohos.hiviewdfx.HiLog;
import ohos.hiviewdfx.HiLogLabel;

public class PayAbilitySlice extends AbilitySlice {
    static final HiLogLabel logLabel =
            new HiLogLabel(HiLog.LOG_APP, 0x00001, "PayAbilitySlice");

    @Override
    public void onStart(Intent intent) {
        super.onStart(intent);
        super.setUIContent(ResourceTable.Layout_ability_pay);

        // 添加点击事件来触发导航
        Text text = (Text) findComponentById(ResourceTable.Id_text_pay);
        text.setClickedListener(listener ->
                present(new MainAbilitySlice(), new Intent()));

        HiLog.info(logLabel, "onStart");
    }

    @Override
    public void onActive() {
        super.onActive();
        HiLog.info(logLabel, "onActive");
    }

    @Override
    public void onForeground(Intent intent) {
        super.onForeground(intent);
        HiLog.info(logLabel, "onForeground");
    }

    @Override
    public void onInactive() {
        HiLog.info(logLabel, "onInactive");
    }

    @Override
    public void onBackground() {
```

```
        HiLog.info(logLabel, "onBackground");
    }

    @Override
    public void onStop() {
        HiLog.info(logLabel, "onStop");
    }

}
```

在PayAbilitySlice类中：

- PayAbilitySlice重写了AbilitySlice的生命周期回调方法；
- 在每个生命周期回调方法中，通过HiLog日志工具将状态进行了输出，以便能够在控制台看到实时的状态。

5.6.3 运行

运行应用，此时能看到控制台输出如下内容：

```
4-23 10:20:27.317 11194-11194/com.waylau.hmos.javaabilityslicenavigation I
00001/MainAbilitySlice: onStart
4-23 10:20:27.325 11194-11194/com.waylau.hmos.javaabilityslicenavigation I
00001/MainAbilitySlice: onActive
```

从上述日志可以看出，MainAbilitySlice已经启动并处于ACTIVE状态。

点击文本“你好，世界”后，可以切换到“你好，支付”。此时能看到控制台输出如下内容：

```
4-23 10:22:34.885 11194-11194/com.waylau.hmos.javaabilityslicenavigation I
00001/MainAbilitySlice: onInactive
4-23 10:22:34.893 11194-11194/com.waylau.hmos.javaabilityslicenavigation I
00001/PayAbilitySlice: onStart
4-23 10:22:34.895 11194-11194/com.waylau.hmos.javaabilityslicenavigation I
00001/PayAbilitySlice: onActive
4-23 10:22:34.895 11194-11194/com.waylau.hmos.javaabilityslicenavigation I
00001/MainAbilitySlice: onBackground
```

从上述日志可以看出，MainAbilitySlice失去了焦点并处于INACTIVE状态，而PayAbilitySlice启动并处于ACTIVE状态。最终MainAbilitySlice进入BACKGROUND状态。

再次点击文本“你好，支付”，可以切换到“你好，世界”。此时，能看到控制台输出如下内容：

```
4-23 10:28:37.006 11194-11194/com.waylau.hmos.javaabilityslicenavigation I
00001/PayAbilitySlice: onInactive
4-23 10:28:37.007 11194-11194/com.waylau.hmos.javaabilityslicenavigation I
00001/MainAbilitySlice: onStart
```

```
4-23 10:28:37.008 11194-11194/com.waylau.hmos.javaabilityslicenavigation I
00001/MainAbilitySlice: onActive
4-23 10:28:37.008 11194-11194/com.waylau.hmos.javaabilityslicenavigation I
00001/PayAbilitySlice: onBackground
```

从上述日志可以看出，PayAbilitySlice失去了焦点并处于INACTIVE状态，而MainAbilitySlice启动并处于ACTIVE状态。最终PayAbilitySlice进入BACKGROUND状态。

单击模拟器的“Back”按钮，返回“你好，支付”界面。此时，能看到控制台输出如下内容：

```
4-23 10:34:16.099 11194-11194/com.waylau.hmos.javaabilityslicenavigation I
00001/MainAbilitySlice: onInactive
4-23 10:34:16.099 11194-11194/com.waylau.hmos.javaabilityslicenavigation I
00001/PayAbilitySlice: onForeground
4-23 10:34:16.102 11194-11194/com.waylau.hmos.javaabilityslicenavigation I
00001/PayAbilitySlice: onActive
4-23 10:34:16.102 11194-11194/com.waylau.hmos.javaabilityslicenavigation I
00001/MainAbilitySlice: onBackground
4-23 10:34:16.102 11194-11194/com.waylau.hmos.javaabilityslicenavigation I
00001/MainAbilitySlice: onStop
```

从上述日志可以看出，MainAbilitySlice失去了焦点并处于INACTIVE状态，而PayAbilitySlice重新回到前台，并处于ACTIVE状态。最终MainAbilitySlice进入BACKGROUND状态，并被销毁。

再次单击模拟器的“Back”按钮，返回“你好，世界”界面。此时，能看到控制台输出如下内容：

```
4-23 10:40:55.901 11194-11194/com.waylau.hmos.javaabilityslicenavigation I
00001/PayAbilitySlice: onInactive
4-23 10:40:55.902 11194-11194/com.waylau.hmos.javaabilityslicenavigation I
00001/MainAbilitySlice: onForeground
4-23 10:40:55.904 11194-11194/com.waylau.hmos.javaabilityslicenavigation I
00001/MainAbilitySlice: onActive
4-23 10:40:55.904 11194-11194/com.waylau.hmos.javaabilityslicenavigation I
00001/PayAbilitySlice: onBackground
4-23 10:40:55.905 11194-11194/com.waylau.hmos.javaabilityslicenavigation I
00001/PayAbilitySlice: onStop
```

从上述日志可以看出，PayAbilitySlice失去了焦点并处于INACTIVE状态，而MainAbilitySlice重新回到前台，并处于ACTIVE状态。最终PayAbilitySlice进入BACKGROUND状态，并被销毁。

再次单击模拟器的“Back”按钮，返回“你好，世界”界面。此时，能看到控制台输出如下内容：

```
4-23 10:42:54.947 11194-11194/com.waylau.hmos.javaabilityslicenavigation I
00001/MainAbilitySlice: onInactive
4-23 10:42:55.749 11194-11194/com.waylau.hmos.javaabilityslicenavigation I
00001/MainAbilitySlice: onBackground
4-23 10:42:55.756 11194-11194/com.waylau.hmos.javaabilityslicenavigation I
00001/MainAbilitySlice: onStop
```

从上述日志可以看出，MainAbilitySlice处于INACTIVE状态，接着进入BACKGROUND状态，最后被销毁。

当单击系统主界面应用图标时，将进入应用。此时，能看到控制台输出如下内容：

```
4-23 10:48:23.960 11194-11194/com.waylau.hmos.javaabilityslicenavigation I
00001/MainAbilitySlice: onStart
4-23 10:48:23.969 11194-11194/com.waylau.hmos.javaabilityslicenavigation I
00001/MainAbilitySlice: onActive
```

从上述日志可以看出，MainAbilitySlice启用并处于ACTIVE状态。

5.7 Service Ability

基于Service模板的Ability，简称为Service，其职责主要用于后台运行任务，比如执行音乐播放、文件下载等，但不提供用户交互界面。Service可由其他应用或Ability启动，即使用户切换到其他应用，Service仍将在后台继续运行。

Service是单实例的。在一个设备上，相同的Service只会存在一个实例。如果多个Ability共用这个实例，只有当与Service绑定的所有Ability都退出后，Service才能够退出。由于Service是在主线程里执行的，因此，如果在Service里面的操作时间过长，开发者必须在Service里创建新的线程来处理，防止造成主线程阻塞，应用程序无响应。

有关线程方面的内容，可以详见“第12章 线程管理”。

5.7.1 创建Service

接下来介绍如何创建一个Service。

1. 继承 Ability

每个Service也都是Ability的子类，需要实现Service相关的生命周期方法。

Ability为Service提供了以下生命周期方法，用户可以重写这些方法来添加自己的处理。

- onStart()：该方法在创建Service的时候调用，用于Service的初始化。在Service的整个生命周期只会调用一次，调用时传入的Intent应为空。该方法在前面章节的Page中已经做了介绍。
- onCommand()：在Service创建完成之后调用，该方法在客户端每次启动该Service时都会调用，用户可以在该方法中做一些调用统计、初始化类的操作。
- onConnect()：在Ability和Service连接时调用，该方法返回IRemoteObject对象，用户可以在该回调函数中生成对应Service的IPC通信通道，以便Ability与Service交互。Ability可以多次连接同一个Service，系统会缓存该Service的IPC通信对象，只有当第一个客户端连接Service时，系统才会调用Service的onConnect方法来生成IRemoteObject对象，而后系统会将同一个RemoteObject对象传递至其他连接同一个Service的所有客户端，而无须再次调用onConnect方法。

- onDisconnect()：在Ability与绑定的Service断开连接时调用。
- onStop()：在Service被销毁时调用。Service应通过实现此方法来清理任何资源，如关闭线程、注册的侦听器等。该方法在前面章节的Page中已经做了介绍。

创建Service的代码示例如下：

```
public class ServiceAbility extends Ability {
    @Override
    public void onStart(Intent intent) {
        super.onStart(intent);
    }

    @Override
    public void onCommand(Intent intent, boolean restart, int startId) {
        super.onCommand(intent, restart, startId);
    }

    @Override
    public IRemoteObject onConnect(Intent intent) {
        super.onConnect(intent);
        return null;
    }

    @Override
    public void onDisconnect(Intent intent) {
        super.onDisconnect(intent);
    }

    @Override
    public void onStop() {
        super.onStop();
    }
}
```

2. 注册 Service

Service也需要在应用配置文件中进行注册，注册类型type需要设置为service。配置内容如下：

```
{
    "module": {
        "abilities": [
            {
                "name": ".ServiceAbility",
```

```
                    "type": "service",
                    "visible": true
                    ...
                }
            ]
            ...
        }
        ...
}
```

5.7.2 启动Service

接下来介绍通过startAbility()启动Service及对应的停止方法。

1. 启动Service

Ability为开发者提供了startAbility()方法来启动另外一个Ability。因为Service也是Ability的一种，开发者同样可以通过将Intent传递给该方法来启动Service。Ability不仅支持启动本地Service，还支持启动远程Service。

开发者可以通过构造包含DeviceId、BundleName与AbilityName的Operation对象来设置目标Service信息。这三个参数的含义如下。

- DeviceId：表示设备ID。如果是本地设备，则可以直接留空；如果是远程设备，可以通过ohos.distributedschedule.interwork.DeviceManager提供的getDeviceList获取设备列表。
- BundleName：表示包名称。
- AbilityName：表示待启动的Ability名称。

启动本地设备Service的代码示例如下：

```
Intent intent = new Intent();
Operation operation = new Intent.OperationBuilder()
        .withDeviceId("")
        .withBundleName("com.huawei.hiworld.himusic")
        .withAbilityName("com.huawei.hiworld.himusic.ServiceAbility")
        .build();
intent.setOperation(operation);
startAbility(intent);
```

启动远程设备Service的代码示例如下：

```
Operation operation = new Intent.OperationBuilder()
        .withDeviceId("deviceId")
        .withBundleName("com.huawei.hiworld.himusic")
```

```
        .withAbilityName("com.huawei.hiworld.himusic.ServiceAbility")
        .withFlags(Intent.FLAG_ABILITYSLICE_MULTI_DEVICE) // 设置支持分布式调度系统多设备启动的标识
        .build();
Intent intent = new Intent();
intent.setOperation(operation);
startAbility(intent);
```

执行上述代码后，Ability将通过startAbility()方法来启动Service。

- 如果Service尚未运行，则系统会先调用onStart()来初始化Service，再回调Service的onCommand()方法来启动Service。
- 如果Service正在运行，则系统会直接回调Service的onCommand()方法来启动Service。

2. 停止 Service

Service一旦创建就会一直保持在后台运行，除非必须回收内存资源，否则系统不会停止或销毁Service。开发者可以在Service中通过terminateAbility()停止本Service或在其他Ability调用stopAbility()来停止Service。

支持停止本地设备Service和停止远程设备Service，使用方法与启动Service一样。一旦调用停止Service的方法，系统便会尽快销毁Service。

5.7.3 连接Service

如果Service需要与Page Ability或其他应用的Service Ability进行交互，则应创建用于连接的Connection。Service支持其他Ability通过connectAbility()方法与其进行连接。

在使用connectAbility()处理回调时，需要传入目标Service的Intent与IAbilityConnection的实例。IAbilityConnection提供了两种方法供开发者实现：

- onAbilityConnectDone()用来处理连接的回调；
- onAbilityDisconnectDone()用来处理断开连接的回调。

连接Service的代码示例如下：

```
// 创建连接回调实例
private IAbilityConnection connection = new IAbilityConnection() {
    // 连接到 Service 的回调
    @Override
    public void onAbilityConnectDone(ElementName elementName,
        IRemoteObject iRemoteObject, int resultCode) {
        // Client 侧需要定义与 Service 侧相同的 IRemoteObject 实现类
        // 开发者获取服务端传过来 IRemoteObject 对象，并从中解析出服务端传过来的信息
    }
```

```
    // 断开与连接的回调
    @Override
    public void onAbilityDisconnectDone(ElementName elementName, int
resultCode) {
    }
};

// 连接 Service
connectAbility(intent, connection);
```

同时，Service侧也需要在onConnect()方法中返回IRemoteObject对象，从而定义与Service进行通信的接口。onConnect()需要返回一个IRemoteObject对象，HarmonyOS提供了IRemoteObject的默认实现，用户可以通过继承LocalRemoteObject来创建自定义的实现类。Service侧把自身的实例返回给调用侧的代码示例如下：

```
// 创建自定义 IRemoteObject 实现类
private class MyRemoteObject extends LocalRemoteObject {
    public MyRemoteObject() {
        super("MyRemoteObject");
    }
}

// 把 IRemoteObject 返回给客户端
@Override
protected IRemoteObject onConnect(Intent intent) {
    return new MyRemoteObject();
}
```

5.7.4 Service Ability生命周期

与Page类似，Service Ability也拥有生命周期，如图5-12所示。

根据调用方法的不同，其生命周期有以下两种路径。

- 启动Service：该Service在其他Ability调用startAbility()时创建，然后保持运行。其他Ability通过调用stopAbility()来停止Service，Service停止后，系统会将其销毁。

- 连接Service：该Service在其他Ability调用connectAbility()时创建，客户端可通过调用disconnectAbility()断开连接。多个客户端可以绑定到相同Service，而且当所有绑定全部取消后，系统即会销毁该Service。

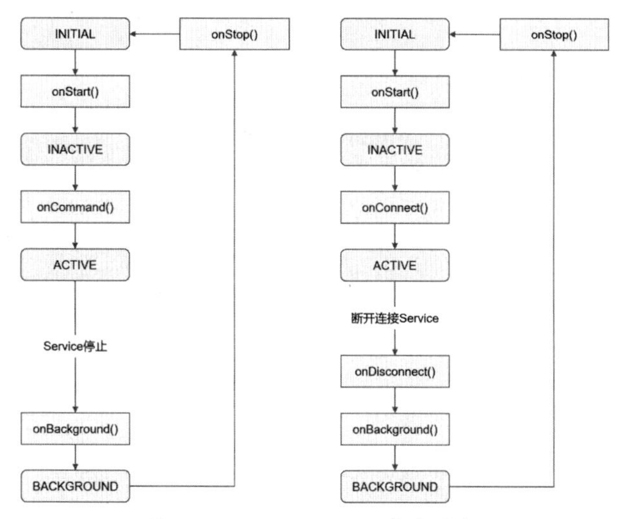

图 5-12　Service Ability 生命周期

5.8 实战：Service Ability生命周期的例子

为了更好理解Service Ability的生命周期，创建一个名为“JavaServiceAbilityLifeCycle”的应用作为演示。

5.8.1　创建Service

在DevEco Studio中，可以按图5-13所示的方式创建一个Service Ability。

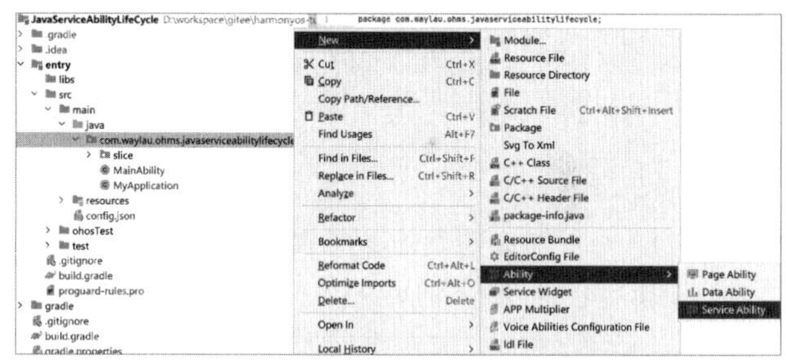

图 5-13　创建一个 Service Ability

根据图5-14所示的引导，我们创建一个名为“TimeServiceAbility”的Service。

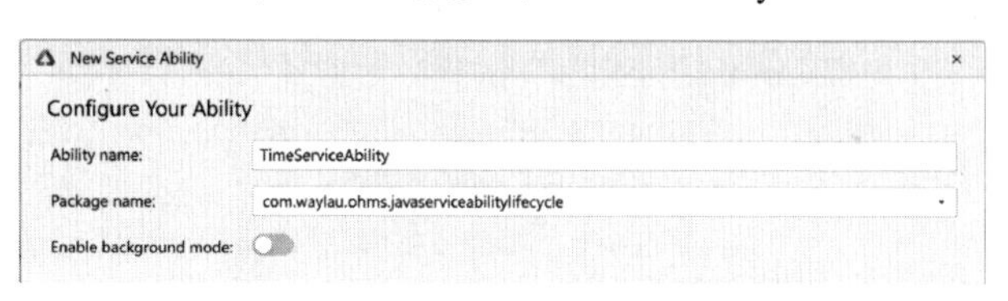

图 5-14　创建了一个 TimeServiceAbility

注意，上述步骤中的“Enable backgroud mode”（后台模式）先不要启用。

在自动创建的TimeServiceAbility的基础上，修改代码如下：

```
package com.waylau.hmos.javaserviceabilitylifecycle;

import ohos.aafwk.ability.Ability;
import ohos.aafwk.content.Intent;
import ohos.rpc.IRemoteObject;
import ohos.hiviewdfx.HiLog;
import ohos.hiviewdfx.HiLogLabel;
```

```
import java.time.LocalDateTime;

public class TimeServiceAbility extends Ability {
    private static final HiLogLabel LOG_LABEL =
            new HiLogLabel(HiLog.LOG_APP, 0x00001, "TimeServiceAbility");

    private TimeRemoteObject timeRemoteObject = new TimeRemoteObject();

    @Override
    public void onStart(Intent intent) {
        HiLog.info(LOG_LABEL, "onStart");
        super.onStart(intent);
    }

    @Override
    public void onBackground() {
        super.onBackground();
        HiLog.info(LOG_LABEL, "onBackground");
    }

    @Override
    public void onStop() {
        super.onStop();
        HiLog.info(LOG_LABEL, "onStop");
    }

    @Override
    public void onCommand(Intent intent, boolean restart, int startId) {
        super.onCommand(intent, restart, startId);
        HiLog.info(LOG_LABEL, "onCommand");
    }

    @Override
    public IRemoteObject onConnect(Intent intent) {
        super.onConnect(intent);
        HiLog.info(LOG_LABEL, "onConnect");

        LocalDateTime now = LocalDateTime.now();
        timeRemoteObject.setTime(now);

        return timeRemoteObject;
    }

    @Override
    public void onDisconnect(Intent intent) {
```

```
        super.onDisconnect(intent);
        HiLog.info(LOG_LABEL, "onDisconnect");
    }
}
```

其中，timeRemoteObject是一个IRemoteObject子类的实例，可以通过timeRemoteObject将当前时间返回给调用侧。稍后我们会介绍TimeRemoteObject类。

同时，在配置文件中会自动新增TimeServiceAbility相关的配置信息。

```
"abilities": [
    {
    "skills": [
        {
        "entities": [
            "entity.system.home"
        ],
        "actions": [
            "action.system.home"
        ]
        }
    ],
    "name": "com.waylau.hmos.javaserviceabilitylifecycle.MainAbility",
    "description": "$string:mainability_description",
    "icon": "$media:icon",
    "label": "$string:entry_MainAbility",
    "launchType": "standard",
    "orientation": "unspecified",
    "visible": true,
    "type": "page"
    },
    // 新增的 TimeServiceAbility
    {
    "name": "com.waylau.hmos.javaserviceabilitylifecycle.TimeServiceAbility",
    "description": "$string:timeserviceability_description",
    "type": "service",
    "backgroundModes": [],
    "icon": "$media:icon"
    }
]
```

5.8.2 创建远程对象

timeRemoteObject是TimeRemoteObject类的实例。TimeRemoteObject类是远程对象，继承自LocalRemoteObject。新增TimeRemoteObject类，代码如下：

```
package com.waylau.hmos.javaserviceabilitylifecycle;

import ohos.aafwk.ability.LocalRemoteObject;

import java.time.LocalDateTime;

public class TimeRemoteObject extends LocalRemoteObject {
    private LocalDateTime time;

    public TimeRemoteObject() {
    }

    public void setTime(LocalDateTime time) {
        this.time = time;
    }

    public LocalDateTime getTime() {
        return time;
    }
}
```

其中，TimeRemoteObject类里面的time 属性是返回给调用方的服务器当前时间。

5.8.3 修改 MainAbilitySlice

修改MainAbilitySlice代码如下：

```
package com.waylau.hmos.javaserviceabilitylifecycle.slice;

import com.waylau.hmos.javaserviceabilitylifecycle.ResourceTable;
import com.waylau.hmos.javaserviceabilitylifecycle.TimeRemoteObject;
import ohos.aafwk.ability.AbilitySlice;
import ohos.aafwk.ability.IAbilityConnection;
import ohos.aafwk.content.Intent;
import ohos.aafwk.content.Operation;
import ohos.agp.components.Text;
import ohos.bundle.ElementName;
import ohos.hiviewdfx.HiLog;
import ohos.hiviewdfx.HiLogLabel;
import ohos.rpc.IRemoteObject;

import java.time.LocalDateTime;

public class MainAbilitySlice extends AbilitySlice {
    private static final HiLogLabel LOG_LABEL =
```

```
            new HiLogLabel(HiLog.LOG_APP, 0x00001, "MainAbilitySlice");
    private TimeRemoteObject timeRemoteObject;

    @Override
    public void onStart(Intent intent) {
        super.onStart(intent);
        super.setUIContent(ResourceTable.Layout_ability_main);

        // 添加点击事件
        Text textStart = (Text) findComponentById(ResourceTable.Id_text_
start);
        textStart.setClickedListener(listener -> {
            // 启动本地服务
            startupLocalService(intent);

            // 连接本地服务
            connectLocalService(intent);
        });

        // 添加点击事件
        Text textStop = (Text) findComponentById(ResourceTable.Id_text_stop);
        textStop.setClickedListener(listener -> {
            // 断开本地服务
            disconnectLocalService(intent);

            // 关闭本地服务
            stopLocalService(intent);
        });


        HiLog.info(LOG_LABEL, "onStart");
    }

    @Override
    public void onActive() {
        super.onActive();
    }

    @Override
    public void onForeground(Intent intent) {
        super.onForeground(intent);
    }

    /**
     * 启动本地服务
     */
```

```
    private void startupLocalService(Intent intent) {
        //Intent intent = new Intent();
        // 构建操作方式
        Operation operation = new Intent.OperationBuilder()
                // 设备 ID
                .withDeviceId("")
                // 包名称
                .withBundleName("com.waylau.hmos.
javaserviceabilitylifecycle")
                // 待启动的 Ability 名称
                .withAbilityName("com.waylau.hmos.
javaserviceabilitylifecycle.TimeServiceAbility")
                .build();
        // 设置操作
        intent.setOperation(operation);
        startAbility(intent);

        HiLog.info(LOG_LABEL, "startupLocalService");
    }

    /**
     * 关闭本地服务
     */
    private void stopLocalService(Intent intent) {
        stopAbility(intent);

        HiLog.info(LOG_LABEL, "stopLocalService");
    }

    // 创建连接回调实例
    private IAbilityConnection connection = new IAbilityConnection() {
        // 连接到 Service 的回调
        @Override
        public void onAbilityConnectDone(ElementName elementName,
                IRemoteObject iRemoteObject, int resultCode) {
            // Client 侧需要定义与 Service 侧相同的 IRemoteObject 实现类
            // 开发者获取服务端传过来 IRemoteObject 对象，并从中解析出服务端传过来的信息
            timeRemoteObject = (TimeRemoteObject) iRemoteObject;

            HiLog.info(LOG_LABEL, "onAbilityConnectDone, time: %{public}s",
                timeRemoteObject.getTime());
        }

        // 断开与连接的回调
        @Override
        public void onAbilityDisconnectDone(ElementName elementName, int
```

```
resultCode) {
            HiLog.info(LOG_LABEL, "onAbilityDisconnectDone");
        }
    };

    /**
     * 连接本地服务
     */
    private void connectLocalService(Intent intent) {
        // 连接 Service
        connectAbility(intent, connection);

        HiLog.info(LOG_LABEL, "connectLocalService");
    }

    /**
     * 断开连接本地服务
     */
    private void disconnectLocalService(Intent intent) {
        // 断开连接 Service
        disconnectAbility(connection);

        HiLog.info(LOG_LABEL, "disconnectLocalService");
    }
}
```

上述代码，在onStart()方法中增加了对Text的事件监听。当点击“textStart”时，会启动本地服务、连接本地服务；当点击“textEnd”时，会断开本地服务、关闭本地服务。

5.8.4 修改ability_main.xml

修改ability_main.xml内容如下：

```
<?xml version="1.0" encoding="utf-8"?>
<DirectionalLayout
    xmlns:ohos="http://schemas.huawei.com/res/ohos"
    ohos:height="match_parent"
    ohos:width="match_parent"
    ohos:orientation="vertical">

    <Text
        ohos:id="$+id:text_start"
        ohos:height="match_content"
        ohos:width="match_content"
        ohos:background_element="$graphic:background_ability_main"
```

```
        ohos:layout_alignment="horizontal_center"
        ohos:text="Start"
        ohos:text_size="50"
        />

    <Text
        ohos:id="$+id:text_stop"
        ohos:height="match_parent"
        ohos:width="match_content"
        ohos:background_element="$graphic:background_ability_main"
        ohos:layout_alignment="horizontal_center"
        ohos:text="End"
        ohos:text_size="50"
        />

</DirectionalLayout>
```

上述代码主要定义了两个 Text，一个用于触发 Start 点击事件，一个用于触发 End 点击事件。

5.8.5 运行

在模拟器中运行应用，此时，能看到控制台输出如下内容：

```
4-23 22:23:13.496 28370-28370/? I 00001/MainAbilitySlice: onStart
```

从上述日志可以看出，MainAbilitySlice 已经启动。

点击文本“Start”后，触发了点击事件。此时，能看到控制台输出如下内容：

```
4-23 22:24:15.739 28370-28370/com.waylau.hmos.javaserviceabilitylifecycle I
00001/MainAbilitySlice: startupLocalService
4-23 22:24:15.745 28370-28370/com.waylau.hmos.javaserviceabilitylifecycle I
00001/MainAbilitySlice: connectLocalService
4-23 22:24:15.748 28370-28370/com.waylau.hmos.javaserviceabilitylifecycle I
00001/TimeServiceAbility: onStart
4-23 22:24:15.750 28370-28370/com.waylau.hmos.javaserviceabilitylifecycle I
00001/TimeServiceAbility: onCommand
4-23 22:24:15.764 28370-28370/com.waylau.hmos.javaserviceabilitylifecycle I
00001/TimeServiceAbility: onConnect
4-23 22:24:15.771 28370-28370/com.waylau.hmos.javaserviceabilitylifecycle I
00001/MainAbilitySlice: onAbilityConnectDone, time: 2020-4-23T22:24:15.769
```

当点击文本“Start”时，会启动本地服务、连接本地服务。而 TimeServiceAbility 也分别执行了 onStart、onCommand 及 onConnect 等生命周期，并将当前时间返回给了 MainAbilitySlice。

点击文本“End”后，触发了点击事件。此时，能看到控制台输出如下内容：

```
4-23 22:26:27.502 28370-28370/com.waylau.hmos.javaserviceabilitylifecycle I
00001/MainAbilitySlice: disconnectLocalService
4-23 22:26:27.505 28370-28370/com.waylau.hmos.javaserviceabilitylifecycle I
00001/MainAbilitySlice: stopLocalService
4-23 22:26:27.508 28370-28370/com.waylau.hmos.javaserviceabilitylifecycle I
00001/TimeServiceAbility: onDisconnect
4-23 22:26:27.513 28370-28370/com.waylau.hmos.javaserviceabilitylifecycle I
00001/TimeServiceAbility: onBackground
4-23 22:26:27.513 28370-28370/com.waylau.hmos.javaserviceabilitylifecycle I
00001/TimeServiceAbility: onStop
```

当点击文本“Start”时，会断开本地服务、关闭本地服务。而TimeServiceAbility也分别执行了onDisconnect、onBackground及onStop等生命周期。

5.9 Data Ability

使用Data模板的Ability，也简称Data，其主要职责是管理其自身应用和其他应用存储数据的访问，并提供与其他应用共享数据的方法。Data既可用于同设备不同应用的数据共享，也支持跨设备不同应用的数据共享。

数据的存储方式多种多样，可以是传统意义上的数据库系统，也可以是本地磁盘上的文件。Data对外提供对数据的增、删、改、查，以及打开文件等接口，这些接口的具体实现由开发者提供。

5.9.1 URI

Data的提供方和使用方都通过URI（Uniform Resource Identifier，统一资源定位符）来标识一个具体的数据，例如数据库中的某个表或磁盘上的某个文件。HarmonyOS的URI基于URI通用标准，具体格式如图5-15所示。

Scheme://[authority]/[path][?query][#fragment]
协议方案名 设备ID 资源路径 查询参数 访问的子资源

图5-15 URI格式

其中，各格式解释如下。

- Scheme：协议方案名，固定为“dataability”，代表Data Ability所使用的协议类型。
- authority：设备ID。如果为跨设备场景，则为目标设备的ID；如果为本地设备场景，则不需要填写。
- path：资源的路径信息，代表特定资源的位置信息。
- query：查询参数。
- fragment：可以用于指示要访问的子资源。

以下是具体的URI示例：

```
// 跨设备场景:
dataability://device_id/com.waylau.hmos.dataabilityhelperaccessfile.
dataability.persondata/person/10

// 本地设备:
dataability:///com.waylau.hmos.dataabilityhelperaccessfile.dataability.
persondata/person/10
```

5.9.2 访问Data

可以通过DataAbilityHelper类来访问当前应用或其他应用提供的共享数据。DataAbilityHelper作为客户端，与提供方的Data进行通信。Data接收到请求后，执行相应的处理，并返回结果。DataAbilityHelper提供了一系列与Data Ability对应的方法。

下面介绍DataAbilityHelper的具体使用步骤。

1. 声明使用权限

如果待访问的Data声明了访问需要权限，则访问此Data需要在配置文件中声明需要此权限。声明请参考权限申请字段说明。代码如下：

```
"reqPermissions": [
    {
        "name": "com.waylau.hmos.dataabilityhelperaccessfile.DataAbility.
DATA"
    }
]
```

2. 创建DataAbilityHelper

DataAbilityHelper为开发者提供了creator()方法来创建DataAbilityHelper实例。该方法为静态方法，有多个重载。最常见的方法是通过传入一个context对象来创建DataAbilityHelper对象。

以下为获取helper对象示例：

```
DataAbilityHelper helper = DataAbilityHelper.creator(this);
```

3. 访问Data Ability

DataAbilityHelper为开发者提供了一系列的接口来访问不同类型的数据，比如文件或是数据库等。

- 访问文件：DataAbilityHelper为开发者提供了FileDescriptor openFile（Uri uri，String mode）方法来操作文件。此方法需要传入两个参数，其中uri用来确定目标资源路径，mode用来指定打开文件的方式，可选方式包含“r”（读）、“w”（写）、“rw”（读写）、“wt”（覆盖写）、“wa”（追加写）、“rwt”（覆盖写且可读）。该方法返回一个目标文件的FD（文件描述符），把文件描述符封装成流，开发者就可以对文件流进行自定义处理。

- 访问数据库：DataAbilityHelper为开发者提供了增、删、改、查及批量处理等方法来操作数据库。

在接下来的章节，将会详细介绍如何使用DataAbilityHelper访问文件和数据库。

5.10 实战：通过DataAbilityHelper访问文件

本节演示如何通过DataAbilityHelper类访问当前应用的文件数据。采用Car设备类型，创建一个名为“JavaDataAbilityHelperAccessFile”的应用。

下面介绍DataAbilityHelper的具体使用步骤。

5.10.1 创建DataAbility

在DevEco Studio中，可以按图5-16所示的方式创建一个Data Ability。

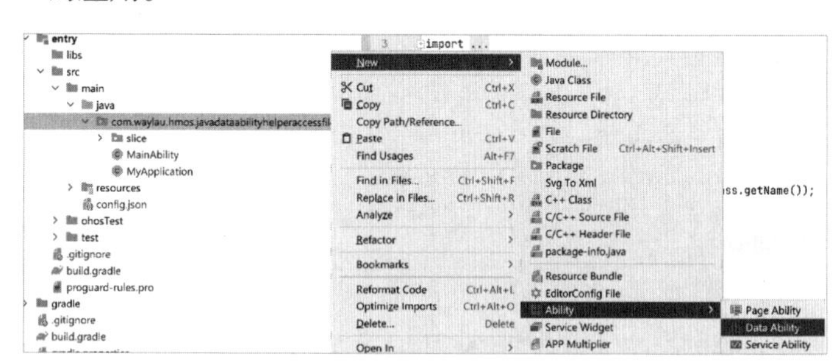

图5-16　创建一个Data Ability

根据图5-17所示的引导，我们创建一个名为“UserDataAbility”的Data。

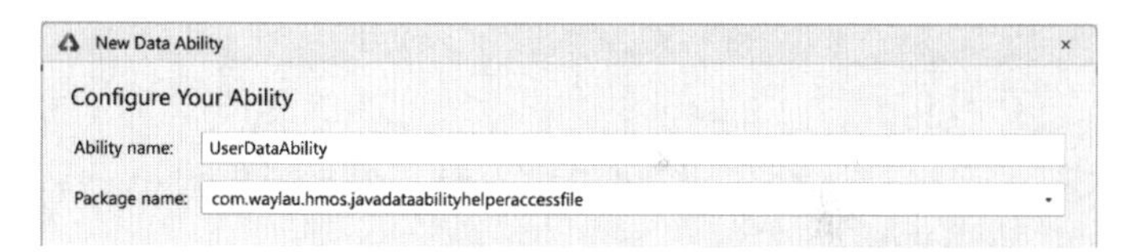

图5-17　创建了一个UserDataAbility

UserDataAbility代码如下：

```
package com.waylau.hmos.javadataabilityhelperaccessfile;

import ohos.aafwk.ability.Ability;
import ohos.aafwk.content.Intent;
import ohos.data.resultset.ResultSet;
import ohos.data.rdb.ValuesBucket;
import ohos.data.dataability.DataAbilityPredicates;
import ohos.hiviewdfx.HiLog;
import ohos.hiviewdfx.HiLogLabel;
import ohos.utils.net.Uri;
import ohos.utils.PacMap;

import java.io.FileDescriptor;

public class UserDataAbility extends Ability {
    private static final HiLogLabel LABEL_LOG = new HiLogLabel(3, 0xD001100,
```

```
"Demo");

    @Override
    public void onStart(Intent intent) {
        super.onStart(intent);
        HiLog.info(LABEL_LOG, "UserDataAbility onStart");
    }

    @Override
    public ResultSet query(Uri uri, String[] columns, DataAbilityPredicates
predicates) {
        return null;
    }

    @Override
    public int insert(Uri uri, ValuesBucket value) {
        HiLog.info(LABEL_LOG, "UserDataAbility insert");
        return 999;
    }

    @Override
    public int delete(Uri uri, DataAbilityPredicates predicates) {
        return 0;
    }

    @Override
    public int update(Uri uri, ValuesBucket value, DataAbilityPredicates
predicates) {
        return 0;
    }

    @Override
    public FileDescriptor openFile(Uri uri, String mode) {
        return null;
    }

    @Override
    public String[] getFileTypes(Uri uri, String mimeTypeFilter) {
        return new String[0];
    }

    @Override
    public PacMap call(String method, String arg, PacMap extras) {
        return null;
    }
```

```
    @Override
    public String getType(Uri uri) {
        return null;
    }
}
```

在创建的时候就生成了一些代码，包括基本的增删改查、打开文件、获取URI类型、获取文件类型，还有一个回调，再加上一个onStart方法，总共是9个。

UserDataAbility自动在配置文件中添加了相应的配置，内容如下：

```
"abilities": [
    {
    "skills": [
        {
        "entities": [
            "entity.system.home"
        ],
        "actions": [
            "action.system.home"
        ]
        }
    ],
    "orientation": "landscape",
    "name": "com.waylau.hmos.javadataabilityhelperaccessfile.MainAbility",
    "icon": "$media:icon",
    "description": "$string:mainability_description",
    "label": "JavaDataAbilityHelperAccessFile",
    "type": "page",
    "launchType": "standard"
    },
    // 新增 UserDataAbility 配置
    {
    "permissions": [
        "com.waylau.hmos.javadataabilityhelperaccessfile.
DataAbilityShellProvider.PROVIDER"
    ],
    "name": "com.waylau.hmos.javadataabilityhelperaccessfile.UserDataAbility",
    "icon": "$media:icon",
    "description": "$string:userdataability_description",
    "type": "data",
    "uri": "dataability://com.waylau.hmos.javadataabilityhelperaccessfile.
UserDataAbility"
    }
],
// 新增 UserDataAbility 所需的权限
```

```
"defPermissions": [
    {
    "name": "com.waylau.hmos.javadataabilityhelperaccessfile.
DataAbilityShellProvider.PROVIDER"
    }
]
```

从上述配置可以看出如下信息。

- type: 类型设置为 data。
- uri: 对外提供的访问路径，全局唯一。
- defPermissions: 访问该 Data Ability 时需要申请的访问权限。

5.10.2 修改 UserDataAbility

由于本示例只是涉及文件，因此修改 UserDataAbility 时，只需重写 onStart 和 openFile 方法：

```
package com.waylau.hmos.javadataabilityhelperaccessfile;

import ohos.aafwk.ability.Ability;
import ohos.aafwk.content.Intent;
import ohos.data.resultset.ResultSet;
import ohos.data.rdb.ValuesBucket;
import ohos.data.dataability.DataAbilityPredicates;
import ohos.global.resource.RawFileEntry;
import ohos.global.resource.Resource;
import ohos.hiviewdfx.HiLog;
import ohos.hiviewdfx.HiLogLabel;
import ohos.rpc.MessageParcel;
import ohos.utils.net.Uri;
import ohos.utils.PacMap;

import java.io.*;
import java.nio.file.Paths;

public class UserDataAbility extends Ability {
    private static final HiLogLabel LABEL_LOG = new HiLogLabel(HiLog.LOG_APP,
0x00001, "UserDataAbility");

    private File targetFile;

    @Override
    public void onStart(Intent intent) {
        super.onStart(intent);
        HiLog.info(LABEL_LOG, "UserDataAbility onStart");
```

```
        try {
            // 初始化目标文件数据
            initFile();
        } catch (IOException e) {
            e.printStackTrace();
        }
    }

    private void initFile() throws IOException {
        // 获取数据目录
        File dataDir = new File(this.getDataDir().toString());
        if(!dataDir.exists()){
            dataDir.mkdirs();
        }

        // 构建目标文件
        targetFile = new File(Paths.get(dataDir.toString(),"users.txt").
toString());

        // 获取源文件
        RawFileEntry rawFileEntry = this.getResourceManager().
getRawFileEntry("resources/rawfile/users.txt");
        Resource resource = rawFileEntry.openRawFile();

        // 新建目标文件
        FileOutputStream fos = new FileOutputStream(targetFile);

        byte[] buffer = new byte[4096];
        int count = 0;

        // 源文件内容写入目标文件
        while((count = resource.read(buffer)) >= 0){
            fos.write(buffer,0,count);
        }

        resource.close();
        fos.close();
    }

    @Override
    public FileDescriptor openFile(Uri uri, String mode) {
        FileDescriptor fd = null;

        try {
            // 获取目标文件 FileDescriptor
```

```
            FileInputStream fileIs = new FileInputStream(targetFile);
            fd = fileIs.getFD();

            HiLog.info(LABEL_LOG, "fd: %{public}s", fd);
        } catch (IOException e) {
            e.printStackTrace();
        }

        // 创建 MessageParcel
        MessageParcel messageParcel = MessageParcel.obtain();

        // 复制 FileDescriptor
        fd = messageParcel.dupFileDescriptor(fd);
        return fd;
    }

    // 为节约篇幅，省略部分代码
}
```

initFile方法用于将源文件写入目标文件。this.getDataDir()方法可以获取数据目录，目标文件最终写入该目录下。

HarmonyOS提供了一个ResourceManager资源管理器，通过该资源管理器可以方便读取resouece目录下的资源文件。其中，RawFileEntry代表rawfile目录下的文件。通过rawFileEntry.openRawFile()方法可以方便获取指定文件。

最后在方法返回前，需要通过MessageParcel对FileDescriptor进行复制。

5.10.3 创建文件

在resouece目录的rawfile目录下，创建一个测试用文件users.txt。该文件的测试内容比较简单，就是一个用户的名字：

```
Way Lau
```

5.10.4 修改MainAbilitySlice

修改MainAbilitySlice的onStart方法，代码如下：

```
package com.waylau.hmos.javadataabilityhelperaccessfile.slice;

import com.waylau.hmos.javadataabilityhelperaccessfile.FileUtils;
import com.waylau.hmos.javadataabilityhelperaccessfile.ResourceTable;
import ohos.aafwk.ability.AbilitySlice;
```

```
import ohos.aafwk.ability.DataAbilityHelper;
import ohos.aafwk.ability.DataAbilityRemoteException;
import ohos.aafwk.content.Intent;
import ohos.agp.components.Text;
import ohos.hiviewdfx.HiLog;
import ohos.hiviewdfx.HiLogLabel;
import ohos.utils.net.Uri;

import java.io.*;

public class MainAbilitySlice extends AbilitySlice {
    private static final HiLogLabel LABEL_LOG = new HiLogLabel(HiLog.LOG_APP,
                                                    0x00001, "MainAbilitySlice");

    @Override
    public void onStart(Intent intent) {
        super.onStart(intent);
        super.setUIContent(ResourceTable.Layout_ability_main);

        // 添加点击事件来触发访问数据
        Text text = (Text) findComponentById(ResourceTable.Id_text_
                                                helloworld);
        text.setClickedListener(listener -> this.getFile());
    }

    private void getFile() {
        DataAbilityHelper helper = DataAbilityHelper.creator(this);

        // 访问数据用的 uri，注意用 3 个斜杠
        Uri uri =
                Uri.parse("dataability:///com.waylau.hmos.
javadataabilityhelperaccessfile.UserDataAbility");

        //  DataAbilityHelper 的 openFile 方法来访问文件
        try {
            FileDescriptor fd = helper.openFile(uri, "r");

            HiLog.info(LABEL_LOG, "fd: %{public}s", fd);
            HiLog.info(LABEL_LOG, "file content: %{public}s", FileUtils.
                    getFileContent(fd));
        } catch (DataAbilityRemoteException | IOException e) {
            e.printStackTrace();
        }
    }
    // 为节约篇幅，省略部分代码
}
```

上述方法中，在Text添加点击事件来触发访问文件的操作。

- getFile方法用于访问文件。借助DataAbilityHelper类的openFile方法访问当前UserDataAbility提供的文件数据。
- FileUtils.getFileContent方法用于将文件内容转为字符串，这样可以在日志里面方便查看文件的具体内容。FileUtils工具类稍后介绍。

注意，上述代码中访问的uri与UserDataAbility在配置文件中添加的uri基本是一致的，唯一的区别是上述代码中访问的uri用了3个斜杠。

5.10.5 创建FileUtils类

FileUtils类是一个工具类，其中getFileContent方法用于将文件内容转为字符串。代码如下：

```
package com.waylau.hmos.javadataabilityhelperaccessfile;

import java.io.FileDescriptor;
import java.io.FileInputStream;
import java.io.IOException;

public class FileUtils {

    /**
     * 输出文件内容
     * @param fd
     * @return 文件内容
     * @throws IOException
     */
    public static String getFileContent(FileDescriptor fd) throws IOException {
        // 根据 FileDescriptor 创建 FileInputStream 对象
        FileInputStream fis = new FileInputStream(fd);

        int b = 0;
        StringBuilder sb = new StringBuilder();
        while((b = fis.read()) != -1){
            sb.append((char)b);
        }
        fis.close();

        return sb.toString();
    }
}
```

5.10.6 运行

运行应用后，点击“你好，世界”触发访问文件的操作，可以看到控制台HiLog输出内容如下：

```
04-24 10:43:22.054 6457-6533/? I 00001/UserDataAbility: UserDataAbility
onStart
04-24 10:43:26.204 6457-6457/com.waylau.hmos.javadataabilityhelperaccessfile
I 00001/UserDataAbility: fd: java.io.FileDescriptor@98b9a7c
04-24 10:43:26.205 6457-6457/com.waylau.hmos.javadataabilityhelperaccessfile
I 00001/MainAbilitySlice: fd: java.io.FileDescriptor@3850b5a
04-24 10:43:26.206 6457-6457/com.waylau.hmos.javadataabilityhelperaccessfile
I 00001/MainAbilitySlice: file content: Way Lau
```

至此，DataAbilityHelper访问文件的示例演示完毕。

5.11 实战：通过DataAbilityHelper访问数据库

上一节演示了如何通过DataAbilityHelper类访问当前应用的文件数据。本节将演示如何通过DataAbilityHelper类访问当前应用的数据库数据。创建一个名为“JavaDataAbilityHelperAccessDatabase”的应用作为演示。

下面介绍DataAbilityHelper的具体使用步骤。

5.11.1 创建DataAbility

在DevEco Studio中，创建一个名为“UserDataAbility”的Data，如图5-18所示。

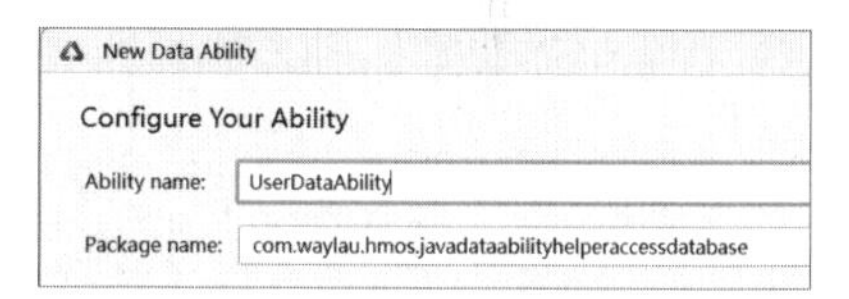

图5-18 创建UserDataAbility

UserDataAbility初始化时代码如下：

```
package com.waylau.hmos.javadataabilityhelperaccessdatabase;

import ohos.aafwk.ability.Ability;
import ohos.aafwk.content.Intent;
import ohos.data.resultset.ResultSet;
import ohos.data.rdb.ValuesBucket;
import ohos.data.dataability.DataAbilityPredicates;
import ohos.hiviewdfx.HiLog;
import ohos.hiviewdfx.HiLogLabel;
import ohos.utils.net.Uri;
import ohos.utils.PacMap;
```

```
import java.io.FileDescriptor;

public class UserDataAbility extends Ability {
    private static final HiLogLabel LABEL_LOG = new HiLogLabel(3, 0xD001100,
"Demo");

    @Override
    public void onStart(Intent intent) {
        super.onStart(intent);
        HiLog.info(LABEL_LOG, "UserDataAbility onStart");
    }

    @Override
    public ResultSet query(Uri uri, String[] columns, DataAbilityPredicates
predicates) {
        return null;
    }

    @Override
    public int insert(Uri uri, ValuesBucket value) {
        HiLog.info(LABEL_LOG, "UserDataAbility insert");
        return 999;
    }

    @Override
    public int delete(Uri uri, DataAbilityPredicates predicates) {
        return 0;
    }

    @Override
    public int update(Uri uri, ValuesBucket value, DataAbilityPredicates
predicates) {
        return 0;
    }

    @Override
    public FileDescriptor openFile(Uri uri, String mode) {
        return null;
    }

    @Override
    public String[] getFileTypes(Uri uri, String mimeTypeFilter) {
        return new String[0];
    }

    @Override
```

```
    public PacMap call(String method, String arg, PacMap extras) {
        return null;
    }

    @Override
    public String getType(Uri uri) {
        return null;
    }
}
```

UserDataAbility 自动在配置文件中添加了相应的配置，内容如下：

```
"abilities": [
    {
    "skills": [
        {
        "entities": [
            "entity.system.home"
        ],
        "actions": [
            "action.system.home"
        ]
        }
    ],
    "name": "com.waylau.hmos.javadataabilityhelperaccessdatabase.MainAbility",
    "description": "$string:mainability_description",
    "icon": "$media:icon",
    "label": "$string:entry_MainAbility",
    "launchType": "standard",
    "orientation": "unspecified",
    "visible": true,
    "type": "page"
    },
    // 新增 UserDataAbility 配置
    {
    "name": "com.waylau.hmos.javadataabilityhelperaccessdatabase.
UserDataAbility",
    "description": "$string:userdataability_description",
    "type": "data",
    "uri": "dataability://com.waylau.hmos.javadataabilityhelperaccessdataba
se.UserDataAbility",
    "icon": "$media:icon"
    }
],
// 新增 UserDataAbility 所需的权限
"defPermissions": [
```

```
    {
    "name": "com.waylau.hmos.javadataabilityhelperaccessdatabase.
DataAbilityShellProvider.PROVIDER"
    }
]
```

从上述配置中可以看出以下信息。

- type: 类型设置为data。
- uri: 对外提供的访问路径，全局唯一。
- permissions: 访问该Data Ability时需要申请的访问权限。

5.11.2 初始化数据库

HarmonyOS关系型数据库对外提供通用的操作接口，底层使用SQLite作为持久化存储引擎，支持SQLite具有的所有数据库特性，包括但不限于事务、索引、视图、触发器、外键、参数化查询和预编译SQL语句。初始化数据库代码如下：

```
import ohos.data.DatabaseHelper;
import ohos.data.rdb.RdbOpenCallback;
import ohos.data.rdb.RdbStore;
import ohos.data.rdb.StoreConfig;

public class UserDataAbility extends Ability {
    private static final HiLogLabel LABEL_LOG =
            new HiLogLabel(HiLog.LOG_APP, 0x00001, "UserDataAbility");
    private static final String DATABASE_NAME = "RdbStoreTest.db";
    private static final String TABLE_NAME = "user_t";
    private RdbStore store = null;

    @Override
    public void onStart(Intent intent) {
        super.onStart(intent);
        HiLog.info(LABEL_LOG, "UserDataAbility onStart");

        // 创建数据库连接，并获取连接对象
        DatabaseHelper helper = new DatabaseHelper(this);
        StoreConfig config = StoreConfig.newDefaultConfig(DATABASE_NAME);
        RdbOpenCallback callback = new RdbOpenCallback() {

            // 初始化数据库
            public void onCreate(RdbStore store) {
                store.executeSql("CREATE TABLE IF NOT EXISTS " + TABLE_NAME +
                        "(user_id INTEGER PRIMARY KEY, user_name TEXT NOT
NULL, user_age INTEGER)");
```

```
            }

            @Override
            public void onUpgrade(RdbStore store, int oldVersion, int
newVersion) {
            }
        };

        store = helper.getRdbStore(config, 1, callback, null);
    }

    // 为节约篇幅，省略部分代码
```

在上述代码中：

- 初始化了一个名为“RdbStoreTest.db”的数据库；
- 创建了一个名为“user_t”的表结构，该表包含user_id、user_name和user_age三个字段，其中user_id为主键；
- 初始化了RdbStore，方便处理后续关系型数据库的管理。

5.11.3 重写query方法

重写UserDataAbility的query方法，代码如下：

```
import ohos.data.rdb.*;
import ohos.data.dataability.DataAbilityUtils;
import ohos.data.dataability.DataAbilityPredicates;

@Override
public ResultSet query(Uri uri, String[] columns, DataAbilityPredicates
predicates) {

    if (store == null) {
        HiLog.error(LABEL_LOG, "failed to query, ormContext is null");
        return null;
    }

    // 查询数据库
    RdbPredicates rdbPredicates = DataAbilityUtils.
createRdbPredicates(predicates, TABLE_NAME);
    ResultSet resultSet = store.query(rdbPredicates, columns);
    if (resultSet == null) {
        HiLog.info(LABEL_LOG, "resultSet is null");
    }
```

```
    // 返回结果
    return resultSet;
}
```

上述代码中：

- DataAbilityUtils工具类将入参DataAbilityPredicates转为了RdbStore所能处理的RdbPredicates；
- 通过RdbStore查询所需要的字段。

5.11.4 重写insert方法

重写UserDataAbility的insert方法，代码如下：

```
@Override
public int insert(Uri uri, ValuesBucket value) {
    HiLog.info(LABEL_LOG, "UserDataAbility insert");

    // 参数校验
    if (store == null) {
        HiLog.error(LABEL_LOG, "failed to insert, ormContext is null");
        return -1;
    }

    // 插入数据库
    long userId = store.insert(TABLE_NAME, value);

    return Integer.valueOf(userId + "");
}
```

上述代码中：

- 通过RdbStore插入表对应的数据；
- RdbStore插入数据后的返回值是该插入数据的主键，即user_id字段值。

5.11.5 重写update方法

重写UserDataAbility的update方法，代码如下：

```
@Override
public int update(Uri uri, ValuesBucket value, DataAbilityPredicates
predicates) {
    if (store == null) {
        HiLog.error(LABEL_LOG, "failed to update, ormContext is null");
        return -1;
    }
```

```
    RdbPredicates rdbPredicates =
            DataAbilityUtils.createRdbPredicates(predicates, TABLE_NAME);
    int index = store.update(value, rdbPredicates);
    return index;
}
```

上述代码中：

- DataAbilityUtils工具类将入参DataAbilityPredicates转为了RdbStore所能处理的RdbPredicates；
- 通过RdbStore更新了对应条件的数据；
- 更新后返回的value值代表更新的数量。

5.11.6 重写delete方法

重写UserDataAbility的delete方法，代码如下：

```
@Override
public int delete(Uri uri, DataAbilityPredicates predicates) {
    if (store == null) {
        HiLog.error(LABEL_LOG, "failed to delete, ormContext is null");
        return -1;
    }

    RdbPredicates rdbPredicates =
        DataAbilityUtils.createRdbPredicates(predicates, TABLE_NAME);
    int value = store.delete(rdbPredicates);
    return value;
}
```

上述代码中：

- DataAbilityUtils工具类将入参DataAbilityPredicates转为了RdbStore所能处理的RdbPredicates；
- 通过RdbStore删除了对应条件的数据；
- 删除后返回的value值代表删除的数量。

5.11.7 修改MainAbilitySlice

修改MainAbilitySlice的onStart方法，代码如下：

```
package com.waylau.hmos.javadataabilityhelperaccessdatabase.slice;

import com.waylau.hmos.javadataabilityhelperaccessdatabase.ResourceTable;
import ohos.aafwk.ability.AbilitySlice;
import ohos.aafwk.ability.DataAbilityHelper;
import ohos.aafwk.ability.DataAbilityRemoteException;
```

```
import ohos.aafwk.content.Intent;
import ohos.agp.components.Text;
import ohos.data.dataability.DataAbilityPredicates;
import ohos.data.rdb.ValuesBucket;
import ohos.data.resultset.ResultSet;
import ohos.hiviewdfx.HiLog;
import ohos.hiviewdfx.HiLogLabel;
import ohos.utils.net.Uri;

public class MainAbilitySlice extends AbilitySlice {
    private static final HiLogLabel LABEL_LOG =
            new HiLogLabel(HiLog.LOG_APP, 0x00001, "MainAbilitySlice");

    @Override
    public void onStart(Intent intent) {
        super.onStart(intent);
        super.setUIContent(ResourceTable.Layout_ability_main);

        // 添加点击事件来触发访问数据
        Text text = (Text) findComponentById(ResourceTable.Id_text_
helloworld);
        text.setClickedListener(listener -> this.doDatabaseAction());
    }

    private void doDatabaseAction() {

        DataAbilityHelper helper = DataAbilityHelper.creator(this);
        // 访问数据用的 uri，注意用 3 个斜杠
        Uri uri =
                Uri.parse("dataability:///com.waylau.hmos.javadataabilityhelp
eraccessdatabase.UserDataAbility");
        String[] columns = {"user_id", "user_name", "user_age"};

        // 查询
        doQuery(helper, uri, columns);

        // 插入
        doInsert(helper, uri, columns);

        // 查询
        doQuery(helper, uri, columns);

        // 更新
        doUpdate(helper, uri, columns);

        // 查询
```

```
        doQuery(helper, uri, columns);

        // 删除
        doDelete(helper, uri, columns);

        // 查询
        doQuery(helper, uri, columns);
    }

    private void doQuery(DataAbilityHelper helper, Uri uri, String[] columns) {
        // 构造查询条件
        DataAbilityPredicates predicates = new DataAbilityPredicates();
        predicates.between("user_id", 101, 200);

        // 进行查询
        ResultSet resultSet = null;
        try {
            resultSet = helper.query(uri, columns, predicates);

        } catch (DataAbilityRemoteException e) {
            e.printStackTrace();
        }


        if (resultSet != null && resultSet.getRowCount() > 0) {
            // 在此处理 ResultSet 中的记录
            while (resultSet.goToNextRow()) {
                // 处理结果
                HiLog.info(LABEL_LOG,
                        "resultSet user_id: %{public}s, user_name: %{public}s,
user_age: %{public}s",
                resultSet.getInt(0), resultSet.getString(1), resultSet.
getInt(2));
            }
        } else {
            HiLog.info(LABEL_LOG, "resultSet is null or row count is 0");
        }
    }

    private void doInsert(DataAbilityHelper helper, Uri uri, String[] columns)
{
        // 构造插入数据
        ValuesBucket valuesBucket = new ValuesBucket();
        valuesBucket.putInteger(columns[0], 101);
        valuesBucket.putString(columns[1], "Way Lau");
        valuesBucket.putInteger(columns[2], 33);
```

```
        try {
            int result = helper.insert(uri, valuesBucket);
            HiLog.info(LABEL_LOG, "insert result: %{public}s", result);
        } catch (DataAbilityRemoteException e) {
            e.printStackTrace();
        }
    }

    private void doUpdate(DataAbilityHelper helper, Uri uri, String[] columns)
{
        // 构造查询条件
        DataAbilityPredicates predicates = new DataAbilityPredicates();
        predicates.equalTo("user_id", 101);

        // 构造更新数据
        ValuesBucket valuesBucket = new ValuesBucket();
        valuesBucket.putInteger(columns[0], 101);
        valuesBucket.putString(columns[1], "Way Lau");
        valuesBucket.putInteger(columns[2], 35);

        try {
            int result = helper.update(uri, valuesBucket, predicates);

            HiLog.info(LABEL_LOG, "update result: %{public}s", result);
        } catch (DataAbilityRemoteException e) {
            e.printStackTrace();
        }
    }


    private void doDelete(DataAbilityHelper helper, Uri uri, String[] columns)
{
        // 构造查询条件
        DataAbilityPredicates predicates = new DataAbilityPredicates();
        predicates.equalTo("user_id", 101);

        try {
            int result = helper.delete(uri, predicates);

            HiLog.info(LABEL_LOG, "delete result: %{public}s", result);
        } catch (DataAbilityRemoteException e) {
            e.printStackTrace();
        }
    }
```

```
    @Override
    public void onActive() {
        super.onActive();
    }

    @Override
    public void onForeground(Intent intent) {
        super.onForeground(intent);
    }
}
```

上述方法中，在Text添加点击事件来触发访问数据库的操作。doDatabaseAction方法用于执行数据库操作，借助DataAbilityHelper类的方法，来访问当前UserDataAbility提供的数据库数据，分别执行了查询、插入、查询、更新、查询、删除和查询动作。

注意，上述代码中访问的uri与UserDataAbility在配置数据库中添加的uri基本是一致的，唯一的区别是上述代码中访问的uri用了3个斜杠。

图5-19　运行应用效果

5.11.8　运行

运行应用后，效果如图5-19所示。

点击“你好，世界”触发了访问数据库的操作，可以看到控制台HiLog输出内容如下：

```
04-24 23:29:45.275 22633-22703/com.waylau.hmos.javadataabilityhelperaccessdat
abase I 00001/UserDataAbility: UserDataAbility onStart
04-24 23:30:38.333 22633-22633/com.waylau.hmos.javadataabilityhelperaccessdat
abase I 00001/MainAbilitySlice: resultSet is null or row count is 0
04-24 23:30:38.334 22633-22633/com.waylau.hmos.javadataabilityhelperaccessdat
abase I 00001/UserDataAbility: UserDataAbility insert
04-24 23:30:38.338 22633-22633/com.waylau.hmos.javadataabilityhelperaccessdat
abase I 00001/MainAbilitySlice: insert result: 101
04-24 23:30:38.348 22633-22633/com.waylau.hmos.javadataabilityhelperaccessda
tabase I 00001/MainAbilitySlice: resultSet user_id: 101, user_name: Way Lau,
user_age: 33
04-24 23:30:38.353 22633-22633/com.waylau.hmos.javadataabilityhelperaccessdat
abase I 00001/MainAbilitySlice: update result: 1
04-24 23:30:38.357 22633-22633/com.waylau.hmos.javadataabilityhelperaccessda
tabase I 00001/MainAbilitySlice: resultSet user_id: 101, user_name: Way Lau,
user_age: 35
04-24 23:30:38.361 22633-22633/com.waylau.hmos.javadataabilityhelperaccessdat
abase I 00001/MainAbilitySlice: delete result: 1
04-24 23:30:38.364 22633-22633/com.waylau.hmos.javadataabilityhelperaccessdat
```

```
abase I 00001/MainAbilitySlice: resultSet is null or row count is 0
```

通过上述运行日志可以验证程序的运行。代码解释如下。

- 首先执行查询，数据库的数据是空的。
- 接着插入一条用户数据，该数据的user_id是101。
- 再次查询，返回user_id是101的用户数据，user_name是Way Lau，user_age是33。
- 执行更新操作。
- 再次查询，可以看到user_id是101的用户数据，user_age被改为了35。
- 执行删除操作。
- 再次执行查询，数据库的数据是空的。

5.12 Intent

在HarmonyOS中，Intent是对象之间传递信息的载体。例如，当一个Ability需要启动另一个Ability时，或者一个AbilitySlice需要导航到另一个AbilitySlice时，可以通过Intent指定哪个Ability需要启动，并传递参数数据。

Intent的构成元素包括Operation与Parameters。

5.12.1 Operation与Parameters

Operation由表5-1所示的属性组成。

表5-1 Operation属性

属性	描述
Action	表示动作，通常使用系统预置Action，应用也可以自定义Action。例如，IntentConstants.ACTION_HOME表示返回桌面动作
Entity	表示类别，通常使用系统预置Entity，应用也可以自定义Entity。例如，Intent.ENTITY_HOME表示在桌面显示图标
Uri	表示Uri描述。如果在Intent中指定了Uri，则Intent将匹配指定的Uri信息，包括scheme、schemeSpecificPart、authority和path信息
Flags	表示处理Intent的方式。例如，Intent.FLAG_ABILITY_CONTINUATION标记在本地的一个Ability是否可以迁移到远端设备继续运行
BundleName	表示包描述。如果在Intent中同时指定了BundleName和AbilityName，则Intent可以直接匹配到指定的Ability

续表

属性	描述
AbilityName	表示待启动的Ability名称。如果在Intent中同时指定了BundleName和AbilityName，则Intent可以直接匹配到指定的Ability
DeviceId	表示运行指定Ability的设备ID

除上述属性之外，开发者也可以通过Parameters传递某些请求所需的额外信息。Parameters是一种支持自定义的数据结构。

当Intent用于发起请求时，根据指定元素的不同，启动分为以下两种类型。

- 如果同时指定了BundleName与AbilityName，则根据Ability的全称（如com.waylau.hmos.PayAbility）来直接启动应用。
- 如果未同时指定BundleName和AbilityName，则根据Operation中的其他属性来启动应用。

5.12.2 根据Ability的全称启动应用

通过构造包含BundleName与AbilityName的Operation对象，可以启动一个Ability，并导航到该Ability。

在“5.8 实战：Service Ability生命周期的例子”的JavaServiceAbilityLifeCycle应用中，启动本地服务的方式就是根据Ability的全称启动应用的一个例子。代码如下：

```
/**
* 启动本地服务
*/
private void startupLocalService(Intent intent) {
    // 构建操作方式
    Operation operation = new Intent.OperationBuilder()
            // 设备 ID
            .withDeviceId("")
            // 包名称
            .withBundleName("com.waylau.hmos.javaserviceabilitylifecycle")
            // 待启动的 Ability 名称
            .withAbilityName("com.waylau.hmos.javaserviceabilitylifecycle.
TimeServiceAbility")
            .build();
    // 设置操作
    intent.setOperation(operation);
    startAbility(intent);
}
```

在上述代码中，通过Intent中的OperationBuilder类构造Operation对象，指定设备标识（空串表示当前设备）、应用包名、Ability名称。

5.12.3 实战：根据Operation的其他属性启动应用

在有些场景下，开发者需要在应用中使用其他应用提供的某种能力，而不关心提供该能力的具体是哪一个应用。例如，开发者需要通过浏览器打开一个链接，而不关心用户最终选择哪一个浏览器应用，则可以通过Operation的其他属性（除BundleName与AbilityName之外的属性）描述需要的能力。如果设备上存在多个应用提供同种能力，系统则弹出候选列表，让用户选择由哪个应用处理请求。以下示例展示使用Intent跨Ability查询天气信息。

创建一个名为“JavaIntentOperationWithAction”的应用作为演示。

1. 创建 Page

通过DevEco Studio创建一个关于天气信息的Page“WeatherAbility”，如图5-20所示。

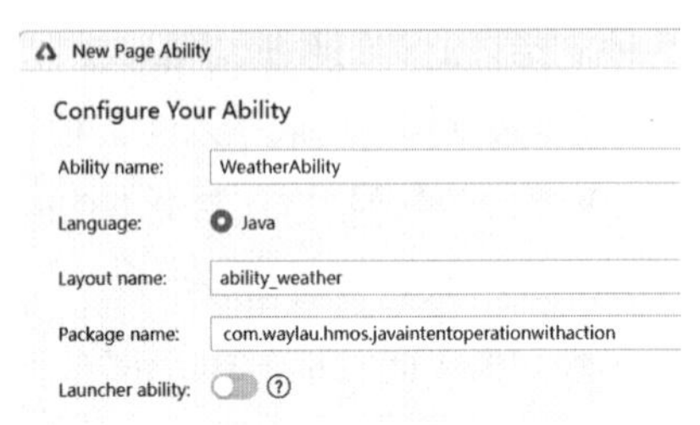

图 5-20　创建Page

自动创建了如下的WeatherAbilitySlice类：

```
package com.waylau.hmos.javaintentoperationwithaction.slice;

import com.waylau.hmos.javaintentoperationwithaction.ResourceTable;
import ohos.aafwk.ability.AbilitySlice;
import ohos.aafwk.content.Intent;

public class WeatherAbilitySlice extends AbilitySlice {
    @Override
    public void onStart(Intent intent) {
        super.onStart(intent);
        super.setUIContent(ResourceTable.Layout_ability_weather);
    }

    @Override
    public void onActive() {
        super.onActive();
    }

    @Override
    public void onForeground(Intent intent) {
        super.onForeground(intent);
    }
}
```

WeatherAbility使用的布局名称是ability_weather。修改ability_weather.xml如下：

```
<?xml version="1.0" encoding="utf-8"?>
<DirectionalLayout
```

```
    xmlns:ohos="http://schemas.huawei.com/res/ohos"
    ohos:height="match_parent"
    ohos:width="match_parent"
    ohos:alignment="center"
    ohos:orientation="vertical">

    <Text
        ohos:id="$+id:text_weather"
        ohos:height="match_parent"
        ohos:width="match_content"
        ohos:background_element="$graphic:background_ability_weather"
        ohos:layout_alignment="horizontal_center"
        ohos:text="Weather"
        ohos:text_size="40vp"
        />

</DirectionalLayout>
```

上述布局会在页面正中显示“Weather”字样。

2. 配置路由

修改配置文件，增加action.weather路由信息，内容如下：

```
"abilities": [
  {
    "skills": [
      {
        "entities": [
          "entity.system.home"
        ],
        "actions": [
          "action.system.home"
        ]
      }
    ],
    "name": "com.waylau.hmos.javaintentoperationwithaction.MainAbility",
    "description": "$string:mainability_description",
    "icon": "$media:icon",
    "label": "$string:entry_MainAbility",
    "launchType": "standard",
    "orientation": "unspecified",
    "visible": true,
    "type": "page"
  },
  {
    "skills": [
```

```
            {
              "actions": [
                "action.weather" // 指定路由的动作
              ]
            }
          ],
          "name": "com.waylau.hmos.javaintentoperationwithaction.WeatherAbility",
          "description": "$string:weatherability_description",
          "icon": "$media:icon",
          "label": "$string:entry_WeatherAbility",
          "launchType": "standard",
          "orientation": "unspecified",
          "type": "page"
        }
      ]
```

上述配置是为了配置路由，以便支持以此action导航到对应的AbilitySlice。

3. 修改 WeatherAbility

重写WeatherAbility类的onActive的方法，代码如下：

```
package com.waylau.hmos.javaintentoperationwithaction;

import com.waylau.hmos.javaintentoperationwithaction.slice.
WeatherAbilitySlice;
import ohos.aafwk.ability.Ability;
import ohos.aafwk.content.Intent;
import ohos.hiviewdfx.HiLog;
import ohos.hiviewdfx.HiLogLabel;

public class WeatherAbility extends Ability {
    private static final HiLogLabel LABEL_LOG =
            new HiLogLabel(HiLog.LOG_APP, 0x00001, "WeatherAbility");
    private final static int CODE = 1;
    private static final String TEMP_KEY = "temperature";

    @Override
    public void onStart(Intent intent) {
        super.onStart(intent);
        super.setMainRoute(WeatherAbilitySlice.class.getName());
    }

    @Override
    protected void onActive() {
        HiLog.info(LABEL_LOG, "before onActive");
```

```
        super.onActive();

        Intent resultIntent = new Intent();
        resultIntent.setParam(TEMP_KEY, "17");

        setResult(CODE, resultIntent); // 暂存返回结果

        HiLog.info(LABEL_LOG, "after onActive");
    }

}
```

上述代码解释如下。

- resultIntent是Intent的示例，用于返回结果。
- 通过Parameters的方式来传递天气信息。Parameters是一种支持自定义的数据结构，通过setParam方法，将温度“17”放到resultIntent中。
- 调用setResult方法暂存返回结果。

4. 修改 MainAbilitySlice

MainAbilitySlice作为请求方，修改代码如下：

```
package com.waylau.hmos.javaintentoperationwithaction.slice;

import com.waylau.hmos.javaintentoperationwithaction.ResourceTable;
import ohos.aafwk.ability.AbilitySlice;
import ohos.aafwk.content.Intent;
import ohos.aafwk.content.Operation;
import ohos.agp.components.Text;
import ohos.hiviewdfx.HiLog;
import ohos.hiviewdfx.HiLogLabel;

public class MainAbilitySlice extends AbilitySlice {
    private static final HiLogLabel LABEL_LOG =
            new HiLogLabel(HiLog.LOG_APP, 0x00001, "MainAbilitySlice");
    private final static int CODE = 1;
    private static final String TEMP_KEY = "temperature";

    @Override
    public void onStart(Intent intent) {
        super.onStart(intent);
        super.setUIContent(ResourceTable.Layout_ability_main);

        // 添加点击事件来触发请求
        Text text = (Text) findComponentById(ResourceTable.Id_text_
helloworld);
```

```
        text.setClickedListener(listener -> this.queryWeather());
    }
    private void queryWeather() {
        HiLog.info(LABEL_LOG, "before queryWeather");

        Intent intent = new Intent();

        Operation operation = new Intent.OperationBuilder()
                .withAction("action.weather")
                .build();
        intent.setOperation(operation);

        // 上述方式等同于 intent.setAction("action.weather");

        startAbilityForResult(intent, CODE);
    }

    @Override
    protected void onAbilityResult(int requestCode, int resultCode, Intent
                                   resultData) {
        HiLog.info(LABEL_LOG, "onAbilityResult");

        switch (requestCode) {
            case CODE:
                HiLog.info(LABEL_LOG, "code 1 result: %{public}s",
                        resultData.getStringParam(TEMP_KEY));
                break;
            default:
                HiLog.info(LABEL_LOG, "defualt result: %{public}s",
                           resultData.getAction());
                break;
        }
    }

    // 为节约篇幅，省略部分代码
}
```

上述代码解释如下。

- Text增加了点击事件以触发请求，并路由到action.weather。
- 重写了onAbilityResult方法，用来接收请求的返回值。注意，resultData获取的返回参数的值是字符串，因此需要用getStringParam方法；如果是整型，则可以使用getIntParam方法。

5. 运行应用

运行应用，界面效果如图5-21所示。

点击“你好，世界”字样，触发路由到“action.weather”的请求，界面效果如图5-22所示。

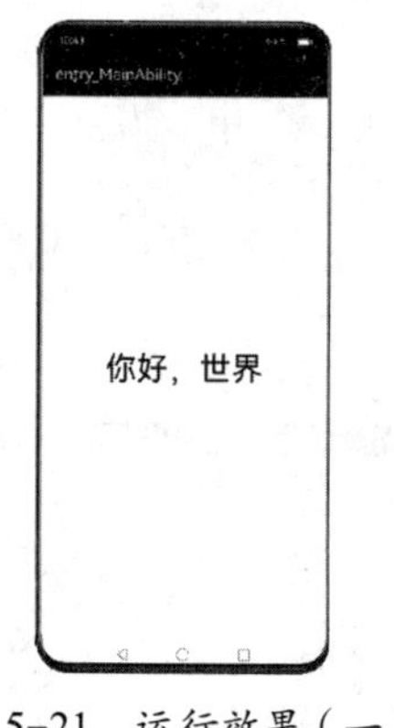

图5-21　运行效果（一）

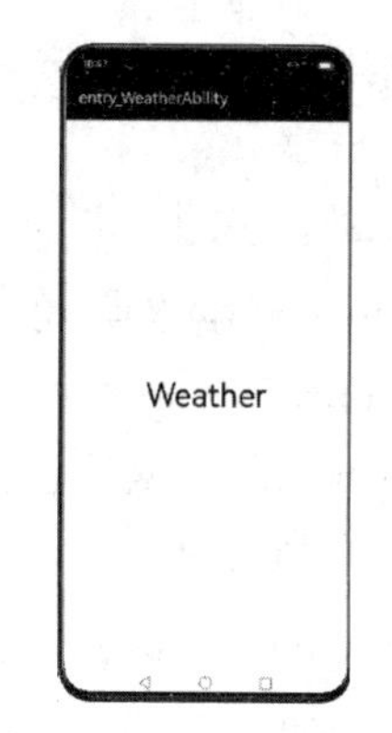

图5-22　运行效果（二）

此时，控制台HiLog输出内容如下：

```
04-24 00:46:11.175 5821-5821/com.waylau.hmos.javaintentoperationwithaction I
00001/MainAbilitySlice: before queryWeather
04-24 00:46:11.374 5821-5821/com.waylau.hmos.javaintentoperationwithaction I
00001/WeatherAbility: before onActive
04-24 00:46:11.375 5821-5821/com.waylau.hmos.javaintentoperationwithaction I
00001/WeatherAbility: after onActive
```

点击模拟器返回按钮，返回“你好，世界”界面。此时，控制台HiLog输出内容如下：

```
04-24 00:49:57.127 5821-5821/com.waylau.hmos.javaintentoperationwithaction I
00001/MainAbilitySlice: onAbilityResult
04-24 00:49:57.127 5821-5821/com.waylau.hmos.javaintentoperationwithaction I
00001/MainAbilitySlice: code 1 result: 17
```

从上述日志文件也可以看出，MainAbilitySlice已经能够拿到WeatherAbility的返回值，温度是“17”。

5.13 实战：Stage模型Ability内页面的跳转和数据传递

在Stage模型下，Ability的数据传递包括Ability内页面的跳转和数据传递、Ability间的数据跳转和数据传递。本节主要讲解Ability内页面的跳转和数据传递。

打开DevEco Studio，创建一个名为“ArkTSPagesRouter”的应用作为演示示例，如图5-23所示。

图5-23　新建应用

5.13.1 新建Ability内页面

初始化应用之后，会生成以下代码。

- 在src/main/ets/entryability目录下，初始会生成一个Ability文件EntryAbility.ts。可以在EntryAbility.ts文件中根据业务需要实现Ability的生命周期回调内容。
- 在src/main/ets/pages目录下，会生成一个Index页面。这也是基于Ability实现的应用的入口页面。可以在Index页面中根据业务需要实现入口页面的功能。

为了实现页面的跳转和数据传递，需要新建一个页面。在src/main/ets/pages目录下，可以按照图5-24所示右击“New”，选择“Page”来新建页面。

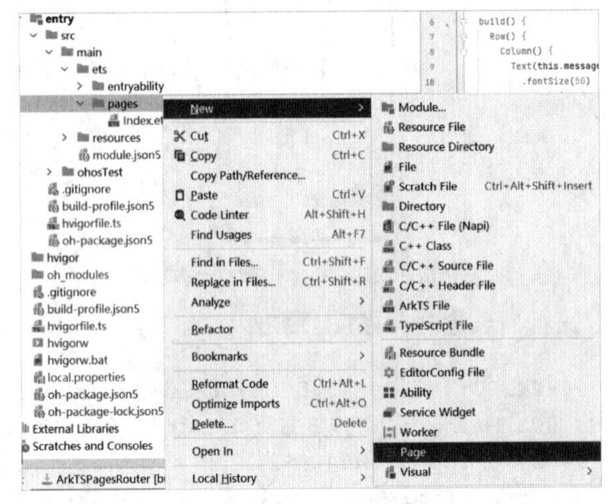

图 5-24 新建页面

新建一个名为“Second”的页面，如图5-25所示。

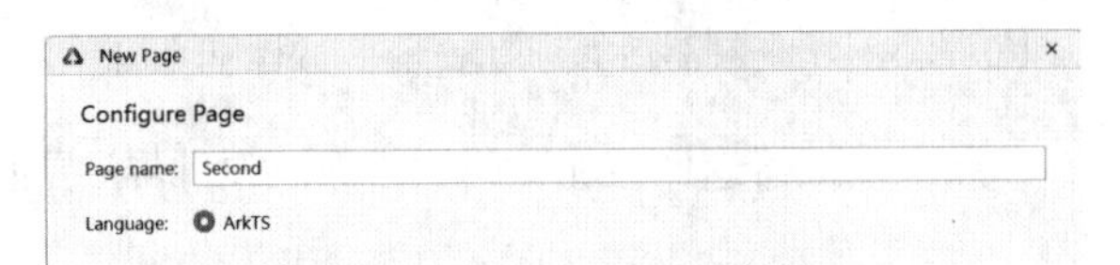

图 5-25 新建一个“Second”页面

“Second”页面创建完成之后，会自动做两个动作。一个动作是在src/main/ets/pages目录下，会创建一个Second.ets的文件。文件内容如下：

```
@Entry
@Component
struct Second {
  @State message: string = 'Hello World'

  build() {
    Row() {
      Column() {
        Text(this.message)
          .fontSize(50)
          .fontWeight(FontWeight.Bold)
      }
      .width('100%')
    }
    .height('100%')
  }
}
```

另外一个动作是将Second页面信息配置到了src/main/resources/base/profile/main_pages.json文

件中。main_pages.json文件内容如下：

```
{
  "src": [
    "pages/Index",
    "pages/Second"
  ]
}
```

分别把Index.ets和Second.ets的message变量值改为“Index页面”和“Second页面”以示区别。

5.13.2 页面跳转及传参

页面间的导航可以通过页面路由router模块来实现。页面路由模块根据页面url找到目标页面，从而实现跳转。通过页面路由模块，可以使用不同的url访问不同的页面，包括跳转到Ability内的指定页面、用Ability内的某个页面替换当前页面、返回上一页面或指定的页面等。通过params来传递参数。

在使用页面路由之前，需要先导入router模块，代码如下所示：

```
// 导入 router 模块
import router from '@ohos.router';
```

下面介绍页面跳转的几种方式，大家根据需要选择一种方式跳转即可。

1. router.pushUrl()

通过调用router.pushUrl()方法跳转到Ability内的指定页面。每调用一次router.pushUrl()方法，就会新建一个页面。默认情况下，页面栈数量会加1，页面栈支持的最大页面数量为32。

当页面栈数量较大或超过32时，可以通过调用router.clear()方法清除页面栈中的所有历史页面，仅保留当前页面作为栈顶页面。

用法示例如下：

```
router.pushUrl({
  url: 'pages/Second',
  params: {
    src: 'Index 页面传来的数据',
  }
})
```

2. router.pushUrl()加mode参数

router.pushUrl()方法新增mode参数，可以将mode参数配置为router.RouterMode.Single单实例模式和router.RouterMode.Standard标准模式。

在单实例模式下，如果目标页面的url在页面栈中已经存在相同url页面，离栈顶最近的同url页面会被移动到栈顶，移动后的页面为新建页，原来的页面仍然存在栈中，页面栈数量不变；如果目标页面的url在页面栈中不存在同url页面，则按标准模式跳转，页面栈数量会加1。

用法示例如下：

```
router.pushUrl({
  url: 'pages/Second',
  params: {
    src: 'Index 页面传来的数据 ',
  }
}, router.RouterMode.Single)
```

3. routerreplaceUrl()

通过调用routerreplaceUrl()方法跳转到Ability内的指定页面，即使用新的页面替换当前页面，并销毁被替换的当前页面，页面栈数量不变。

用法示例如下：

```
routerreplaceUrl({
  url: 'pages/Second',
  params: {
    src: 'Index 页面传来的数据 ',
  }
})
```

4. routerreplaceUrl() 加 mode 参数

routerreplaceUrl()方法新增了mode参数，可以将mode参数配置为router.RouterMode.Single单实例模式和router.RouterMode.Standard标准模式。

在单实例模式下，如果目标页面的url在页面栈中已经存在同url页面，离栈顶最近的同url页面会被移动到栈顶，替换当前页面，并销毁被替换的当前页面，移动后的页面为新建页，页面栈数量会减1；如果目标页面的url在页面栈中不存在同url页面，则按标准模式跳转，页面栈数量不变。

用法示例如下：

```
routerreplaceUrl({
  url: 'pages/Second',
  params: {
    src: 'Index 页面传来的数据 ',
  }
}, router.RouterMode.Single)
```

最后，在Index.ets文件中添加按钮以触发跳转。Index.ets代码如下：

```
// 导入 router 模块
import router from '@ohos.router';
```

```
@Entry
@Component
struct Index {
  @State message: string = 'Index 页面'

  build() {
    Row() {
      Column() {
        Text(this.message)
          .fontSize(50)
          .fontWeight(FontWeight.Bold)

        // 添加按钮，触发跳转
        Button('跳转')
          .fontSize(40)
          .onClick(() => {
            router.pushUrl({
              url: 'pages/Second',
              params: {
                src: 'Index 页面传来的数据',
              }
            });
          })
      }
      .width('100%')
    }
    .height('100%')
  }
}
```

5.13.3 参数接收

通过调用router.getParams()方法获取Index页面传递过来的自定义路由参数。代码如下：

```
import router from '@ohos.router';

@Entry
@Component
struct Second {
  // 获取 Index 页面传递过来的自定义路由参数
  @State src: string = router.getParams()?.['src'];
  // 为节约篇幅，省略部分代码
}
```

可以调用router.back()方法返回上一个页面。

最终，完整Second.ets的代码如下：

```
// 导入 router 模块
import router from '@ohos.router';

@Entry
@Component
struct Second {
  @State message: string = 'Second 页面 '

  // 获取 Index 页面传递过来的自定义路由参数
  @State src: string = router.getParams()?.['src'];

  build() {
    Row() {
      Column() {
        Text(this.message)
          .fontSize(50)
          .fontWeight(FontWeight.Bold)

        // 显示传参的内容
        Text(this.src)
          .fontSize(30)

        // 添加按钮，触发返回
        Button(' 返回 ')
          .fontSize(40)
          .onClick(() => {
            router.back();
          })
      }
      .width('100%')
    }
    .height('100%')
  }
}
```

5.13.4 运行

运行项目后，初始化界面如图5-26所示。

在Index页面中点击“跳转”后，即可从Index页面跳转到Second页面，并在Second页面中接收参数和进行页面刷新展示，界面效果如图5-27所示。

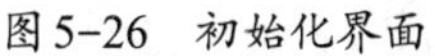
图5-26　初始化界面

图5-27　Second 页面

在Second页面点击“返回”后，则会回到如图5-26所示的Index页面。

以上就是完整的页面跳转及传参、接收参数的过程。

5.14 Want概述

在Stage模型中，Want是对象间信息传递的载体，可以用于应用组件间的信息传递。而在FA模型中，Intent是与之有相同概念的类。

5.14.1 Want的用途

Want的使用场景之一是作为startAbility的参数，其包含指定的启动目标，以及启动时需携带的相关数据，如bundleName和abilityName字段分别指明目标Ability所在应用的包名及对应包内的Ability名称。当AbilityA启动AbilityB并需要传入一些数据给AbilityB时，Want可以作为一个数据载体将数据传给AbilityB，如图5-28所示。

图5-28　Want用法示意

5.14.2 Want的类型

Want的类型主要分为显式和隐式。

1. 显式Want

在启动Ability时指定了abilityName和bundleName的Want称为显式Want。

当有明确处理请求的对象时，通过提供目标Ability所在应用的包名信息（bundleName），并在Want内指定abilityName，便可启动目标Ability。显式Want通常在启动当前应用开发中某个已知Ability时被用到。示例如下：

```
let want = {
    deviceId: '',
    bundleName: 'com.example.myapplication',
    abilityName: 'calleeAbility',
};
```

2. 隐式 Want

在启动Ability时未指定abilityName的Want称为隐式Want。

当请求处理的对象不明确，如开发者希望在当前应用中使用其他应用提供的某个能力（通过skills定义），而不关心提供该能力的具体应用时，可以使用隐式Want。例如，使用隐式Want描述需要打开一个链接的请求，而不关心具体通过哪个应用打开，系统将匹配声明支持该请求的所有应用。当未匹配到支持的应用时，系统将弹窗说明无法打开；当仅匹配到一个时，系统将自动拉起对应应用；当匹配到多个时，系统将弹出候选列表，由用户选择拉起哪个应用。示例如下：

```
let want = {
    action: 'ohos.want.action.search',
    entities: [ 'entity.system.browsable' ],
    uri: 'https://www.test.com:8080/query/student',
    type: 'text/plain',
};
```

其中，action表示调用方要执行的通用操作（如查看、分享、应用详情）。在隐式Want中，可定义该字段，配合uri或parameters来表示对数据要执行的操作（如打开，可查看该uri数据）。例如，当uri为一段网址，action为ohos.want.action.viewData时，则表示匹配可查看该网址的Ability。在Want内声明action字段，表示希望被调用方应用支持声明的操作。在被调用方应用配置文件skills字段内声明actions，表示该应用支持声明操作。

常见action如下。

- ACTION_HOME：启动应用入口组件的动作，需要和ENTITY_HOME配合使用。系统桌面应用图标就是显式的入口组件，点击也是启动入口组件；入口组件可以配置多个。
- ACTION_CHOOSE：选择本地资源数据，例如联系人、相册等。系统一般对不同类型的数据提供对应的Picker应用，例如联系人和图库。
- ACTION_VIEW_DATA：查看数据，当使用网址uri时，则表示显示该网址对应的内容。
- ACTION_VIEW_MULTIPLE_DATA：发送多个数据记录的操作。

entities表示目标Ability的类别信息（如浏览器、视频播放器），在隐式Want中是对action的补充。在隐式Want中，开发者可定义该字段来过滤匹配应用的类别，如必须是浏览器。在Want内声明entities字段，表示希望被调用方应用属于声明的类别。在被调用方应用配置文件skills字段内声明entites，表示该应用支持的类别。

常用entities有以下几种。

- ENTITY_DEFAULT：默认类别无实际意义。
- ENTITY_HOME：主屏幕有图标点击入口类别。
- ENTITY_BROWSABLE：指示浏览器类别。

所有action和entities定义在wantConstant模块中。

5.14.3 Want参数属性

Want参数属性说明如表5-2所示。

表5-2 Want参数属性说明

名称	读写属性	类型	必填	描述
deviceId	只读	string	否	表示目标Ability所在设备ID。如果未设置该字段，则表明本设备
bundleName	只读	string	否	表示目标Ability所在应用名称
moduleName	只读	string	否	表示目标Ability所属的模块名称
abilityName	只读	string	否	表示目标Ability名称。如果未设置该字段，则该Want为隐式。如果在Want中同时指定了bundleName、moduleName和abilityName，则Want可以直接匹配到指定的Ability
uri	只读	string	否	表示携带的数据，一般配合type使用，指明待处理的数据类型。如果在Want中指定了uri，则Want将匹配指定的uri信息，包括scheme、schemeSpecificPart、authority和path信息
type	只读	string	否	表示携带数据类型，使用MIME类型规范。例如："text/plain""image/*"等
action	只读	string	否	表示要执行的通用操作（如查看、分享、应用详情）。在隐式Want中，可定义该字段，配合uri或parameters来表示对数据要执行的操作（如打开，则查看该uri数据）。例如，当uri为一段网址，action为ohos.want.action.viewData时，则表示匹配可查看该网址的Ability
entities	只读	Array<string>	否	表示目标Ability额外的类别信息（如浏览器、视频播放器），在隐式Want中是对action的补充。在隐式Want中，可定义该字段来过滤匹配Ability类别，如必须是浏览器。例如，在action字段的举例中，可存在多个应用声明支持查看网址的操作，其中有的应用为普通社交应用，有的为浏览器应用，可通过entity.system.browsable过滤掉非浏览器的其他应用
flags	只读	number	否	表示处理Want的方式。例如，通过wantConstant.Flags.FLAG_ABILITY_CONTINUATION，表示是否以设备间迁移方式启动Ability

续表

名称	读写属性	类型	必填	描述
parameters	只读	{[key: string]: any}	否	此参数用于传递自定义数据，通过用户自定义的键值对进行数据填充，具体支持的数据类型如Want API所示

5.15 实战：通过显式Want启动Ability

本节演示如何通过显式Want拉起应用内的一个指定Ability组件。

打开DevEco Studio，创建一个名为“ArkTSWantStartAbility”的应用作为演示示例。

5.15.1 新建Ability内页面

初始化应用之后，在原有的代码基础上，需要新建一个页面。在src/main/ets/pages目录下新建一个名为“Second”的页面。

对Second.ets文件中的message变量值进行修改，最终文件内容如下：

```
@Entry
@Component
struct Second {
  // 修改变量值为Second
  @State message: string = 'Second'

  build() {
    Row() {
      Column() {
        Text(this.message)
          .fontSize(50)
          .fontWeight(FontWeight.Bold)
      }
      .width('100%')
    }
    .height('100%')
  }
}
```

5.15.2 新建Ability

在原有的代码基础上，需要新建一个Ability。在src/main/ets目录下通过右击“New->Ability”

新建一个名为“SecondAbility”的Ability。

创建完成之后，会自动在module.json5文件中添加该Ability的信息。内容如下：

```
{
  "name": "SecondAbility",
  "srcEntry": "./ets/secondability/SecondAbility.ts",
  "description": "$string:SecondAbility_desc",
  "icon": "$media:icon",
  "label": "$string:SecondAbility_label",
  "startWindowIcon": "$media:icon",
  "startWindowBackground": "$color:start_window_background"
}
```

此时，在src/main/ets目录下会初始化一个secondability目录，并在secondability目录下生成一个SecondAbility.ts文件。修改该SecondAbility.ts文件，将pages/Index改为pages/Second。最终文件内容如下：

```
onWindowStageCreate(windowStage: window.WindowStage) {
  // Main window is created, set main page for this ability
  hilog.info(0x0000, 'testTag', '%{public}s', 'Ability onWindowStageCreate');

  // 加载Second页面
  windowStage.loadContent('pages/Second', (err, data) => {
    if (err.code) {
      hilog.error(0x0000, 'testTag', 'Failed to load the content. Cause:
%{public}s', JSON.stringify(err) ?? '');
      return;
    }
    hilog.info(0x0000, 'testTag', 'Succeeded in loading the content. Data:
%{public}s', JSON.stringify(data) ?? '');
  });
}
```

上述修改是为了当启动SecondAbility时能够展示Second页面。

5.15.3 使用显式Want启动Ability

在Index.ets文件中添加按钮以触发执行启动Ability。Index.ets代码修改如下：

```
// 导入common
import common from '@ohos.app.ability.common';

@Entry
@Component
struct Index {
```

```
  @State message: string = 'Hello World'

  build() {
    Row() {
      Column() {
        Text(this.message)
          .fontSize(50)
          .fontWeight(FontWeight.Bold)

        // 添加按钮，启动 Ability
        Button('启动')
          .fontSize(40)
          .onClick(this.explicitStartAbility) // 显示启动 Ability
      }
      .width('100%')
    }
    .height('100%')
  }

  // 显示启动 Ability
  async explicitStartAbility() {
    try {
      // 在启动 Ability 时指定了 abilityName 和 bundleName
      let want = {
        deviceId: "",
        bundleName: "com.waylau.hmos.arktswantstartability",
        abilityName: "SecondAbility"
      };
      let context = getContext(this) as common.UIAbilityContext;
      await context.startAbility(want);
      console.info(`explicit start ability succeed`);
    } catch (error) {
      console.info(`explicit start ability failed with ${error.code}`);
    }
  }
}
```

与上述代码中一样，abilityName 可在 module.json5 文件中查看。注意，bundleName 需要与对应 app.json5 文件中的 bundleName 一致，因此修改 app.json5 文件中的 bundleName。内容如下：

```
{
  "app": {
    "bundleName": "com.waylau.hmos.arktswantstartability", // 修改 bundleName
    "vendor": "example",
    "versionCode": 1000000,
    "versionName": "1.0.0",
```

```
    "icon": "$media:app_icon",
    "label": "$string:app_name"
  }
}
```

5.15.4 运行

运行项目后，初始化界面如图5-29所示。

在Index页面中，单击“启动”后，此时启动了SecondAbility，并展示了Second页面，界面效果如图5-30所示。

图5-29 初始化界面

图5-30 Second页面

以上就是完整的通过显式Want启动Ability的过程。

5.16 实战：通过隐式Want打开应用管理

本节演示如何通过隐式Want打开应用管理。

打开DevEco Studio，创建一个名为“ArkTSWantOpenManageApplications”的应用作为演示示例。

5.16.1 使用隐式Want启动Ability

在Index.ets文件中添加按钮以触发执行启动Ability。Index.ets代码如下：

```
// 导入 common
import common from '@ohos.app.ability.common';

@Entry
@Component
struct Index {
  build() {
    Row() {
      Column() {
        // 添加按钮，启动 Ability
        Button('启动')
          .fontSize(40)
          .onClick(this.implicitStartAbility) // 隐示启动 Ability
      }
      .width('100%')
    }
```

```
      .height('100%')
    }

  // 隐示启动 Ability
  async implicitStartAbility() {
    try {
      let want = {
        // 调用应用管理
        "action": "ohos.settings.manage.applications"
      }
      let context = getContext(this) as common.UIAbilityContext;
      await context.startAbility(want)
      console.info(`implicit start ability succeed`)
    } catch (error) {
      console.info(`implicit start ability failed with ${error.code}`)
    }
  }
}
```

上述implicitStartAbility()方法通过指定action为ohos.settings.manage.applications，从而实现隐式启动应用管理。

5.16.2 运行

运行项目后，初始化界面如图5-31所示。

在Index页面中，点击“启动”后，此时启动了应用管理并展示了其页面，界面效果如图5-32所示。

图5-31 初始化界面

图5-32 应用管理页面

以上就是完整的通过隐式Want启动Ability的过程。

第6章 安全管理

本章介绍 HarmonyOS 应用的安全管理机制。

6.1 权限基本概念

HarmonyOS的权限包含以下基本概念。

（1）应用沙盒：系统利用内核保护机制来识别和隔离应用资源，可将不同的应用隔离开，保护应用自身和系统免受恶意应用的攻击。在默认情况下，应用间不能彼此交互，而且对系统的访问会受到限制。例如，如果应用A（一个单独的应用）尝试在没有权限的情况下读取应用B的数据或调用系统的能力拨打电话，操作系统会阻止此类行为，因为应用A没有被授予相应的权限。

（2）应用权限：由于系统通过沙盒机制管理各个应用，在默认规则下，应用只能访问有限的系统资源。但应用为了扩展功能，需要访问自身沙盒之外的系统或其他应用的数据（包括用户个人数据）或能力；系统或应用也必须以明确的方式对外提供接口来共享其数据或能力。为了保证这些数据或能力不会被不当或恶意使用，就需要有一种访问控制机制来保护，这就是应用权限。应用权限是程序访问操作某种对象的许可。权限在应用层面要求明确定义且经用户授权，以便系统化地规范各类应用程序的行为准则与权限许可。

（3）权限保护的对象：权限保护的对象可以分为数据和能力。

- 数据包含个人数据（如照片、通讯录、日历、位置等）、设备数据（如设备标识、相机、麦克风等）、应用数据。
- 能力包括设备能力（如打电话、发短信、联网等）、应用能力（如弹出悬浮框、创建快捷方式等）等。

（4）权限开放范围：权限开放范围指一个权限能被哪些应用申请。按可信程度从高到低的顺序，不同权限开放范围对应的应用可分为：系统服务、系统应用、系统预置特权应用、同签名应用、系统预置普通应用、持有权限证书的后装应用、其他普通应用，开放范围依次扩大。

（5）敏感权限：涉及访问个人数据（如照片、通讯录、日历、本机号码、短信等）和操作敏感能力（如相机、麦克风、拨打电话、发送短信等）的权限。

（6）应用核心功能：一个应用可能提供了多种功能，其中应用为满足用户的关键需求而提供的功能，称为应用的核心功能。这是一个相对宽泛的概念，主要用来辅助描述用户权限授权的预期。用户选择安装一个应用，通常是被应用的核心功能所吸引。比如，导航类应用，定位导航就是这种应用的核心功能；媒体类应用，播放及媒体资源管理就是其核心功能，这些功能所需要的权限，用户在安装时内心已经倾向于授予（否则就不会去安装）。与核心功能相对应的是辅助功能，这些功能所需要的权限，需要向用户清晰说明目的、场景等信息，由用户授权。有些功能既不属于核心功能，也不是辅助功能，那么这些功能就是多余功能，这些功能所需要的权限通常被用户禁止。

（7）最小必要权限：保障应用某一服务类型正常运行所需要的应用权限的最小集，一旦缺少将导致该类型服务无法实现或无法正常运行的应用权限。

6.2 权限运作机制

系统所有应用均在应用沙盒内运行。在默认情况下，应用只能访问有限的系统资源。这些限制是通过DAC（Discretionary Access Control，自主访问控制）、MAC（Mandatory Access Control，强制访问控制）及本节描述的应用权限机制等多种不同的形式实现的。因为应用要实现某些功能就必须访问系统或其他应用的数据或操作某些器件，所以需要系统或其他应用能提供接口，考虑到安全，就需要对这些接口采用一种限制措施，这就是被称为“应用权限”的安全机制。

接口的提供涉及其权限的命名和分组、对外开放的范围、被授予的应用，以及用户的参与和体验。应用权限管理模块的目的就是负责管理由接口提供方（访问客体）、接口使用方（访问主体）、系统（包括云侧和端侧）和用户等共同参与的整个流程，保证受限接口在约定好的规则下被正常使用，避免接口被滥用而导致用户、应用和设备受损。

6.3 权限约束与限制

HarmonyOS权限在使用时，应考虑以下约束与限制。

- 同一应用自定义权限个数不能超过1024个。
- 同一应用申请权限个数不能超过1024个。
- 为避免与系统权限名冲突，应用自定义权限名不能以ohos开头，且权限名长度不能超过256个字符。
- 自定义权限授予方式不能为user_grant。
- 自定义权限开放范围不能为restricted。

6.4 应用权限列表

HarmonyOS根据接口所涉数据的敏感程度或所涉能力的安全威胁影响，定义了不同开放范围与授权方式的权限来保护数据。

6.4.1 权限分类

当前权限的开放范围分为以下几种。

- all：所有应用可用。

- signature：平台签名应用可用。
- privileged：预制特权应用可用。
- restricted：证书可控应用可用。

应用在使用对应服务的能力或数据时，需要申请对应权限。

- 已在config.json文件中声明的非敏感权限，会在应用安装时自动授予，该类权限的授权方式为系统授权（system_grant）。
- 敏感权限需要应用动态申请，通过运行时发送弹窗的方式请求用户授权，该类权限的授权方式为用户授权（user_grant）。

当应用调用服务时，服务会对应用进行权限检查，如果没有对应权限则无法使用该服务。

当前仅介绍对所有应用开发的HarmonyOS的应用权限。

6.4.2 敏感权限

敏感权限的申请需要按照动态申请流程向用户申请授权。

敏感权限包括以下几种。

- ohos.permission.LOCATION：允许应用在前台运行时获取位置信息。如果应用在后台运行时也要获取位置信息，则需要同时申请ohos.permission.LOCATION_IN_BACKGROUND权限。
- ohos.permission.LOCATION_IN_BACKGROUND：允许应用在后台运行时获取位置信息，需要同时申请ohos.permission.LOCATION权限。
- ohos.permission.CAMERA：允许应用使用相机拍摄照片和录制视频。
- ohos.permission.MICROPHONE：允许应用使用麦克风进行录音。
- ohos.permission.READ_CALENDAR：允许应用读取日历信息。
- ohos.permission.WRITE_CALENDAR：允许应用在设备上添加、移除或修改日历活动。
- ohos.permission.ACTIVITY_MOTION：允许应用读取用户当前的运动状态。
- ohos.permission.READ_HEALTH_DATA：允许应用读取用户的健康数据。
- ohos.permission.DISTRIBUTED_DATASYNC：允许不同设备间的数据交换。
- ohos.permission.DISTRIBUTED_DATA：允许应用使用分布式数据的能力。
- ohos.permission.MEDIA_LOCATION：允许应用访问用户媒体文件中的地理位置信息。
- ohos.permission.READ_MEDIA：允许应用读取用户外部存储中的媒体文件信息。
- ohos.permission.WRITE_MEDIA：允许应用读写用户外部存储中的媒体文件信息。

6.4.3 非敏感权限

非敏感权限不涉及用户的敏感数据或危险操作，仅需在config.json中声明，应用安装后即被授权。

非敏感权限包括以下几种。

- ohos.permission.GET_NETWORK_INFO：允许应用获取数据网络信息。
- ohos.permission.GET_WIFI_INFO：允许获取WLAN信息。
- ohos.permission.USE_BLUETOOTH：允许应用查看蓝牙的配置。
- ohos.permission.DISCOVER_BLUETOOTH：允许应用配置本地蓝牙，并允许其查找远端设备且与之配对连接。
- ohos.permission.SET_NETWORK_INFO：允许应用控制数据网络。
- ohos.permission.SET_WIFI_INFO：允许配置WLAN设备。
- ohos.permission.SPREAD_STATUS_BAR：允许应用以缩略图方式呈现在状态栏。
- ohos.permission.INTERNET：允许使用网络socket。
- ohos.permission.MODIFY_AUDIO_SETTINGS：允许应用程序修改音频设置。
- ohos.permission.RECEIVER_STARTUP_COMPLETED：允许应用接收设备启动完成广播。
- ohos.permission.RUNNING_LOCK：允许申请休眠运行锁，并执行相关操作。
- ohos.permission.ACCESS_BIOMETRIC：允许应用使用生物识别能力进行身份认证。
- ohos.permission.RCV_NFC_TRANSACTION_EVENT：允许应用接收卡模拟交易事件。
- ohos.permission.COMMONEVENT_STICKY：允许发布黏性公共事件的权限。
- ohos.permission.SYSTEM_FLOAT_WINDOW：提供显示悬浮窗的能力。
- ohos.permission.VIBRATE：允许应用程序使用马达。
- ohos.permission.USE_TRUSTCIRCLE_MANAGER：允许调用设备间认证能力。
- ohos.permission.USE_WHOLE_SCREEN：允许通知携带一个全屏IntentAgent。
- ohos.permission.SET_WALLPAPER：允许设置静态壁纸。
- ohos.permission.SET_WALLPAPER_DIMENSION：允许设置壁纸尺寸。
- ohos.permission.REARRANGE_MISSIONS：允许调整任务栈。
- ohos.permission.CLEAN_BACKGROUND_PROCESSES：允许根据包名清理相关后台进程。
- ohos.permission.KEEP_BACKGROUND_RUNNING：允许Service Ability在后台继续运行。
- ohos.permission.GET_BUNDLE_INFO：允许查询其他应用的信息。
- ohos.permission.ACCELEROMETER：允许应用程序读取加速度传感器的数据。
- ohos.permission.GYROSCOPE：允许应用程序读取陀螺仪传感器的数据。
- ohos.permission.MULTIMODAL_INTERACTIVE：允许应用订阅语音或手势事件。
- ohos.permission.radio.ACCESS_FM_AM：允许用户获取收音机相关服务。
- ohos.permission.NFC_TAG：允许应用读写Tag卡片。
- ohos.permission.NFC_CARD_EMULATION：允许应用实现卡模拟功能。
- ohos.permission.DISTRIBUTED_DEVICE_STATE_CHANGE：允许获取分布式组网内设备的状态变化。

- ohos.permission.GET_DISTRIBUTED_DEVICE_INFO：允许获取分布式组网内的设备列表和设备信息。

6.4.4 受限开放的权限

受限开放的权限通常是不允许第三方应用申请的。如果有特殊场景需要使用，请提供相关申请材料到应用市场申请相应权限证书。如果应用未申请相应的权限证书，却试图在config.json文件中声明此类权限，将会导致应用安装失败。另外，由于此类权限涉及用户敏感数据或危险操作，当应用申请到权限证书后，还需按照动态申请权限的流程向用户申请授权。

受限开放权限包括以下两种。

- ohos.permission.READ_CONTACTS：允许应用读取联系人数据。
- ohos.permission.WRITE_CONTACTS：允许应用添加、移除和更改联系人数据。

6.5 FA模型应用权限开发流程

HarmonyOS支持开发者自定义权限来保护能力或接口，同时开发者也可申请权限来访问受权限保护的对象。

在前面已经介绍了应用权限的一些使用方式，本节就详细总结一下应用权限的开发流程。

6.5.1 权限申请

开发者需要在config.json文件中的reqPermissions字段中声明所需要的权限。示例如下：

```
{
    "reqPermissions": [
        {
            "name": "ohos.permission.CAMERA",
            "reason": "$string:permreason_camera",
            "usedScene":
            {
                "ability": ["com.mycamera.Ability", "com.mycamera.
AbilityBackground"],
                "when": "always"
            }
        },{
        // 为节约篇幅，省略部分代码
        }
```

```
    ]
}
```

权限申请格式采用数组格式，可支持同时申请多个权限，权限个数最多不能超过1024个。

reqPermissions权限申请字段说明如表6-1所示。

表6-1 reqPermissions权限申请字段说明

键	值说明	类型	取值范围	默认值	规则约束
name	必填。填写需要使用的权限名称	字符串	自定义	无	未填写时，解析失败
reason	可选。当申请的权限为user_grant权限时，此字段必填	描述申请权限的原因	字符串	显示文字长度不能超过256个字节	空
usedScene	可选。当申请的权限为user_grant权限时，此字段必填	描述权限使用的场景和时机。场景类型有：ability、when（调用时机）。可配置多个ability	ability：字符串数组；when：字符串	ability：ability的名称；when：inuse（使用时）、always（始终）	ability：空；when：inuse

如果声明使用的权限的grantMode是system_grant，则权限会在应用安装的时候被自动授予。

如果声明使用的权限的grantMode是user_grant，则必须经用户手动授权（用户在弹框中授权或进入权限设置界面授权）才可使用。用户会看到reason字段中填写的理由，来帮助用户决定是否给予授权。

注意：对于授权方式为user_grant的权限，每一次执行需要这一权限的操作时，都需要检查自身是否有该权限。当自身具有权限时，才可继续执行，否则应用需要请求用户授予权限。示例参见动态申请权限开发步骤。

6.5.2 自定义权限

开发者需要在config.json文件中的defPermissions字段中自定义所需的权限。示例如下：

```
{
    "defPermissions": [
        {
            "name": "com.myability.permission.MYPERMISSION",
            "grantMode": "system_grant",
            "availableScope": ["signature"]
        }, {
        // 为节约篇幅，省略部分代码
        }
    ]
}
```

权限定义格式采用数组格式，可支持同时定义多个权限，自定义的权限个数最多不能超过1024个。

defPermissions权限定义字段说明如表6-2所示。

表6-2　defPermissions权限定义字段说明

键	值说明	类型	取值范围	默认值	规则约束
name	必填。权限名称。为最大可能避免重名，采用反向域公司名+应用名+权限名组合	字符串	自定义	无	第三方应用不允许填写系统存在的权限，否则安装失败。未填写解析失败。权限名长度不能超过256个字符
grantMode	必填。权限授予方式	字符串	user_grant（用户授权）；system_grant（系统授权）	system_grant	未填值或填写取值范围以外的值时，自动赋予默认值；不允许第三方应用填写user_grant，填写后会自动赋予默认值
availableScope	选填。权限限制范围。不填则表示此权限对所有应用开放	字符串数组	signature；privileged；restricted	空	填写取值范围以外的值时，权限限制范围不生效。由于第三方应用并不在restricted的范围内，很少会出现权限定义者不能访问自身定义的权限的情况，所以不允许第三方应用填写restricted
label	选填。权限的简短描述。若未填写，则使用到简短描述的地方时由权限名取代	字符串	自定义	空	需要多语种适配
description	选填。权限的详细描述。若未填写，则使用到详细描述的地方时由label取代	字符串	自定义	空	需要多语种适配

权限授予方式字段说明如表6-3所示。

表6-3　权限授予方式字段说明

授予方式(grantMode)	说明	自定义权限是否可指定该级别	取值样例
system_grant	在config.json里面声明，安装后系统自动授予	是	GET_NETWORK_INFO、GET_WIFI_INFO
user_grant	在config.json里面声明，并在使用时动态申请，用户授权后才可使用	否，如自定义则强制修改为system_grant	CAMERA、MICROPHONE

权限限制范围字段说明如表6-4所示。

表6-4 权限限制范围字段说明

权限范围(availableScope)	说明	自定义权限是否可指定该级别	取值样例
restricted	需要开发者向华为申请后才能被使用的特殊权限	否	ANSWER_CALL、READ_CALL_LOG、RECEIVE_SMS
signature	权限定义方和使用方的签名一致。需在config.json里面声明后，由权限管理模块负责签名校验，一致后方可使用	是	对应用（或Ability）操作的系统接口上由系统定义的及应用自定义的权限。例如：find某Ability，连接某Ability
privileged	预置在系统版本中的特权应用可申请的权限	是	SET_TIME、MANAGE_USER_STORAGE

6.5.3 访问权限控制

1. Ability的访问权限控制

在config.json中填写abilities到permissions字段，表示只有拥有该权限的应用可访问此Ability。下面的例子表明只有拥有ohos.permission.CAMERA权限的应用可以访问此ability：

```
"abilities": [
    {
        "name": ".MainAbility",
        "description": "$string:description_main_ability",
        "icon": "$media:hiworld.png",
        "label": "HiCamera",
        "launchType": "standard",
        "orientation": "portrait",
        "visible": false,
        "permissions": [
            "ohos.permission.CAMERA"
        ],
    }
]
```

其中，permissions用以表示此ability受哪个权限保护，只有拥有此权限的应用才可访问此ability。目前仅支持填写一个权限名，若填写多个权限名，仅第一个权限名称有效。

2. Ability接口的访问权限控制

在Ability实现中，如需要针对特定接口进行访问控制，可在服务侧的接口实现中，主动通过

verifyCallingPermission、verifyCallingOrSelfPermission来检查访问者是否拥有所需要的权限。示例如下：

```
if (verifyCallingPermission("ohos.permission.CAMERA") != IBundleManager.
PERMISSION_GRANTED) {
    // 调用者无权限，做错误处理
}
    // 调用者权限校验通过，开始提供服务
```

6.5.4 接口说明

应用权限接口有以下几种。

- verifyPermission(String permissionName, int pid, int uid)：查询指定PID、UID的应用是否已被授予某权限。
- verifyCallingPermission(String permissionName)：查询IPC跨进程调用的调用方的进程是否已被授予某权限。
- verifySelfPermission(String permissionName)：查询自身进程是否已被授予某权限。
- verifyCallingOrSelfPermission(String permissionName)：当有远端调用时，检查远端是否有权限，否则检查自身是否拥有权限。
- canRequestPermission(String permissionName)：向系统权限管理模块查询某权限是否不再弹框授权了。
- requestPermissionsFromUser (String[] permissions, int requestCode)：向系统权限管理模块申请权限。（接口可支持一次申请多个。若下一步操作涉及多个敏感权限，可以这么用，其他情况建议不要这么用。因为弹框还是按权限组一个个去弹框，耗时比较长，最好用到哪个权限就去申请哪个。）
- onRequestPermissionsFromUserResult (int requestCode, String[] permissions, int[] grantResults)：调用requestPermissionsFromUser后的应答接口。

6.5.5 动态申请权限开发步骤

动态申请权限开发步骤如下。

1. 声明所需要的权限

在config.json文件中声明所需要的权限。示例如下：

```
{
    "reqPermissions": [
        {
            "name": "ohos.permission.CAMERA",
            "reason": "$string:permreason_camera",
```

```
            "usedScene": {
                "ability": ["com.mycamera.Ability", "com.mycamera.
AbilityBackground"],
                "when": "always"}
        }, {
        // 为节约篇幅，省略部分代码
        }
    ]
}
```

2. 检查是否已被授予该权限

继承Ability，使用ohos.app.Context.verifySelfPermission接口查询应用是否已被授予该权限。

- 如果已被授予权限，可以结束权限申请流程。
- 如果未被授予权限，继续执行下一步。

3. 查询是否可动态申请

使用canRequestPermission查询是否可动态申请。

- 如果不可动态申请，说明已被用户或系统永久禁止授权，可以结束权限申请流程。
- 如果可动态申请，继续执行下一步。

4. 动态申请权限

使用requestPermissionFromUser动态申请权限，通过回调函数接收授予结果。样例代码如下：

```
if (verifySelfPermission("ohos.permission.CAMERA")
        != IBundleManager.PERMISSION_GRANTED) {
    // 应用未被授予权限
    if (canRequestPermission("ohos.permission.CAMERA")) {
        // 是否可以申请弹框授权（首次申请或用户未选择禁止且不再提示）
        requestPermissionsFromUser(
                new String[] { "ohos.permission.CAMERA" } , MY_PERMISSIONS_
REQUEST_CAMERA);
    } else {
        // 显示应用需要权限的理由，提示用户进入设置授权
    }
} else {
    // 权限已被授予
}

@Override
public void onRequestPermissionsFromUserResult (int requestCode, String[]
permissions, int[] grantResults) {
    switch (requestCode) {
        case MY_PERMISSIONS_REQUEST_CAMERA: {
            // 匹配requestPermissions的requestCode
```

```
            if (grantResults.length > 0
                && grantResults[0] == IBundleManager.PERMISSION_GRANTED) {
                // 权限被授予
                // 注意：因时间差导致接口权限检查时有无权限
                // 所以对那些因无权限而抛异常的接口进行异常捕获处理
            } else {
                // 权限被拒绝
            }
            return;
        }
    }
}
```

6.6 Stage模型访问控制开发步骤

如果应用需要获取目标权限，那么需要先进行权限申请。

- 权限申请：开发者需要在配置文件中声明目标权限。
- 权限授权：如果目标权限是system_grant类型，开发者在进行权限申请后，系统会在安装应用时自动为其进行权限预授予，开发者不需要做其他操作即可使用权限。如果目标权限是user_grant类型，开发者在进行权限申请后，应用在运行时触发动态弹窗，请求用户授权。

6.6.1 权限申请

应用需要在工程配置文件中对需要的权限逐个声明，没有在配置文件中声明的权限，应用将无法获得授权。

使用Stage模型的应用，需要在module.json5文件中声明权限。示例如下：

```
{
  "module" : {
    "requestPermissions":[
      {
        "name" : "ohos.permission.PERMISSION1",
        "reason": "$string:reason",
        "usedScene": {
          "abilities": [
            "FormAbility"
          ],
          "when":"inuse"
        }
```

```
      },
      {
        "name" : "ohos.permission.PERMISSION2",
        "reason": "$string:reason",
        "usedScene": {
          "abilities": [
            "FormAbility"
          ],
          "when":"always"
        }
      }
    ]
  }
}
```

配置文件标签说明如下。

- name：权限名称。
- reason：当申请的权限为user_grant权限时，此字段必填，描述申请权限的原因。
- usedScene：当申请的权限为user_grant权限时，此字段必填，描述权限使用的场景和时机。
- ability：标识需要使用到该权限的Ability，标签为数组形式。
- when：标识权限使用的时机，值为inuse/always，表示仅允许前台使用/前后台都可使用。

6.6.2 权限授权

经过前期的权限声明步骤，应用在安装过程中，系统会对system_grant类型的权限进行权限预授权，而user_grant类型权限则需要用户进行手动授权。

所以，应用在调用受ohos.permission.PERMISSION2权限保护的接口前，需要先校验应用是否已经获取该权限。

如果校验结果显示应用已经获取了该权限，那么应用可以直接访问该目标接口；否则，应用需要通过动态弹框先申请用户授权，并根据授权结果进行相应处理。

6.7 实战：访问控制授权

本节演示访问控制授权申请的流程。为了演示该功能，创建一个名为“ArkTSUserGrant”的应用。

6.7.1 场景介绍

本示例代码假设应用因为应用核心功能诉求，需要申请权限ohos.permission.INTERNET和权限ohos.permission.CAMERA。其中部分情况，说明如下。

- 应用的APL等级为normal。
- 权限ohos.permission.INTERNET的权限等级为normal，权限类型为system_grant。
- 权限ohos.permission.CAMERA的权限等级为system_basic, 权限类型为user_grant。

在当前场景下，应用申请的权限包括user_grant权限，对于这部分user_grant权限，可以先通过权限校验，判断当前调用者是否具备相应权限。

当权限校验结果显示当前应用尚未被授予该权限时，再通过动态弹框授权方式给用户提供手动授权入口。

6.7.2 声明访问的权限

在module.json5文件中声明权限，配置如下：

```
{
  "module": {
    // 为节约篇幅，省略部分代码

    // 声明所需要的权限
    "requestPermissions":[
      {
        "name" : "ohos.permission.INTERNET"
      },
      {
        "name" : "ohos.permission.CAMERA"
      }
    ]
  }
}
```

6.7.3 申请授权user_grant权限

经过前期的权限声明步骤，应用在安装过程中，系统会对system_grant类型的权限进行权限预授权，而user_grant类型权限则需要用户进行手动授权。

所以，应用在调用受ohos.permission.CAMERA权限保护的接口前，需要先校验应用是否已经获取该权限。

如果校验结果显示应用已经获取了该权限，那么应用可以直接访问该目标接口；否则，应用需要通过动态弹框先申请用户授权，并根据授权结果进行相应处理，处理方式可参考访问控制开发概述。

修改EntryAbility.ets的onWindowStageCreate方法，在windowStage.loadContent方法之前添加如下内容：

```
// 导入abilityAccessCtrl, Permissions
import abilityAccessCtrl, { Permissions } from '@ohos.abilityAccessCtrl';

onWindowStageCreate(windowStage: window.WindowStage): void {
  // Main window is created, set main page for this ability
  hilog.info(0x0000, 'testTag', '%{public}s', 'Ability onWindowStageCreate');

  // 权限校验
  let context = this.context;
  let atManager = abilityAccessCtrl.createAtManager();
  let permissions: Array<Permissions> = ["ohos.permission.CAMERA"];

  // requestPermissionsFromUser会判断权限的授权状态
  atManager.requestPermissionsFromUser(context, permissions).then((data) => {
    let grantStatus: Array<number> = data.authResults;
    let length: number = grantStatus.length;
    for (let i = 0; i < length; i++) {
      if (grantStatus[i] === 0) {
        // 用户同意授权
        windowStage.loadContent('pages/Index', (err, data) => {
          if (err.code) {
            hilog.error(0x0000, 'testTag', 'Failed to load the content.
Cause: %{public}s', JSON.stringify(err) ?? '');
            return;
          }
          hilog.info(0x0000, 'testTag', 'Succeeded in loading the content.
Data: %{public}s', JSON.stringify(data) ?? '');
        });
      } else {
        // 用户拒绝授权
        return;
      }
    }
    // 授权成功
  }).catch((err) => {
    console.error(`requestPermissionsFromUser failed, code is ${err.code},
message is ${err.message}`);
  })
}
```

上述代码演示了请求用户授予权限的开发步骤：

- 获取 ability 的上下文 context。
- 调用 requestPermissionsFromUser 接口请求权限。在运行过程中，该接口会根据应用是否已获得目标权限，决定是否拉起动态弹框请求用户授权。
- 根据 requestPermissionsFromUser 接口返回值判断是否已获取目标权限。如果当前已经获取权限，则可以继续正常访问目标接口。

6.7.4 运行

运行应用，显示的界面效果如图 6-1 所示。

上述界面提示让用户授权。当用户点击“仅使用期间允许”或“允许本次使用”时，代表同意授权，授权成功后界面执行加载，如图 6-2 所示。

当用户点击“禁止”时，代表不同意授权，则界面不会执行加载，最终效果如图 6-3 所示。

图 6-1 提示授权

图 6-2 授权成功后界面执行加载

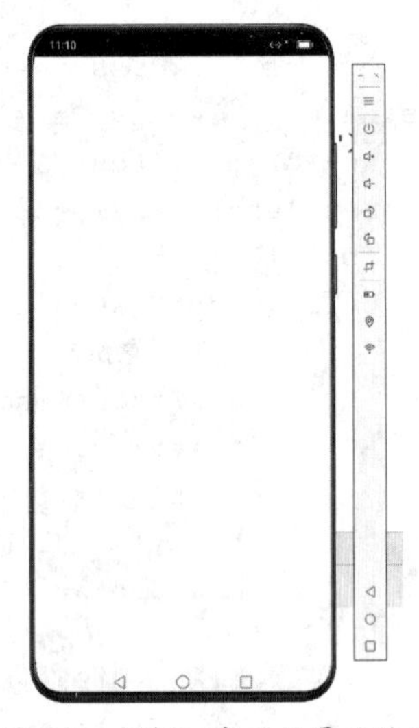

图 6-3 授权不成功后界面效果

6.8 生物特征识别约束与限制

生物特征识别运作机制包含以下约束与限制。

- 当前版本提供的生物特征识别能力只包含人脸识别，且只支持本地认证，不提供认证界面。
- 要求设备上具备摄像器件，且人脸图像像素大于 100×100。
- 要求设备上具有 TEE（Trusted Execution Environment，可信执行环境），人脸特征信息高强度加密保存在 TEE 中。
- 对于面部特征相似的人、面部特征不断发育的儿童，人脸特征匹配率有所不同。如果对此担忧，可考虑其他认证方式。

6.9 生物特征识别开发流程

当前生物特征识别支持2D人脸识别、3D人脸识别，可应用于设备解锁、应用登录、支付等身份认证场景。

6.9.1 接口说明

BiometricAuthentication类提供了生物认证的相关方法，包括检测认证能力、认证和取消认证等，用户可以通过人脸等生物特征信息进行认证操作。在执行认证前，需要检查设备是否支持该认证能力，具体指认证类型、安全级别和是否本地认证。如果不支持，需要考虑使用其他认证能力。

生物特征识别开放能力接口有以下几种。

- getInstance(Ability ability)：获取BiometricAuthentication的单例对象。
- checkAuthenticationAvailability(AuthType type, SecureLevel level, boolean isLocalAuth)：检测设备是否具有生物认证能力。
- execAuthenticationAction(AuthType type, SecureLevel level, boolean isLocalAuth,boolean isAppAuthDialog, SystemAuthDialogInfo information)：调用者使用该方法进行生物认证。可以使用自定义的认证界面，也可以使用系统提供的认证界面。当使用系统认证界面时，调用者可以自定义提示语。该方法直到认证结束才返回认证结果。
- getAuthenticationTips()：获取生物认证过程中的提示信息。
- cancelAuthenticationAction()：取消生物认证操作。
- setSecureObjectSignature(Signature sign)：设置需要关联认证结果的Signature对象，在进行认证操作后，如果认证成功则Signature对象被授权可以使用。设置前Signature对象需要正确初始化，且配置为认证成功才能使用。
- getSecureObjectSignature()：在认证成功后，可通过该方法获取已授权的Signature对象。如果未设置过Signature对象，则返回null。
- setSecureObjectCipher(Cipher cipher)：设置需要关联认证结果的Cipher对象，在进行认证操作后，如果认证成功则Cipher对象被授权可以使用。设置前Cipher对象需要正确初始化，且配置为认证成功才能使用。
- getSecureObjectCipher()：在认证成功后，可通过该方法获取已授权的Cipher对象。如果未设置过Cipher对象，则返回null。
- setSecureObjectMac(Mac mac)：设置需要关联认证结果的Mac对象，在进行认证操作后，如果认证成功则Mac对象被授权可以使用。设置前Mac对象需要正确初始化，且配置为认证成功才能使用。

- getSecureObjectMac()：在认证成功后，可通过该方法获取已授权的Mac对象。如果未设置过Mac对象，则返回null。

6.9.2 开发准备

开发前请完成以下准备工作。

- 在应用配置权限文件中，增加ohos.permission.ACCESS_BIOMETRIC的权限声明。
- 在使用生物特征识别认证能力的代码文件中，增加导入ohos.biometrics.authentication.BiometricAuthentication。

6.9.3 开发过程

开发过程大致如下。

1. 获取BiometricAuthentication的单例对象

代码示例如下：

```
BiometricAuthentication  biometricAuthentication =
    BiometricAuthentication.getInstance(MainAbility.mAbility);
```

2. 检测设备是否具有生物认证能力

- 2D人脸识别建议使用SECURE_LEVEL_S2；
- 3D人脸识别建议使用SECURE_LEVEL_S3。

代码示例如下：

```
int retChkAuthAvb =
    biometricAuthentication.checkAuthenticationAvailability(
        BiometricAuthentication.AuthType.AUTH_TYPE_BIOMETRIC_FACE_ONLY,
        BiometricAuthentication.SecureLevel.SECURE_LEVEL_S2, true);
```

3. 设置需要关联认证结果（可选）

设置需要关联认证结果的Signature对象、Cipher对象或Mac对象，代码示例如下：

```
// 定义一个Signature对象sign;
biometricAuthentication.setSecureObjectSignature(sign);

// 定义一个Cipher对象cipher;
biometricAuthentication.setSecureObjectCipher(cipher);
```

```
// 定义一个 Mac 对象 mac;
biometricAuthentication.setSecureObjectMac(mac);
```

4. 执行认证操作

在新线程里面执行认证操作，避免阻塞其他操作，代码示例如下：

```
new Thread(new Runnable() {
    @Override
    public void run() {
        int retExcAuth;
        retExcAuth = biometricAuthentication.execAuthenticationAction(
            BiometricAuthentication.AuthType.AUTH_TYPE_BIOMETRIC_FACE_ONLY,
            BiometricAuthentication.SecureLevel.SECURE_LEVEL_S2, true, false,
null);
    }
}).start();
```

5. 获得认证过程中的提示信息

获得认证过程中的提示信息，代码示例如下：

```
AuthenticationTips tips = biometricAuthentication.getAuthenticationTips();
```

6. 认证成功后获取结果对象（可选）

认证成功后获取已设置的Signature对象、Cipher对象或Mac对象，代码示例如下：

```
Signature sign = biometricAuthentication.getSecureObjectSignature();

Cipher cipher = biometricAuthentication.getSecureObjectCipher();

Mac mac = biometricAuthentication.getSecureObjectMac();
```

7. 认证过程中取消认证

认证过程中取消认证，代码示例如下：

```
int ret = biometricAuthentication.cancelAuthenticationAction();
```

第7章

Ability公共事件与通知

HarmonyOS 通过 CES（Common Event Service，公共事件服务）为应用程序提供订阅、发布、退订公共事件的能力；通过 ANS（Advanced Notification Service，高级通知服务）为应用程序提供发布通知的能力。

7.1 公共事件与通知概述

在应用里面，往往会有事件和通知。比如，给朋友手机发了一条信息，未读信息会在手机的通知栏给出提示。

在HarmonyOS里面，系统给应用发送提示一般分为两种方式，即公共事件和通知。

7.1.1 公共事件和通知

公共事件可分为系统公共事件和自定义公共事件。

- 系统公共事件：系统将收集到的事件信息，根据系统策略发送给订阅该事件的用户程序。例如：用户可感知亮灭屏事件，系统关键服务发送的系统事件（如USB插拔、网络连接、系统升级等）。
- 自定义公共事件：应用自定义一些公共事件用来处理业务逻辑。

IntentAgent封装了一个指定行为的Intent，可以通过IntentAgent启动Ability和发送公共事件。应用如果需要接收公共事件，需要订阅相应的事件。

通知提供应用的即时消息或通信消息，用户可以直接删除或点击通知触发进一步的操作。比如，收到一条未读信息的通知，则可以选择删除该通知，或者点击该通知进入短信应用查看该信息。

7.1.2 约束与限制

1. 公共事件的约束与限制

- 目前公共事件仅支持动态订阅。部分系统事件需要具有指定的权限，具体的权限见API参考。
- 目前公共事件订阅不支持多用户。
- ThreadMode表示线程模型，目前仅支持HANDLER模式，即在当前UI线程上执行回调函数。
- deviceId用来指定订阅本地公共事件还是远端公共事件。deviceId为null、空字符串或本地设备deviceId时，表示订阅本地公共事件，否则表示订阅远端公共事件。

2. 通知的约束与限制

- 通知目前支持六种样式：普通文本、长文本、图片、社交、多行文本和媒体样式。创建通知时必须包含其中一种样式。
- 通知支持快捷回复。

3. IntentAgent 的限制

使用IntentAgent启动Ability时，Intent必须指定Ability的包名和类名。

7.2 公共事件服务

每个应用都可以订阅自己感兴趣的公共事件，订阅成功且公共事件发布后，系统会将其发送给应用。这些公共事件可能来自系统、其他应用和应用自身。HarmonyOS提供了一套完整的API，支持用户订阅、发送和接收公共事件。发送公共事件需要借助CommonEventData对象，接收公共事件需要继承CommonEventSubscriber类并实现onReceiveEvent回调函数。

7.2.1 接口说明

公共事件相关基础类包含以下几种。

- CommonEventData。
- CommonEventPublishInfo。
- CommonEventSubscribeInfo。
- CommonEventSubscriber。
- CommonEventManager。

上述基础类之间的关系如图7-1所示。

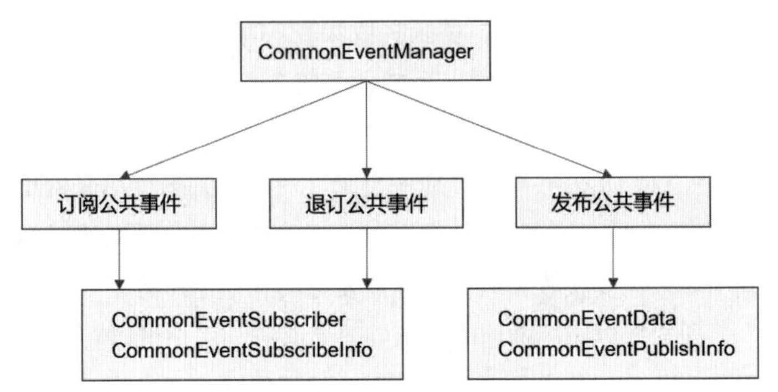

图7-1 基础类之间的关系

1. CommonEventData

CommonEventData封装公共事件相关信息，用于在发布、分发和接收时处理数据。在构造CommonEventData对象时，相关参数需要注意以下事项。

- code为有序公共事件的结果码，data为有序公共事件的结果数据，仅用于有序公共事件场景。
- intent不允许为空，否则发布公共事件失败。

2. CommonEventPublishInfo

CommonEventPublishInfo封装公共事件发布相关属性、限制等信息，包括公共事件类型（有序或黏性）、接收者权限等。

- 有序公共事件：主要场景是多个订阅者有依赖关系或对处理顺序有要求，例如，高优先级订阅者可修改公共事件内容或处理结果，包括终止公共事件处理；或低优先级订阅者依赖高优先级的处理结果等。有序公共事件的订阅者可以通过CommonEventSubscribeInfo.setPriority()方法指定优先级，缺省为0，优先级范围[-1000, 1000]，值越大优先级越高。
- 黏性公共事件：指公共事件的订阅动作是在公共事件发布之后进行，订阅者也能收到的公共事件类型。主要场景是由公共事件服务记录某些系统状态，如蓝牙、WLAN、充电等事件和状态。不使用黏性公共事件机制时，应用可以通过直接访问系统服务获取该状态；在状态变化时，系统服务、硬件需要提供类似于observer等方式通知应用。发布黏性公共事件可以通过setSticky()方法设置，发布黏性公共事件需要申ohos.permission.COMMONEVENT_STICKY权限。

3. CommonEventSubscribeInfo

CommonEventSubscribeInfo封装公共事件订阅相关信息，比如优先级、线程模式、事件范围等。

线程模式（ThreadMode）：设置订阅者的回调方法所执行的线程模式。主要有HANDLER、POST、ASYNC、BACKGROUND四种模式。

- HANDLER：在Ability的主线程上执行。
- POST：在事件分发线程执行。
- ASYNC：在一个新创建的异步线程执行。
- BACKGROUND：在后台线程执行。

截至目前，线程模式只支持HANDLER模式。

4. CommonEventSubscriber

CommonEventSubscriber封装公共事件订阅者及相关参数。

CommonEventSubscriber.AsyncCommonEventResult类处理有序公共事件异步执行。目前订阅者只能通过调用CommonEventManager的subscribeCommonEvent()进行订阅。

5. CommonEventManager

CommonEventManager是为应用提供订阅、退订和发布公共事件的静态接口类。主要接口如下。

- 发布公共事件：publishCommonEvent(CommonEventData event)。
- 发布公共事件指定发布信息：publishCommonEvent(CommonEventData event, CommonEventPublishInfo publishinfo)。
- 发布有序公共事件，指定发布信息和最后一个接收者：publishCommonEvent(CommonEventData event, CommonEventPublishInfo publishinfo, CommonEventSubscriber resultSubscriber)。
- 订阅公共事件：subscribeCommonEvent(CommonEventSubscriber subscriber)。
- 退订公共事件：unsubscribeCommonEvent(CommonEventSubscriber subscriber)。

7.2.2 发布公共事件

开发者可以发布4种公共事件：无序的公共事件、携带权限的公共事件、有序的公共事件、黏性公共事件。

1. 发布无序的公共事件

发布无序的公共事件时，构造CommonEventData对象，设置Intent，通过构造operation对象把需要发布的公共事件信息传入intent对象，然后调用 CommonEventManager.publishCommonEvent(CommonEventData) 接口发布公共事件。

示例代码如下：

```
try {
    Intent intent = new Intent();
    Operation operation = new Intent.OperationBuilder()
            .withAction("com.my.test")
            .build();
    intent.setOperation(operation);
    CommonEventData eventData = new CommonEventData(intent);
    CommonEventManager.publishCommonEvent(eventData);
} catch (RemoteException e) {
    HiLog.info(LABEL, "publishCommonEvent occur exception.");
}
```

2. 发布携带权限的公共事件

发布携带权限的公共事件时，构造CommonEventPublishInfo对象，设置订阅者的权限。订阅者在config.json中申请所需的权限，各字段含义详见权限定义字段说明。

示例如下：

```
{
    "reqPermissions": [{
        "name": "com.example.MyApplication.permission",
        "reason": "get right",
        "usedScene": {
           "ability": [
            ".MainAbility"
           ],
           "when": "inuse"
        }
    }, {
    ...
    }]
}
```

发布携带权限的公共事件示例代码如下：

```
Intent intent = new Intent();
Operation operation = new Intent.OperationBuilder()
        .withAction("com.my.test")
        .build();
intent.setOperation(operation);
CommonEventData eventData = new CommonEventData(intent);
CommonEventPublishInfo publishInfo = new CommonEventPublishInfo();
String[] permissions = {"com.example.MyApplication.permission" };
publishInfo.setSubscriberPermissions(permissions); // 设置权限
```

```
try {
    CommonEventManager.publishCommonEvent(eventData, publishInfo);
} catch (RemoteException e) {
    HiLog.info(LABEL, "publishCommoneEvent occur exception.");
}
```

3. 发布有序的公共事件

发布有序的公共事件时，构造CommonEventPublishInfo对象，通过setOrdered(true)指定公共事件属性为有序公共事件，也可以指定一个最后的公共事件接收者。

示例代码如下：

```
CommonEventSubscriber resultSubscriber = new MyCommonEventSubscriber();
CommonEventPublishInfo publishInfo = new CommonEventPublishInfo();
publishInfo.setOrdered(true); // 设置属性为有序公共事件
try {
    // 指定 resultSubscriber 为有序公共事件最后一个接收者。
    CommonEventManager.publishCommonEvent(eventData, publishInfo,
resultSubscriber);
} catch (RemoteException e) {
    HiLog.info(LABEL, "publishCommoneEvent occur exception.");
}
```

4. 发布黏性公共事件

发布黏性公共事件时，构造CommonEventPublishInfo对象，通过setSticky(true)指定公共事件属性为黏性公共事件。

发布者首先在config.json中申请发布黏性公共事件所需的权限，配置如下：

```
{
    "reqPermissions": [{
        "name": "ohos.permission.COMMONEVENT_STICKY",
        "reason": "get right",
        "usedScene": {
            "ability": [
             ".MainAbility"
            ],
            "when": "inuse"
        }
    }, {
    ...
    }]
}
```

发布黏性公共事件代码如下：

```
CommonEventPublishInfo publishInfo = new CommonEventPublishInfo();
publishInfo.setSticky(true); // 设置属性为黏性公共事件
try {
    CommonEventManager.publishCommonEvent(eventData, publishInfo);
} catch (RemoteException e) {
    HiLog.info(LABEL, "publishCommoneEvent occur exception.");
}
```

7.2.3 订阅公共事件

订阅公共事件时，首先创建CommonEventSubscriber派生类，在onReceiveEvent()回调函数中处理公共事件。示例代码如下：

```
class MyCommonEventSubscriber extends CommonEventSubscriber {
    MyCommonEventSubscriber(CommonEventSubscribeInfo info) {
        super(info);
    }
    @Override
    public void onReceiveEvent(CommonEventData commonEventData) {
    }
}
```

注意：此处不能执行耗时操作，否则会阻塞UI线程，产生用户点击没有反应等异常。

接着构造MyCommonEventSubscriber对象，调用CommonEventManager.subscribeCommonEvent()接口进行订阅。示例代码如下：

```
String event = "com.my.test";
MatchingSkills matchingSkills = new MatchingSkills();
matchingSkills.addEvent(event); // 自定义事件
matchingSkills.addEvent(CommonEventSupport.COMMON_EVENT_SCREEN_ON); // 亮屏事件
CommonEventSubscribeInfo subscribeInfo = new CommonEventSubscribeInfo(matchin
gSkills);
MyCommonEventSubscriber subscriber = new MyCommonEventSubscriber(subscribeIn
fo);
try {
    CommonEventManager.subscribeCommonEvent(subscriber);
} catch (RemoteException e) {
    HiLog.info(LABEL, "subscribeCommonEvent occur exception.");
}
```

如果订阅拥有指定权限应用发布的公共事件，发布者需要在config.json中申请权限，各字段含义详见权限申请字段说明。代码如下：

```
"reqPermissions": [
```

```
    {
        "name": "ohos.abilitydemo.permission.PROVIDER",
        "reason": "get right",
        "usedScene": {
            "ability": ["com.huawei.hmi.ivi.systemsetting.MainAbility"],
            "when": "inuse"
        }
    }
]
```

如果订阅的公共事件是有序的，可以调用setPriority()指定优先级。代码如下：

```
String event = "com.my.test";
MatchingSkills matchingSkills = new MatchingSkills();
matchingSkills.addEvent(event ); // 自定义事件

CommonEventSubscribeInfo subscribeInfo = new CommonEventSubscribeInfo(matchin
gSkills);
subscribeInfo.setPriority(100); // 设置优先级,优先级取值范围[-1000,1000],值默认为0。
MyCommonEventSubscriber subscriber = new MyCommonEventSubscriber(subscribeIn
fo);
try {
     CommonEventManager.subscribeCommonEvent(subscriber);
} catch (RemoteException e) {
     HiLog.info(LABEL, "subscribeCommonEvent occur exception.");
}
```

最后，针对在onReceiveEvent中不能执行耗时操作的限制，可以使用CommonEventSubscriber的goAsyncCommonEvent()来实现异步操作，函数返回后仍保持该公共事件活跃，且执行完成后必须调用AsyncCommonEventResult.finishCommonEvent()来结束。示例代码如下：

```
// 创建新线程，将耗时的操作放到新的线程上执行
EventRunner runner = EventRunner.create();

//MyEventHandler 为 EventHandler 的派生类，在不同线程间分发和处理事件和 Runnable 任务
MyEventHandler myHandler = new MyEventHandler(runner);

@Override
public void onReceiveEvent(CommonEventData commonEventData){
    final AsyncCommonEventResult result = goAsyncCommonEvent();

    Runnable task = new Runnable() {
        @Override
        public void run() {
            ......          // 待执行的操作，由开发者定义
            result.finishCommonEvent(); // 调用 finish 结束异步操作
```

```
        }
    };
    myHandler.postTask(task);
}
```

7.2.4 退订公共事件

在 Ability 的 onStop() 中调用 CommonEventManager.unsubscribeCommonEvent() 方法来退订公共事件。调用后，之前订阅的所有公共事件均被退订。示例代码如下：

```
try {
    CommonEventManager.unsubscribeCommonEvent(subscriber);
 } catch (RemoteException e) {
    HiLog.info(LABEL, "unsubscribeCommonEvent occur exception.");
 }
```

7.3 实战：公共事件服务发布事件

本节演示使用公共事件服务发布事件的例子。为了演示该功能，创建一个名为 “CommonEvent Publisher” 的应用，作为公共事件的发布者。

7.3.1 修改 ability_main.xml

修改 ability_main.xml 文件如下：

```
<?xml version="1.0" encoding="utf-8"?>
<DirectionalLayout
    xmlns:ohos="http://schemas.huawei.com/res/ohos"
    ohos:height="match_parent"
    ohos:width="match_parent"
    ohos:alignment="center"
    ohos:orientation="vertical">

    <Text
        ohos:id="$+id:text_publish_event"
        ohos:height="match_content"
        ohos:width="match_content"
        ohos:background_element="$graphic:background_ability_main"
        ohos:layout_alignment="horizontal_center"
        ohos:text="Publish Event"
```

```
        ohos:text_size="40vp"
        />

</DirectionalLayout>
```

上述修改只是将显示文本内容改为“Publish Event”。

7.3.2 修改MainAbilitySlice

在初始化应用时，应用已经包含一个主AbilitySlice，即MainAbilitySlice。对MainAbilitySlice进行修改，代码如下：

```
package com.waylau.hmos.commoneventpublisher.slice;

import com.waylau.hmos.commoneventpublisher.ResourceTable;
import ohos.aafwk.ability.AbilitySlice;
import ohos.aafwk.content.Intent;
import ohos.aafwk.content.Operation;
import ohos.agp.components.Text;
import ohos.event.commonevent.CommonEventData;
import ohos.event.commonevent.CommonEventManager;
import ohos.event.commonevent.CommonEventPublishInfo;
import ohos.hiviewdfx.HiLog;
import ohos.hiviewdfx.HiLogLabel;
import ohos.rpc.RemoteException;

public class MainAbilitySlice extends AbilitySlice {
    private static final String TAG = MainAbilitySlice.class.getSimpleName();
    private static final HiLogLabel LABEL_LOG =
            new HiLogLabel(HiLog.LOG_APP, 0x00001, TAG);

    private static final String EVENT_PERMISSION =
            "com.waylau.hmos.commoneventpublisher.PERMISSION";
    private static final String EVENT_NAME =
            "com.waylau.hmos.commoneventpublisher.EVENT";
    private static final int EVENT_CODE = 1;
    private static final String EVENT_DATA = "Welcome to waylau.com";
    private int index = 0; // 递增的序列

    @Override
    public void onStart(Intent intent) {
        super.onStart(intent);
        super.setUIContent(ResourceTable.Layout_ability_main);

        // 添加点击事件来触发
```

```
        Text text = (Text) findComponentById(ResourceTable.Id_text_publish_event);
        text.setClickedListener(listener -> publishEvent());
    }

    private void publishEvent() {
        HiLog.info(LABEL_LOG, "before publishEvent");

        Intent intent = new Intent();
        Operation operation = new Intent.OperationBuilder()
                .withAction(EVENT_NAME) // 设置事件名称
                .build();
        intent.setOperation(operation);

        index++;
        CommonEventData eventData =
                new CommonEventData(intent, EVENT_CODE, EVENT_DATA +
                                        " times " + index);
        CommonEventPublishInfo publishInfo = new CommonEventPublishInfo();
        publishInfo.setOrdered(true); // 设置属性为有序公共事件

        String[] permissions = {EVENT_PERMISSION};
        publishInfo.setSubscriberPermissions(permissions); // 设置权限
        try {
            CommonEventManager.publishCommonEvent(eventData, publishInfo);
        } catch (RemoteException e) {
            HiLog.info(LABEL_LOG, "publishCommonEvent occur exception.");
        }

        HiLog.info(LABEL_LOG, "end publishEvent, event data %{}s", eventData);
    }

    @Override
    public void onActive() {
        super.onActive();
    }

    @Override
    public void onForeground(Intent intent) {
        super.onForeground(intent);
    }
}
```

在MainAbilitySlice类中：

- 在Text中增加了点击事件，以便触发发送事件的方法；
- 通过CommonEventManager.publishCommonEvent发送了事件，该事件是一个携带权限信息的事件；

- 每个事件的内容都不同，会附加一个唯一的索引；
- setOrdered(true)用于设置属性为有序公共事件。

7.3.3 运行

运行应用之后，点击“Publish Event”文本内容，可以看到控制台日志输出内容如下：

```
04-24 23:25:21.030 5676-5676/com.waylau.hmos.commoneventpublisher I 00001/
MainAbilitySlice: [5922962461070bb, 0, 0] before publishEvent
04-24 23:25:21.034 5676-5676/com.waylau.hmos.commoneventpublisher I 00001/
MainAbilitySlice: [5922962461070bb, 0, 0] end publishEvent, event data
<private>
04-24 23:25:24.591 5676-5676/com.waylau.hmos.commoneventpublisher I 00001/
MainAbilitySlice: [592298246490263, 0, 0] before publishEvent
04-24 23:25:24.593 5676-5676/com.waylau.hmos.commoneventpublisher I 00001/
MainAbilitySlice: [592298246490263, 0, 0] end publishEvent, event data
<private>
04-24 23:25:25.910 5676-5676/com.waylau.hmos.commoneventpublisher I 00001/
MainAbilitySlice: [59229a2465ddee2, 0, 0] before publishEvent
04-24 23:25:25.914 5676-5676/com.waylau.hmos.commoneventpublisher I 00001/
MainAbilitySlice: [59229c24671693d, 0, 0] end publishEvent, event data
<private>
04-24 23:25:27.094 5676-5676/com.waylau.hmos.commoneventpublisher I 00001/
MainAbilitySlice: [59229c24671693d, 0, 0] before publishEvent
04-24 23:25:27.121 5676-5676/com.waylau.hmos.commoneventpublisher I 00001/
MainAbilitySlice: [59229c24671693d, 0, 0] end publishEvent, event data
<private>
04-24 23:25:27.959 5676-5676/com.waylau.hmos.commoneventpublisher I 00001/
MainAbilitySlice: [59229e2467e9d8d, 0, 0] before publishEvent
04-24 23:25:27.962 5676-5676/com.waylau.hmos.commoneventpublisher I 00001/
MainAbilitySlice: [59229e2467e9d8d, 0, 0] end publishEvent, event data
<private>
```

可以看到，每次点击都会发送一次事件。

7.4 实战：公共事件服务订阅事件

本节演示使用公共事件服务订阅事件的例子。为了演示该功能，创建一个名为“CommonEvent Subscriber”的应用，作为公共事件的订阅者。

7.4.1 修改 ability_main.xml

修改ability_main.xml文件如下：

```
<?xml version="1.0" encoding="utf-8"?>
<DirectionalLayout
    xmlns:ohos="http://schemas.huawei.com/res/ohos"
    ohos:height="match_parent"
    ohos:width="match_parent"
    ohos:alignment="center"
    ohos:orientation="vertical">

    <Text
        ohos:id="$+id:text_eventsubscriber"
        ohos:height="match_content"
        ohos:width="match_content"
        ohos:background_element="$graphic:background_ability_main"
        ohos:layout_alignment="horizontal_center"
        ohos:text="EventSubscriber"
        ohos:text_size="40vp"
        />

</DirectionalLayout>
```

上述修改只是将显示文本内容改为“EventSubscriber”。

7.4.2 创建 CommonEventSubscriber

创建一个名为“WelcomeCommonEventSubscriber”的事件订阅者，该类继承了CommonEvent Subscriber。

代码如下：

```
package com.waylau.hmos.commoneventsubscriber;

import ohos.event.commonevent.CommonEventData;
import ohos.event.commonevent.CommonEventSubscribeInfo;
import ohos.event.commonevent.CommonEventSubscriber;
import ohos.hiviewdfx.HiLog;
import ohos.hiviewdfx.HiLogLabel;

public class WelcomeCommonEventSubscriber extends CommonEventSubscriber {
    private static final String TAG = WelcomeCommonEventSubscriber.class.
getSimpleName();
    private static final HiLogLabel LABEL_LOG =
```

```
            new HiLogLabel(HiLog.LOG_APP, 0x00001, TAG);

    public WelcomeCommonEventSubscriber(CommonEventSubscribeInfo info) {
        super(info);
    }

    @Override
    public void onReceiveEvent(CommonEventData commonEventData) {
        HiLog.info(LABEL_LOG, "receive event data %{public}s",
commonEventData);
    }

}
```

在上述代码中，当接收到事件时，只是简单地将事件内容在日志中输出。

7.4.3 修改MainAbility

在初始化应用时，应用已经包含一个MainAbility。对MainAbility进行修改，代码如下：

```
package com.waylau.hmos.commoneventsubscriber.slice;

import com.waylau.hmos.commoneventsubscriber.ResourceTable;
import com.waylau.hmos.commoneventsubscriber.WelcomeCommonEventSubscriber;
import ohos.aafwk.ability.AbilitySlice;
import ohos.aafwk.content.Intent;
import ohos.event.commonevent.CommonEventManager;
import ohos.event.commonevent.CommonEventSubscribeInfo;
import ohos.event.commonevent.MatchingSkills;
import ohos.hiviewdfx.HiLog;
import ohos.hiviewdfx.HiLogLabel;
import ohos.rpc.RemoteException;


public class MainAbilitySlice extends AbilitySlice {
    private static final String TAG = MainAbilitySlice.class.getSimpleName();
    private static final HiLogLabel LABEL_LOG =
            new HiLogLabel(HiLog.LOG_APP, 0x00001, TAG);
    private static final String EVENT_NAME =
            "com.waylau.hmos.commoneventpublisher.EVENT";
    @Override
    public void onStart(Intent intent) {
        super.onStart(intent);
        super.setUIContent(ResourceTable.Layout_ability_main);
```

```
        // 订阅事件
        subscribeEvent();
    }

    private void subscribeEvent() {
        MatchingSkills matchingSkills = new MatchingSkills();
        matchingSkills.addEvent(EVENT_NAME);
        CommonEventSubscribeInfo subscribeInfo = new CommonEventSubscribeInfo
(matchingSkills);
        subscribeInfo.setPriority(100); // 设置优先级

        WelcomeCommonEventSubscriber subscriber = new WelcomeCommonEventSubsc
riber(subscribeInfo);
        try {
            CommonEventManager.subscribeCommonEvent(subscriber);
        } catch (RemoteException e) {
            HiLog.error(LABEL_LOG, "%{public}s", "subscribeEvent
remoteException.");
        }
    }

    @Override
    public void onActive() {
        super.onActive();
    }

    @Override
    public void onForeground(Intent intent) {
        super.onForeground(intent);
    }
}
```

在MainAbility类中：

- 在该类启动时，会执行subscribeEvent()方法；
- subscribeEvent()方法内容会通过CommonEventManager.subscribeCommonEvent订阅指定的事件；
- setPriority(100)用于设置优先级。

7.4.4 修改配置文件

修改配置文件，增加了权限的申请。代码如下：

```
"abilities": [
    // 为节约篇幅，省略部分代码
```

```
],
"reqPermissions": [{
    "name": "com.waylau.hmos.commoneventpublisher.PERMISSION"
}]
```

7.4.5 运行

先运行CommonEventSubscriber应用，而后运行前一节所介绍的CommonEventPublisher应用，并点击CommonEventPublisher应用的“Publish Event”文本内容，可以看到控制台日志输出内容如下：

```
04-24 23:25:21.030 5676-5676/com.waylau.hmos.commoneventpublisher I 00001/
MainAbilitySlice: [5922962461070bb, 0, 0] before publishEvent
04-24 23:25:21.034 5676-5676/com.waylau.hmos.commoneventpublisher I 00001/
MainAbilitySlice: [5922962461070bb, 0, 0] end publishEvent, event data
<private>
04-24 23:25:24.591 5676-5676/com.waylau.hmos.commoneventpublisher I 00001/
MainAbilitySlice: [592298246490263, 0, 0] before publishEvent
04-24 23:25:24.593 5676-5676/com.waylau.hmos.commoneventpublisher I 00001/
MainAbilitySlice: [592298246490263, 0, 0] end publishEvent, event data
<private>
04-24 23:25:25.910 5676-5676/com.waylau.hmos.commoneventpublisher I 00001/
MainAbilitySlice: [59229a2465ddee2, 0, 0] before publishEvent
04-24 23:25:25.914 5676-5676/com.waylau.hmos.commoneventpublisher I 00001/
MainAbilitySlice: [59229c24671693d, 0, 0] end publishEvent, event data
<private>
04-24 23:25:27.094 5676-5676/com.waylau.hmos.commoneventpublisher I 00001/
MainAbilitySlice: [59229c24671693d, 0, 0] before publishEvent
04-24 23:25:27.121 5676-5676/com.waylau.hmos.commoneventpublisher I 00001/
MainAbilitySlice: [59229c24671693d, 0, 0] end publishEvent, event data
<private>
04-24 23:25:27.959 5676-5676/com.waylau.hmos.commoneventpublisher I 00001/
MainAbilitySlice: [59229e2467e9d8d, 0, 0] before publishEvent
04-24 23:25:27.962 5676-5676/com.waylau.hmos.commoneventpublisher I 00001/
MainAbilitySlice: [59229e2467e9d8d, 0, 0] end publishEvent, event data
<private>
```

可以看到，每次点击CommonEventPublisher应用的“Publish Event”文本都会发送一次事件，而CommonEventSubscriber应用都能收到相应的事件。控制台日志输出内容如下：

```
04-24 23:25:21.045 6331-6331/com.waylau.hmos.commoneventsubscriber I 00001/
WelcomeCommonEventSubscriber: [5922962461070bb, 1c49898, 11634a3] receive
event data CommonEventData[ code = 1 data = Welcome to waylau.com times 11]
04-24 23:25:24.596 6331-6331/com.waylau.hmos.commoneventsubscriber I 00001/
```

```
WelcomeCommonEventSubscriber: [592298246490263, 3605d95, 34382dc] receive
event data CommonEventData[ code = 1 data = Welcome to waylau.com times 12]
04-24 23:25:25.915 6331-6331/com.waylau.hmos.commoneventsubscriber I 00001/
WelcomeCommonEventSubscriber: [59229a2465ddee2, 2d2d0af, 154d37a] receive
event data CommonEventData[ code = 1 data = Welcome to waylau.com times 13]
04-24 23:25:27.121 6331-6331/com.waylau.hmos.commoneventsubscriber I 00001/
WelcomeCommonEventSubscriber: [59229c24671693d, 3f6d897, 6b8c64] receive
event data CommonEventData[ code = 1 data = Welcome to waylau.com times 14]
04-24 23:25:27.967 6331-6331/com.waylau.hmos.commoneventsubscriber I 00001/
WelcomeCommonEventSubscriber: [59229e2467e9d8d, 25be7ee, 31f35d8] receive
event data CommonEventData[ code = 1 data = Welcome to waylau.com times 15]
```

7.5 高级通知服务

大家对手机的通知功能应该不会陌生。HarmonyOS也提供了通知功能，即在一个应用的UI界面之外显示的消息，主要用来提醒用户有来自某个应用的信息。当应用向系统发出通知时，它将先以图标的形式显示在通知栏中，用户可以下拉通知栏查看通知的详细信息。常见的使用场景如下。

- 显示接收到短消息、即时消息等。
- 显示应用的推送消息，如广告、版本更新等。
- 显示当前正在进行的事件，如播放音乐、导航、下载等。

7.5.1 接口说明

通知相关基础类包含以下几种。

- NotificationSlot。
- NotificationRequest。
- NotificationHelper。

上述基础类之间的关系如图7-2所示。

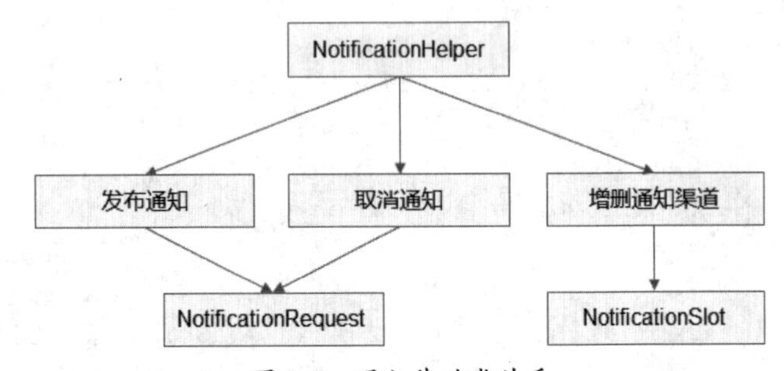

图7-2　通知基础类关系

1. NotificationSlot

NotificationSlot可以对提示音、振动、锁屏显示和重要级别等进行设置。一个应用可以创建一个或多个NotificationSlot，在发送通知时，通过绑定不同的NotificationSlot实现不同用途。

NotificationSlot需要先通过NotificationHelper的addNotificationSlot(NotificationSlot)方法发布后，通知才能绑定使用。所有绑定该NotificationSlot的通知在发布后都具备相应的特性，对象在创建后，将无法更改这些设置，对于是否启动相应设置，用户有最终控制权。

不指定NotificationSlot时，当前通知会使用默认的NotificationSlot，默认的NotificationSlot优先级为LEVEL_DEFAULT。

NotificationSlot的级别目前支持如下几种，顺序由低到高。

- LEVEL_NONE：表示通知不发布。
- LEVEL_MIN：表示通知可以发布，但是不显示在通知栏，不自动弹出，无提示音。该级别不适用于前台服务的场景。
- LEVEL_LOW：表示通知可以发布且显示在通知栏，不自动弹出，无提示音。
- LEVEL_DEFAULT：表示通知发布后可在通知栏显示，不自动弹出，触发提示音。
- LEVEL_HIGH：表示通知发布后可在通知栏显示，自动弹出，触发提示。

2. NotificationRequest

NotificationRequest用于设置具体的通知对象，包括设置通知的属性，如通知的分发时间、小图标、大图标、自动删除等参数，以及设置具体的通知类型，如普通文本、长文本等。

通知的常用属性包括以下几种。

- 通知分组：对于同一类型的通知，比如电子邮件，可以放在一个群组内展示。
- 小图标、大图标：为分别通过NotificationRequest的setLittleIcon(PixelMap)、setBigIcon(PixelMap)设置的小图标、大图标。
- 显示时间戳：通知除了显示时间戳外，还可以显示计时器功能，包含正计时和倒计时。通知通过NotificationRequest的setCreateTime(Long)、setShowCreateTime(boolean)设置并显示时间戳；通知通过NotificationRequest的setShowStopwatch(boolean)显示计时器功能；通知通过NotificationRequest的setShowStopwatch(boolean)、setCountdownTimer(boolean)显示倒计时功能。
- 进度条：主要用于播放音乐、下载等场景。通知通过NotificationRequest的setProgressBar(int, int, boolean)显示进度条。
- 从通知启动Ability：点击通知栏的通知，可以通过启动Ability触发新的事件。通知通过NotificationRequest的setIntentAgent(IntentAgent)设置IntentAgent后，点击通知栏上发布的通知，将触发通知中的IntentAgent承载的事件。IntentAgent的设置请参考IntentAgent开发指导。
- 通知设置ActionButton：通过点击通知按钮，可以触发按钮承载的事件。通过NotificationRequest的addActionButton(NotificationActionButton)附加按钮，点击按钮后可以触发相关的事件，具体事件内容如何设置需要参考NotificationActionButton。
- 通知设置ComponentProvider：通过ComponentProvider设置自定义的布局。通过NotificationRequest的setCustomView(ComponentProvider)配置自定义布局，替代系统布局，具体布局信息如何设置需要参考ComponentProvider。

目前支持6种通知类型，具体如下。

- 普通文本NotificationNormalContent：通知的标题，通过NotificationRequest的setTitle(String)方法设置；通知的内容，通过NotificationRequest的setText(String)方法设置。
- 长文本NotificationLongTextContent：为长文本的内容，通过setLongText(String) 设置，文本长度最大支持1024个字符。

- 图片 NotificationPictureContent：为图片通知样式的图片。
- 多行 NotificationMultiLineContent：为折叠状态下的多行通知样式的标题，通过 NotificationMultiLineContent 的 setTitle(String) 方法设置；为折叠状态下的多行通知样式的内容，通过 NotificationMultiLineContent 的 setText(String) 方法设置；为展开状态下的多行通知样式的标题，通过 NotificationMultiLineContent 的 setExpandedTitle(String) 方法设置；为展开状态下的多行通知样式的内容，通过 NotificationMultiLineContent 的 addSingleLine(String) 方法设置。
- 社交 NotificationConversationalContent：为社交通知样式的标题，通过 NotificationConversationalContent 的 setConversationTitle(String) 方法设置；为社交通知样式中的消息内容，通过 NotificationConversationalContent 的 addConversationalMessage(ConversationalMessage) 方法设置。
- 媒体 NotificationMediaContent：为媒体通知样式的标题，通过 NotificationMediaContent 的 setTitle(String) 方法设置；为媒体通知样式中的消息内容，通过 NotificationMediaContent 的 setText(String) 方法设置；为媒体通知样式对应的多媒体按钮，具备控制音频媒体的用途，通过 NotificationMediaContent 的 setAVToken(AVToken)、NotificationMediaContent 的 setShownActions(int[]) 方法设置。

3. NotificationHelper

NotificationHelper 封装了发布、更新、删除通知等静态方法，主要接口如下。

- 发布一条通知：publishNotification(NotificationRequest request)。
- 发布一条带 TAG 的通知：publishNotification(String tag, NotificationRequest)。
- 取消指定的通知：cancelNotification(int notificationId)。
- 取消指定的带 TAG 的通知：cancelNotification(String tag, int notificationId)。
- 取消之前发布的所有通知：cancelAllNotifications()。
- 创建一个 NotificationSlot：addNotificationSlot(NotificationSlot slot)。
- 获取 NotificationSlo：getNotificationSlot(String slotId)。
- 删除一个 NotificationSlot：removeNotificationSlot(String slotId)。
- 获取当前应用发的活跃通知：getActiveNotifications()。
- 获取系统中当前应用发的活跃通知的数量：getActiveNotificationNums()。
- 设置通知的角标：setNotificationBadgeNum(int num)。
- 设置当前应用中活跃状态通知的数量在角标显示：setNotificationBadgeNum()。

7.5.2 创建 NotificationSlot

NotificationSlot 可以设置公共通知的震动、锁屏模式、重要级别等，并通过调用 NotificationHelper.addNotificationSlot() 发布 NotificationSlot 对象。

示例代码如下：

```
 // 创建 notificationSlot 对象
NotificationSlot slot =
    new NotificationSlot("slot_001","slot_default", NotificationSlot.LEVEL_
MIN);

slot.setDescription("NotificationSlotDescription");
slot.setEnableVibration(true); // 设置振动提醒
slot.setLockscreenVisibleness(NotificationRequest.VISIBLENESS_TYPE_PUBLIC);//
设置锁屏模式
slot.setEnableLight(true); // 设置开启呼吸灯提醒
slot.setLedLightColor(Color.RED.getValue());// 设置呼吸灯的提醒颜色

try {
   NotificationHelper.addNotificationSlot(slot);
} catch (RemoteException ex) {
   HiLog.warn(LABEL, "addNotificationSlot occur exception.");
}
```

7.5.3 发布通知

发布通知分为以下几个步骤。

1. 构建 NotificationRequest 对象

应用发布通知前，通过NotificationRequest的setSlotId()方法与NotificationSlot绑定，使该通知在发布后都具备该对象的特征。示例代码如下：

```
int notificationId = 1;
NotificationRequest request = new NotificationRequest(notificationId);
request.setSlotId(slot.getId());
```

2. 设置通知内容

调用setContent()设置通知的内容，示例代码如下：

```
String title = "Welcome";
String text = "Welcome to waylau.com!";
NotificationNormalContent content = new NotificationNormalContent();
content.setTitle(title)
       .setText(text);
NotificationContent notificationContent = new NotificationContent(content);
request.setContent(notificationContent); // 设置通知的内容
```

3. 发送通知

调用publishNotification()发送通知。示例代码如下：

```
try {
   NotificationHelper.publishNotification(request);
} catch (RemoteException ex) {
   HiLog.warn(LABEL, "publishNotification occur exception.");
}
```

7.5.4 取消通知

取消通知分为取消指定单条通知和取消所有通知，应用只能取消自己发布的通知。

1. 取消指定的单条通知

调用cancelNotification()取消指定的单条通知，示例代码如下：

```
int notificationId = 1;
try {
    NotificationHelper.cancelNotification(notificationId);
} catch (RemoteException ex) {
    HiLog.warn(LABEL, "cancelNotification occur exception.");
}
```

2. 取消所有通知

调用cancelAllNotifications()取消所有通知，示例代码如下：

```
try {
    NotificationHelper.cancelAllNotifications();
} catch (RemoteException ex) {
    HiLog.warn(LABEL, "cancelAllNotifications occur exception.");
}
```

7.6 实战：通知发布与取消

本节演示如何实现通知发布与取消的例子。为了演示该功能，创建一个名为“Notification”的应用，用于通知发布与取消。

7.6.1 修改ability_main.xml

修改ability_main.xml文件如下：

```
<?xml version="1.0" encoding="utf-8"?>
<DirectionalLayout
```

```
    xmlns:ohos="http://schemas.huawei.com/res/ohos"
    ohos:height="match_parent"
    ohos:width="match_parent"
    ohos:alignment="center"
    ohos:orientation="vertical">

    <Text
        ohos:id="$+id:text_publish_notification"
        ohos:height="match_content"
        ohos:width="match_content"
        ohos:background_element="$graphic:background_ability_main"
        ohos:layout_alignment="horizontal_center"
        ohos:text="Publish Notification"
        ohos:text_size="40vp"
        />

    <Text
        ohos:id="$+id:text_cancel_notification"
        ohos:height="match_parent"
        ohos:width="match_content"
        ohos:background_element="$graphic:background_ability_main"
        ohos:layout_alignment="horizontal_center"
        ohos:text="Cancel Notification"
        ohos:text_size="40vp"
        />
</DirectionalLayout>
```

上述修改将显示两端文本内容“Publish Notification”与“Cancel Notification”。在模拟中运行，效果图7-3所示。

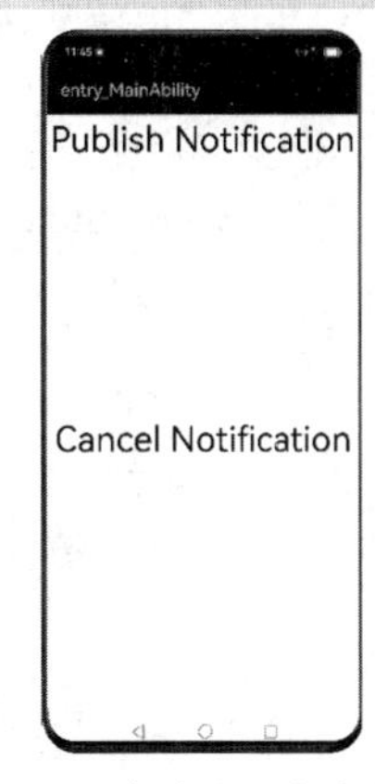

图7-3　修改显示文本内容

7.6.2　修改MainAbilitySlice

在初始化应用时，应用已经包含一个主AbilitySlice，即MainAbilitySlice。对MainAbilitySlice进行修改，代码如下：

```
package com.waylau.hmos.notification.slice;

import com.waylau.hmos.notification.ResourceTable;
import ohos.aafwk.ability.AbilitySlice;
import ohos.aafwk.content.Intent;
import ohos.agp.components.Text;
```

```
import ohos.agp.utils.Color;
import ohos.event.notification.NotificationHelper;
import ohos.event.notification.NotificationRequest;
import ohos.event.notification.NotificationSlot;
import ohos.hiviewdfx.HiLog;
import ohos.hiviewdfx.HiLogLabel;
import ohos.rpc.RemoteException;

public class MainAbilitySlice extends AbilitySlice {
    private static final String TAG = MainAbilitySlice.class.getSimpleName();
    private static final HiLogLabel LABEL_LOG =
            new HiLogLabel(HiLog.LOG_APP, 0x00001, TAG);

    private int notificationId = 0; // 递增的序列

    @Override
    public void onStart(Intent intent) {
        super.onStart(intent);
        super.setUIContent(ResourceTable.Layout_ability_main);
        // 添加点击事件来触发
        Text textPublishNotification = (Text) findComponentById(ResourceTable.
            Id_text_publish_notification);
        textPublishNotification.setClickedListener(listener ->
            publishNotification());


        // 添加点击事件来触发
        Text textCancelNotification = (Text) findComponentById(ResourceTable.
            Id_text_cancel_notification);
        textCancelNotification.setClickedListener(listener ->
            cancelNotification());
    }

    private void publishNotification() {
        HiLog.info(LABEL_LOG, "before publishNotification");

        // 创建 notificationSlot 对象
        NotificationSlot slot =
                new NotificationSlot("slot_001", "slot_default",
                                     NotificationSlot.LEVEL_HIGH);

        slot.setLevel(NotificationSlot.LEVEL_HIGH);           // 设置提醒级别
        slot.setDescription("NotificationSlotDescription"); // 设置提示内容
        slot.setEnableVibration(true); // 设置振动提醒
        slot.setLockscreenVisibleness(NotificationRequest.VISIBLENESS_TYPE_
                                       PUBLIC);// 设置锁屏模式
```

```
        slot.setEnableLight(true); // 设置开启呼吸灯提醒
        slot.setLedLightColor(Color.RED.getValue());// 设置呼吸灯的提醒颜色

        try {
            NotificationHelper.addNotificationSlot(slot);

            String title = "title";
            String text = "There is a normal notification content.";
            NotificationRequest.NotificationNormalContent content =
                    new NotificationRequest.NotificationNormalContent();
            content.setTitle(title)
                    .setText(text);
            notificationId++;
            NotificationRequest request = new NotificationRequest(notificationId);

            NotificationRequest.NotificationContent notificationContent =
                    new NotificationRequest.NotificationContent(content);
            request.setContent(notificationContent); // 设置通知的内容
            request.setSlotId(slot.getId());

            NotificationHelper.publishNotification(request);
        } catch (RemoteException ex) {
            HiLog.warn(LABEL_LOG, "publishNotification occur exception.");
        }

        HiLog.info(LABEL_LOG, "end publishNotification");
    }

    private void cancelNotification() {
        HiLog.info(LABEL_LOG, "before cancelNotification");
        try {
            NotificationHelper.cancelNotification(notificationId);
        } catch (RemoteException ex) {
            HiLog.warn(LABEL_LOG, "cancelNotification occur exception.");
        }

        HiLog.info(LABEL_LOG, "end cancelNotification");
    }


    @Override
    public void onActive() {
        super.onActive();
    }

    @Override
```

```
    public void onForeground(Intent intent) {
        super.onForeground(intent);
    }
}
```

在MainAbilitySlice类中：

- 在Text中增加了点击事件，以便触发发送事件的方法；
- publishNotification方法用于触发发布通知，而cancelNotification方法用于触发取消通知；
- 通过NotificationHelper.publishNotification发送通知；
- 通过NotificationHelper.cancelNotification取消通知。

点击文本内容“Publish Notification”触发发布通知，在模拟器中可以看到，通知内容效果如图7-4所示。

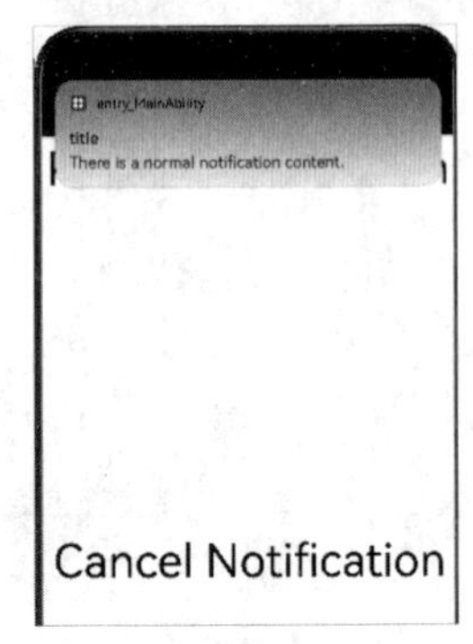

图7-4　通知内容

7.7 实战：Stage模型的订阅、发布、取消公共事件

前面几节都是介绍在FA模型下如何订阅、发布事件，本节主要演示如何实现公共事件的订阅、发布和取消操作。

打开DevEco Studio，创建一个名为“ArkTSCommonEventService”的应用作为演示示例。

7.7.1 添加按钮

在Index.ets的Text组件下添加4个按钮。代码如下：

```
// 创建订阅者
Button(('创建订阅者'), { type: ButtonType.Capsule })
    .fontSize(40)
    .fontWeight(FontWeight.Medium)
    .margin({ top: 10, bottom: 10 })
    .onClick(() => {
    this.createSubscriber()
    })

// 订阅事件
Button(('订阅事件'), { type: ButtonType.Capsule })
    .fontSize(40)
    .fontWeight(FontWeight.Medium)
    .margin({ top: 10, bottom: 10 })
    .onClick(() => {
    this.subscriberCommonEvent()
```

```
    })

// 发送事件
Button(('发送事件'), { type: ButtonType.Capsule })
    .fontSize(40)
    .fontWeight(FontWeight.Medium)
    .margin({ top: 10, bottom: 10 })
    .onClick(() => {
    this.publishCommonEvent()
    })

// 发送事件
Button(('取消订阅'), { type: ButtonType.Capsule })
    .fontSize(40)
    .fontWeight(FontWeight.Medium)
    .margin({ top: 10, bottom: 10 })
    .onClick(() => {
    this.unsubscribeCommonEvent()
    })
```

其中4个按钮分别设置了onClick点击事件，分别来触发创建订阅者、订阅事件、发送事件及取消订阅的操作。界面效果如图7-5所示。

图7-5　界面效果

7.7.2　添加Text显示接收的事件

为了能显示接收到的事件的信息，在4个按钮下面添加一个Text组件。代码如下：

```
// 用于接收事件数据
@State eventData: string = ''

// ……

// 接收到的事件数据
Text(this.eventData)
    .fontSize(50)
    .fontWeight(FontWeight.Bold)
```

Text组件的显示内容通过@State绑定了eventData变量。当eventData变量变化时，Text的显示内容也会实时更新。

7.7.3　设置按钮的点击事件方法

设置4个按钮的点击事件，方法如下：

```
// 导入公共事件管理器
import commonEventManager from '@ohos.commonEventManager';

// 为节约篇幅，省略部分代码

// 用于保存创建成功的订阅者对象，后续使用其完成订阅及退订的动作
private subscriber = null

private createSubscriber() {
    if (this.subscriber) {
        this.message = "subscriber already created";
    } else {
        commonEventManager.createSubscriber({ // 创建订阅者
        events: ["testEvent"] // 指定订阅的事件名称
        }, (err, subscriber) => { // 创建结果的回调
        if (err) {
            this.message = "create subscriber failure"
        } else {
            this.subscriber = subscriber; // 创建订阅成功
            this.message = "create subscriber success";
        }
        })
    }
}

private subscriberCommonEvent() {
    if (this.subscriber) {
        // 根据创建的 subscriber 开始订阅事件
        commonEventManager.subscribe(this.subscriber, (err, data) => {
        if (err) {
            // 异常处理
            this.eventData = "subscribe event failure: " + err;
        } else {
            // 接收到事件
            this.eventData = "subscribe event success: " +
                                JSON.stringify(data);
        }
        })
    } else {
        this.message = "please create subscriber";
    }
}

private publishCommonEvent() {
    // 发布公共事件
    commonEventManager.publish("testEvent", (err) => { // 结果回调
```

```
        if (err) {
        this.message = "publish event error: " + err;
        } else {
        this.message = "publish event with data success";
        }
    })
}

private unsubscribeCommonEvent() {
    if (this.subscriber) {
        commonEventManager.unsubscribe(this.subscriber, (err) => {
                                                            // 取消订阅事件
        if (err) {
            this.message = "unsubscribe event failure: " + err;
        } else {
            this.subscriber = null;
            this.message = "unsubscribe event success";
        }
        })
    } else {
        this.message = "already subscribed";
    }
}
```

subscriber是作为订阅者的变量。createSubscriber()方法用于创建订阅者；subscriberCommonEvent()方法用于订阅事件；publishCommonEvent()方法用于发布公共事件；unsubscribeCommonEvent()方法用于取消订阅。

7.7.4 运行

创建订阅者界面效果如图7-6所示。

点击“订阅事件”及“发送事件”按钮后，界面效果如图7-7所示。

可以看到订阅者已经能够正确接收事件，并将事件的信息显示在页面上。

点击“取消订阅”按钮后界面效果如图7-8所示。

图7-6 创建订阅者

图7-7 发送事件

图7-8 取消订阅

第8章

用ArkUI开发UI

本章重点介绍以 ArkTS 语言为核心的 ArkUI 框架的使用方法。

8.1 ArkUI概述

ArkUI（方舟开发框架），是一套构建HarmonyOS应用界面的UI开发框架，它提供了极简的UI语法与包括UI组件、动画机制、事件交互等在内的UI开发基础设施，以满足应用开发者的可视化界面开发需求。

8.1.1 ArkUI基本概念

ArkUI基本概念主要包括两部分。

- 组件：组件是界面搭建与显示的最小单位。开发者通过多种组件的组合，构建出满足自身应用诉求的完整界面。

- 页面：page页面是ArkUI最小的调度分割单位。开发者可以将应用设计为多个功能页面，每个页面进行单独的文件管理，并通过页面路由API完成页面间的调度管理，以实现应用内功能的解耦。

我们以“5.13 实战：Stage模型Ability内页面的跳转和数据传递”中的Index.ets代码为例：

```
// 导入 router 模块
import router from '@ohos.router';

@Entry
@Component
struct Index {
  @State message: string = 'Index 页面'

  build() {
    Row() {
      Column() {
        Text(this.message)
          .fontSize(50)
          .fontWeight(FontWeight.Bold)

        // 添加按钮，触发跳转
        Button('跳转')
          .fontSize(40)
          .onClick(() => {
            router.push({
              url: 'pages/Second',
              params: {
                src: 'Index 页面传来的数据',
              }
```

```
                });
            })
        }
        .width('100%')
      }
      .height('100%')
  }
}
```

在上述代码中，Index和Second就是页面，而Row、Column、Text、Button等都是ArkUI的组件。

Index和Second这两个页面，通过页面路由API完成页面间的调度管理，以实现应用内功能的解耦。

8.1.2 ArkUI主要特征

ArkUI的主要特征如下。

- UI组件：ArkUI内置了丰富的多态组件，包括Image、Text、Button等基础组件，可包含一个或多个子组件的容器组件，满足开发者自定义绘图需求的绘制组件，以及提供视频播放能力的媒体组件等。其中“多态”是指组件针对不同类型设备进行了设计，提供了在不同平台上的样式适配能力。同时，ArkUI也支持用户自定义组件。
- 布局：UI界面设计离不开布局的参与。ArkUI提供了多种布局方式，不仅保留了经典的弹性布局能力，也提供了列表、宫格、栅格布局和适应多分辨率场景开发的原子布局能力。
- 动画：ArkUI对于UI界面的美化，除了组件内置动画效果外，也提供了属性动画、转场动画和自定义动画能力。
- 绘制：ArkUI提供了多种绘制能力，以满足开发者的自定义绘图需求，支持绘制形状、颜色填充、绘制文本、变形与裁剪、嵌入图片等。
- 交互事件：ArkUI提供了多种交互能力，以满足应用在不同平台通过不同输入设备进行UI交互响应的需求，默认适配了触摸手势、遥控器按键输入、键鼠输入，同时提供了相应的事件回调，以便开发者添加交互逻辑。
- 平台API通道：ArkUI提供了API扩展机制，可通过该机制对平台能力进行封装，提供风格统一的JS接口。
- 两种开发范式：ArkUI针对不同的应用场景及不同技术背景的开发者提供了两种开发范式，分别是基于ArkTS的声明式开发范式（简称“声明式开发范式”）和兼容JS的类Web开发范式（简称“类Web开发范式”）。

8.1.3 JS、TS、ArkTS、ArkUI、ArkCompiler之间的联系

JS（JavaScript的简写）、TS（TypeScript的简写）和ArkTS都是开发语言，其中TS是JS的超集，

而ArkTS在TS的基础上，扩展了声明式UI、状态管理等相应的能力，让开发者可以以更简洁、更自然的方式开发高性能应用。ArkTS会结合应用开发和运行的需求持续演进，包括但不限于引入分布式开发范式、并行和并发能力增强、类型系统增强等方面的语言特性。因此三者的关系如图8-1所示。

ArkUI是一套构建分布式应用界面的声明式UI开发框架。它使用极简的UI信息语法、丰富的UI组件，以及实时界面预览工具，帮助提升HarmonyOS应用界面开发效率。只需使用一套ArkTS API，就能在多个HarmonyOS设备上实现生动而流畅的用户界面体验。

ArkCompiler（方舟编译器）是华为自研的统一编程平台，包含编译器、工具链、运行时等关键部件，支持高级语言在多种芯片平台的编译与运行，并支撑应用和服务运行在手机、个人计算机、平板、电视、汽车和智能穿戴等多种设备上的需求。ArkCompiler会把ArkTS/TS/JS编译为方舟字节码，运行时直接运行方舟字节码，并且ArkCompiler使用多种混淆技术提供更高强度的混淆与保护，确保HarmonyOS应用包中装载的是多重混淆后的字节码。ArkCompiler所处结构如图8-2所示。

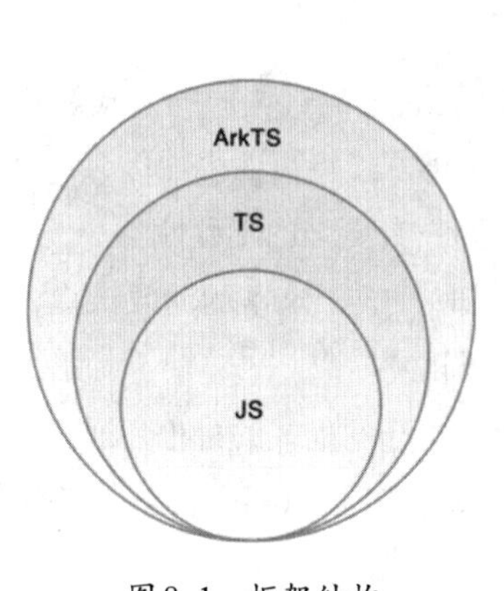

图8-1 框架结构

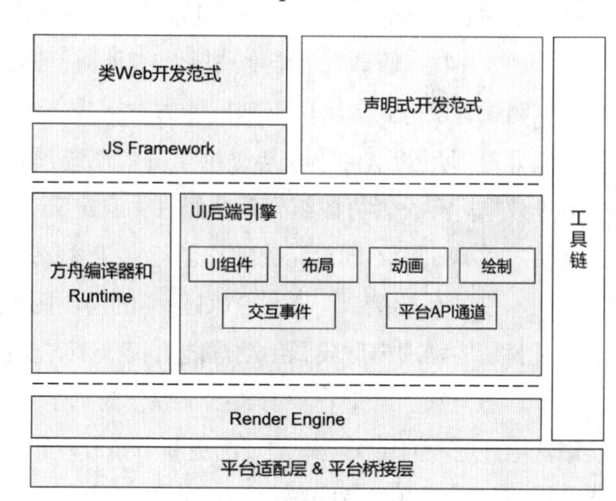

图8-2 ArkCompiler所处结构

8.2 声明式开发范式

ArkUI是一套开发极简、高性能、跨设备应用的UI开发框架，支持开发者高效地构建跨设备应用UI界面。

8.2.1 声明式开发范式与类Web开发范式

声明式开发范式是采用基于TypeScript进行声明式UI语法扩展而来的ArkTS语言，从组件、动画和状态管理三个维度提供了UI绘制能力。声明式开发范式更接近自然语义的编程方式，让开发者直观地描述UI界面，不必关心框架如何实现UI绘制和渲染，实现极简高效开发。因此，声明式开发范式适合复杂度较大、团队合作度较高的程序。

类Web开发范式是采用经典的HML、CSS、JavaScript三段式的开发方式，使用HML标签文件进行布局搭建，使用CSS文件进行样式描述，使用JavaScript文件进行逻辑处理。UI组件与数据之间通过单向数据绑定的方式建立关联，当数据发生变化时，UI界面自动触发刷新。该开发方式更接近Web前端开发者的使用习惯，便于快速将已有的Web应用改造成ArkUI应用。因此，类Web开发范式适合界面较简单的中小型应用和卡片。

本章示例推荐采用声明式开发范式。

8.2.2 声明式开发范式的基础能力

使用基于ArkTS的声明式开发范式的ArkUI，采用更接近自然语义的编程方式，让开发者可以直观地描述UI界面，不必关心框架如何实现UI绘制和渲染，实现极简高效开发。开发框架不仅从组件、动效和状态管理三个维度来提供UI能力，还提供了系统能力接口，实现系统能力的极简调用。

声明式开发范式具备以下基础能力。

- 开箱即用的组件：框架提供丰富的系统预置组件，可以通过链式调用的方式设置系统组件的渲染效果。开发者可以组合系统组件为自定义组件，通过这种方式将页面组件化为一个个独立的UI单元，实现页面不同单元的独立创建、开发和复用，使页面具有更强的工程性。
- 丰富的动效接口：提供SVG标准的绘制图形能力，同时开放了丰富的动效接口，开发者可以通过封装的物理模型或调用动画能力接口来实现自定义动画轨迹。
- 状态与数据管理：状态数据管理作为基于ArkTS的声明式开发范式的特色，通过功能不同的装饰器给开发者提供了清晰的页面更新渲染流程和管道。状态管理包括UI组件状态和应用程序状态，两者协作可以使开发者完整地构建整个应用的数据更新和UI渲染。
- 系统能力接口：ArkUI还封装了丰富的系统能力接口，开发者可以通过简单的接口调用，实现从UI设计到系统能力调用的极简开发。

8.2.3 声明式开发范式的整体架构

声明式开发范式的整体架构如图8-3所示，内容包括以下几个部分。

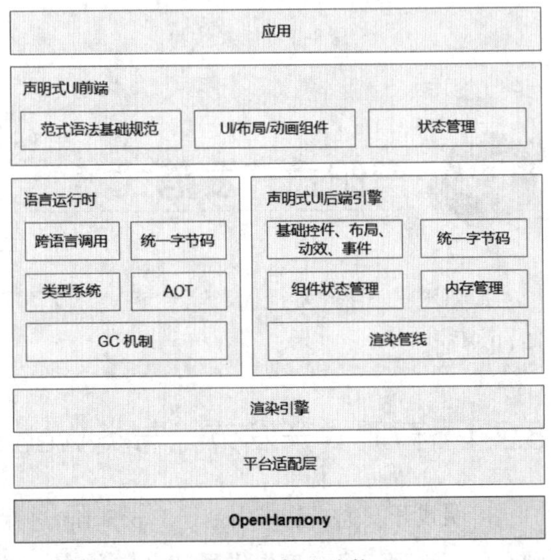

图8-3 整体结构

- 声明式UI前端：提供了UI开发范式的基础语言规范，并提供内置的UI、布局和动画组件，以及多种状态管理机制，为应用开发者提供一系列接口支持。

• 语言运行时：选用方舟语言运行时，提供了针对UI范式语法的解析能力、跨语言调用支持的能力和TS语言高性能运行环境。

• 声明式UI后端引擎：后端引擎提供了兼容不同开发范式的UI渲染管线，提供多种基础控件、布局计算、动效、交互事件，以及状态管理和绘制能力。

• 渲染引擎：提供了高效的绘制能力，将渲染管线收集的渲染指令绘制到屏幕。

• 平台适配层：提供了对系统平台的抽象接口，具备接入不同系统的能力，如系统渲染管线、生命周期调度等。

8.2.4 声明式开发范式的基本组成

声明式开发范式的基本组成如图8-4所示，内容包括以下几个方面。

• 装饰器：用来装饰类、结构体、方法及变量，赋予其特殊的含义。图8-4所示的示例中的@Entry、@Component、@State都是装饰器。具体而言，@Component表示这是个自定义组件；@Entry则表示这是个入口组件；@State表示组件中的状态变量，此状态变化会引起UI变更。

• 自定义组件：可复用的UI单元，可组合其他组件，如图8-4中被@Component装饰的struct Hello。

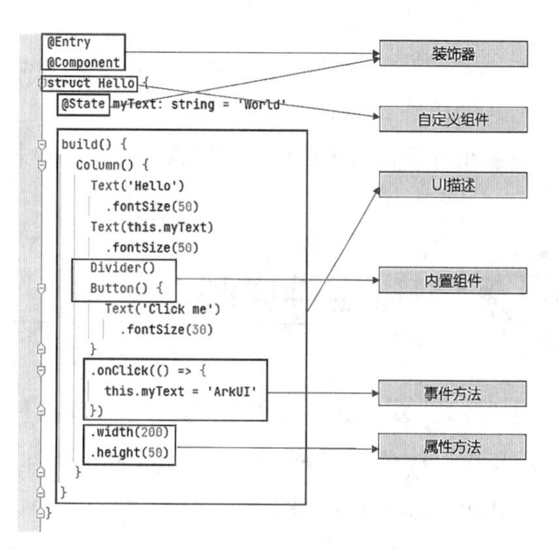

图8-4　声明式开发范式的基本组成

• UI描述：用声明式的方式来描述UI的结构，如图8-4中build()方法内部的代码块。

• 内置组件：框架中默认内置的基础和布局组件，可直接被开发者调用，如图8-4中的Divider、Button。

• 事件方法：用于添加组件对事件的响应逻辑，统一通过事件方法进行设置，如图8-4中跟随在Button后面的onClick()。

• 属性方法：用于组件属性的配置，统一通过属性方法进行设置，如图8-4中的width()、height()等，可通过链式调用的方式设置多项属性。

8.3 常用组件

组件是构建页面的核心，每个组件通过对数据和方法的简单封装，实现独立的可视、可交互功能单元。组件之间相互独立，随取随用，也可以在需求相同的地方重复使用。

声明式开发范式目前包括如下组件。

- 基础组件：Blank、Button、Checkbox、CheckboxGroup、DataPanel、DatePicker、Divider、Gauge、Image、ImageAnimator、LoadingProgress、Marquee、Navigation、PatternLock、Progress、QRCode、Radio、Rating、RichText、ScrollBar、Search、Select、Slider、Span、Stepper、StepperItem、Text、TextArea、TextClock、TextInput、TextPicker、TextTimer、TimePicker、Toggle、Web、XComponent。
- 容器组件：AlphabetIndexer、Badge、Column、ColumnSplit、Counter、Flex、GridContainer、GridCol、GridRow、Grid、GridItem、List、ListItem、Navigator、Panel、Refresh、RelativeContainer、Row、RowSplit、Scroll、SideBarContainer、Stack、Swiper、Tabs、TabContent。
- 媒体组件：Video。
- 绘制组件：Circle、Ellipse、Line、Polyline、Polygon、Path、Rect、Shape。
- 画布组件：Canvas。

这些组件的详细用法可以查阅API文档。本书后续也会对常用的组件做进一步的使用介绍。

8.4 基础组件详解

声明式开发范式目前可供选择的基础组件有Blank、Button、Checkbox、CheckboxGroup、DataPanel、DatePicker、Divider、Gauge、Image、ImageAnimator、LoadingProgress、Marquee、Navigation、PatternLock、Progress、QRCode、Radio、Rating、RichText、ScrollBar、Search、Select、Slider、Span、Stepper、StepperItem、Text、TextArea、TextClock、TextInput、TextPicker、TextTimer、TimePicker、Toggle、Web等。

本节演示如何使用声明式开发范式的部分基础组件。相关示例可以在ArkUIBasicComponents应用中找到。

8.4.1 Blank

Blank是空白填充组件，在容器主轴方向上，空白填充组件具有自动填充容器空余部分的能力。

需要注意的是，Blank组件仅当其父组件为Row/Column，且父组件设置了宽度时才生效。以下示例展示了Blank父组件Row未设置宽度及设置了宽度的效果对比。

示例如下：

```
// Blank 父组件 Row 未设置宽度时，子组件间无空白填充
Row() {
   Text('Left Space').fontSize(24)
```

```
    Blank()
    Text('Right Space').fontSize(24)
}

// Blank父组件Row设置了宽度时，子组件间以空白填充
Row() {
    Text('Left Space').fontSize(24)
    Blank()
    Text('Right Space').fontSize(24)
}.width('100%')
```

界面效果如图8-5所示，第一行Row由于未设置宽度，导致Blank未生效。

Blank支持color属性，用来设置空白填充的颜色。示例如下：

```
Row() {
    Text('Left Space').fontSize(24)

    // 设置空白填充的填充颜色
    Blank().color(Color.Yellow)

    Text('Right Space').fontSize(24)
}.width('100%')
```

在上述示例中，Blank组件设置了黄色作为空白填充，界面效果如图8-6所示。

图8-5　Blank组件效果

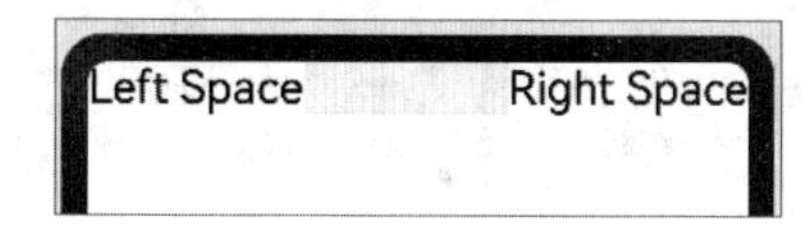

图8-6　Blank组件设置了黄色作为空白填充的效果

8.4.2 Button

Button是按钮组件，可快速创建不同样式的按钮。以下是两个按钮示例：

```
// 一个基本的按钮，设置要显示的文字
Button('01')

// 设置边框半径、背景色、宽度
Button('02').borderRadius(8).backgroundColor(0x317aff).width(90)
```

其中，第一个按钮是一个最为基本的按钮，仅设置要显示的文字“01”；第二个按钮则设置了边框半径、背景色、宽度。两个按钮界面效果如图8-7所示。

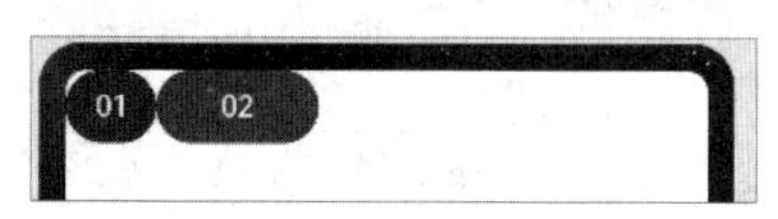

图8-7　Button组件效果

Button组件支持通过type属性来设置按钮显示样式。示例如下：

```
// 胶囊型按钮（圆角默认为高度的一半）
```

```
Button('03', { type: ButtonType.Capsule }).width(90)

// 圆形按钮。
Button('04', { type: ButtonType.Circle}).width(90)

// 普通按钮（默认不带圆角）。
Button('05', { type: ButtonType.Normal}).width(90)
```

上述三个按钮的样式分别是Capsule（胶囊型）、Circle（圆形）、Normal（普通），界面效果如图8-8所示。

图8-8　Button组件显示样式效果

Button组件支持包含子组件，示例如下：

```
// 可以包含子组件。但是文字就不会显示了
Button('06', { type: ButtonType.Normal }){
    LoadingProgress().width(20).height(20).color(0xFFFFFF)
}.width(90)

// 可以包含单个子组件。文字就用 Text 组件来显示
Button({ type: ButtonType.Capsule, stateEffect: true }) {
    Row() {
        LoadingProgress().width(20).height(20).margin({ left: 12
}).color(0xFFFFFF)
        Text('06').fontSize(12).fontColor(0xffffff).margin({ left: 5, right:
12 })
    }.alignItems(VerticalAlign.Center).width(90).height(40)
}.backgroundColor(0x317aff)
```

上述第一个按钮包括LoadingProgress组件。需要注意的是，包含子组件之后，原本按钮上的文字“06”就不会显示了。如果想显示文字，可以参考第二个按钮的设置方式，增加一个Text组件。界面效果如图8-9所示。

图8-9　Button组件包含子组件效果

8.4.3　Checkbox

Checkbox是多选框组件，通常用于某选项的打开或关闭。示例如下：

```
// 设置多选框名称、多选框的群组名称。
Checkbox({ name: 'checkbox1', group: 'checkboxGroup' })
    .select(true) // 设置默认选中
    .selectedColor(0xed6f21) // 设置选中颜色
    .onChange((value: boolean) => { // 设置选中事件
        console.info('Checkbox1 change is ' + value)
    })

Checkbox({ name: 'checkbox2', group: 'checkboxGroup' })
```

```
    .select(false)
    .selectedColor(0x39a2db)
    .onChange((value: boolean) => {
        console.info('Checkbox2 change is ' + value)
    })
```

Checkbox在实例化时主要用于设置多选框名称、多选框的群组名称，并支持通过select、selectedColor来设置是否选中、选中颜色等属性。

上述示例界面效果如图8-10所示。

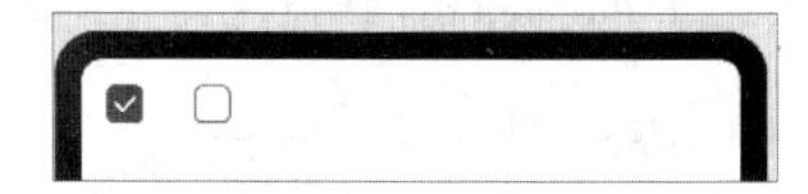
图8-10　Checkbox组件效果

Checkbox支持onChange事件，当Checkbox选中状态发生变化时，会触发该回调。当value为true时，表示已选中；当value为false时，表示未选中。

8.4.4　CheckboxGroup

CheckboxGroup是多选框群组，用于控制多选框全选或不全选状态。示例如下：

```
Row() {
CheckboxGroup({ group: 'checkboxGroup' })
Text(' 全要 ').fontSize(20)
}

Row() {
    Checkbox({ name: 'checkbox1', group: 'checkboxGroup' })
    Text(' 可乐 ').fontSize(20)
}

Row() {
    Checkbox({ name: 'checkbox2', group: 'checkboxGroup' })
    Text(' 鸡翅 ').fontSize(20)
}
```

checkbox1和checkbox2都属于同一个checkboxGroup。当只选中组中的一个Checkbox组件时（不全选），界面效果如图8-11所示。当CheckboxGroup组件全选时，界面效果如图8-12所示。

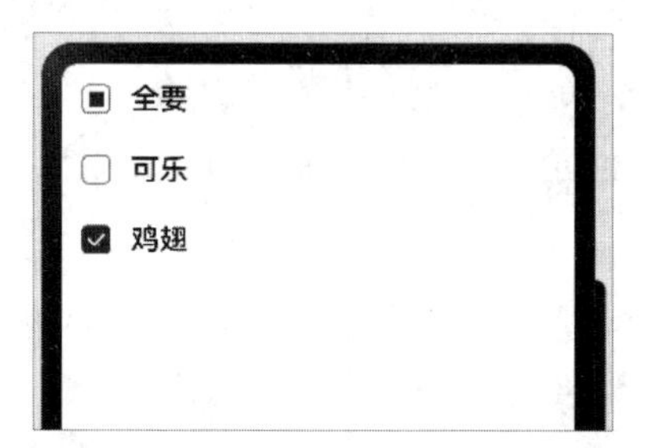

图8-11　CheckboxGroup组件不全选效果

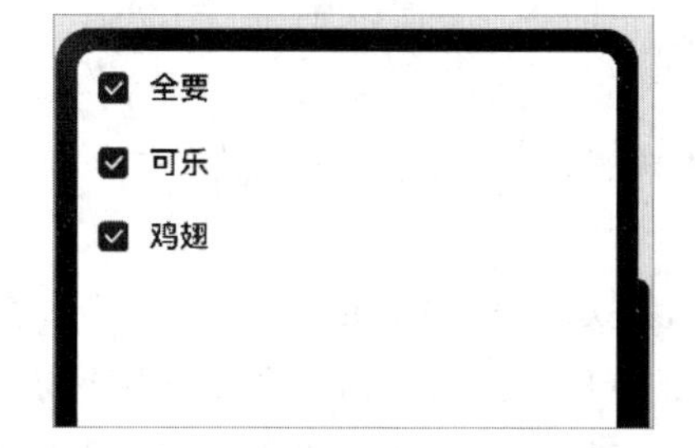

图8-12　CheckboxGroup组件全选效果

8.4.5 DataPanel

DataPanel是数据面板组件，用于将多个数据占比情况使用占比图进行展示。

DataPanel主要支持以下两类。

- Line：线型数据面板。
- Circle：环形数据面板。

DataPanel示例如下：

```
private dataPanelValues: number[] = [11, 3, 10, 2, 36, 4, 7, 22, 5]

build() {
    Column() {
        // 环形数据面板
        DataPanel({ values: this.dataPanelValues, max: 100, type:
DataPanelType.Circle }).width(350).height(350)

        // 线型数据面板
        DataPanel({ values: this.dataPanelValues, max: 100, type:
DataPanelType.Line }).width(350).height(50)
    }
    .height('100%')
}
```

在上述示例中，DataPanel主要有三个参数，其中values是数据值列表，最大支持9个数据；max表示数据的最大值；type就是类型。界面效果如图8-13所示。

图8-13　DataPanel组件效果

8.4.6 DatePicker

DatePicker是选择日期的滑动选择器组件。以下是一个DatePicker的基本示例：

```
DatePicker({
    start: new Date('1970-1-1'), // 指定选择器的起始日期。 默认值: Date('1970-1-1')
    end: new Date('2100-1-1'), // 指定选择器的结束日期。 默认值: Date('2100-12-
31')
    selected: new Date('2023-02-14'), // 设置选中项的日期。默认值: 当前系统日期
})
```

在上述示例中，DatePicker主要有三个参数，其中start是指定选择器的起始日期；end是指定选择器的结束日期；selected是设置选中项的日期。如果三个参数不设置就会使用默认值。

上述示例的DatePicker界面效果如图8-14所示。

DatePicker支持农历。可以通过设置lunar属性来设置日期是否显示农历。

- true：展示农历。
- false：不展示农历。默认值为false。

DatePicker在选择日期时会触发onChange事件。以下是示例：

图8-14　DatePicker组件效果

```
DatePicker({
    start: new Date('1970-1-1'), // 指定选择器的起始日期。 默认值：Date('1970-1-1')
    end: new Date('2100-1-1'), // 指定选择器的结束日期。 默认值：Date('2100-12-31')
    selected: new Date('2023-02-15'), // 设置选中项的日期。默认值：当前系统日期
    }).lunar(true) // 设置农历
    .onChange((value: DatePickerResult) => { // 选择日期时触发该事件
        console.info('select current date is: ' + JSON.stringify(value))
    })
```

在上述示例中，DatePicker主设置了农历，同时监听onChange事件。上述示例的DatePicker界面效果如图8-15所示。

图8-15　DatePicker组件效果

8.4.7　Divider

Divider是分隔器组件，分隔不同内容块/内容元素。以下是示例：

```
Text(' 我是天 ').fontSize(29)
Divider()
Text(' 我是地 ').fontSize(29)
```

在上述示例中，Divider在两个Text组件之间形成了一条分割线，界面效果如图8-16所示。

在默认情况下，Divider是水平的，但也可以通过vertical属性来设置为垂直。以下是示例：

```
Text(' 我是天 ').fontSize(29)
// 设置垂直
Divider().vertical(true).height(100)
Text(' 我是地 ').fontSize(29)
```

在上述示例中，Divider在两个Text组件之间形成了一条垂直分割线，界面效果如图8-17所示。

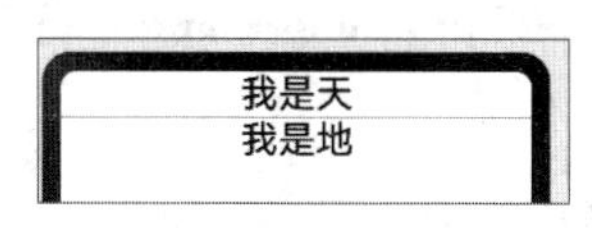

图8-16　Divider组件效果

图8-17　Divider组件垂直效果

Divider还可以通过以下属性来设置样式。

- color：分割线颜色。

- strokeWidth：分割线宽度。默认值为1。
- lineCap：分割线的端点样式。默认值是LineCapStyle.Butt。

以下是设置了样式的Divider示例：

```
Text(' 我是天 ').fontSize(29)
// 设置样式
Divider()
    .strokeWidth(15)  // 宽度
    .color(0x2788D9)  // 颜色
    .lineCap(LineCapStyle.Round) // 端点样式
Text(' 我是地 ').fontSize(29)
```

在上述示例中，界面效果如图8-18所示。

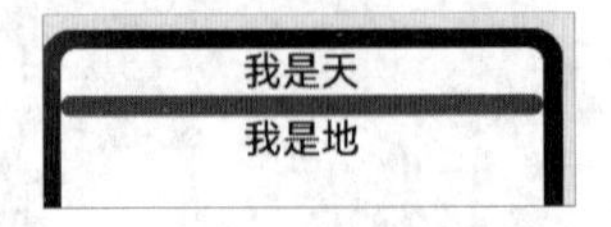

图8-18　Divider组件样式效果

8.4.8　Gauge

Gauge是数据量规图表组件，用于将数据展示为环形图表。以下是Gauge示例：

```
// value 值的设置，默认的 min 和 max 为 0 ～ 100，角度范围默认 0 ～ 360
// 参数中设置当前值为 75
Gauge({ value: 75 })
    .width(200).height(200)
    // 设置量规图的颜色，支持分段颜色设置。
    .colors([[0x317AF7, 1], [0x5BA854, 1], [0xE08C3A, 1], [0x9C554B, 1]])
```

在上述示例中，colors是一个颜色数组，表示该量规图由四段颜色组成。参数value是量规图的当前数据值，即图中指针指向位置。界面效果如图8-19所示。上述value值也可以在属性中进行设置。如果属性和参数都设置时，以参数为准。以下是Gauge示例：

```
// 参数设置当前值为 75，属性设置值为 25，属性设置优先级高
Gauge({ value: 75 })
    .value(25) // 属性和参数都设置时以属性为准
    .width(200).height(200)
    .colors([[0x317AF7, 1], [0x5BA854, 1], [0xE08C3A, 1], [0x9C554B, 1]])
```

在上述示例中，参数设置当前值为75，属性设置值为25，属性设置优先级高，因此Gauge的最终value是25。界面效果如图8-20所示。

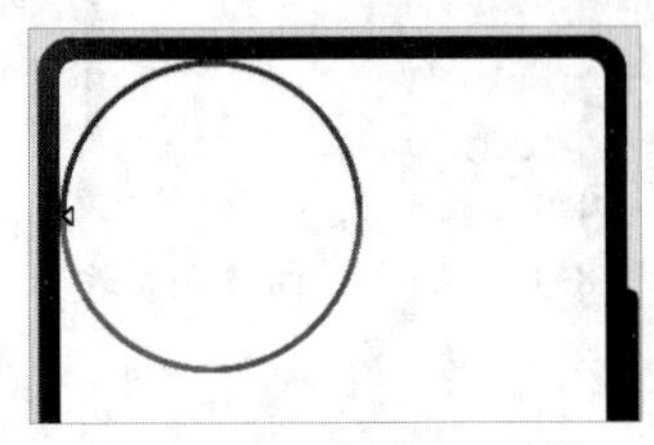

图8-19　Gauge组件效果

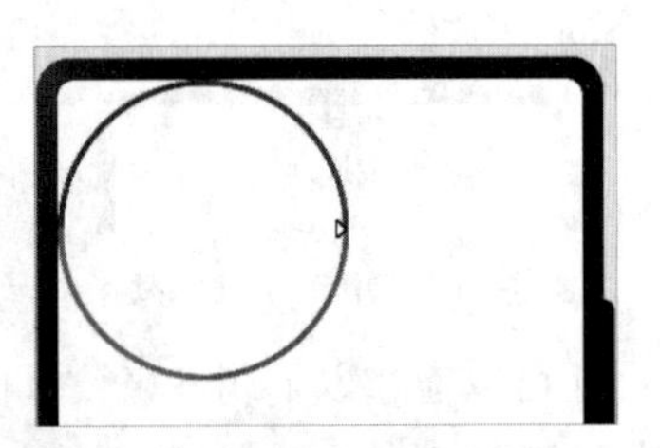

图8-20　Gauge组件属性和参数都设置的效果

Gauge组件还有其他的一些属性设置，具体如下。

- startAngle：设置起始角度位置，时钟0点为0度，顺时针方向为正角度。默认值为0。
- endAngle：设置终止角度位置，时钟0点为0度，顺时针方向为正角度。默认值为360。
- strokeWidth：设置环形量规图的环形厚度。

以下是一个设置了210到150度、厚度为20的Gauge示例：

```
// 210--150 度环形图表
Gauge({ value: 70})
    .startAngle(210) // 起始角度
    .endAngle(150)   // 终止角度
    .colors([[0x317AF7, 0.1], [0x5BA854, 0.2], [0xE08C3A, 0.3], [0x9C554B, 
0.4]])
    .strokeWidth(20) // 环形厚度
    .width(200)
    .height(200)
```

在上述示例中，界面效果如图8-21所示。

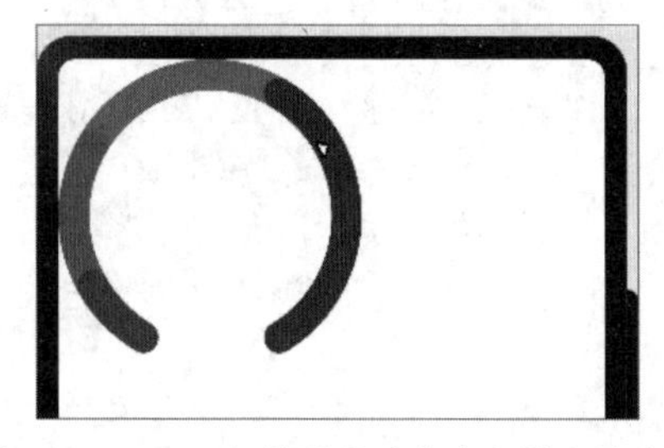

图8-21　Gauge组件设置了角度和厚度的效果

8.4.9 Image

Image是图片组件，支持本地图片和网络图片的渲染展示。

以下是Image组件使用本地图片的示例：

```
// 使用本地图片的示例
// 图片资源在 base/media 目录下
Image($r('app.media.waylau_181_181'))
    .width(180).height(180)
```

在上述示例中，图片资源waylau_181_181.jpg放置在base/media目录下。界面效果如图8-22所示。

图8-22　Image组件使用本地图片的效果

以下展示Image采用网络图片的过程。

首选需要在module.json5文件中声明使用网络的权限ohos.permission.INTERNET。示例如下。

```
{
  "module": {
    // ...
    "requestPermissions": [
      {
        "name": "ohos.permission.INTERNET"
      }
    ]
}
```

其次是编写请求网络图片的方法。示例如下：

```
// http 请求网络图片需要导入的包
import http from '@ohos.net.http';
import imageModule from '@ohos.multimedia.image'

@Entry
@Component
struct Index {

  // 先创建一个 PixelMap 状态变量用于接收网络图片
  @State imagePixelMap: PixelMap = undefined

// 网络图片请求方法
  private httpRequest() {
    let httpRequest = http.createHttp();

    httpRequest.request(
      "https://waylau.com/images/showmethemoney-sm.jpg",   // 网络图片地址
      (error, data) => {
        if(error) {
          console.log("error code: " + error.code + ", msg: " + error.message)
        } else {
          let code = data.responseCode
          if(http.ResponseCode.OK == code) {
            // @ts-ignore
            let imageSource = imageModule.createImageSource(data.result)
            let options = {alphaType: 0,          // 透明度
              editable: false,                    // 是否可编辑
              pixelFormat: 3,                     // 像素格式
              scaleMode: 1,                       // 缩略值
              size: {height: 281, width: 207}}  // 创建图片大小
            imageSource.createPixelMap(options).then((pixelMap) => {
              this.imagePixelMap = pixelMap
            })
          } else {
            console.log("response code: " + code);
          }
        }
      }
    )
  }

  // ...
}
```

上述httpRequest方法需要导入http及image的包。请求到网络图片资源后，会转为一个

PixelMap对象imagePixelMap。

最后再将imagePixelMap复制到Image组件中即可。代码如下：

```
// 使用网络图片的示例
Button(" 获取网络图片 ")
    .onClick(() => {
        // 请求网络资源
        this.httpRequest();
    })
Image(this.imagePixelMap).width(207).height(281)
```

在上述示例中，通过Button来点击触发httpRequest方法。界面效果如图8-23所示。

图8-23 Image组件使用网络图片的效果

8.4.10 ImageAnimator

ImageAnimator提供帧动画组件来实现逐帧播放图片的能力，可以配置需要播放的图片列表，每张图片可以配置时长。示例代码如下：

```
// 按钮控制动画的播放和暂停
Button(' 播放 ').width(100).padding(5).onClick(() => {
  this.animationStatus = AnimationStatus.Running
}).margin(5)
Button(' 暂停 ').width(100).padding(5).onClick(() => {
  this.animationStatus = AnimationStatus.Paused
}).margin(5)

// images 设置设置图片帧信息集合
// 每一帧的帧信息 (ImageFrameInfo) 包含图片路径、图片大小、图片位置和图片播放时长信息，
ImageAnimator()
  .images([
    {
      src: $r('app.media.book01'), // 图片路径
      duration: 500, // 播放时长
      width: 240,    // 图片大小
      height: 350,
      top: 0,        // 图片位置
      left: 0
    },
    {
      src: $r('app.media.book02'),
      duration: 500,
      width: 240,
      height: 350,
```

```
      top: 0,
      left: 170
    },
    {
      src: $r('app.media.book03'),
      duration: 500,
      width: 240,
      height: 350,
      top: 120,
      left: 170
    },
    {
      src: $r('app.media.book04'),
      duration: 500,
      width: 240,
      height: 350,
      top: 120,
      left: 0
    }
  ])
  .state(this.animationStatus)
  .reverse(false) // 是否逆序播放
  .fixedSize(false) // 是否固定大小
  .preDecode(2) // 是否启用预解码
  .iterations(-1) // 循环播放次数
  .width(240)
  .height(350)
  .margin({ top: 100 })
```

在上述示例中，通过Button来点击触发播放或暂停方法，界面效果如图8-24至8-27所示。

图 8-24　ImageAnimator 播放第 1 帧图的效果

图 8-25　ImageAnimator 播放第 2 帧图的效果

图 8-26　ImageAnimator 播放第 3 帧图的效果

图 8-27　ImageAnimator 播放第 4 帧图的效果

8.4.11 LoadingProgress

LoadingProgress是用于显示加载动效的组件。示例代码如下：

```
// 显示加载动效
LoadingProgress()
  .color(Color.Red) // 设置为红色
```

在上述示例中，通过color来设置LoadingProgress的颜色为红色。界面效果如图8-28所示。

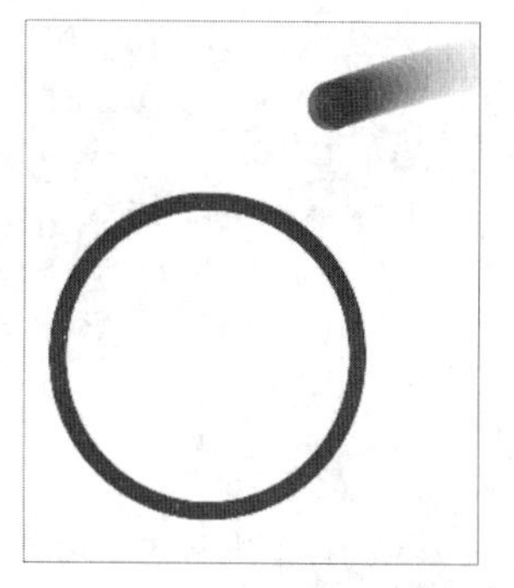

图8-28 LoadingProgress组件效果

8.4.12 Marquee

Marquee是跑马灯组件，用于滚动展示一段单行文本，仅当文本内容宽度超过跑马灯组件宽度时滚动。示例代码如下：

```
// 文本内容宽度未超过跑马灯组件宽度，不滚动
Marquee({
  start: true, // 控制跑马灯是否进入播放状态
  step: 12, // 滚动动画文本滚动步长。默认值：6，单位 vp
  loop: -1, // 循环次数，-1 为无限循环
  fromStart: true, // 设置文本从头开始滚动或反向滚动
  src: "HarmonyOS 也称为鸿蒙系统 "
}).fontSize(20)

// 文本内容宽度超过了跑马灯组件宽度，滚动
Marquee({
  start: true, // 控制跑马灯是否进入播放状态
  step: 12, // 滚动动画文本滚动步长。默认值：6，单位 vp
  loop: -1, // 循环次数，-1 为无限循环
  fromStart: true, // 设置文本从头开始滚动或反向滚动
  src: " 在传统的单设备系统能力基础上，HarmonyOS 提出了基于同一套系统能力、适配多种终端形态的分布式理念。"
}).fontSize(20)
```

在上述示例中，第1个Marquee的文本内容宽度未超过跑马灯组件宽度，因此不滚动；第2个Marquee的文本内容宽度超过了跑马灯组件宽度，因此会滚动。界面效果如图8-29所示。

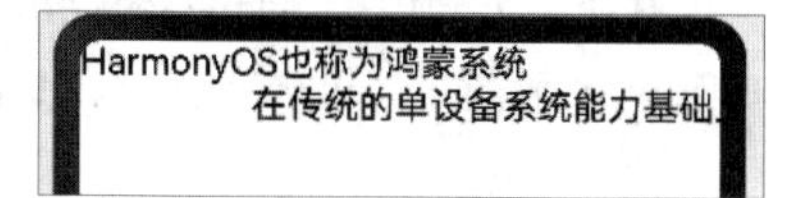

图8-29 Marquee组件效果

8.4.13 Navigation

Navigation组件一般作为Page页面的根容器，通过属性设置来展示页面的标题、工具栏、菜单。示例代码如下：

```
// 自定义一个 Toolbar 组件
@Builder NavigationToolbar() {
  Row() {
      Text("首页").fontSize(25).margin({ left: 70 })
      Text("+").fontSize(25).margin({ left: 70 })
      Text("我").fontSize(25).margin({ left: 70 })
  }
}

// Navigation 使用自定义的 NavigationToolbar 组件
Navigation() {
    Flex() {
    }
  }
  .toolBar(this.NavigationToolbar) // 使用自定义一个 Toolbar 组件
```

在上述示例中，通过@Builder来构造一个名为“Navigation Toolbar”的Toolbar组件，而后在Navigation的属性toolBar属性中设置该Toolbar组件，界面效果如图8-30所示。

图8-30　Navigation组件效果

8.4.14　PatternLock

PatternLock是图案密码锁组件，以九宫格图案的方式输入密码，用于密码验证场景。手指在PatternLock组件区域按下时开始进入输入状态，手指离开屏幕时结束输入状态完成密码输入。

```
PatternLock()
  .sideLength(200)              // 设置组件的宽度和高度（宽高相同）
  .circleRadius(9)              // 设置宫格中圆点的半径
  .pathStrokeWidth(18)          // 设置连线的宽度。设置为 0 或负数等非法值时连线不显示
  .activeColor('#B0C4DE')       // 设置宫格圆点在“激活”状态的填充颜色
  .selectedColor('#228B22')     // 设置宫格圆点在“选中”状态的填充颜色
  .pathColor('#90EE90')         // 设置连线的颜色
  .backgroundColor('#F5F5F5') // 背景颜色
  .autoReset(true) // 设置在完成密码输入后再次在组件区域按下时是否重置组件状态
```

在上述示例中，通过sideLength、circleRadius等属性设置PatternLock的样式。初始状态界面效果如图8-31所示。输入密码后界面效果如图8-32所示。

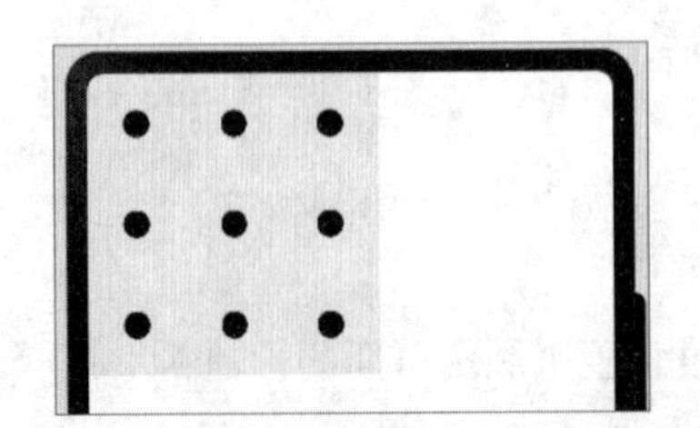
图8-31　PatternLock组件初始状态效果

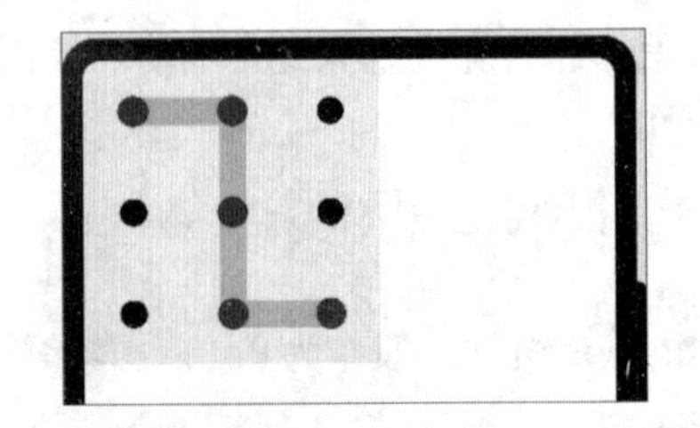
图8-32　PatternLock组件输入密码后效果

8.4.15 Progress

Progress为进度条组件，用于显示内容加载或操作处理等进度。

Progress主要有以下参数。

- value：指定当前进度值。
- total：指定进度总长。
- type：指定进度条类型ProgressType。

其中，ProgressType主要有以下5种。

①Linear：线性样式。

②Ring：环形无刻度样式，环形圆环逐渐显示至完全填充效果。

③Eclipse：圆形样式，显示类似月圆月缺的进度展示效果，从月牙逐渐变化至满月。

④ScaleRing：环形有刻度样式，显示类似时钟刻度形式的进度展示效果。

⑤Capsule：胶囊样式，头尾两端圆弧处的进度展示效果与Eclipse相同；中段处的进度展示效果与Linear相同。

以下是5种ProgressType的具体示例：

```
Progress({ value: 20, total: 100, type: ProgressType.
Linear }).width(150).margin({ top: 10 })
Progress({ value: 20, total: 100, type: ProgressType.
Ring }).width(150).margin({ top: 10 })
Progress({ value: 20, total: 100, type: ProgressType.
Eclipse }).width(150).margin({ top: 10 })
Progress({ value: 20, total: 100, type: ProgressType.
ScaleRing }).width(150).margin({ top: 10 })
Progress({ value: 20, total: 100, type: ProgressType.
Capsule }).width(40).margin({ top: 10 })
```

在上述示例中，界面效果如图8-33所示。

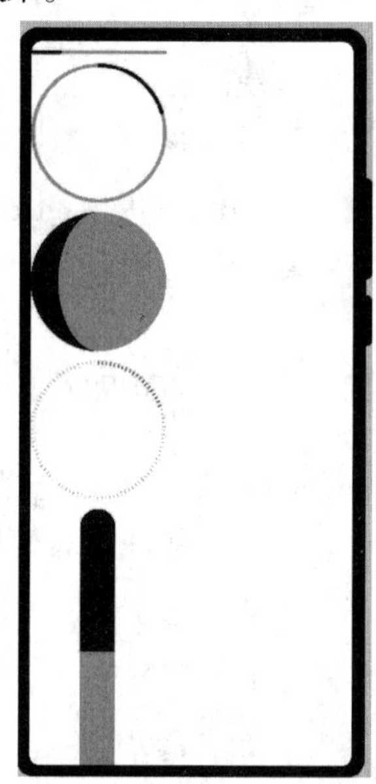

图8-33　Progress组件效果

8.4.16 QRCode

QRCode是用于显示单个二维码的组件。以下是附了黄码的具体示例：

```
QRCode("https://waylau.comn")
  .width(360).height(360)          // 大小
  .backgroundColor(Color.Orange) // 颜色
```

在上述示例中，QRCode会自动将“https://waylau.comn” URL链接转为二维码的图片，并且根据backgroundColor将二维码设置为黄码。界面效果如图8-34所示。

图8-34　Progress组件效果

8.4.17 Radio

Radio是单选框，提供相应的用户交互选择项。当前单选框的所属群组名称，相同group的Radio只能有一个被选中。

以下是一组Radio的具体示例：

```
Radio({ value: 'Radio1', group: 'radioGroup' })
  .checked(false) // 默认不选中
  .height(50)
  .width(50)
Radio({ value: 'Radio2', group: 'radioGroup' })
  .checked(true) // 默认选中
  .height(50)
  .width(50)
Radio({ value: 'Radio2', group: 'radioGroup' })
  .checked(false) // 默认不选中
  .height(50)
  .width(50)
```

在上述示例中，checked属性用来配置Radio是否会被默认选中。界面效果如图8-35所示。

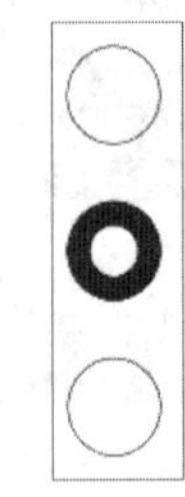

图8-35　Radio组件效果

8.4.18 Rating

Rating是提供在给定范围内选择评分的组件。

以下是一组Rating的具体示例：

```
// 设置初始星数为1，可以操作
Rating({ rating: 1, indicator: false })
  .stars(5)       // 设置评星总数。默认值：5
  .stepSize(0.5) // 操作评级的步长。默认值：0.5
  .onChange((value: number) => {
    //...
  })
```

在上述示例中，Rating构造函数接收两个参数：rating是初始星数；indicator指示是否仅作为指示器使用，不可操作。Rating属性有stars、stepSize等，还可以通过onChange监听Rating选择的星数。界面效果如图8-36所示。

图8-36　Rating组件效果

8.4.19 RichText

RichText是富文本组件，可以解析并显示HTML格式文本。示例如下：

```
RichText('<h1 style="text-align: center;">h1标题</h1>' +
  '<h1 style="text-align: center;"><i>h1斜体</i></h1>' +
  '<h1 style="text-align: center;"><u>h1下划线</u></h1>' +
  '<h2 style="text-align: center;">h2标题</h2>' +
  '<h3 style="text-align: center;">h3标题</h3>' +
  '<p style="text-align: center;">p常规</p><hr/>' +
  '<div style="width: 500px;height: 500px;border: 1px solid;margin: 0auto;">'
+
  '<p style="font-size: 35px;text-align: center;font-weight: bold; color:
rgb(24,78,228)">字体大小35px, 行高45px</p>' +
  '<p style="background-color: #e5e5e5;line-height: 45px;font-size: 35px;text-
indent: 2em;">' +
  '<p>这是一段文字这是一段文字这是一段文字这是一段文字这是一段文字这是一段文字这是一段文字
这是一段文字这是一段文字</p>')
```

在上述示例中，界面效果如图8-37所示。

图8-37　RichText组件效果

注意：示例效果请以真机或虚拟机运行为准，当前IDE预览器不支持RichText的显示。

8.4.20　ScrollBar

滚动条组件ScrollBar，用于配合可滚动组件使用，如List、Grid、Scroll等。

ScrollBar实例化构造函数为ScrollBar(value: { scroller: Scroller, direction?: ScrollBarDirection, state?: BarState })，这些参数说明如下。

- Scroller：可滚动组件的控制器。用于与可滚动组件进行绑定。
- ScrollBarDirection：滚动条的方向，控制可滚动组件对应方向的滚动。默认值是ScrollBarDirection.Vertical。
- BarState：滚动条状态。默认值是BarState.Auto。

ScrollBar示例如下：

```
// 可滚动组件的控制器
private scroller: Scroller = new Scroller()

private dataScroller: number[] = [0, 1, 2, 3, 4, 5, 6, 7, 8, 9]


Stack({ alignContent: Alignment.End }) {
  // 定义了可滚动组件 Scroll
  Scroll(this.scroller) {
    Flex({ direction: FlexDirection.Column }) {
      ForEach(this.arr, (item) => {
        Row() {
          Text(item.toString())
            .width('90%')
            .height(100)
            .backgroundColor('#3366CC')
            .borderRadius(15)
            .fontSize(16)
            .textAlign(TextAlign.Center)
            .margin({ top: 5 })
        }
      }, item => item)
    }.margin({ left: 52 })
  }
  .scrollBar(BarState.Off)
  .scrollable(ScrollDirection.Vertical)

  // 定义了滚动条组件 ScrollBar
  ScrollBar({ scroller: this.scroller, direction: ScrollBarDirection.Vertical,
state: BarState.Auto }) {
    // 定义 Text 作为滚动条的样式
    Text()
      .width(30)
      .height(100)
      .borderRadius(10)
      .backgroundColor('#C0C0C0')
  }.width(30).backgroundColor('#ededed')
}
```

上述示例定义了可滚动组件Scroll及滚动条组件ScrollBar。在ScrollBar子组件中定义Text作为滚动条的样式。可滚动组件Scroll及滚动条组件ScrollBar通过Scroller进行绑定，且只有当两者方向相同时，才能联动，ScrollBar与可滚动组件Scroll仅支持一对一绑定。

上述示例界面效果如图8-38所示。

8.4.21 Search

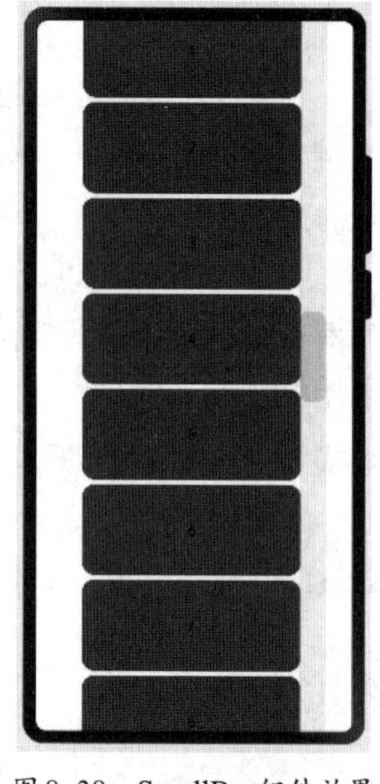
图8-38 ScrollBar组件效果

Search是搜索框组件，适用于浏览器的搜索内容输入框等应用场景。

Search示例如下：

```
Search({ placeholder: '输入内容...'})
  // 搜索按钮的文字
  .searchButton('搜索')
  .width(300)
  .height(80)
  // 提示文本样式
  .placeholderColor(Color.Grey)
  // 提示文本字体大小
  .placeholderFont({ size: 24, weight: 400 })
  // 搜索框文字字体大小
  .textFont({ size: 24, weight: 400 })
```

上述示例定义了Search组件及搜索按钮的文字、提示文本样式、字体大小等。界面效果如图8-39所示。

图8-39 Search组件效果

Search还支持以下事件。

- onSubmit(callback: (value: string) => void)：点击搜索图标、搜索按钮或按下软键盘搜索按钮时，触发该回调。
- onChange(callback: (value: string) => void)：输入内容发生变化时，触发该回调。

在上述事件中，value是指当前搜索框中输入的文本内容。

8.4.22 Select

Select用于提供下拉选择菜单，可以让用户在多个选项之间选择。

Select示例如下：

```
// 设置下拉列表值和图标
Select([{ value: 'Java 核心编程', icon: $r('app.media.book01') },
  { value: '轻量级 Java EE 企业应用开发实战', icon: $r('app.media.book02') },
  { value: '鸿蒙 HarmonyOS 手机应用开发实战', icon: $r('app.media.book03') },
  { value: 'Node.js+Express+MongoDB+Vue.js 全栈开发实战', icon: $r('app.media.
book04') }])
  .selected(2)          // 选中的下拉列表索引
  .value('老卫作品集') // 下拉按钮本身的文本内容
  .font({ size: 16, weight: 500 }) // 下拉按钮本身的文本样式
  .fontColor('#182431') // 下拉按钮本身的文本颜色
  .selectedOptionFont({ size: 16, weight: 400 }) // 下拉菜单选中项的文本样式
```

```
.optionFont({ size: 16, weight: 400 }) // 下拉菜单项的文本样式
```

上述示例定义了Select组件，构造函数是一个SelectOption数组。SelectOption分为value及icon属性，分别用来定义下拉框的文字及图标。

Select组件还可以设置默认选中的下拉列表索引、下拉按钮本身的文本内容、下拉按钮本身的文本样式、下拉按钮本身的文本颜色、下拉菜单选中项的文本样式、下拉菜单项的文本样式等。上述示例界面效果如图8-40所示。

图 8-40　Select组件效果

Search还支持事件onSelect(callback: (index: number, value?: string) => void)。其中index是选中项的索引，value是选中项的值。

8.4.23　Slider

Slider是滑动条组件，通常用于快速调节设置值，如音量调节、亮度调节等应用场景。

Slider示例如下：

```
// 设置垂直的 Slider
Slider({
  value: 40,
  step: 10,
  style: SliderStyle.InSet,   // 滑块在滑轨上
  direction: Axis.Vertical    // 方向
})
  .showSteps(true) // 设置显示步长刻度值
  .height('50%')

// 设置水平的 Slider
Slider({
  value: 40,
  min: 0,
  max: 100,
  style: SliderStyle.OutSet   //滑块在滑轨内
})
  .blockColor('#191970')      // 设置滑块的颜色
  .trackColor('#ADD8E6')      // 设置滑轨的背景颜色
  .selectedColor('#4169E1')   // 设置滑轨的已滑动部分颜色
  .showTips(true) // 设置气泡提示
  .width('50%')
```

上述示例定义了两个Slider组件，一个是垂直的，另一个是水平的，由direction参数决定。参数style用来设置滑块是否在滑轨内。

Slider组件还包括以下属性。

- blockColor：设置滑块的颜色。
- trackColor：设置滑轨的背景颜色。
- selectedColor：设置滑轨的已滑动部分颜色。
- showSteps：设置当前是否显示步长刻度值。
- showTips：设置滑动时是否显示百分比气泡提示。
- trackThickness：设置滑轨的粗细。

上述示例界面效果如图8-41所示。

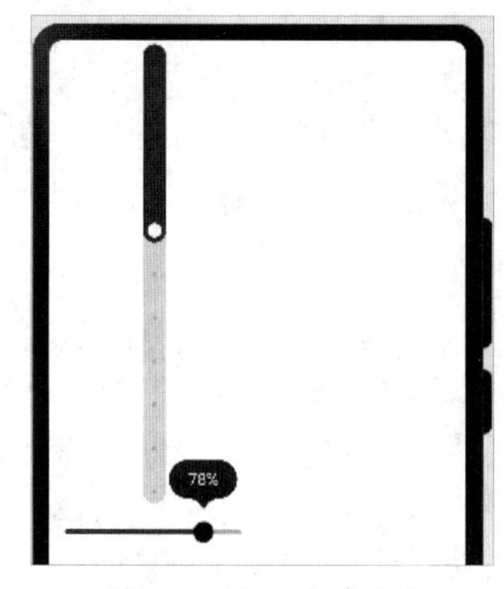

图8-41　Slider组件效果

8.4.24　Span

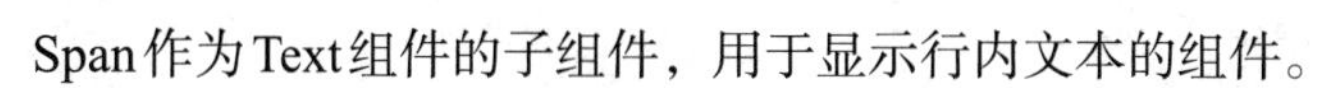

Span作为Text组件的子组件，用于显示行内文本的组件。

Span组件主要包括以下属性。

- decoration：设置文本装饰线样式及其颜色。
- letterSpacing：设置文本字符间距。取值小于0，字符聚集重叠；取值大于0且随着数值变大，字符间距越来越大，稀疏分布。
- textCase：设置文本大小写。

Span示例如下：

```
// 文本添加横线
Text() {
  Span('文本添加横线').decoration({ type: TextDecorationType.Underline, color: 
Color.Red }).fontSize(24)
}

// 文本添加划掉线
Text() {
  Span('文本添加划掉线')
    .decoration({ type: TextDecorationType.LineThrough, color: Color.Red })
    .fontSize(24)
}

// 文本添加上划线
Text() {
  Span('文本添加上划线').decoration({ type: TextDecorationType.Overline, color: 
Color.Red }).fontSize(24)
}

// 文本字符间距
Text() {
  Span('文本字符间距')
```

```
    .letterSpacing(10)
    .fontSize(24)
}

// 文本转化为小写 LowerCase
Text() {
  Span('文本转化为小写 LowerCase').fontSize(24)
    .textCase(TextCase.LowerCase)
    .decoration({ type: TextDecorationType.None })
}

// 文本转化为大写 UpperCase
Text() {
  Span('文本转化为小写 UpperCase').fontSize(24)
    .textCase(TextCase.UpperCase)
    .decoration({ type: TextDecorationType.None })
}
```

上述示例界面效果如图 8-42 所示。

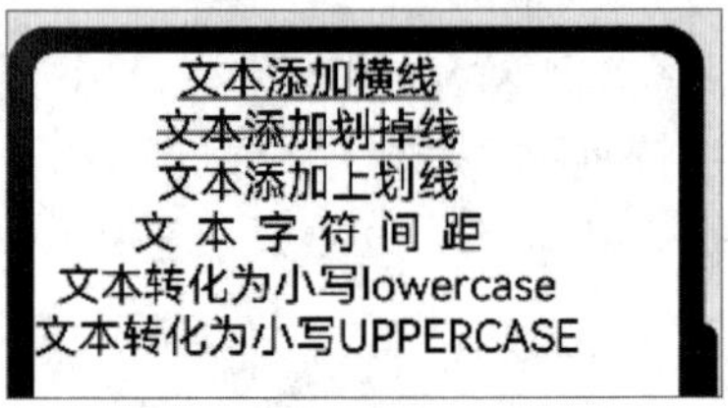

图 8-42　Span 组件效果

8.4.25　Stepper 与 StepperItem

Stepper 是步骤导航器组件，适用于引导用户按照步骤完成任务的导航场景。而 StepperItem 是用作 Stepper 组件的页面子组件。

Stepper 与 StepperItem 示例如下：

```
Stepper({
  // 设置 Stepper 当前显示 StepperItem 的索引值
  index: 0
}) {
  // 第 1 页
  StepperItem() {
    Text('第 1 页').fontSize(34)
  }
  .nextLabel('下一页')

  // 第 2 页
  StepperItem() {
    Text('第 2 页').fontSize(34)
  }
  .nextLabel('下一页')
  .prevLabel('上一页')
```

```
// 第3页
StepperItem() {
  Text('第3页').fontSize(34)
}
.prevLabel('上一页')
```

上述示例设置Stepper当前显示StepperItem的索引值为0，即显示第1页的内容。后续定义了3个StepperItem页面。

上述示例第1页界面效果如图8-43所示。

点击“下一页”会切换到第2页界面，效果如图8-44所示。

继续点击“下一页”会切换到第3页界面，效果如图8-45所示。

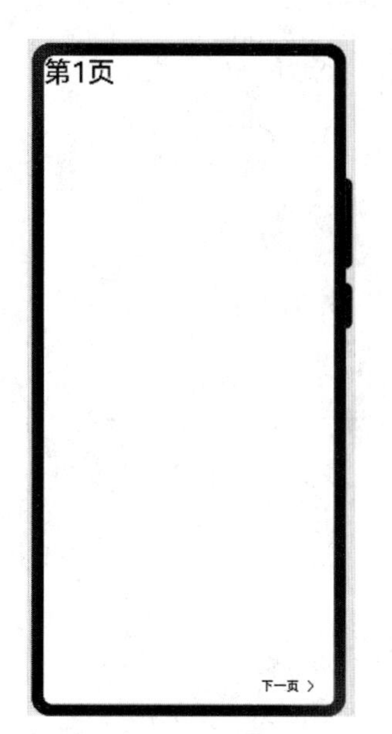

图8-43　Stepper组件第1页效果

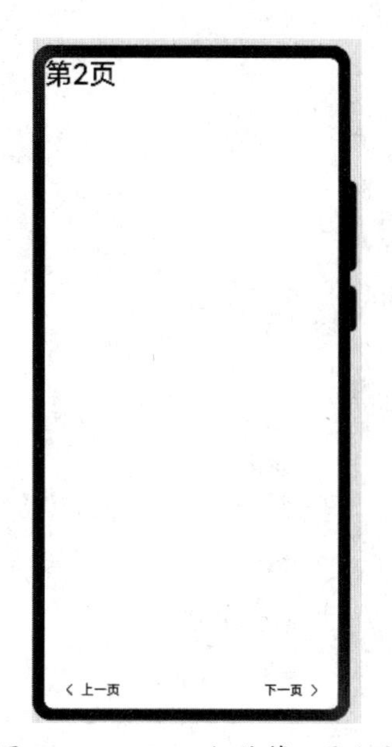

图8-44　Stepper组件第2页效果

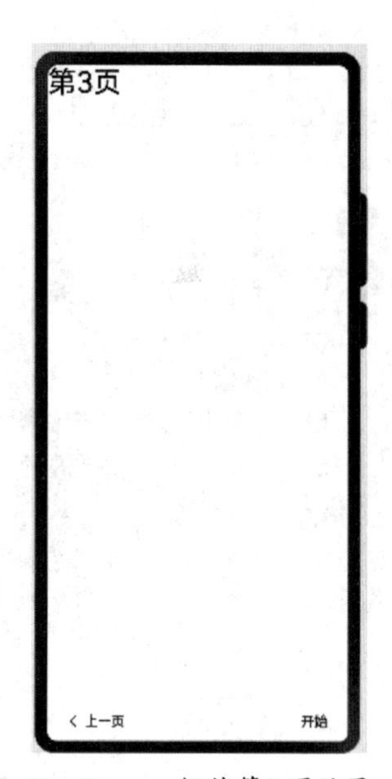

图8-45　Stepper组件第3页效果

8.4.26　Text

Text是显示一段文本的组件，可以包含Span子组件。

Text包含以下属性。

- textAlign：设置文本在水平方向的对齐方式。
- textOverflow：设置文本超长时的显示方式。默认值是TextOverflow.Clip。
- maxLines：设置文本的最大行数。默认值是Infinity。
- lineHeight：设置文本的文本行高，设置值不大于0时不限制文本行高，自适应字体大小，Length为number类型时单位为fp。
- decoration：设置文本装饰线样式及其颜色。
- baselineOffset：设置文本基线的偏移量，默认值为0。

- letterSpacing：设置文本字符间距。
- minFontSize：设置文本最小显示字号。需配合maxFontSize及maxline或布局大小限制使用，单独设置不生效。
- maxFontSize：设置文本最大显示字号。需配合minFontSize及maxline或布局大小限制使用，单独设置不生效。
- textCase：设置文本大小写。默认值是TextCase.Normal。
- copyOption：组件支持设置文本是否可复制粘贴。默认值是CopyOptions.None。

Text示例如下：

```
// 单行文本
// 红色单行文本居中
Text('红色单行文本居中').fontSize(24)
  .fontColor(Color.Red) // 红色
  .textAlign(TextAlign.Center) // 居中
  .width('100%')
// 单行文本对齐左侧
Text('单行文本对齐左侧').fontSize(24)
  .textAlign(TextAlign.Start) // 对齐左侧
  .width('100%')
// 单行文本带边框对齐右侧
Text('单行文本带边框对齐右侧')
  .fontSize(24)
  .textAlign(TextAlign.End) // 对齐右侧
  .border({ width: 1 }) // 边宽
  .padding(10)
  .width('100%')
// 多行文本
// 超出 maxLines 截断内容展示
Text('寒雨连江夜入吴，平明送客楚山孤。洛阳亲友如相问，一片冰心在玉壶。')
  .textOverflow({ overflow: TextOverflow.None }) // 超出截断内容
  .maxLines(2) //  最多显示 2 行
  .fontSize(24)
  .border({ width: 1 })
  .padding(10)
// 超出 maxLines 展示省略号
Text('寒雨连江夜入吴，平明送客楚山孤。洛阳亲友如相问，一片冰心在玉壶。')
  .textOverflow({ overflow: TextOverflow.Ellipsis }) // 超出展示省略号
  .maxLines(2)
  .fontSize(24)
  .border({ width: 1 })
  .padding(10)
Text('寒雨连江夜入吴，平明送客楚山孤。洛阳亲友如相问，一片冰心在玉壶。')
  .textOverflow({ overflow: TextOverflow.Ellipsis }) // 超出展示省略号
  .maxLines(2)
```

```
  .fontSize(24)
  .border({ width: 1 })
  .padding(10)
  .lineHeight(50) // 设置文本的文本行高
```

上述示例界面效果如图8-46所示。

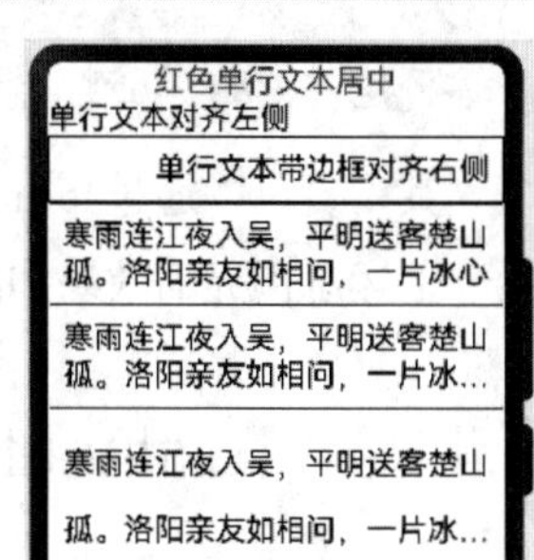

图8-46　Text组件效果

8.4.27　TextArea

TextArea是多行文本输入框组件，当输入的文本内容超过组件宽度时会自动换行显示。

TextArea支持以下属性。

- placeholderColor：设置placeholder文本颜色。
- placeholderFont：设置placeholder文本样式。
- textAlign：设置文本在输入框中的水平对齐方式。
- caretColor：设置输入框光标颜色。
- inputFilter：通过正则表达式设置输入过滤器。匹配表达式的输入允许显示，不匹配的输入将被过滤。仅支持单个字符匹配，不支持字符串匹配。
- copyOption：设置输入的文本是否可复制。

TextArea示例如下：

```
TextArea({
  // 设置无输入时的提示文本
  placeholder: '寒雨连江夜入吴，平明送客楚山孤。洛阳亲友如相问，一片冰心在玉壶。'
})
  .placeholderFont({ size: 24, weight: 400 })  // 设置placeholder文本样式
  .width(336)
  .height(100)
  .margin(20)
  .fontSize(16)
  .fontColor('#182431')
  .backgroundColor('#FFFFFF')
```

上述示例界面效果如图8-47所示。

图8-47　TextArea组件效果

8.4.28　TextClock

TextClock组件通过文本将当前系统时间显示在设备上。支持不同时区的时间显示，最高精度到秒级。

TextClock示例如下：

```
// 普通的 TextClock 示例
TextClock().margin(20).fontSize(30)

// 带日期格式化的 TextClock 示例
TextClock().margin(20).fontSize(30)
  .format('yyyyMMdd hh:mm:ss') // 日期格式化
```

其中，可以通过format属性设置显示时间格式。

上述示例界面效果如图 8-48 所示。

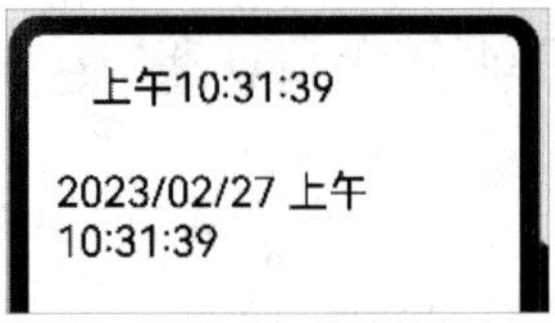

上午10:31:39

2023/02/27 上午
10:31:39

图 8-48　TextClock组件效果

8.4.29　TextInput

TextInput是单行文本输入框组件。

TextInput示例如下：

```
// 文本输入框
TextInput({ placeholder: '请输入 ...'}) // 设置无输入时的提示文本
  .placeholderColor(Color.Grey) // 设置 placeholder 文本颜色
  .placeholderFont({ size: 14, weight: 400 }) // 设置 placeholder 文本样式
  .caretColor(Color.Blue) // 设置输入框光标颜色
  .width(300)
  .height(40)
  .margin(20)
  .fontSize(24)
  .fontColor(Color.Black)

// 密码输入框
TextInput({ placeholder: '请输入密码 ...' })
  .width(300)
  .height(40)
  .margin(20)
  .fontSize(24)
  .type(InputType.Password) // 密码类型
  .maxLength(9) //  设置文本的最大输入字符数
  .showPasswordIcon(true)  // 输入框末尾的图标显示
```

TextInput常见的属性包括以下几种。

- type：设置输入框类型。默认值是InputType.Normal。
- placeholderColor：设置placeholder文本颜色。
- placeholderFont：设置placeholder文本样式。
- enterKeyType：设置输入法回车键类型，目前仅支持默认类型显示。
- caretColor：设置输入框光标颜色。
- maxLength：设置文本的最大输入字符数。

- showPasswordIcon：密码输入模式时，输入框末尾的图标是否显示。默认值是true。

上述示例界面效果如图8-49所示。

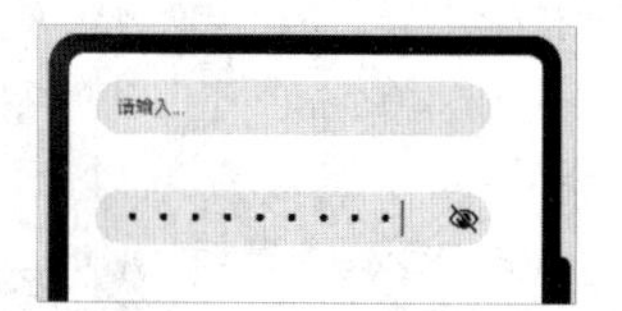

图8-49　TextInput组件效果

8.4.30　TextPicker

TextPicker是滑动选择文本内容的组件。

TextPicker示例如下：

```
// 文本输入框
TextPicker({
  // 选择器的数据选择列表
  range: ['Java 核心编程', '轻量级 Java EE 企业应用
开发实战', '鸿蒙 HarmonyOS 手机应用开发实战', 'Node.
js+Express+MongoDB+Vue.js 全栈开发实战'],
  // 设置默认选中项在数组中的索引值。默认值是 0
  selected: 1
}).defaultPickerItemHeight(30)// 设置 Picker 各选择项的
高度
```

从上述示例看出，参数range用于设置选择器的数据选择列表；selected用于设置默认选中项在数组中的索引值；defaultPickerItemHeight属性用于设置Picker各选择项的高度。

上述示例界面效果如图8-50所示。

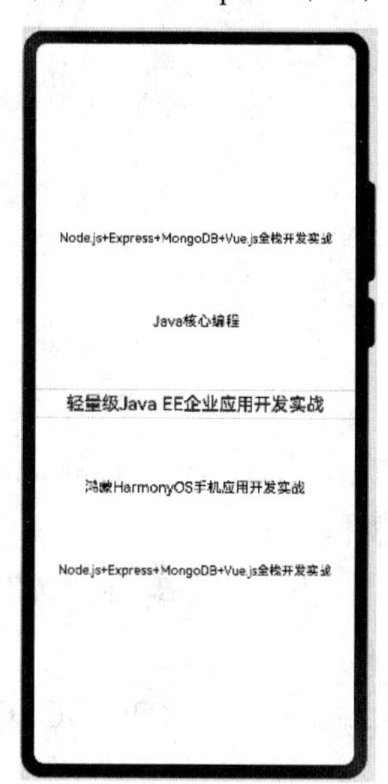

图8-50　TextPicker组件效果

8.4.31　TextTimer

TextTimer是通过文本显示计时信息并控制其计时器状态的组件。TextTimer组件支持绑定一个控制器TextTimerController用来控制文本计时器。

TextTimer示例如下：

```
// TextTimer 组件的控制器
private textTimerController: TextTimerController = new TextTimerController()


// 定义 TextTimer 组件
TextTimer({ controller: this.textTimerController,
  isCountDown: true,       // 是否倒计时。默认值 false
  count: 30000 })          // 倒计时时间，单位为毫秒
  .format('mm:ss.SS')      // 格式化
  .fontColor(Color.Black) // 字体颜色
  .fontSize(50)            // 字体大小
```

```
// 控制按钮
Row() {
  Button("开始").onClick(() => {
    this.textTimerController.start()
  })
  Button("暂停").onClick(() => {
    this.textTimerController.pause()
  })
  Button("重置").onClick(() => {
    this.textTimerController.reset()
  })
}
```

从上述示例看出，TextTimer绑定一个控制器TextTimerController，设置了倒计时30秒。通过TextTimerController的start()、pause()、reset()实现对计时器状态的控制。

上述示例界面效果如图8-51所示。

图8-51　TextTimer组件效果

8.4.32　TimePicker

TimePicker是滑动选择时间的组件。

TimePicker示例如下：

```
TimePicker()
  .useMilitaryTime(true) // 设置为 24 小时制
```

从上述示例看出，TimePicker可以设置useMilitaryTime属性来实现展示时间是否为24小时制。上述示例界面效果如图8-52所示。

图8-52　TimePicker组件效果

8.4.33　Toggle

Toggle组件提供勾选框样式、状态按钮样式及开关样式。仅当ToggleType为Button时，可包含子组件。

Toggle组件构造函数参数主要有以下两个。

- typ：开关类型。可以是Checkbox、Button、Switch。
- isOn：开关是否打开。默认值是false。

Toggle组件还可以设置以下属性。

- selectedColor：设置组件打开状态的背景颜色。
- switchPointColor：设置Switch类型的圆形滑块颜色。

Toggle示例如下：

```
// 关闭的 Switch 类型
Toggle({ type: ToggleType.Switch, isOn: false })
  .size({ width: 40, height: 40 })   // 设置大小
  .selectedColor('#007DFF')          // 设置组件打开状态的背景颜色
  .switchPointColor('#FFFFFF')       // 设置 Switch 类型的圆形滑块颜色

// 打开的 Switch 类型
Toggle({ type: ToggleType.Switch, isOn: true })
  .size({ width: 40, height: 40 })   // 设置大小
  .selectedColor('#007DFF')          // 设置组件打开状态的背景颜色
  .switchPointColor('#FFFFFF')       // 设置 Switch 类型的圆形滑块颜色

// 关闭的 Checkbox 类型
Toggle({ type: ToggleType.Checkbox, isOn: false })
  .size({ width: 40, height: 40 })   // 设置大小
  .selectedColor('#007DFF')          // 设置组件打开状态的背景颜色

// 打开的 Checkbox 类型
Toggle({ type: ToggleType.Checkbox, isOn: true })
  .size({ width: 40, height: 40 })   // 设置大小
  .selectedColor('#007DFF')          // 设置组件打开状态的背景颜色

// 关闭的 Button 类型
Toggle({ type: ToggleType.Button, isOn: false })
  .size({ width: 40, height: 40 })   // 设置大小
  .selectedColor('#007DFF')          // 设置组件打开状态的背景颜色

// 打开的 Button 类型
Toggle({ type: ToggleType.Button, isOn: true })
  .size({ width: 40, height: 40 })   // 设置大小
  .selectedColor('#007DFF')          // 设置组件打开状态的背景颜色
```

上述示例界面效果如图8-53所示。

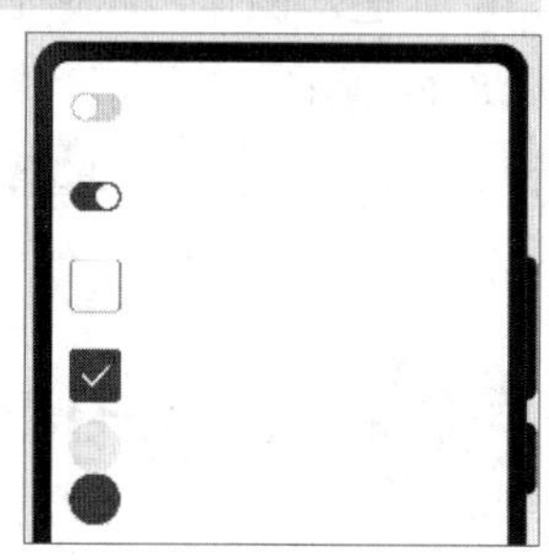

图8-53　Toggle组件效果

8.4.34 Web

Web组件是提供具有网页显示能力的组件。需要注意的是，在访问在线网页时需添加网络权限ohos.permission.INTERNET。修改module.json5文件如下：

```
{
  "module": {
    "name": "entry",
```

```
    "type": "entry",
    "description": "$string:module_desc",
    "mainElement": "EntryAbility",
    "deviceTypes": [
      "phone",
      "tablet"
    ],
    "deliveryWithInstall": true,
    "installationFree": false,
    "pages": "$profile:main_pages",
    "abilities": [
      // 为节约篇幅，省略部分代码
    ],
    // 添加相关权限
    "requestPermissions": [
      {
        "name": "ohos.permission.INTERNET"
      }
    ]
  }
}
```

Web组件示例如下：

```
// Web 组件控制器需要导入的包
import web_webview from '@ohos.web.webview'

private webviewController: web_webview.
WebviewController = new web_webview.
WebviewController()

Web({ src: 'https://waylau.com', controller: this.
webviewController })
```

上述示例显示了来自https://waylau.com 网页的界面效果，如图8-54所示。

有关Web组件的更多内容，还会在“第15章 网络管理”深入探讨。

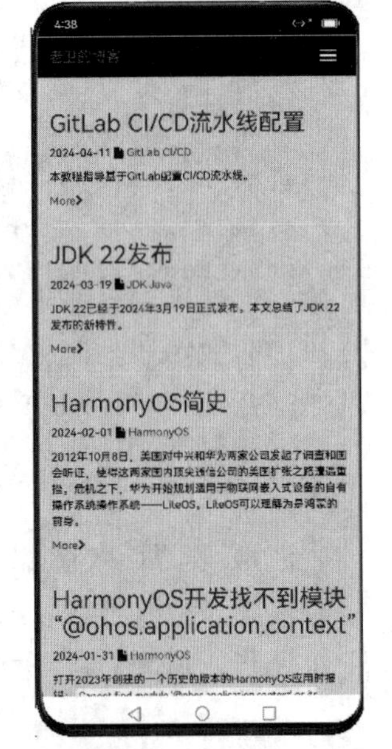

图 8-54　Web组件效果

8.5 容器组件详解

声明式开发范式目前可供选择的容器组件有Column、Row、ColumnSplit、RowSplit、Flex、Grid、GridItem、GridRow、GridCol、List、ListItem、ListItemGroup、AlphabetIndexer、Badge、Counter、Navigator、Panel、Refresh、RelativeContainer、Scroll、SideBarContainer、Stack、Swiper、

Tabs、TabContent。

本节演示如何使用这些容器组件，相关示例可以在ArkUIContainerComponents应用中找到。

8.5.1 Column和Row

Column和Row是最常用的容器组件。其中，Column是沿垂直方向布局的容器，Row是沿水平方向布局的容器。

Column和Row的构造函数都有space参数，表示元素间的间距。

Column和Row都包含属性alignItems和justifyContent，用来设置子组件的对齐格式。

所不同的是，针对Column而言，alignItems是设置子组件在水平方向上的对齐格式，默认值是HorizontalAlign.Center；justifyContent是设置子组件在垂直方向上的对齐格式，默认值是FlexAlign.Start。

而Row则相反，alignItems是设置子组件在垂直方向上的对齐格式，默认值是VerticalAlign.Center；而justifyContent是设置子组件在水平方向上的对齐格式，默认值是FlexAlign.Start。

Column和Row示例如下：

```
Column() {
  // 设置子组件水平方向的间距为 5
  Row({ space: 5 }) {
    Row().width('30%').height(50).backgroundColor(0xAFEEEE)
    Row().width('30%').height(50).backgroundColor(0x00FFFF)
  }.width('90%').height(107).border({ width: 1 })

  // 设置子元素垂直方向对齐方式
  Row() {
    Row().width('30%').height(50).backgroundColor(0xAFEEEE)
    Row().width('30%').height(50).backgroundColor(0x00FFFF)
  }.width('90%').alignItems(VerticalAlign.Bottom).height('15%').border({
width: 1 })

  Row() {
    Row().width('30%').height(50).backgroundColor(0xAFEEEE)
    Row().width('30%').height(50).backgroundColor(0x00FFFF)
  }.width('90%').alignItems(VerticalAlign.Center).height('15%').border({
width: 1 })

  // 设置子元素水平方向对齐方式
  Row() {
    Row().width('30%').height(50).backgroundColor(0xAFEEEE)
    Row().width('30%').height(50).backgroundColor(0x00FFFF)
  }.width('90%').border({ width: 1 }).justifyContent(FlexAlign.End)
```

```
  Row() {
    Row().width('30%').height(50).backgroundColor(0xAFEEEE)
    Row().width('30%').height(50).backgroundColor(0x00FFFF)
  }.width('90%').border({ width: 1 }).justifyContent(FlexAlign.Center)
}
```

上述示例界面效果如图 8-55 所示。

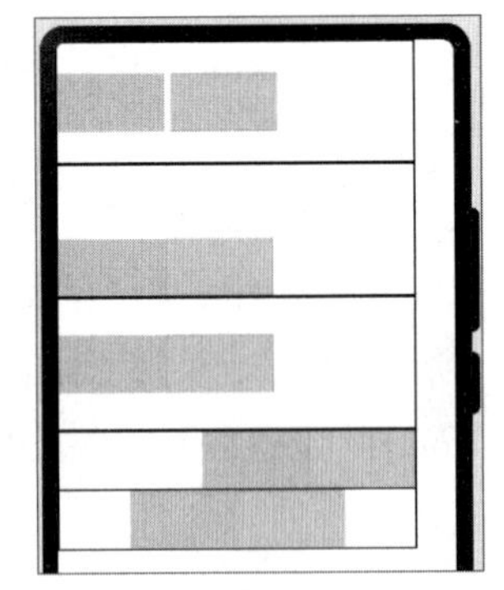
图 8-55　Column 和 Row 组件效果

8.5.2　ColumnSplit 和 RowSplit

ColumnSplit 和 RowSplit 在每个子组件之间插入一根分割线。其中 ColumnSplit 是横向的分割线，RowSplit 是纵向的分割线。

ColumnSplit 和 RowSplit 示例如下：

```
// 纵向的分割线
RowSplit() {
  Text('1').width('10%').height(400).backgroundColor(0xF5DEB3).
      textAlign(TextAlign.Center)
  Text('2').width('10%').height(400).backgroundColor(0xD2B48C).
      textAlign(TextAlign.Center)
  Text('3').width('10%').height(400).backgroundColor(0xF5DEB3).
      textAlign(TextAlign.Center)
  Text('4').width('10%').height(400).backgroundColor(0xD2B48C).
      textAlign(TextAlign.Center)
  Text('5').width('10%').height(400).backgroundColor(0xF5DEB3).
      textAlign(TextAlign.Center)
}
.resizeable(true) // 可拖动
.width('90%').height(400)

// 横向的分割线
ColumnSplit() {
  Text('1').width('100%').height(50).backgroundColor(0xF5DEB3).
      textAlign(TextAlign.Center)
  Text('2').width('100%').height(50).backgroundColor(0xD2B48C).
      textAlign(TextAlign.Center)
  Text('3').width('100%').height(50).backgroundColor(0xF5DEB3).
      textAlign(TextAlign.Center)
  Text('4').width('100%').height(50).backgroundColor(0xD2B48C).
      textAlign(TextAlign.Center)
  Text('5').width('100%').height(50).backgroundColor(0xF5DEB3).
      textAlign(TextAlign.Center)
```

```
}
.resizeable(true) // 可拖动
.width('90%').height('60%')
```

ColumnSplit和RowSplit还可以设置resizeable属性，表示分割线是否可以拖动。上述示例界面效果如图8-56所示。

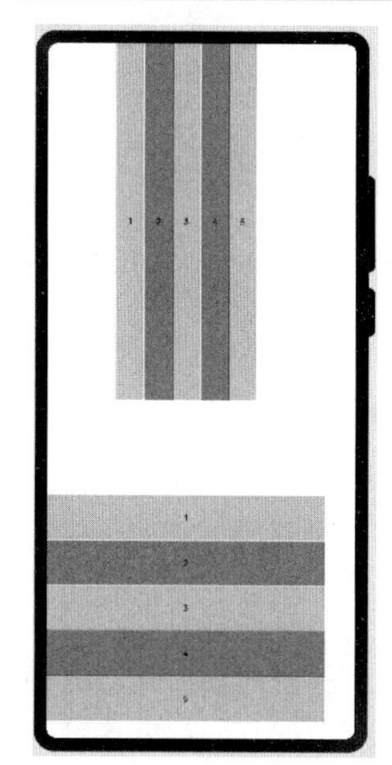

图 8-56　ColumnSplit和RowSplit组件效果

8.5.3　Flex

Flex是以弹性方式布局子组件的容器组件。

标准Flex布局容器包含以下参数。

- direction：子组件在Flex容器上排列的方向，即主轴的方向。
- wrap：Flex容器是单行/列还是多行/列排列。
- justifyContent：子组件在Flex容器主轴上的对齐格式。
- alignItems：子组件在Flex容器交叉轴上的对齐格式。
- alignContent：交叉轴中有额外的空间时，多行内容的对齐方式。仅在wrap为Wrap或WrapReverse时生效。

Flex示例如下：

```
// 主轴方向为 FlexDirection.Row
Flex({ direction: FlexDirection.Row }) {
  Text('1').width('20%').height(50).backgroundColor(0xF5DEB3)
  Text('2').width('20%').height(50).backgroundColor(0xD2B48C)
  Text('3').width('20%').height(50).backgroundColor(0xF5DEB3)
  Text('4').width('20%').height(50).backgroundColor(0xD2B48C)
}
.height('40%')
.width('90%')
.padding(10)
.backgroundColor(0xAFEEEE)

// 主轴方向为 FlexDirection.Column
Flex({ direction: FlexDirection.Column }) {
  Text('1').width('20%').height(50).backgroundColor(0xF5DEB3)
  Text('2').width('20%').height(50).backgroundColor(0xD2B48C)
  Text('3').width('20%').height(50).backgroundColor(0xF5DEB3)
  Text('4').width('20%').height(50).backgroundColor(0xD2B48C)
}
```

```
.height('40%')
.width('90%')
.padding(10)
.backgroundColor(0xAFEEEE)
```

上述示例界面效果如图8-57所示。

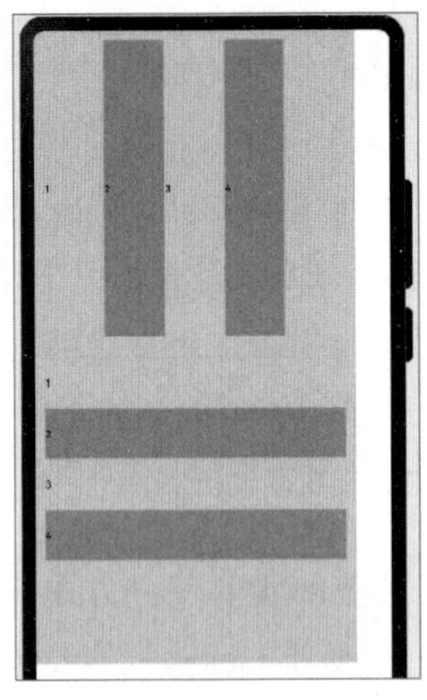

图8-57 Flex组件效果

8.5.4 Grid和GridItem

Grid网格容器，由行和列分割的单元格所组成，通过指定GridItem所在的单元格做出各种各样的布局。

Grid和GridItem示例如下：

```
private numberArray: String[] = ['0', '1', '2', '3', '4']

Grid() {
  ForEach(this.numberArray, (day: string) => {
    ForEach(this.numberArray, (day: string) => {
      GridItem() {
        Text(day)
          .fontSize(16)
          .backgroundColor(0xF9CF93)
          .width('100%')
          .height('100%')
          .textAlign(TextAlign.Center)
      }
    }, day => day)
  }, day => day)
}
.columnsTemplate('1fr 1fr 1fr 1fr 1fr') // 设置当前网格布局列的数量
.rowsTemplate('1fr 1fr 1fr 1fr 1fr')     // 设置当前网格布局行的数量
.columnsGap(10) // 设置列与列的间距
.rowsGap(10)     // 设置行与行的间距
.width('90%')
.backgroundColor(0xFAEEE0)
.height(300)
```

在上述示例中，columnsTemplate用来设置当前网格布局列的数量，不设置时默认为1列。例如，“1fr 1fr 2fr”是将父组件分为3列，将父组件允许的宽分为4等份，第一列占1份，第二列占1份，第三列占2份。 同理，rowsTemplate是用来设置当前网格布局行的数量，不设置时默认为1行。例如，“1fr 1fr 2fr”是将父组件分为三行，将父组件允许的高分为4等份，第一行占1份，第二行占1份，第三行占2份。

上述示例界面效果如图8-58所示。

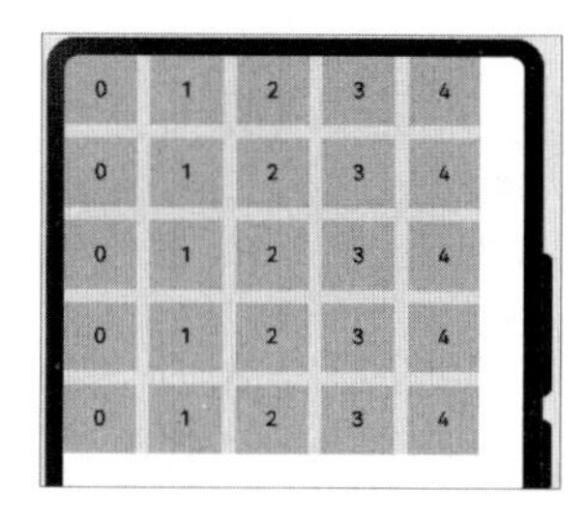

图8-58　Grid和GridItem组件效果

8.5.5　GridRow和GridCol

GridRow是栅格容器组件，仅可以和栅格子组件GridCol在栅格布局场景中使用。

GridRow和GridCol示例如下：

```
private bgColors: Color[] = [Color.Red, Color.Orange, Color.Yellow, Color.
Green, Color.Pink, Color.Grey, Color.Blue, Color.Brown]


GridRow({
  columns: 5, // 设置布局列数
  gutter: { x: 5, y: 20 }, // 栅格布局间距，x代表水平方向，y代表垂直方向
  breakpoints: { value: ["400vp", "600vp", "800vp"], // 断点发生变化时触发回调
    reference: BreakpointsReference.WindowSize },
  direction: GridRowDirection.Row // 栅格布局排列方向
}) {
  ForEach(this.bgColors, (color) => {
    GridCol({ span: { xs: 1, sm: 2, md: 3, lg: 4 } }) {
      Row().width("100%").height("80vp")
    }.borderColor(color).borderWidth(2)
  })
}.width("100%").height("100%")
```

GridRow参数如下。

- gutter：栅格布局间距，x代表水平方向。
- columns：设置布局列数。
- breakpoints：设置断点值的断点数列及基于窗口或容器尺寸的相应参照。
- direction：栅格布局排列方向。

上述示例界面效果如图8-59所示。

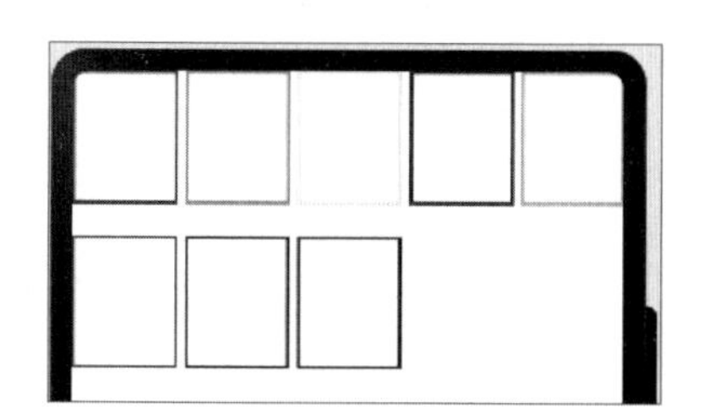
图8-59　GridRow和GridCol组件效果

8.5.6　List、ListItem和ListItemGroup

List是列表，包含一系列相同宽度的列表项。适合连续、多行呈现同类数据，例如图片和文本。

List可以包含ListItem、ListItemGroup子组件。ListItem用来展示列表具体item，必须配合List来使用。ListItemGroup组件用来展示列表item分组，宽度默认充满List组件，必须配合List组件使用。

以下是一个List、ListItem和ListItemGroup的示例：

```
private timetableListItemGroup: any = [
```

```
  {
    title:'星期一',
    projects:['语文', '数学', '英语']
  },
  {
    title:'星期二',
    projects:['物理', '化学', '生物']
  },
  {
    title:'星期三',
    projects:['历史', '地理', '政治']
  },
  {
    title:'星期四',
    projects:['美术', '音乐', '体育']
  }
]

List({ space: 2 }) {
  ForEach(this.timetableListItemGroup, (item) => {
    ListItemGroup() {
      ForEach(item.projects, (project) => {
        ListItem() {
          Text(project)
            .width("100%").height(30).fontSize(20)
            .textAlign(TextAlign.Center)
        }
      }, item => item)
    }
    .borderRadius(20)
    .divider({ strokeWidth: 2, color: 0xDCDCDC }) // 每行之间的分界线
  })
}
.width('100%')
```

上述示例界面效果如图 8-60 所示。

图 8-60　List、ListItem 和 ListItemGroup 组件效果

8.5.7　AlphabetIndexer

AlphabetIndexer 是可以与容器组件联动用于按逻辑结构快速定位容器显示区域的组件。

AlphabetIndexer 构造函数接受以下两个参数。

- arrayValue：字母索引字符串数组，不可设置为空。

- selected：初始选中项索引值，若超出索引值范围，则取默认值0。

以下是一个AlphabetIndexer的示例：

```
Row() {
  List({ space: 10, initialIndex: 0 }) {
    ForEach(this.alphabetIndexerArrayA, (item) => {
      ListItem() {
        Text(item)
          .width('80%')
          .height('5%')
          .fontSize(20)
          .textAlign(TextAlign.Center)
      }
    }, item => item)

    ForEach(this.alphabetIndexerArrayB, (item) => {
      ListItem() {
        Text(item)
          .width('80%')
          .height('5%')
          .fontSize(20)
          .textAlign(TextAlign.Center)
      }
    }, item => item)

    ForEach(this.alphabetIndexerArrayC, (item) => {
      ListItem() {
        Text(item)
          .width('80%')
          .height('5%')
          .fontSize(20)
          .textAlign(TextAlign.Center)
      }
    }, item => item)

    ForEach(this.alphabetIndexerArrayL, (item) => {
      ListItem() {
        Text(item)
          .width('80%')
          .height('5%')
          .fontSize(20)
          .textAlign(TextAlign.Center)
      }
    }, item => item)
  }
  .width('50%')
```

```
  .height('100%')

  AlphabetIndexer({ arrayValue: this.alphabetIndexerArrayValue, selected: 0 })
    .selectedColor(0xFFFFFF) // 选中项文本颜色
    .popupColor(0xFFFAF0)     // 弹出框文本颜色
    .selectedBackgroundColor(0xCCCCCC)   // 选中项背景颜色
    .popupBackground(0xD2B48C)             // 弹出框背景颜色
    .usingPopup(true) // 是否显示弹出框
    .selectedFont({ size: 16, weight: FontWeight.Bolder }) // 选中项字体样式
    .popupFont({ size: 30, weight: FontWeight.Bolder }) // 弹出框内容的字体样式
    .itemSize(28)      // 每一项的尺寸大小
    .alignStyle(IndexerAlign.Left) // 弹出框在索引条右侧弹出
    .onRequestPopupData((index: number) => {
      if (this.alphabetIndexerArrayValue[index] == 'A') {
        return this.alphabetIndexerArrayA // 当选中 A 时，弹出框里面的提示文本列表显
示 A 对应的列表 arrayA，选中 B、C、L 时同理
      } else if (this.alphabetIndexerArrayValue[index] == 'B') {
        return this.alphabetIndexerArrayB
      } else if (this.alphabetIndexerArrayValue[index] == 'C') {
        return this.alphabetIndexerArrayC
      } else if (this.alphabetIndexerArrayValue[index] == 'L') {
        return this.alphabetIndexerArrayL
      } else {
        return [] // 选中其余字母项时，提示文本列表为空
      }
    })
  }
```

上述示例界面效果如图 8-61 所示。

图 8-61　AlphabetIndexer 组件效果

8.5.8 Badge

Badge是可以附加在单个组件上用于信息标记的容器组件。

Badge构造函数主要由以下四个参数组成。

- count：设置提醒消息数。
- position：设置提示点显示位置。
- maxCount：最大消息数，超过最大消息时仅显示maxCount。
- style：可设置Badge组件样式，支持设置文本颜色、尺寸、圆点颜色和尺寸。

其中，position可以设置三种情况。

①RightTop：圆点显示在右上角。

②Right：圆点显示在右侧纵向居中。

③Left：圆点显示在左侧纵向居中。

以下是一个Badge的示例：

```
// 如果不设置position，默认是在右上显示红点
Badge({
  value: '',
  style: { badgeSize: 16, badgeColor: '#FA2A2D' }
}) {
  Image($r('app.media.ic_user_portrait'))
    .width(40)
    .height(40)
}
.width(40)
.height(40)

// 在右侧显示“New”
Badge({
  value: 'New',
  position: BadgePosition.Right,
  style: { badgeSize: 16, badgeColor: '#FA2A2D' }
}) {
  Text('我的消息').width(170).height(40).fontSize(40).fontColor('#182431')
}.width(170).height(40)

// 在右侧显示“数字
Badge({
  value: '1',
  position: BadgePosition.Right,
  style: { badgeSize: 16, badgeColor: '#FA2A2D' }
}) {
  Text('我的消息').width(170).height(40).fontSize(40).fontColor('#182431')
}.width(170).height(40)
```

上述示例界面效果如图8-62所示。

图8-62　Badge组件效果

8.5.9 Counter

Counter是计数器组件，提供相应的增加或减少的计数操作。

以下是一个Counter的示例：

```
Counter() {
  Text(this.counterValue.toString())
}.margin(100)
// 监听数值增加事件
.onInc(() => {
```

```
  this.counterValue++
})
// 监听数值减少事件。
.onDec(() => {
  this.counterValue--
})
```

上述示例界面效果如图 8-63 所示。

图 8-63　Counter 组件效果

8.5.10　Navigator

Navigator 是路由容器组件，提供路由跳转能力。

Navigator 的构造函数参数主要有以下两个。

- target：指定跳转目标页面的路径。
- type：指定路由方式。默认值是 NavigationType.Push。

其中路由方式主要有以下三类。

①Push：跳转到应用内的指定页面。

②Replace：用应用内的某个页面替换当前页面，并销毁被替换的页面。

③Back：返回上一页面或指定的页面。

以下是 Navigator 示例。假设有 NavigatorExample.ets、DetailExample.ets、BackExample.ets 三个文件。

NavigatorExample.ets 文件如下：

```
@Entry
@Component
struct NavigatorExample {
  @State active: boolean = false
  @State Text: object = {name: 'news'}

  build() {
    Flex({ direction: FlexDirection.Column, alignItems: ItemAlign.Start,
           justifyContent: FlexAlign.SpaceBetween }) {
      Navigator({ target: 'pages/DetailExample', type: NavigationType.Push }) {
        Text('Go to ' + this.Text['name'] + ' page')
          .width('100%').textAlign(TextAlign.Center)
      }.params({ text: this.Text }) // 传参数到 Detail 页面

      Navigator() {
        Text('Back to previous page').width('100%').textAlign(TextAlign.
                                                             Center)
      }.active(this.active)
      .onClick(() => {
```

```
        this.active = true
      })
    }.height(150).width(350).padding(35)
  }
}
```

DetailExample.ets文件如下：

```
import router from '@ohos.router'

@Entry
@Component
struct DetailExample {
  // 接收NavigatorExample.ets的传参
  @State text: any = router.getParams()['text']

  build() {
    Flex({ direction: FlexDirection.Column, alignItems: ItemAlign.Start,
           justifyContent: FlexAlign.SpaceBetween }) {
      Navigator({ target: 'pages/BackExample', type: NavigationType.Push }) {
        Text('Go to back page').width('100%').height(20)
      }

      Text('This is ' + this.text['name'] + ' page')
        .width('100%').textAlign(TextAlign.Center)
    }
    .width('100%').height(200).padding({ left: 35, right: 35, top: 35 })
  }
}
```

BackExample.ets文件如下：

```
@Entry
@Component
struct BackExample {
  build() {
    Column() {
      Navigator({ target: 'pages/NavigatorExample', type: NavigationType.Back
}) {
        Text('Return to Navigator Page').width('100%').textAlign(TextAlign.
                                                                    Center)
      }
    }.width('100%').height(200).padding({ left: 35, right: 35, top: 35 })
  }
}
```

注意，上述三个页面需要在main_pages.json文件中进行注册。代码如下：

```
{
  "src": [
    "pages/Index",

     // 注册页面
    "pages/NavigatorExample",
    "pages/BackExample",
    "pages/DetailExample"
  ]
}
```

通过点击页面上的文本实现三个页面之间的切换，如图8-64～图8-66所示。

图8-64 NavigatorExample.ets 页面效果　　图8-65 DetailExample.ets 页面效果　　图8-66 BackExample.ets 页面效果

8.5.11 Panel

Panel是可滑动面板，提供一种轻量的内容展示窗口，方便在不同尺寸中切换。

Panel示例如下：

```
Panel(true) { // 展示日程
  Column() {
    Text('今日历程').fontSize(20)

    Text('1、Java 核心编程').fontSize(20)
    Text('2、轻量级 Java EE 企业应用开发实战').fontSize(20)
    Text('3、鸿蒙 HarmonyOS 手机应用开发实战').fontSize(20)
    Text('4、Node.js+Express+MongoDB+Vue.js 全栈开发实战').fontSize(20)
  }
}
.type(PanelType.Foldable).mode(PanelMode.Half)
.dragBar(true)   // 默认开启
.halfHeight(500) // 默认一半
```

上述示例指定了PanelMode.Half状态下的窗口高度，默认为屏幕尺寸的一半。界面效果如图8-67所示。

8.5.12 Refresh

Refresh是可以进行页面下拉操作并显示刷新动效的容器组件，主要包含以下参数。

- refreshing：当前组件是否正在刷新。该参数支持$$双向绑定变量。
- offset：刷新组件静止时距离父组件顶部的距离。默认值是16，单位是vp。
- friction：下拉摩擦系数，取值范围为0到100，默认值是62。

图8-67 Panel效果

Refresh示例如下：

```
Refresh({ refreshing: true,   // 当前组件是否正在刷新
  offset: 120,                // 新组件静止时距离父组件顶部的距离
  friction: 100 }) {          // 下拉摩擦系数，取值范围为 0 到 100，默认值是 62
  Text(' 下拉刷新 ')
    .fontSize(30)
    .margin(10)
}
```

上述示例界面效果如图8-68所示。

图8-68 Refresh效果

8.5.13 RelativeContainer

RelativeContainer是相对布局组件，用于复杂场景中元素对齐的布局。容器内子组件区分水平方向、垂直方向，具体如下。

- 水平方向为left、middle、right，对应容器的HorizontalAlign.Start、HorizontalAlign.Center、HorizontalAlign.End。
- 垂直方向为top、center、bottom，对应容器的VerticalAlign.Top、VerticalAlign.Center、VerticalAlign.Bottom。

RelativeContainer示例如下：

```
RelativeContainer() {
  Row()
    .width(100)
    .height(100)
    .backgroundColor('#FF3333')
    .alignRules({
      top: { anchor: '__container__', align: VerticalAlign.Top },  // 以父容器为
锚点，竖直方向顶头对齐
      middle: { anchor: '__container__', align: HorizontalAlign.Center }  //
```

```
以父容器为锚点，水平方向居中对齐
    })
    .id('row1')  // 设置锚点为 row1

  Row() {
    Image($r('app.media.icon'))
  }
  .height(100).width(100)
  .alignRules({
    top: { anchor: 'row1', align: VerticalAlign.Bottom },  // 以 row1 组件为锚点，
竖直方向底端对齐
    left: { anchor: 'row1', align: HorizontalAlign.Start }  // 以 row1 组件为锚点，
水平方向开头对齐
  })
  .id('row2')  // 设置锚点为 row2

  Row()
    .width(100)
    .height(100)
    .backgroundColor('#FFCC00')
    .alignRules({
      top: { anchor: 'row2', align: VerticalAlign.Top }
    })
    .id('row3')  // 设置锚点为 row3

  Row()
    .width(100)
    .height(100)
    .backgroundColor('#FF9966')
    .alignRules({
      top: { anchor: 'row2', align: VerticalAlign.Top },
      left: { anchor: 'row2', align: HorizontalAlign.End },
    })
    .id('row4')  // 设置锚点为 row4

  Row()
    .width(100)
    .height(100)
    .backgroundColor('#FF66FF')
    .alignRules({
      top: { anchor: 'row2', align: VerticalAlign.Bottom },
      middle: { anchor: 'row2', align: HorizontalAlign.Center }
    })
    .id('row5')  // 设置锚点为 row5
}
.width(300).height(300)
.border({ width: 2, color: '#6699FF' })
```

在上述示例中，子组件可以将容器或其他子组件设为锚点，参与相对布局的容器内组件必须设置id，不设置id的组件不显示，容器id固定为“__container__”。界面效果如图8-69所示。

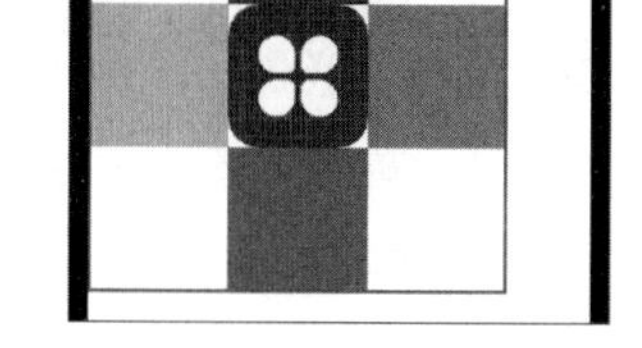

图8-69　RelativeContainer效果

8.5.14　Scroll

Scroll是可滚动的容器组件，当子组件的布局尺寸超过父组件的尺寸时，内容可以滚动。

Scroll示例如下：

```
// 与 Scroller 绑定
Scroll(new Scroller()) {
  Column() {
    ForEach(this.numberArray, (item) => {
      Text(item.toString())
        .width('90%')
        .height(250)
        .backgroundColor(0xFFFFFF)
        .borderRadius(15)
        .fontSize(26)
        .textAlign(TextAlign.Center)
        .margin({ top: 10 })
    }, item => item)
  }.width('100%')
}
.scrollable(ScrollDirection.Vertical)  // 滚动方向纵向
.scrollBar(BarState.On)       // 滚动条常驻显示
.scrollBarColor(Color.Gray)  // 滚动条颜色
.scrollBarWidth(40) // 滚动条宽度
.edgeEffect(EdgeEffect.None)
.onScroll((xOffset: number, yOffset: number) => {
  console.info(xOffset + ' ' + yOffset)
})
.onScrollEdge((side: Edge) => {
  console.info('To the edge')
})
.onScrollEnd(() => {
  console.info('Scroll Stop')
}).backgroundColor(0xDCDCDC)
```

在上述示例中，Scroll与Scroller进行了绑定，以此控制容器组件的滚动。界面效果如图8-70所示。

8.5.15 SideBarContainer

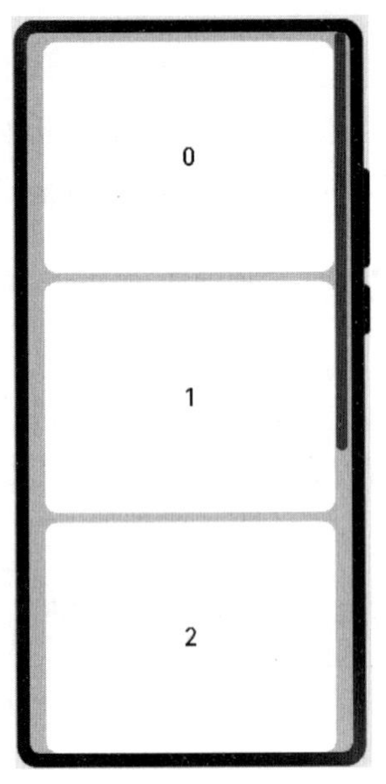

图 8-70 Scroll 效果

SideBarContainer是提供侧边栏可以显示和隐藏的侧边栏容器，通过子组件定义侧边栏和内容区，第一个子组件表示侧边栏，第二个子组件表示内容区。

SideBarContainer示例如下：

```
SideBarContainer(SideBarContainerType.Embed) {
  Column() {
    Text('菜单 1').fontSize(25)
    Text('菜单 2').fontSize(25)
  }.width('100%')
  .justifyContent(FlexAlign.SpaceEvenly)
  .backgroundColor('#19000000')

  Column() {
    Text('内容 1').fontSize(25)
    Text('内容 2').fontSize(25)
  }
}
```

上述示例界面效果如图 8-71 所示。

点击左上角菜单可以显示侧边栏，界面效果如图 8-72 所示。

图 8-71 SideBarContainer 效果

图 8-72 SideBarContainer 显示侧边栏效果

8.5.16 Stack

Stack是堆叠容器，子组件按照顺序依次入栈，后一个子组件覆盖前一个子组件。

Stack示例如下：

```
// 第一层组件
Text('第一层')
  .width('90%')
  .height('100%')
  .backgroundColor(Color.Grey)
  .align(Alignment.Top)
  .fontSize(40)

// 第二层组件
Text('第二层')
  .width('70%')
  .height('60%')
  .backgroundColor(Color.Orange)
  .align(Alignment.Top)
  .fontSize(40)
}.width('100%').height(400).margin({ top: 5 })
```

在上述示例中，第二层组件盖在了第一层组件上面，界面效果如图8-73所示。

图8-73 Stack效果

8.5.17 Swiper

Swiper为滑块视图容器，提供子组件滑动轮播显示的能力。

Swiper示例如下：

```
Swiper() {
  Image($r('app.media.book01'))
    .width(280).height(380)
  Image($r('app.media.book02'))
    .width(280).height(380)
  Image($r('app.media.book03'))
    .width(280).height(380)
  Image($r('app.media.book04'))
    .width(280).height(380)
}
.cachedCount(2)      // 设置预加载子组件个数
.index(1)            // 设置当前在容器中显示的子组件的索引值
.autoPlay(true)      // 子组件是否自动播放，自动播放状态下，导航点不可操作
.interval(4000)      // 使用自动播放时播放的时间间隔，单位为毫秒
```

```
.indicator(true) // 是否启用导航点指示器
.loop(true) // 是否开启循环
.duration(1000) // 子组件切换的动画时长，单位为毫秒
.itemSpace(0) // 设置子组件与子组件之间间隙
.curve(Curve.Linear) // 设置 Swiper 的动画曲线
```

上述示例界面效果如图8-74所示，会自动播放图片。

图 8-74 Swiper 效果

8.5.18 Tabs和TabContent

Tabs是通过页签进行内容视图切换的容器组件，每个页签对应一个内容视图 TabContent。

Tabs主要包括以下三个参数。

- barPosition：设置Tabs的页签位置。默认值是BarPosition.Start。
- index：设置初始页签索引。默认值是0。
- controller：设置Tabs控制器。

Tabs示例如下：

```
Tabs({ barPosition: BarPosition.Start, // 设置 Tabs 的页签位置
    controller: new TabsController() // 设置 Tabs 控制器
  }) {
    TabContent() {
      Column().width('100%').height('100%').backgroundColor(Color.Orange)
    }.tabBar(' 首页 ')

    TabContent() {
      Column().width('100%').height('100%').backgroundColor(Color.Blue)
    }.tabBar(' 商城 ')

    TabContent() {
      Column().width('100%').height('100%').backgroundColor(Color.Red)
    }.tabBar(' 直播 ')
  }
  .vertical(false) // 设置为 false 是为横向 Tabs，设置为 true 时为纵向 Tabs
  .barMode(BarMode.Fixed) // TabBar 布局模式
  .barWidth(360) // TabBar 的宽度值
  .barHeight(56) // TabBar 的高度值
  .animationDuration(400) //TabContent 滑动动画时长
  .width(360)
  .height(296)
  .margin({ top: 52 })
```

上述示例界面效果如图8-75所示。

8.6 媒体组件详解

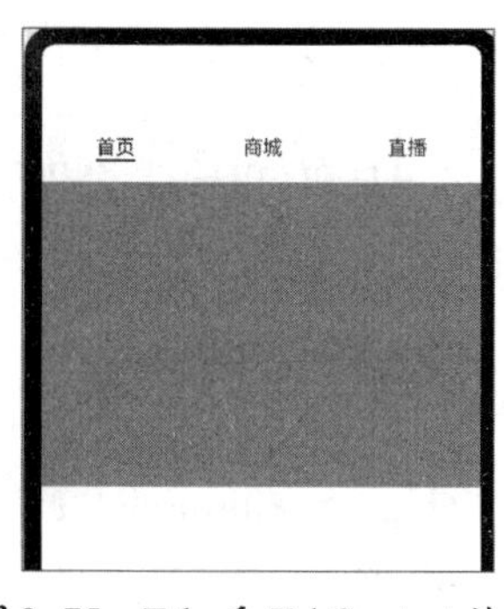

图 8-75　Tabs 和 TabContent 效果

声明式开发范式目前可供选择的媒体组件只有Video。

本节演示如何使用Video组件。相关示例，可以在ArkUI MediaComponents应用中找到。

Video是用于播放视频文件并控制其播放状态的组件。如果使用网络视频，则需要申请权限ohos.permission.INTERNET。

Video的参数主要有以下几个。

- src：视频播放源的路径，支持本地视频路径和网络路径。支持在resources下面的video或rawfile文件夹里放置媒体资源。支持dataability://的路径前缀，用于访问通过Data Ability提供的视频路径。视频支持的格式有MP4、MKV、WebM、TS。
- currentProgressRate：视频播放倍速。取值仅支持0.75，1.0，1.25，1.75，2.0。
- previewUri：视频未播放时的预览图片路径。
- controller：设置视频控制器。

Video示例如下：

```
Video({
  src: $rawfile('video_11.mp4'),        // 视频播放源的路径
  previewUri: $r('app.media.book01'), // 视频未播放时的预览图片路径
  currentProgressRate: 0.75, // 视频播放倍速
  controller: new VideoController()
}).width(400).height(600)
  .autoPlay(true) // 自动播放
  .controls(true) // 显示视频控制器
```

上述示例界面效果如图8-76所示。

8.7 绘制组件详解

声明式开发范式目前可供选择的绘制组件有Circle、Ellipse、Line、Polyline、Polygon、Path、Rect、Shape等。

本节演示如何使用绘制组件。相关示例可以在ArkUIDrawing Components应用中找到。

图 8-76　Video 组件效果

8.7.1 Circle 和 Ellipse

Circle 和 Ellipse 是分别用于绘制圆形和椭圆的组件。

Circle 和 Ellipse 的参数如下。

- width：宽度。
- height：高度。

Circle 和 Ellipse 的参数属性如下。

- fill：设置填充区域颜色。默认值是 Color.Black。
- fillOpacity：设置填充区域透明度。默认值是 1。
- stroke：设置边框颜色。不设置时，默认没有边框。
- strokeDashArray：设置边框间隙。默认值是 []。
- strokeDashOffset：边框绘制起点的偏移量。默认值是 0。
- strokeLineCap：设置边框端点绘制样式。默认值是 LineCapStyle.Butt。
- strokeLineJoin：设置边框拐角绘制样式。默认值是 LineJoinStyle.Miter。
- strokeMiterLimit：设置斜接长度与边框宽度比值的极限值。默认值是 4。
- strokeOpacity：设置边框透明度。默认值是 1。
- strokeWidth：设置边框宽度。默认值是 1。
- antiAlias：是否开启抗锯齿效果。默认值是 true。

Circle 和 Ellipse 示例如下：

```
// 绘制一个直径为 150 的圆
Circle({ width: 150, height: 150 })

// 绘制一个直径为 150、线条为红色虚线的圆环（宽高设置不一致时以短边为直径）
Circle()
  .width(150)
  .height(200)
  .fillOpacity(0)      // 设置填充区域透明度
  .strokeWidth(3)      //设置边框宽度
  .stroke(Color.Red)   //设置边框颜色
  .strokeDashArray([1, 2]) //设置边框间隙

// 绘制一个 150 * 80 的椭圆
Ellipse({ width: 150, height: 50 })

// 绘制一个 150 * 100 、线条为蓝色的椭圆环
Ellipse()
  .width(150)
  .height(50)
  .fillOpacity(0)      // 设置填充区域透明度
```

```
  .strokeWidth(3)           //设置边框宽度
  .stroke(Color.Red)        //设置边框颜色
  .strokeDashArray([1, 2]) //设置边框间隙
```

上述示例界面效果如图8-77所示。

Circle和Ellipse的最为重要的区别在于，即便Circle的width、height设置的内容不一样，仍然会以两者最短的边为直径。

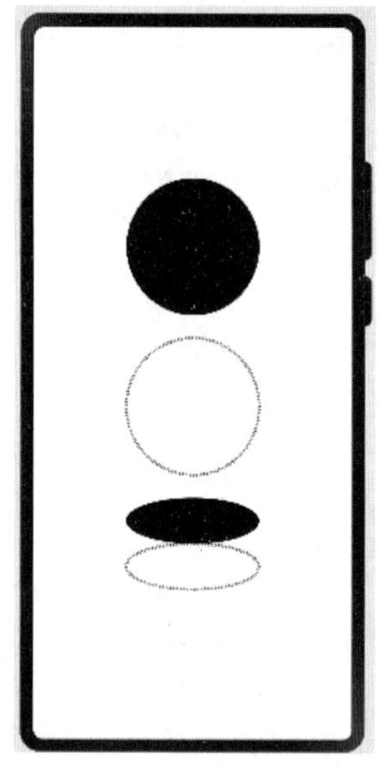

图8-77 Circle和Ellipse效果

8.7.2 Line

Line是用于绘制直线的组件。

Line的参数如下。

- width：宽度。
- height：高度。

Line的参数属性如下。

- startPoint：直线起点坐标点（相对坐标），单位是vp。
- endPoint：直线终点坐标点（相对坐标），单位是vp。
- fill：设置填充区域颜色。默认值是Color.Black。
- fillOpacity：设置填充区域透明度。默认值是1。
- stroke：设置边框颜色。不设置时，默认没有边框。
- strokeDashArray：设置边框间隙。默认值是[]。
- strokeDashOffset：边框绘制起点的偏移量。默认值是0。
- strokeLineCap：设置边框端点绘制样式。默认值是LineCapStyle.Butt。
- strokeLineJoin：设置边框拐角绘制样式。默认值是LineJoinStyle.Miter。
- strokeMiterLimit：设置斜接长度与边框宽度比值的极限值。默认值是4。
- strokeOpacity：设置边框透明度。默认值是1。
- strokeWidth：设置边框宽度。默认值是1。
- antiAlias：是否开启抗锯齿效果。默认值是true。

Line示例如下：

```
// 线条绘制的起止点坐标均是相对于Line组件本身绘制区域的坐标
Line()
  .startPoint([0, 0])
  .endPoint([50, 100])
  .backgroundColor('#F5F5F5')
Line()
  .width(200)
  .height(200)
  .startPoint([50, 50])
```

```
  .endPoint([150, 150])
  .strokeWidth(5)
  .stroke(Color.Orange)
  .strokeOpacity(0.5)
  .backgroundColor('#F5F5F5')

// 当坐标点设置的值超出 Line 组件的宽高范围时，线条会画出组件绘制区域
Line({ width: 50, height: 50 })
  .startPoint([0, 0])
  .endPoint([100, 100])
  .strokeWidth(3)
  .strokeDashArray([10, 3])
  .backgroundColor('#F5F5F5')

// strokeDashOffset 用于定义关联虚线 strokeDashArray 数组渲染时的偏移
Line({ width: 50, height: 50 })
  .startPoint([0, 0])
  .endPoint([100, 100])
  .strokeWidth(3)
  .strokeDashArray([10, 3])
  .strokeDashOffset(5)
  .backgroundColor('#F5F5F5')
```

上述示例界面效果如图 8-78 所示。

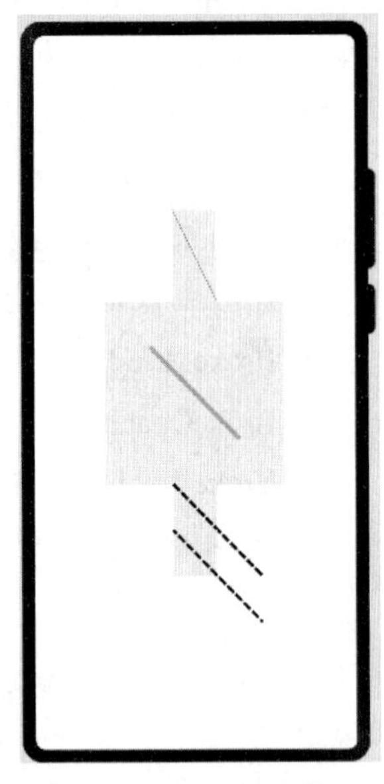

图 8-78　Line 效果

8.7.3　Polyline

Polyline 是用于绘制折线直线的组件。

Polyline 的参数如下。

- width：宽度。
- height：高度。

Polyline 的参数属性如下。

- points：折线经过坐标点列表。
- fill：设置填充区域颜色。默认值是 Color.Black。
- fillOpacity：设置填充区域透明度。默认值是 1。
- stroke：设置边框颜色。不设置时，默认没有边框。
- strokeDashArray：设置边框间隙。默认值是 []。
- strokeDashOffset：边框绘制起点的偏移量。默认值是 0。
- strokeLineCap：设置边框端点绘制样式。默认值是 LineCapStyle.Butt。
- strokeLineJoin：设置边框拐角绘制样式。默认值是 LineJoinStyle.Miter。
- strokeMiterLimit：设置斜接长度与边框宽度比值的极限值。默认值是 4。

- strokeOpacity：设置边框透明度。默认值是1。
- strokeWidth：设置边框宽度。默认值是1。
- antiAlias：是否开启抗锯齿效果。默认值是true。

Polyline示例如下：

```
// 在 100 * 100 的矩形框中绘制一段折线，起点 (0, 0)，经过 (20,60)，到达终点 (100, 100)
Polyline({ width: 100, height: 100 })
  .points([[0, 0], [20, 60], [100, 100]])
  .fillOpacity(0)
  .stroke(Color.Blue)
  .strokeWidth(3)

// 在 100 * 100 的矩形框中绘制一段折线，起点 (20, 0)，经过 (0,100)，到达终点 (100, 90)
Polyline()
  .width(100)
  .height(100)
  .fillOpacity(0)
  .stroke(Color.Red)
  .strokeWidth(8)
  .points([[20, 0], [0, 100], [100, 90]])
    // 设置折线拐角处为圆弧
  .strokeLineJoin(LineJoinStyle.Round)
    // 设置折线两端为半圆
  .strokeLineCap(LineCapStyle.Round)
```

上述示例界面效果如图8-79所示。

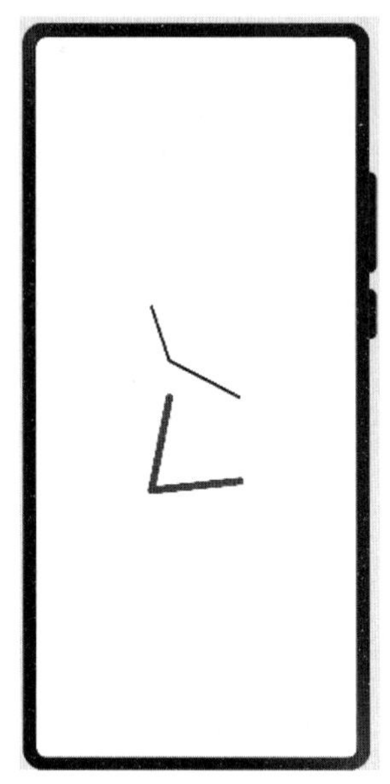

图8-79　Polyline效果

8.7.4 Polygon

Polygon是用于绘制多边形的组件。

Polygon的参数如下。

- width：宽度。
- height：高度。

Polygon的参数属性如下。

- points：折线经过坐标点列表。
- fill：设置填充区域颜色。默认值是Color.Black。
- fillOpacity：设置填充区域透明度。默认值是1。
- stroke：设置边框颜色。不设置时，默认没有边框。
- strokeDashArray：设置边框间隙。默认值是[]。
- strokeDashOffset：边框绘制起点的偏移量。默认值是0。
- strokeLineCap：设置边框端点绘制样式。默认值是LineCapStyle.Butt。

- strokeLineJoin：设置边框拐角绘制样式。默认值是 LineJoinStyle.Miter。
- strokeMiterLimit：设置斜接长度与边框宽度比值的极限值。默认值是 4。
- strokeOpacity：设置边框透明度。默认值是 1。
- strokeWidth：设置边框宽度。默认值是 1。
- antiAlias：是否开启抗锯齿效果。默认值是 true。

Polygon 示例如下：

```
// 在 100 * 100 的矩形框中绘制一个三角形，起点 (0, 0)，经过 (50, 100)，终点 (100, 0)
Polygon({ width: 100, height: 100 })
  .points([[0, 0], [50, 100], [100, 0]])
  .fill(Color.Green)
  .stroke(Color.Transparent)

// 在 100 * 100 的矩形框中绘制一个四边形，起点 (0, 0)，经过 (0, 100) 和 (100, 100)，终点 (100, 0)
Polygon()
  .width(100)
  .height(100)
  .points([[0, 0], [0, 100], [100, 100], [100, 0]])
  .fillOpacity(0)
  .strokeWidth(5)
  .stroke(Color.Blue)

// 在 100 * 100 的矩形框中绘制一个五边形,起点 (50, 0),依次经过 (0, 50)、(20, 100) 和 (80, 100)，终点 (100, 50)
Polygon()
  .width(100)
  .height(100)
  .points([[50, 0], [0, 50], [20, 100], [80, 100], [100, 50]])
  .fill(Color.Red)
  .fillOpacity(0.6)
  .stroke(Color.Transparent)
```

上述示例界面效果如图 8-80 所示。

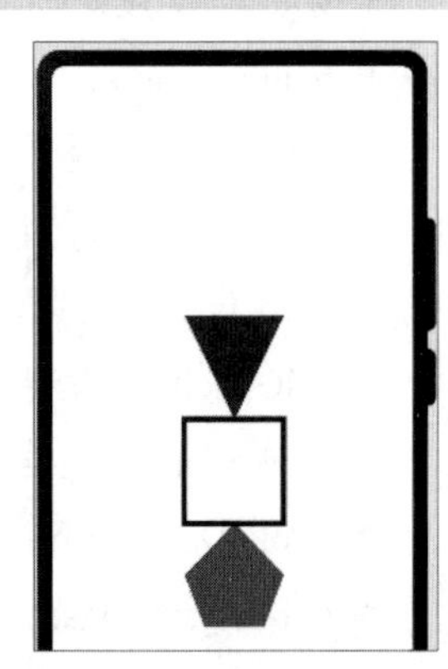

图 8-80 Polygon 效果

8.7.5 Path

Path 是根据绘制路径生成封闭的自定义形状。

Path 的参数如下。

- width：宽度。
- height：高度。
- commands：路径绘制的命令字符串。默认值是"（两个单引号）。

Path的参数属性如下。

- commands：路径绘制的命令字符串，单位为px。
- fill：设置填充区域颜色。默认值是Color.Black。
- fillOpacity：设置填充区域透明度。默认值是1。
- stroke：设置边框颜色。不设置时，默认没有边框。
- strokeDashArray：设置边框间隙。默认值是[]。
- strokeDashOffset：边框绘制起点的偏移量。默认值是0。
- strokeLineCap：设置边框端点绘制样式。默认值是LineCapStyle.Butt。
- strokeLineJoin：设置边框拐角绘制样式。默认值是LineJoinStyle.Miter。
- strokeMiterLimit：设置斜接长度与边框宽度比值的极限值。默认值是4。
- strokeOpacity：设置边框透明度。默认值是1。
- strokeWidth：设置边框宽度。默认值是1。
- antiAlias：是否开启抗锯齿效果。默认值是true。

commands支持的绘制命令如下。

- M：在给定的 (x, y) 坐标处开始一个新的子路径。例如，“M 0 0”表示将（0, 0）点作为新子路径的起始点。
- L：从当前点到给定的 (x, y) 坐标画一条线，该坐标成为新的当前点。例如，“L 50 50”表示绘制当前点到（50, 50）点的直线，并将（50, 50）点作为新子路径的起始点。
- H：从当前点绘制一条水平线，等效于将y坐标指定为0的L命令。例如，“H 50”表示绘制当前点到（50, 0）点的直线，并将（50, 0）点作为新子路径的起始点。
- V：从当前点绘制一条垂直线，等效于将x坐标指定为0的L命令。例如，“V 50”表示绘制当前点到（0, 50）点的直线，并将（0, 50）点作为新子路径的起始点。
- C：使用 (x1, y1) 作为曲线起点的控制点，(x2, y2) 作为曲线终点的控制点，从当前点到 (x, y) 绘制三次贝塞尔曲线。例如，“C100 100 250 100 250 200”表示绘制当前点到（250, 200）点的三次贝塞尔曲线，并将（250, 200）点作为新子路径的起始点。
- S：(x2, y2) 作为曲线终点的控制点，绘制从当前点到 (x, y) 绘制三次贝塞尔曲线。若前一个命令是C或S，则起点控制点是上一个命令的终点控制点相对于起点的映射。例如，“C100 100 250 100 250 200 S400 300 400 200”第二段贝塞尔曲线的起点控制点为（250, 300）。如果没有前一个命令，或前一个命令不是C或S，则第一个控制点与当前点重合。
- Q：使用 (x1, y1) 作为控制点，从当前点到 (x, y) 绘制二次贝塞尔曲线。例如，“Q400 50 600 300”表示绘制当前点到（600, 300）点的二次贝塞尔曲线，并将（600, 300）点作为新子路径的起始点。
- T：绘制从当前点到 (x, y) 绘制二次贝塞尔曲线。若前一个命令是Q或T，则控制点是上一个命令的终点控制点相对于起点的映射。例如，“Q400 50 600 300 T1000 300”第二段贝塞尔曲线的控制点为（800, 350）。如果没有前一个命令，或前一个命令不是Q或T，则第一个控制点与当前点重合。

- A：从当前点到 (x, y) 绘制一条椭圆弧。椭圆的大小和方向由两个半径 (rx, ry) 和x-axis-rotation定义，指示整个椭圆相对于当前坐标系如何旋转（以度为单位）。 large-arc-flag 和 sweep-flag确定弧的绘制方式。
- Z：通过将当前路径连接回当前子路径的初始点来关闭当前子路径。

Path示例如下：

```
// 绘制一条长 900px，宽 3vp 的直线
Path()
  .height(10)
  .commands('M0 0 L600 0')
  .stroke(Color.Black)
  .strokeWidth(3)

// 绘制直线图形
Path()
  .commands('M100 0 L200 240 L0 240 Z')
  .fillOpacity(0)
  .stroke(Color.Black)
  .strokeWidth(3)
Path()
  .commands('M0 0 H200 V200 H0 Z')
  .fillOpacity(0)
  .stroke(Color.Black)
  .strokeWidth(3)
Path()
  .commands('M100 0 L0 100 L50 200 L150 200 L200 100 Z')
  .fillOpacity(0)
  .stroke(Color.Black)
  .strokeWidth(3)

// 绘制弧线图形
Path()
  .commands("M0 300 S100 0 240 300 Z")
  .fillOpacity(0)
  .stroke(Color.Black)
  .strokeWidth(3)
Path()
  .commands('M0 150 C0 100 140 0 200 150 L100 300 Z')
  .fillOpacity(0)
  .stroke(Color.Black)
  .strokeWidth(3)
Path()
  .commands('M0 100 A30 20 20 0 0 200 100 Z')
  .fillOpacity(0)
  .stroke(Color.Black)
```

```
  .strokeWidth(3)
```

上述示例界面效果如图8-81所示。

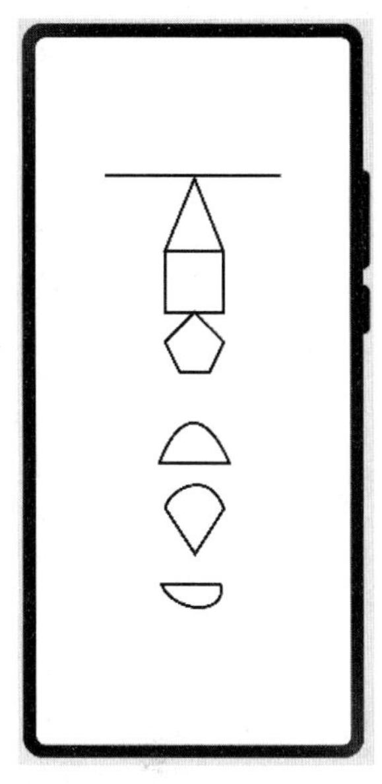

图8-81 Path效果

8.7.6 Rect

Rect是绘制矩形的组件。

Rect的参数如下。

- width：宽度。
- height：高度。
- radius：圆角半径，支持分别设置四个角的圆角度数。
- radiusWidth：圆角宽度。
- radiusHeight：圆角高度。

Rect的参数属性如下。

- radiusWidth：圆角的宽度。仅设置宽时，宽高一致。
- radiusHeight：圆角的高度。仅设置高时，宽高一致。
- radius：圆角半径大小。
- fill：设置填充区域颜色。默认值是Color.Black。
- fillOpacity：设置填充区域透明度。默认值是1。
- stroke：设置边框颜色。不设置时，默认没有边框。
- strokeDashArray：设置边框间隙。默认值是[]。
- strokeDashOffset：边框绘制起点的偏移量。默认值是0。
- strokeLineCap：设置边框端点绘制样式。默认值是LineCapStyle.Butt。
- strokeLineJoin：设置边框拐角绘制样式。默认值是LineJoinStyle.Miter。
- strokeMiterLimit：设置斜接长度与边框宽度比值的极限值。默认值是4。
- strokeOpacity：设置边框透明度。默认值是1。
- strokeWidth：设置边框宽度。默认值是1。
- antiAlias：是否开启抗锯齿效果。默认值是true。

Rect示例如下：

```
// 绘制 90% * 50 矩形
Rect({ width: '90%', height: 50 })
  .fill(Color.Pink)
  .stroke(Color.Transparent)

// 绘制 90% * 50 的矩形框
Rect()
  .width('90%')
```

```
  .height(50)
  .fillOpacity(0)
  .stroke(Color.Red)
  .strokeWidth(3)

// 绘制 90% * 80 的矩形，圆角宽高分别为 40、20
Rect({ width: '90%', height: 80 })
  .radiusHeight(20)
  .radiusWidth(40)
  .fill(Color.Pink)
  .stroke(Color.Transparent)

// 绘制 90% * 80 的矩形，圆角宽高为 20
Rect({ width: '90%', height: 80 })
  .radius(20)
  .fill(Color.Pink)
  .stroke(Color.Transparent)

// 绘制 90% * 50 矩形，左上圆角宽高 40，右上圆角宽高 20，右下圆角宽高 40，左下圆角宽高 20
Rect({ width: '90%', height: 80 })
  .radius([[40, 40], [20, 20], [40, 40], [20, 20]])
  .fill(Color.Pink)
  .stroke(Color.Transparent)
```

上述示例界面效果如图8-82所示。

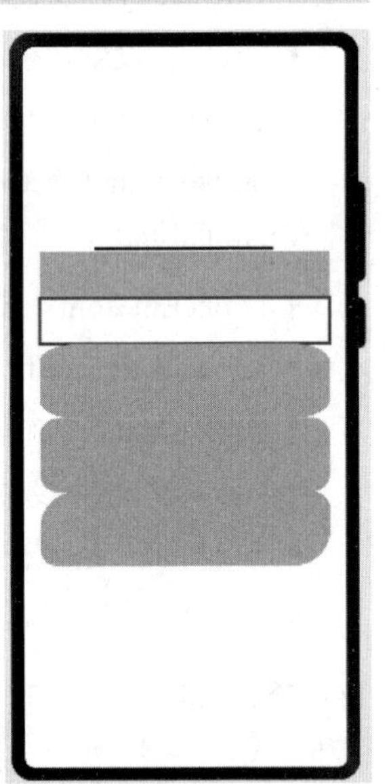

图8-82 Rect效果

8.7.7 Shape

Shape是绘制组件的父组件，父组件中会描述所有绘制组件均支持的通用属性。

- 绘制组件使用Shape作为父组件时，实现类似SVG的效果。
- 绘制组件单独使用时，用于在页面上绘制指定的图形。

Shape的参数为value，可将图形绘制在指定的PixelMap对象中；若未设置，则在当前绘制目标中进行绘制。

Shape的参数属性如下。

- viewPort：形状的视口。
- fill：设置填充区域颜色。默认值是Color.Black。
- fillOpacity：设置填充区域透明度。默认值是1。
- stroke：设置边框颜色。不设置时，默认没有边框。
- strokeDashArray：设置边框间隙。默认值是[]。
- strokeDashOffset：边框绘制起点的偏移量。默认值是0。

- strokeLineCap：设置边框端点绘制样式。默认值是 LineCapStyle.Butt。
- strokeLineJoin：设置边框拐角绘制样式。默认值是 LineJoinStyle.Miter。
- strokeMiterLimit：设置斜接长度与边框宽度比值的极限值。默认值是4。
- strokeOpacity：设置边框透明度。默认值是1。
- strokeWidth：设置边框宽度。默认值是1。
- antiAlias：是否开启抗锯齿效果。默认值是true。
- mesh：设置mesh效果。第一个参数为长度(column + 1) × (row + 1) × 2的数组，它记录扭曲后的位图各个顶点位置；第二个参数为mesh矩阵列数column；第三个参数为mesh矩阵行数row。

Shape示例如下：

```
// 在Shape的(-2, 118)点绘制一个 300 * 10 直线路径，颜色0x317AF7，边框颜色黑色，宽度4，间隙20，向左偏移10，线条两端样式为半圆，拐角样式圆角，抗锯齿(默认开启)
Shape() {
  Rect().width(300).height(50)
  Ellipse().width(300).height(50).offset({ x: 0, y: 60 })
  Path().width(300).height(10).commands('M0 0 L900 0').offset({ x: 0, y: 120
})
}
.viewPort({ x: -2, y: -2, width: 304, height: 130 })
.fill(0x317AF7)
.stroke(Color.Black)
.strokeWidth(4)
.strokeDashArray([20])
.strokeDashOffset(10)
.strokeLineCap(LineCapStyle.Round)
.strokeLineJoin(LineJoinStyle.Round)
.antiAlias(true)

// 分别在Shape的(0, 0)、(-5, -5)点绘制一个 300 * 50 带边框的矩形，可以看出之所以将视口的起始位置坐标设为负值，是因为绘制的起点默认为线宽的中点位置，因此要让边框完全显示则需要让视口偏移半个线宽
Shape() {
  Rect().width(300).height(50)
}
.viewPort({ x: 0, y: 0, width: 320, height: 70 })
.fill(0x317AF7)
.stroke(Color.Black)
.strokeWidth(10)

// 在Shape的(0, -5)点绘制一条直线路径，颜色0xEE8443，线条宽度10，线条间隙20
Shape() {
  Path().width(300).height(10).commands('M0 0 L900 0')
}
```

```
.viewPort({ x: 0, y: -5, width: 300, height: 20 })
.stroke(0xEE8443)
.strokeWidth(10)
.strokeDashArray([20])

// 在 Shape 的 (0, -5) 点绘制一条直线路径，颜色 0xEE8443，线条宽度 10，线条间隙 20，向左偏移
10
Shape() {
  Path().width(300).height(10).commands('M0 0 L900 0')
}
.viewPort({ x: 0, y: -5, width: 300, height: 20 })
.stroke(0xEE8443)
.strokeWidth(10)
.strokeDashArray([20])
.strokeDashOffset(10)

// 在 Shape 的 (0, -5) 点绘制一条直线路径，颜色 0xEE8443，线条宽度 10，透明度 0.5
Shape() {
  Path().width(300).height(10).commands('M0 0 L900 0')
}
.viewPort({ x: 0, y: -5, width: 300, height: 20 })
.stroke(0xEE8443)
.strokeWidth(10)
.strokeOpacity(0.5)

// 在 Shape 的 (0, -5) 点绘制一条直线路径，颜色 0xEE8443，线条宽度 10，线条间隙 20，线条两端
样式为半圆
Shape() {
  Path().width(300).height(10).commands('M0 0 L900 0')
}
.viewPort({ x: 0, y: -5, width: 300, height: 20 })
.stroke(0xEE8443)
.strokeWidth(10)
.strokeDashArray([20])
.strokeLineCap(LineCapStyle.Round)

// 在 Shape 的 (-80, -5) 点绘制一个封闭路径，颜色 0x317AF7，线条宽度 10，边框颜色
0xEE8443，拐角样式锐角（默认值）
Shape() {
  Path().width(200).height(60).commands('M0 0 L400 0 L400 150 Z')
}
.viewPort({ x: -80, y: -5, width: 310, height: 90 })
.fill(0x317AF7)
.stroke(0xEE8443)
.strokeWidth(10)
.strokeLineJoin(LineJoinStyle.Miter)
```

```
.strokeMiterLimit(5)
```

上述示例界面效果如图8-83所示。

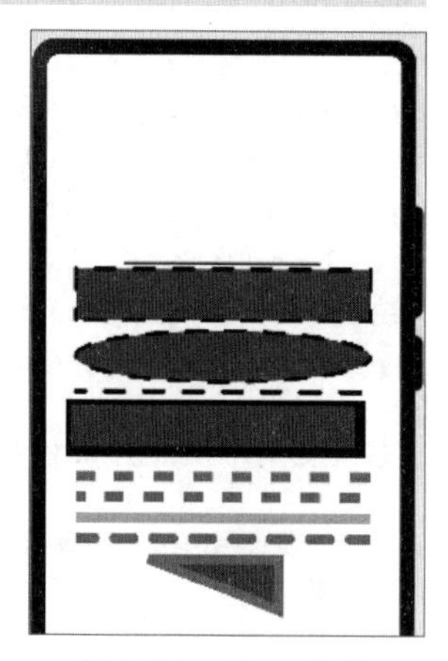

图 8-83　Shape 效果

8.8 画布组件详解

声明式开发范式目前可供选择的画布组件为Canvas。与Canvas配合使用的还有CanvasRenderingContext2D、CanvasGradient、ImageBitmap、ImageData、OffscreenCanvasRenderingContext2D、Path2D等对象。

以下是Canvas示例：

```
private renderingContextSettings: RenderingContextSettings = new
RenderingContextSettings(true)

// 使用 RenderingContext 在 Canvas 组件上进行绘制，绘制对象可以是矩形、文本、图片等
private canvasRenderingContext2D: CanvasRenderingContext2D = new
CanvasRenderingContext2D(this.renderingContextSettings)


Canvas(this.canvasRenderingContext2D)
  .width('100%')
  .height('100%')
  // onReady 是 Canvas 组件初始化完成时的事件回调，该事件之后 Canvas 组件宽高确定且可获取
  .onReady(() => {
    // 绘制矩形
    this.canvasRenderingContext2D.fillRect(0, 30, 100, 100)
  })
```

上述示例通过CanvasRenderingContext2D实例化一个Canvas，而后通过CanvasRenderingContext2D的fillRect绘制了一个矩形。

上述示例界面效果如图8-84所示。

图 8-84　Canvas绘制矩形效果

以下是通过CanvasRenderingContext2D绘制贝赛尔曲线的示例：

```
// 绘制贝赛尔曲线。
this.canvasRenderingContext2D.beginPath()
this.canvasRenderingContext2D.moveTo(170, 10)
this.canvasRenderingContext2D.bezierCurveTo(20, 100, 200, 100, 200, 20)
this.canvasRenderingContext2D.stroke()
```

上述示例通过CanvasRenderingContext2D绘制了贝赛尔曲线，界面效果如图8-85所示。

以下是通过Canvas绘制渐变对象的示例：

```
// 绘制渐变对象。
var grad = this.canvasRenderingContext2D.createLinearGradient(150, 0, 300, 100)
grad.addColorStop(0.0, 'red')
grad.addColorStop(0.5, 'white')
grad.addColorStop(1.0, 'green')
this.canvasRenderingContext2D.fillStyle = grad
this.canvasRenderingContext2D.fillRect(200, 0, 100, 100)
```

上述示例通过CanvasRenderingContext2D来绘制渐变对象，界面效果如图8-85所示。

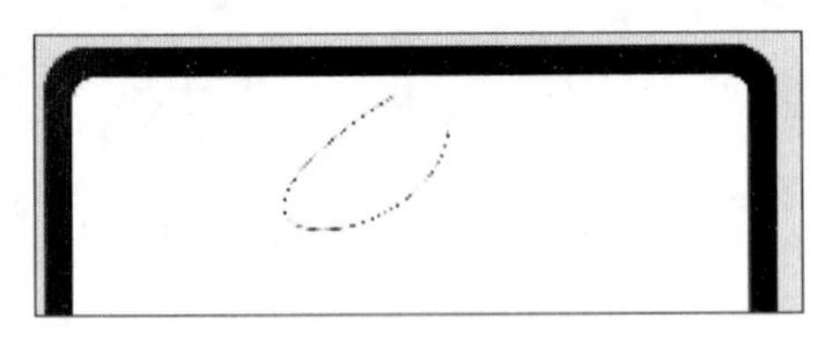

图8-85　通过CanvasRenderingContext2D绘制贝赛尔曲线效果

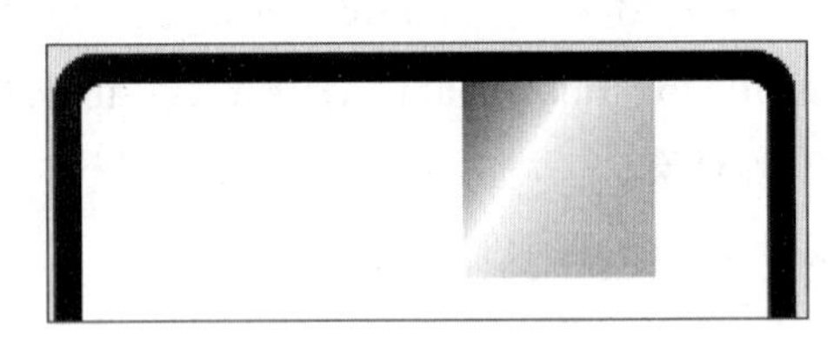

图8-86　通过CanvasRenderingContext2D绘制渐变对象效果

8.9　常用布局

ArkUI常用布局主要分为两大类：自适应布局和响应式布局。

8.9.1　自适应布局

自适应布局包含以下4类。

1. 线性布局

线性布局（LinearLayout）是开发中最常用的布局。线性布局的子组件在线性方向上（水平方向和垂直方向）依次排列。

通过线性容器Row和Column实现线性布局。Column容器内子组件按照垂直方向排列；Row容器内子组件按照水平方向排列。

线性布局的排列方向由所选容器组件决定。根据不同的排列方向，选择使用Row或Column容器创建线性布局，通过调整space、alignItems、justifyContent属性调整子组件的间距、水平和垂直方向的对齐方式。

- 通过space参数设置主轴（排列方向）上子组件的间距，达到各子组件在排列方向上的等间距效果。
- 通过alignItems属性设置子组件在交叉轴（排列方向的垂直方向）上的对齐方式。且在各类尺寸屏幕中，表现一致。其中，交叉轴为垂直方向时，取值为VerticalAlign类型；为水平方向时，取值为HorizontalAlign类型。

- 通过justifyContent属性设置子组件在主轴（排列方向）上的对齐方式，实现布局的自适应均分能力，取值为FlexAlign类型。

2. 层叠布局

层叠布局（StackLayout）用于在屏幕上预留一块区域来显示组件中的元素，提供元素可以重叠的布局。通过层叠容器Stack实现容器中的子元素依次入栈，后一个子元素覆盖前一个子元素显示。

层叠布局可以设置子元素在容器内的对齐方式，支持九种对齐方式：TopStart（左上）、Top（上中）、TopEnd（右上）、Start（左）、Center（中）、End（右）、BottomStart（左下）、Bottom（中下）、BottomEnd（右下）。

3. 弹性布局

弹性布局（Flex布局）是自适应布局中使用最为灵活的布局。弹性布局提供一种更加有效的方式对容器中的子组件进行排列、对齐和空白空间分配。弹性布局包括以下概念。

- 容器：Flex组件作为Flex布局的容器，用于设置布局相关属性。
- 子组件：Flex组件内的子组件自动成为布局的子组件。
- 主轴：Flex组件布局方向的轴线，子组件默认沿着主轴排列。主轴开始的位置称为主轴起始端，结束位置称为主轴终点端。
- 交叉轴：垂直于主轴方向的轴线。交叉轴起始的位置称为主轴首部，结束位置称为交叉轴尾部。

弹性布局示意图如图8-87所示。

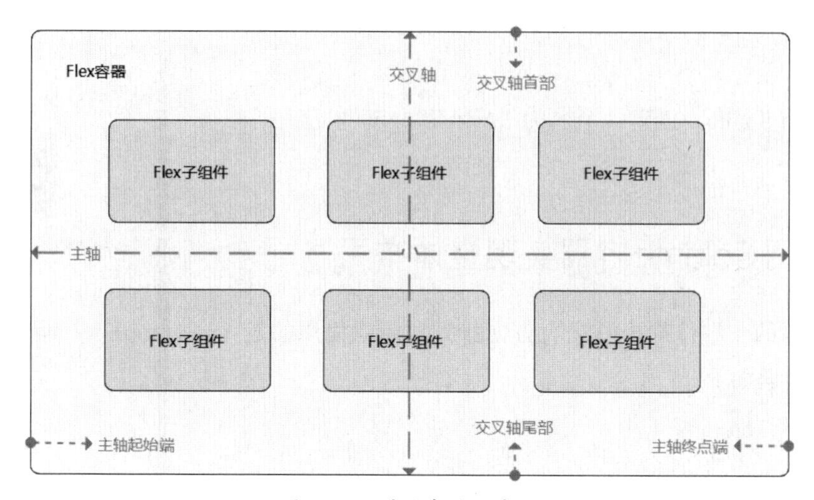

图8-87　弹性布局示意图

4. 网格布局

网格布局（GridLayout）是自适应布局中一种重要的布局，具备较强的页面均分能力、子组件占比控制能力。通过Grid容器组件和子组件GridItem，实现Grid用于设置网格布局相关参数，GridItem定义子组件相关特征。

8.9.2 响应式布局

响应式布局包含以下两类。

1. 栅格布局

栅格系统作为一种辅助布局的定位工具，在平面设计和网站设计中都起到了很好的作用，对移动设备的界面设计有较好的借鉴作用。

栅格组件GridRow和GridCol提供了更灵活、全面的栅格系统实现方案。GridRow为栅格容器组件，只与栅格子组件GridCol在栅格布局场景中使用。

2. 媒体查询

媒体查询（Media Query）作为响应式设计的核心，在移动设备上应用十分广泛。它根据不同设备类型或同设备不同状态修改应用的样式。

8.10 实战：使用ArkUI实现“登录”界面

本节主要介绍在App应用中常见的“登录”页面的实现。本示例“登录”页面使用Column容器组件布局，由Image、TextInput、Button、Text等基础组件构成。最终的界面效果如图8-88所示。

图8-88 界面效果图

打开DevEco Studio，创建一个名为“ArkUILogin”的应用作为演示示例。

8.10.1 使用Column容器实现整体布局

“登录”页面的子组件都按照垂直方向排列，因此使用的是Column容器。代码如下：

```
@Entry
@Component
struct Index {
  build() {
    // 子组件都按照垂直方向排列
    Row() {
```

```
        // 子组件都按照垂直方向排列
        Column() {

        }
        .width('100%')
      }
      .height('100%'))
    }
  }
```

上述width('100%')代表设置了容器的宽度为100%。

8.10.2 使用Image组件实现标志展示

"登录"页面的标志是图片，因此使用Image组件来实现。代码如下：

```
// 头像
Image($r('app.media.waylau_181_181'))
  .height(108)
  .width(108)
```

其中，Image设置了宽、高，Image所引用的图片资源waylau_181_181.jpg放置在src/main/resources/base/media目录下。

8.10.3 使用TextInput组件实现账号密码的输入

用户注册登录需要输入账号、密码等信息，因此可以采用HarmonyOS ArkUI里面的TextInput组件作为账号和密码输入框。代码如下：

```
// 账号输入框
TextInput({placeholder:'请输入账号'})
  .width(320)
  .height(50)
  .borderRadius(8)
  .backgroundColor('#f9f9f9')
  .margin({top:'10'})
  .fontSize(27)
  .type(InputType.Normal) // 输入框类型：平常

// 密码输入框
TextInput({placeholder:'请输入密码'})
  .width(320)
  .height(50)
  .borderRadius(8)
```

```
  .backgroundColor('#f9f9f9')
  .margin({top:'10'})
  .fontSize(27)
  .type(InputType.Password) // 输入框类型：密码
```

TextInput在初始化的时候，可以设置placeholder指定输入框的提示信息。

与账号的输入框不同，密码输入框多了一个type属性。其中type的值设置为InputType.Password用以表明该输入框是一个密码输入框。密码输入框会对输入框的内容进行隐私保护处理，比如用户输入的是“abcabc”，那么在密码输入框里面会变成“******”。

8.10.4　使用Button组件实现登录按钮

“登录”页面在输入框的下方增加一个Button组件，以实现登录按钮。代码如下：

```
// 登录按钮
Button('登录', { type: ButtonType.Normal })
  .width('320')
  .height('50')
  .borderRadius(8)
  .backgroundColor('#ffd0da')
  .margin({ top: '10' })
  .fontSize(24)
```

type方法用于指定登录按钮的样式，本例ButtonType.Normal为默认样式。

8.10.5　使用Text组件实现注册按钮

在“登录”按钮的下方增加一个Text组件，以实现注册按钮。代码如下：

```
// 注册按钮
Text('注册')
  .fontColor(Color.Black)
  .margin({ top: '10' })
  .fontSize(24)
```

fontColor方法用于指定字体颜色，本例Color.Black为黑色样式。

8.10.6　完整代码

最终，整个例子的完整示例代码如下：

```
@Entry
@Component
struct Index {
```

```
@State message: string = 'Hello World'

build() {
  Row() {
    // 子组件都按照垂直方向排列
    Column() {
      // 头像
      Image($r('app.media.waylau_181_181'))
        .height(108)
        .width(108)

      // 账号输入框
      TextInput({ placeholder: '请输入账号' })
        .width(320)
        .height(50)
        .borderRadius(8)
        .backgroundColor('#f9f9f9')
        .margin({ top: '10' })
        .fontSize(27)
        .type(InputType.Normal) // 输入框类型：平常

      // 密码输入框
      TextInput({ placeholder: '请输入密码' })
        .width(320)
        .height(50)
        .borderRadius(8)
        .backgroundColor('#f9f9f9')
        .margin({ top: '10' })
        .fontSize(27)
        .type(InputType.Password) // 输入框类型：密码

      // 登录按钮
      Button('登录', { type: ButtonType.Normal })
        .width('320')
        .height('50')
        .borderRadius(8)
        .backgroundColor('#ffd0da')
        .margin({ top: '10' })
        .fontSize(24)

      // 注册按钮
      Text('注册')
        .fontColor(Color.Black)
        .margin({ top: '10' })
        .fontSize(24)
    }
```

```
        .width('100%')
      }
      .height('100%')
    }
  }
```

8.11 实战：使用ArkUI实现“计算器”

本节主要介绍如何使用ArkUI实现一个“计算器”应用。内容涉及UI布局、事件响应、状态管理、自定义组件等，相当于对ArkUI的一个综合应用。“计算器”最终的界面效果如图8-89所示。

打开DevEco Studio，选择一个Empty Ability工程模板，创建一个名为“ArkUICalculator”的工程作为演示示例。

图8-89 “计算器”效果图

8.11.1 新增Calculator.ets文件

在src>main>ets目录中下，创建一个名为“Calculator.ets”的文件。该文件主要实现“计算器”的核心计算逻辑。代码如下：

```
/**
 * 计算器计算逻辑
 */
export class Calculator {

}
```

8.11.2 实现递归运算

在上述文件中添加recursiveCompute方法。代码如下：

```
/**
 * 计算器计算逻辑
 */
export class Calculator {

  /**
   * 递归计算直至完成，一次计算一对数，从左往右，乘除法优先于加减法
```

```
   * @param split
   * 例: split = ['1.1', '-', '0.1', '+', '2', '×', '3', '÷', '4']
   * 第1次: split = ['1.1', '-', '0.1', '+', '6', '÷', '4']
   * 第2次: split = ['1.1', '-', '0.1', '+', '1.5']
   * 第3次: split = ['1', '+', '1.5']
   * 第4次: split = ['2.5']
   */
  private static recursiveCompute(split: string[]): string[] {
    var symbolIndex // 符号索引
    // 先寻找乘除符号
    for (var i = 0;i < split.length; i++) {
      if (split[i].match(RegExp('^(×|÷)$')) != null) {
        symbolIndex = i
        break
      }
    }
    // 若没找到乘除符号，则寻找加减符号
    if (symbolIndex == null) {
      for (var j = 0;j < split.length; j++) {
        if (split[j].match(RegExp('^(\\+|-)$')) != null) {
          symbolIndex = j
          break
        }
      }
    }
    if (symbolIndex == null) { // 若没找到运算符号，表明计算结束，返回结果
      return split
    } else { // 若找到运算符号，运算后继续寻找运算
      var num1 = +split[symbolIndex-1]
      var symbol = split[symbolIndex]
      var num2 = +split[symbolIndex+1]
      var result = 0
      switch (symbol) {
        case '+':
          result = num1 + num2
          break
        case '-':
          result = num1 - num2
          break
        case '×':
          result = num1 * num2
          break
        case '÷':
          result = num1 / num2
          break
      }
```

```
      split = split.slice(0, symbolIndex - 1).concat(`${result}`).
concat(split.slice(symbolIndex + 2))
      return Calculator.recursiveCompute(split)
    }
  }
}
```

recursiveCompute方法是一个递归计算的方法，主要用于实现算式的递归运算。比如，输入如下的字符串数组：

```
['1.1', '-', '0.1', '+', '2', '×', '3', '÷', '4']
```

根据四则运算的法则，会先计算“乘除”再计算“加减”，因此，第一次会先执行“'2', '×', '3'”，运算结果如下：

```
['1.1', '-', '0.1', '+', '6', '÷', '4']
```

同理，第二次的运算结果如下：

```
['1.1', '-', '0.1', '+', '1.5']
```

第三次的运算结果如下：

```
['1', '+', '1.5']
```

第四次的运算结果如下：

```
['2.5']
```

至此，递归运算结束。

8.11.3　实现输入字符串转为字符串数组

在上述Calculator.ets文件中添加calculate方法，用以实现输入字符串转为字符串数组。代码如下：

```
public static calculate(input: string): string {
  // 先将百分数转为小数
  input = input.replace(RegExp(`(((\\d*\\.\\d*)|(\\d+))%)`, 'g'), s =>
String(Number(s.replace(/%/, '')) / 100)) // input = '1.1-0.1+2×3÷4'
  // 要将 input 分割为数与运算符，分割节点的索引储存在 splitIndex
  var splitIndex = [0]
  for (var i = 1;i < input.length; i++) {
    if (input[i].match(RegExp('(\\+|-|×|÷)')) != null) {
      splitIndex.push(i)
      splitIndex.push(i + 1)
      i++
```

```
    }
  }
  splitIndex.push(input.length) // splitIndex = [0, 3, 4, 7, 8, 9, 10, 11,
12, 13]
  // 分割 input 为数与运算符，储存在 split
  var split = []
  for (var j = 0;j < splitIndex.length - 1; j++) {
    split.push(input.substring(splitIndex[j], splitIndex[j+1]))
  }
  // split = ['1.1', '-', '0.1', '+', '2', '×', '3', '÷', '4']
  return Calculator.recursiveCompute(split)[0] // 递归计算直至完成
}
```

8.11.4 新增CalculatorButtonInfo.ets文件

在src>main>ets目录中下，创建一个名为“CalculatorButtonInfo.ets”的文件。该文件主要表示“计算器”的按钮样式信息。代码如下：

```
// 按钮样式信息
export class CalculatorButtonInfo {
  text: string       // 按钮上的文字
  textColor: number // 文字的颜色
  bgColor: number    // 按钮背景颜色

  constructor(text: string, textColor: number = Color.Black, bgColor: number
= Color.White) {
    this.text = text
    this.textColor = textColor
    this.bgColor = bgColor
  }
}
```

其中，text是按钮上的文字；textColor是文字的颜色；bgColor是按钮背景颜色。

8.11.5 实现CalculatorButton组件

“计算器”按钮CalculatorButton组件的实现方式如下：

```
// 导入 CalculatorButtonInfo
import { CalculatorButtonInfo } from '../CalculatorButtonInfo';


// 构造计算器按钮
@Builder CalculatorButton(btnInfo: CalculatorButtonInfo) { // 计算器按钮组件
```

```
  GridItem() {
    Text(btnInfo.text) // 文本
      .fontSize(50)
      .fontWeight(FontWeight.Bold)
      .width('100%')
      .height('100%')
      .textAlign(TextAlign.Center)
      .borderRadius(100) // 圆角
      .fontColor(btnInfo.textColor) // 字体颜色
      .backgroundColor(btnInfo.bgColor) // 背景颜色
  }
  .forceRebuild(false)
  .onClick(() => this.onClickBtn(btnInfo.text))
  .rowStart(btnInfo.text == '=' ? 4 : null)
  .rowEnd(btnInfo.text == '=' ? 5 : null) // 等于按钮占两格，其他按钮默认
}
```

在上述代码中，CalculatorButton组件的实现主要是对GridItem做了封装。通过传入的CalculatorButtonInfo来实现“计算器”按钮的显示样式的个性化。

同时，CalculatorButton也设置了点击事件，以触发onClickBtn方法。onClickBtn方法代码如下：

```
// 导入 Calculator
import { Calculator } from '../Calculator';

@State input: string = '' // 输入内容

// 点击计算器按钮
onClickBtn = (text: string) => {
  switch (text) {
    case 'C': // 清空所有输入
      this.input = ''
      break
    case '←': // 删除输入的最后一个字符
      if (this.input.length > 0) {
        this.input = this.input.substring(0, this.input.length - 1)
      }
      break
    case '=': // 计算结果
      this.input = Calculator.calculate(this.input)
      break
    default: // 输入内容
      this.input += text
      break
  }
}
```

上述onClickBtn方法会将“计算器”按钮所点击的对应文字进行拼接，并最终调用Calculator.calculate来执行计算。

8.11.6 构造整体页面

现在将CalculatorButton组件进行组装，成为一个完整的“计算器”界面。代码如下：

```
@Entry
@Component
struct Index {
  private BTN_INFO_ARRAY: CalculatorButtonInfo[] = [ // 所有按钮样式信息
    new CalculatorButtonInfo('C', Color.Blue),
    new CalculatorButtonInfo('÷', Color.Blue),
    new CalculatorButtonInfo('×', Color.Blue),
    new CalculatorButtonInfo('←', Color.Blue),
    new CalculatorButtonInfo('7'),
    new CalculatorButtonInfo('8'),
    new CalculatorButtonInfo('9'),
    new CalculatorButtonInfo('-', Color.Blue),
    new CalculatorButtonInfo('4'),
    new CalculatorButtonInfo('5'),
    new CalculatorButtonInfo('6'),
    new CalculatorButtonInfo('+', Color.Blue),
    new CalculatorButtonInfo('1'),
    new CalculatorButtonInfo('2'),
    new CalculatorButtonInfo('3'),
    new CalculatorButtonInfo('=', Color.White, Color.Blue),
    new CalculatorButtonInfo('%'),
    new CalculatorButtonInfo('0'),
    new CalculatorButtonInfo('.')
  ]

  build() {
    Stack({ alignContent: Alignment.Bottom }) {
      Column() {
        // 输入显示区
        Text(this.input.length == 0 ? '0' : this.input) // 输入内容，若没有内容显
示0
          .width('100%')
          .padding(10)
          .textAlign(TextAlign.End)
          .fontSize(46)

        // 按键区
        Grid() {
```

```
            // 遍历生成按钮
            ForEach(this.BTN_INFO_ARRAY, btnInfo => this.
CalculatorButton(btnInfo))
          }
          .columnsTemplate('1fr 1fr 1fr 1fr') // 按钮比重分配
          .rowsTemplate('1fr 1fr 1fr 1fr 1fr')
          .columnsGap(2) // 按钮间隙
          .rowsGap(2)
          .width('100%')
          .aspectRatio(1) // 长宽比
        }
      }.width('100%').height('100%').backgroundColor(Color.Gray)
    }

  // ...
}
```

整体的“计算器”界面分为上下两部分。上部分为输入显示区，主要是采用Text实现；下部分为按键区，通过Grid结合CalculatorButtonInfo来实现按键格子，每个CalculatorButtonInfo就是一个按键，所有按键的样式定义在BTN_INFO_ARRAY数组中。

8.11.7 运行

可以在预览器中直接运行该应用，界面效果如图8-90所示。

最终计算结果界面效果如图8-91所示。

图8-90 “计算器”计算过程效果图

图8-91 “计算器”计算结果效果图

第9章

用Java开发UI

Java UI 框架提供了用于创建用户界面（UI）的各类组件，包括一些常用的组件和常用的布局。用户可通过组件进行交互操作，并获得响应。

9.1 用Java开发UI概述

在前面的应用开发过程中，我们已经初步接触了Java UI编程。以JavaAbilitySliceNavigation应用为例，在开发一个Page Ability时，往往需要涉及AbilitySlice和ability_main.xml文件的修改，这其实就是Java UI编程的一部分。

应用的Ability在屏幕上将显示一个用户界面，该界面用来显示所有可被用户查看和交互的内容。应用中所有的用户界面元素都是由Component和ComponentContainer对象构成的。Component是绘制在屏幕上的一个对象，用户能与之交互；ComponentContainer是一个用于容纳其他Component和ComponentContainer对象的容器。

Java UI框架提供了一部分Component和ComponentContainer的具体子类，即创建用户界面（UI）的各类组件，包括一些常用的组件（比如文本、按钮、图片、列表等）和常用的布局（比如DirectionalLayout和DependentLayout）。用户可通过组件进行交互操作，并获得响应。所有的UI操作都应该在主线程进行设置。

9.1.1 组件和布局

用户界面元素统称为组件，组件根据一定的层级结构进行组合形成布局。组件在未被添加到布局中时，既无法显示也无法交互，因此一个用户界面至少包含一个布局。在UI框架中，具体的布局类通常以XXLayout命名，完整的用户界面是一个布局，用户界面中的一部分也可以是一个布局。布局中容纳Component与ComponentContainer对象。

9.1.2 Component和ComponentContainer

Component用于提供内容显示，是界面中所有组件的基类，开发者可以给Component设置事件处理回调来创建一个可交互的组件。Java UI框架提供了一些常用的界面元素，也可称之为组件，组件一般直接继承Component或它的子类，如Text、Image等。

ComponentContainer作为容器容纳Component或ComponentContainer对象，并对它们进行布局。Java UI框架提供了一些标准布局功能的容器，它们继承自ComponentContainer，一般以“Layout”结尾，如DirectionalLayout、DependentLayout等。

图9-1展示了Component和ComponentContainer的结构组成。

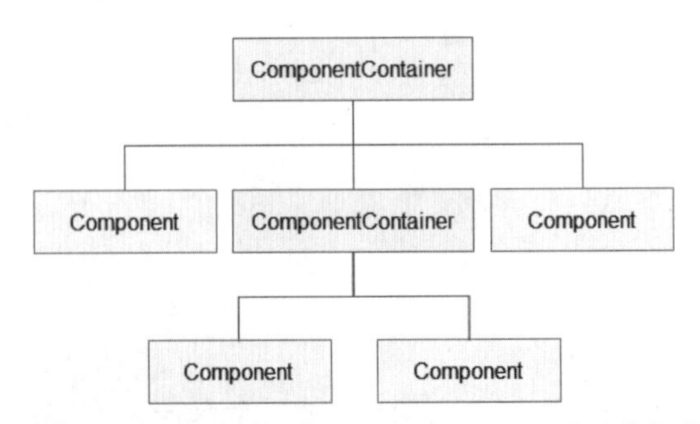

图9-1　Component和ComponentContainer的结构组成

9.1.3 LayoutConfig

每种布局都根据自身特点提供LayoutConfig，供子Component设定布局属性和参数，指定布局属性可以对子Component在布局中的显示效果进行约束。例如，width和height是最基本的布局属性，它们指定了组件的大小。

图9-2展示了LayoutConfig在布局中的作用。

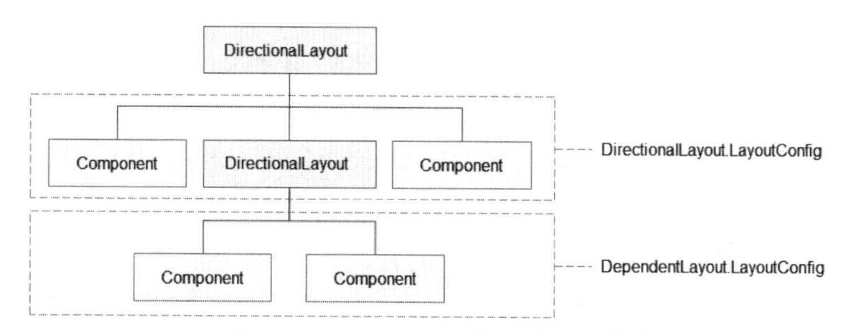

图9-2　LayoutConfig在布局中的作用

9.1.4 组件树

正如图9-1所展示的那样，布局把Component和ComponentContainer以树状的层级结构进行组织，这样的一个布局就称为组件树。组件树的特点是仅有一个根组件，其他组件有且仅有一个父节点，组件之间的关系受到父节点的规则约束。

9.2 组件与布局

正如前面章节所介绍的那样，HarmonyOS提供了Ability和AbilitySlice两个基础类。一个Page Ability可以由一个或多个AbilitySlice构成，AbilitySlice主要用于承载单个页面的具体逻辑实现和界面UI，是应用显示、运行和跳转的最小单元。AbilitySlice通过setUIContent为界面设置布局，示例如下：

```
public class MainAbilitySlice extends AbilitySlice {

    @Override
    public void onStart(Intent intent) {
        super.onStart(intent);

        // 设置布局
        super.setUIContent(ResourceTable.Layout_ability_main);
    }
```

```
    // 为节约篇幅，省略部分代码
}
```

在上述代码中，ResourceTable.Layout_ability_main即为界面组件树的根节点。

9.2.1 编写布局的方式

组件需要进行组合，并添加到界面的布局中。在Java UI框架中，提供了两种编写布局的方式。

- 在代码中创建布局：用代码创建Component和ComponentContainer对象，为这些对象设置合适的布局参数和属性值，并将Component添加到ComponentContainer中，从而创建出完整界面。
- 在XML中声明UI布局：按层级结构来描述Component和ComponentContainer的关系，给组件节点设定合适的布局参数和属性值，代码中可直接加载生成此布局。

这两种方式创建出的布局没有本质差别，在XML中声明布局，在加载后同样可在代码中对该布局进行修改。在前面章节中所介绍的实战案例，多是在XML中声明布局，这也是本书所推崇的布局方式。

9.2.2 组件分类

根据组件的功能，可以将组件分为布局类、显示类、交互类三类。表9-1总结了各个组件的分类。

表9-1　组件分类

组件类别	组件名称	功能描述
布局类	PositionLayout、DirectionalLayout、StackLayout、DependentLayout、TableLayout、AdaptiveBoxLayout	提供了不同布局规范的组件容器，例如以单一方向排列的DirectionalLayout、以相对位置排列的DependentLayout、以确切位置排列的PositionLayout等。
显示类	Text、Image、Clock、TickTimer、ProgressBar	提供了单纯的内容显示，例如用于文本显示的Text，用于图像显示的Image等。
交互类	TextField、Button、Checkbox、RadioButton/RadioContainer、Switch、ToggleButton、Slider、Rating、ScrollView、TabList、ListContainer、PageSlider、PageFlipper、PageSliderIndicator、Picker、TimePicker、DatePicker、SurfaceProvider、ComponentProvider	提供了具体场景下与用户交互响应的功能，例如Button提供了点击响应功能，Slider提供了进度选择功能等。

框架提供的组件使应用界面开发更加便利。

9.3 实战：通过XML创建布局

本节演示如何通过XML来创建DirectionalLayout布局。为了演示该功能，创建一个名为“DirectionalLayoutWithXml”的应用。

XML声明布局的方式是非常简便直观的。在初始化DirectionalLayoutWithXml应用时，已经为应用创建了一个默认的界面布局，即MainAbilitySlice和ability_main.xml。

9.3.1 理解XML布局文件

每一个Component和ComponentContainer对象大部分属性都支持在XML中进行设置，它们都有各自的XML属性列表。某些属性仅适用于特定的组件，例如，只有Text支持text_color属性，不支持该属性的组件如果添加了该属性，该属性则会被忽略。具有继承关系的组件子类将继承父类的属性列表，Component作为组件的基类，拥有各个组件常用的属性，比如ID、布局参数等。

下面是初始化DirectionalLayoutWithXml应用时ability_main.xml文件的内容：

```
<?xml version="1.0" encoding="utf-8"?>
<DirectionalLayout
    xmlns:ohos="http://schemas.huawei.com/res/ohos"
    ohos:height="match_parent"
    ohos:width="match_parent"
    ohos:orientation="vertical">

    <Text
        ohos:id="$+id:text_helloworld"
        ohos:height="match_parent"
        ohos:width="match_content"
        ohos:background_element="$graphic:background_ability_main"
        ohos:layout_alignment="horizontal_center"
        ohos:text="Hello World"
        ohos:text_size="50"
    />

</DirectionalLayout>
```

接下来详细介绍ID和布局参数。

1. ID

在上述配置中，ohos:id="$+id:text_helloworld"就是在XML中使用此格式声明的一个对开发者友好的ID，它会在编译过程中转换成一个常量。尤其是在DependentLayout布局中，组件之间需要描述相对位置关系，描述时要通过ID来指定对应组件。

布局中的组件通常要设置独立的ID，以便在程序中查找该组件。如果布局中有不同组件设置了相同的ID，在通过ID查找组件时会返回查找到的第一个组件，因此尽量保证在所要查找的布局中为组件设置独立的ID值，避免出现与预期不符合的问题。

比如，在Notification应用中，就为不同的Text设置了不同的ID，代码如下：

```
<?xml version="1.0" encoding="utf-8"?>
<DirectionalLayout
    xmlns:ohos="http://schemas.huawei.com/res/ohos"
    ohos:height="match_parent"
    ohos:width="match_parent"
    ohos:alignment="center"
    ohos:orientation="vertical">

    <Text
        ohos:id="$+id:text_publish_notification"
        ohos:height="match_content"
        ohos:width="match_content"
        ohos:background_element="$graphic:background_ability_main"
        ohos:layout_alignment="horizontal_center"
        ohos:text="Publish Notification"
        ohos:text_size="40vp"
        />

    <Text
        ohos:id="$+id:text_cancel_notification"
        ohos:height="match_parent"
        ohos:width="match_content"
        ohos:background_element="$graphic:background_ability_main"
        ohos:layout_alignment="horizontal_center"
        ohos:text="Cancel Notification"
        ohos:text_size="40vp"
        />
</DirectionalLayout>
```

2. 布局参数

在上述配置中，ohos:width和ohos:height都是布局参数，在XML中它们的取值如下。

- 具体的数值：10（以像素为单位）、10vp（以屏幕相对像素为单位）。
- match_parent：表示组件大小将扩展为父组件允许的最大值，它将占据父组件方向上的剩余大小。
- match_content：表示组件大小与它的内容占据的大小范围相适应。

9.3.2 创建XML布局文件

如果要新建XML布局文件，则可以在DevEco Studio的“Project”窗口，打开“entry > src >

main > resources > base”，右击“layout”文件夹，选择“New > File”，来创建布局文件。比如，在本例中，我们将它命名为“ability_pay.xml”。

当然，另外一种快捷的创建布局文件的方式是，直接复制ability_main.xml的内容来修改。

打开新创建的ability_pay.xml布局文件，修改其中的内容，对布局和组件的属性和层级进行描述。代码如下：

```
<?xml version="1.0" encoding="utf-8"?>
<DirectionalLayout
    xmlns:ohos="http://schemas.huawei.com/res/ohos"
    ohos:height="match_parent"
    ohos:width="match_parent"
    ohos:alignment="center"
    ohos:orientation="vertical">
    <Text
        ohos:id="$+id:text_pay"
        ohos:height="match_content"
        ohos:width="match_content"
        ohos:background_element="$graphic:background_ability_main"
        ohos:layout_alignment="horizontal_center"
        ohos:text="Show me the money"
        ohos:text_size="38vp"
        />

</DirectionalLayout>
```

在预览器中，可以对上述布局进行实时预览，如图9-3所示。

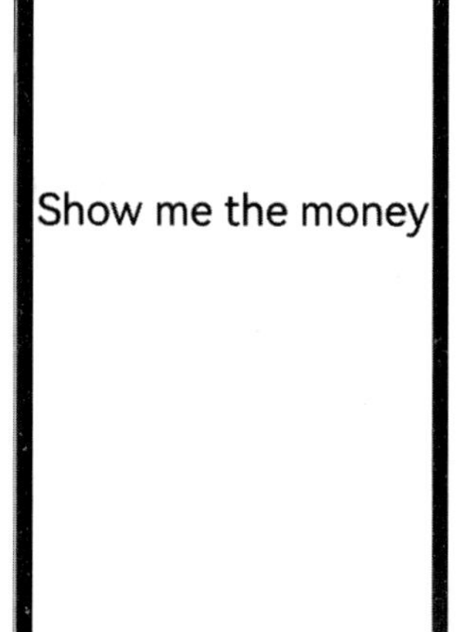

图9-3　对ability_pay.xml布局文件的预览

9.3.3　加载XML布局

在代码中需要加载XML布局，并添加为根布局或作为其他布局的子Component。代码如下：

```
package com.waylau.hmos.directionallayoutwithxml.
slice;

import com.waylau.hmos.directionallayoutwithxml.
ResourceTable;
import ohos.aafwk.ability.AbilitySlice;
import ohos.aafwk.content.Intent;
import ohos.agp.colors.RgbColor;
import ohos.agp.components.Text;
import ohos.agp.components.element.ShapeElement;

public class PayAbilitySlice extends AbilitySlice {
```

```
    @Override
    public void onStart(Intent intent) {
        super.onStart(intent);

        // 加载 XML 布局作为根布局
        super.setUIContent(ResourceTable.Layout_ability_pay);

        // 获取组件
        Text textPay = (Text) findComponentById(ResourceTable.Id_text_pay);

        // 设置组件的属性
        ShapeElement background = new ShapeElement();
        background.setRgbColor(new RgbColor(0, 125, 255));
        background.setCornerRadius(25);
        textPay.setBackground(background);
    }

    @Override
    public void onActive() {
        super.onActive();
    }

    @Override
    public void onForeground(Intent intent) {
        super.onForeground(intent);
    }
}
```

上述代码解释如下。

- 通过setUIContent的方式来加载XML布局。
- 通过findComponentById方法来获取组件。
- 组件可以重新设置属性。在上述例子中，我们设置了文本的背景。

9.3.4 显示XML布局

那么，如何来显示PayAbilitySlice所设置的布局呢？

一种方式是用“5.4.4 实现AbilitySlice之间的路由和导航”一节所介绍的导航的方式，从MainAbilitySlice导航到PayAbilitySlice。

另外一种更为简单的方式是，直接将PayAbilitySlice设置为主的AbilitySlice。代码如下：

```
package com.waylau.hmos.directionallayoutwithxml;

import com.waylau.hmos.directionallayoutwithxml.slice.MainAbilitySlice;
```

```
import com.waylau.hmos.directionallayoutwithxml.slice.PayAbilitySlice;
import ohos.aafwk.ability.Ability;
import ohos.aafwk.content.Intent;

public class MainAbility extends Ability {
    @Override
    public void onStart(Intent intent) {
        super.onStart(intent);
        //super.setMainRoute(MainAbilitySlice.class.getName());

        super.setMainRoute(PayAbilitySlice.class.getName());
    }
}
```

这样，运行应用后，可以看到界面显示，如图9-4所示。

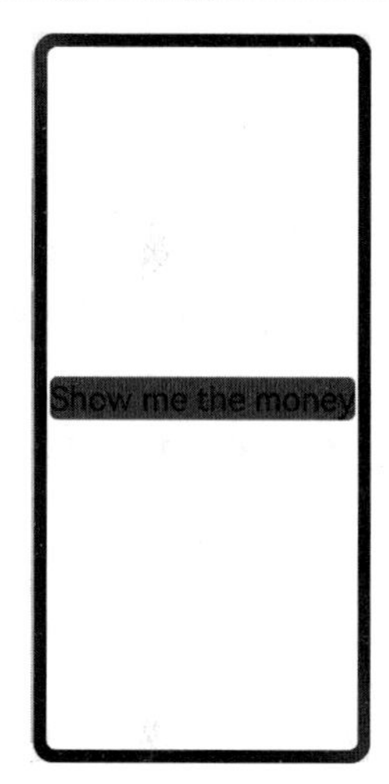

图9-4　应用主界面显示效果

9.4 实战：通过Java创建布局

如果有Java Swing或Java AWT编程经验的话，那么对于用Java语言通过代码方式来创建布局就不会陌生。本节演示如何通过Java来创建布局。为了演示该功能，创建一个名为“DirectionalLayoutWithJava”的应用。

在初始化DirectionalLayoutWithJava应用时，已经为应用创建了一个默认的界面布局，即MainAbilitySlice和ability_main.xml。我们需要再创建一个新的DirectionalLayout布局。

9.4.1 新建AbilitySlice

创建一个新的AbilitySlice，我们命名为“PayAbilitySlice”。PayAbilitySlice需要继承AbilitySlice，代码如下：

```
package com.waylau.hmos.directionallayoutwithjava.slice;

import ohos.aafwk.ability.AbilitySlice;

public class PayAbilitySlice extends AbilitySlice {
    package com.waylau.hmos.directionallayoutwithjava.slice;

import ohos.aafwk.ability.AbilitySlice;
import ohos.aafwk.content.Intent;
```

```
public class PayAbilitySlice extends AbilitySlice {
    @Override
    public void onStart(Intent intent) {
        super.onStart(intent);
    }

    @Override
    public void onActive() {
        super.onActive();
    }

    @Override
    public void onForeground(Intent intent) {
        super.onForeground(intent);
    }

}
```

上述代码重写了onStart、onActive、onForeground方法。

9.4.2 创建布局

在PayAbilitySlice的onStart方法中，创建布局并使用。代码如下：

```
@Override
public void onStart(Intent intent) {
    super.onStart(intent);

    // 声明布局
    DirectionalLayout directionalLayout = new DirectionalLayout(getConte
xt());

    // 设置布局大小
    directionalLayout.setWidth(ComponentContainer.LayoutConfig.MATCH_PARENT);
    directionalLayout.setHeight(ComponentContainer.LayoutConfig.MATCH_
PARENT);

    // 设置布局属性
    directionalLayout.setOrientation(Component.VERTICAL);
    directionalLayout.setAlignment(LayoutAlignment.CENTER);

    // 将布局添加到组件树中
    setUIContent(directionalLayout);
}
```

上述代码解释如下。

- 声明了布局。
- 设置了布局的大小和属性。
- 将布局添加到组件树中。

这样，一个布局就创建完成了。

点击预览器，可以对布局进行预览，如图9-5所示。

图9-5　预览布局

9.4.3　在布局中添加组件

光有布局，那么界面只会显示一片空白，此时，需要在布局中添加组件。代码如下：

```
// 声明 Text 组件
Text textPay = new Text(getContext());
textPay.setText("Show me the money");
textPay.setTextSize(38, Text.TextSizeType.VP);
textPay.setId(1);

// 设置组件的属性
ShapeElement background = new ShapeElement();
background.setRgbColor(new RgbColor(0, 125, 255));
background.setCornerRadius(25);
textPay.setBackground(background);

// 为组件添加对应布局的布局属性
DirectionalLayout.LayoutConfig layoutConfig =
        new DirectionalLayout.LayoutConfig(
            ComponentContainer.LayoutConfig.MATCH_CONTENT,
            ComponentContainer.LayoutConfig.MATCH_CONTENT);
layoutConfig.alignment = LayoutAlignment.HORIZONTAL_CENTER;
textPay.setLayoutConfig(layoutConfig);

// 将组件添加到布局中（视布局需要对组件设置布局属性进行约束）
directionalLayout.addComponent(textPay);
```

上述代码解释如下。

- 声明Text组件。
- 设置组件的属性。在上述例子中，我们设置了文本的背景。
- 为组件添加对应布局的布局属性。
- 将组件添加到布局中。

点击预览器，可以对布局进行预览，如图9-6所示。

9.4.4 显示布局

那么如何来显示PayAbilitySlice所设置的布局呢？

一种方式是用“5.4 实战：多个AbilitySlice间的路由和导航”一节所介绍的导航的方式，从MainAbilitySlice导航到PayAbilitySlice。

另外一种更为简单的方式是，直接将PayAbilitySlice设置为主的AbilitySlice。代码如下：

图9-6 预览布局

```
package com.waylau.hmos.directionallayoutwithjava;

import com.waylau.hmos.directionallayoutwithjava.slice.MainAbilitySlice;
import com.waylau.hmos.directionallayoutwithjava.slice.PayAbilitySlice;
import ohos.aafwk.ability.Ability;
import ohos.aafwk.content.Intent;

public class MainAbility extends Ability {
    @Override
    public void onStart(Intent intent) {
        super.onStart(intent);
        //super.setMainRoute(MainAbilitySlice.class.getName());

        super.setMainRoute(PayAbilitySlice.class.getName());
    }
}
```

这样，运行应用后，可以看到界面显示，如图9-7所示。

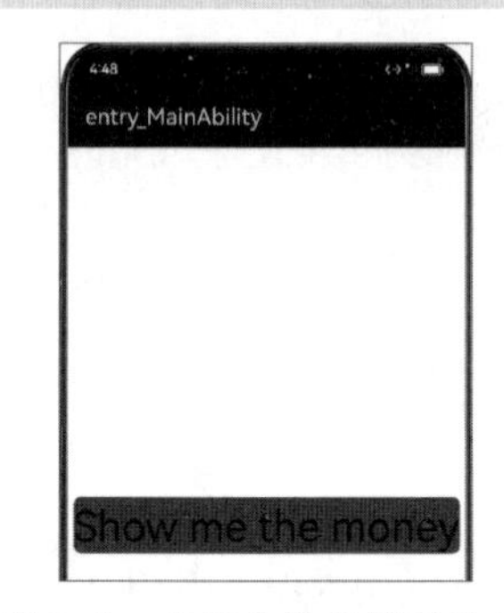

图9-7 应用主界面显示效果

9.5 实战：常用显示类组件——Text

常用显示类组件包括Text、Image、ProgressBar等，这些组件一般提供单纯的内容显示，如用于文本显示的Text，用于图像显示的Image等。本节介绍Text组件的用法。

Text是在前面章节中介绍最多的组件。Text是用来显示字符串的组件，在界面上显示为一块文本区域。Text作为一个基本组件，有很多扩展，常见的有按钮组件Button，文本编辑组件TextField。

创建一个名为“Text”的应用，来作为演示。

9.5.1 设置背景

以下是创建Text应用时产生的ability_main.xml文件，内容如下：

```
<?xml version="1.0" encoding="utf-8"?>
<DirectionalLayout
    xmlns:ohos="http://schemas.huawei.com/res/ohos"
    ohos:height="match_parent"
    ohos:width="match_parent"
    ohos:alignment="center"
    ohos:orientation="vertical">

    <Text
        ohos:id="$+id:text_helloworld"
        ohos:height="match_content"
        ohos:width="match_content"
        ohos:background_element="$graphic:background_ability_main"
        ohos:layout_alignment="horizontal_center"
        ohos:text="$string:mainability_HelloWorld"
        ohos:text_size="40vp"
        />

</DirectionalLayout>
```

在上述文件中，ohos:background_element可以用来配置常用的背景，如常见的文本背景、按钮背景。上述配置引用了background_ability_main.xml文件里面的内容，该文件放置在graphic目录下。

background_ability_main.xml文件内容如下：

```
<?xml version="1.0" encoding="UTF-8" ?>
<shape xmlns:ohos="http://schemas.huawei.com/res/ohos"
       ohos:shape="rectangle">
    <solid
        ohos:color="#FFFFFF"/>
</shape>
```

修改上述配置，可以设置Text背景的效果。修改后的background_ability_main.xml文件内容如下：

```
<?xml version="1.0" encoding="UTF-8" ?>
<shape xmlns:ohos="http://schemas.huawei.com/res/ohos"
       ohos:shape="rectangle">
    <solid
        ohos:color="#007CFD"/><!-- 设置背景色 -->
</shape>
```

最终效果如图9-8所示。

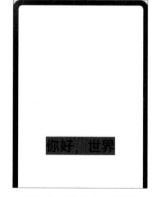

图9-8　界面显示效果

9.5.2 设置字体大小和颜色

为了演示字体大小和颜色的设置过程，通过DevEco Studio创建名为“ColorSizeAbility”的Page，则会自动创建如图9-9所示的4个文件：ColorSizeAbilitySlice、ColorSizeAbility、background_ability_color_size.xml和ability_color_size.xml。

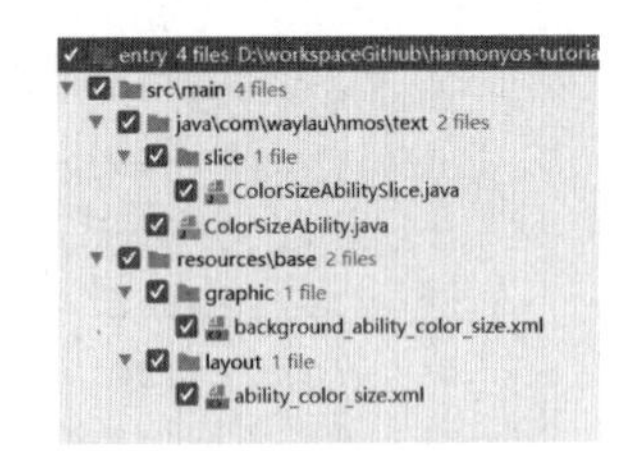

图9-9 创建一个Page

修改ability_color_size.xml内容如下：

```
<?xml version="1.0" encoding="utf-8"?>
<DirectionalLayout
    xmlns:ohos="http://schemas.huawei.com/res/ohos"
    ohos:height="match_parent"
    ohos:width="match_parent"
    ohos:orientation="vertical">

    <Text
        ohos:id="$+id:text_color_size"
        ohos:height="match_parent"
        ohos:width="match_content"
        ohos:background_element="$graphic:background_ability_color_size"
        ohos:layout_alignment="horizontal_center"
        ohos:text="Hello World"
        ohos:text_size="40vp"
        ohos:text_color="#0000FF"
        ohos:left_margin="15vp"
        ohos:bottom_margin="15vp"
        ohos:right_padding="15vp"
        ohos:left_padding="15vp"
        />

</DirectionalLayout>
```

图9-10展示了预览器显示的设置字体大小和颜色之后的效果。

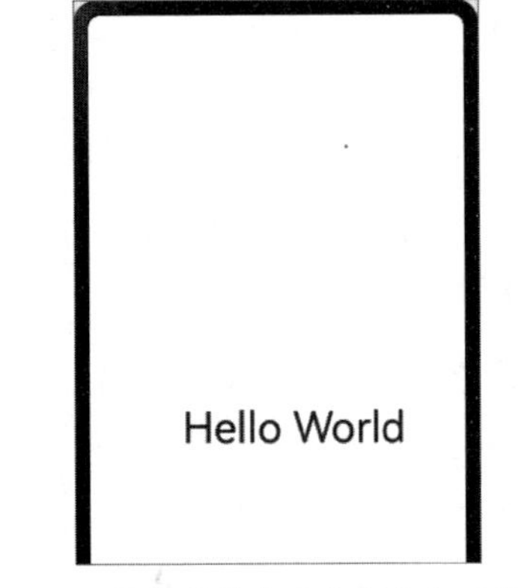

图9-10 设置字体大小和颜色之后的效果

9.5.3 设置字体风格和字重

为了演示设置字体风格和字重的过程，通过DevEco Studio创建名为“ItalicWeightAbility”的Page，则会自动创建以下4个文件：ItalicWeightAbility、ItalicWeightAbilitySlice、ability_italic_weight.xml和background_ability_italic_weight.xml。

修改ability_italic_weight.xml内容如下：

```
<?xml version="1.0" encoding="utf-8"?>
<DirectionalLayout
    xmlns:ohos="http://schemas.huawei.com/res/ohos"
    ohos:height="match_parent"
    ohos:width="match_parent"
    ohos:orientation="vertical">

    <Text
        ohos:id="$+id:text_italic_weight"
        ohos:height="match_parent"
        ohos:width="match_content"
        ohos:background_element="$graphic:background_ability_italic_weight"
        ohos:layout_alignment="horizontal_center"
        ohos:text="Hello World"
        ohos:text_size="40vp"
        ohos:text_color="#0000FF"
        ohos:italic="true"
        ohos:text_weight="700"
        ohos:text_font="serif"
        ohos:left_margin="15vp"
        ohos:bottom_margin="15vp"
        ohos:right_padding="15vp"
        ohos:left_padding="15vp"
        />

</DirectionalLayout>
```

图9-11展示了预览器显示的设置italic字体风格和字重之后的效果。

图9-11　设置italic字体风格和字重之后的效果

9.5.4　设置文本对齐方式

为了演示设置文本对齐方式的过程，通过DevEco Studio创建名为“AlignmentAbility”的Page，则会自动创建以下4个文件：AlignmentAbility、AlignmentAbilitySlice、ability_alignment.xml和background_ability_alignment.xml。

修改ability_alignment.xml内容如下：

```
<?xml version="1.0" encoding="utf-8"?>
<DirectionalLayout
    xmlns:ohos="http://schemas.huawei.com/res/ohos"
```

```
    ohos:height="match_parent"
    ohos:width="match_parent"
    ohos:orientation="vertical">

    <Text
        ohos:id="$+id:text_alignment"
        ohos:background_element="$graphic:background_ability_alignment"
        ohos:text="Hello World"
        ohos:width="300vp"
        ohos:height="100vp"
        ohos:text_size="40vp"
        ohos:text_color="#0000FF"
        ohos:italic="true"
        ohos:text_weight="700"
        ohos:text_font="serif"
        ohos:left_margin="15vp"
        ohos:bottom_margin="15vp"
        ohos:right_padding="15vp"
        ohos:left_padding="15vp"
        ohos:text_alignment="horizontal_center|bottom"
        />

</DirectionalLayout>
```

图 9-12 展示了预览器显示的设置文本对齐方式之后的效果。

图 9-12　设置文本对齐方式之后的效果

9.5.5　设置文本换行和最大显示行数

为了演示设置文本换行和最大显示行数的过程，通过DevEco Studio创建名为“LinesAbility”的Page，则会自动创建以下4个文件：LinesAbility、LinesAbilitySlice、ability_lines.xml和background_ability_lines.xml。

修改ability_lines.xml内容如下：

```
<?xml version="1.0" encoding="utf-8"?>
<DirectionalLayout
    xmlns:ohos="http://schemas.huawei.com/res/ohos"
    ohos:height="match_parent"
    ohos:width="match_parent"
    ohos:alignment="center"
    ohos:orientation="vertical">

    <Text
```

```
        ohos:id="$+id:text_lines"
        ohos:background_element="$graphic:background_ability_lines"
        ohos:layout_alignment="horizontal_center"
        ohos:text="Hello World"
        ohos:width="300"
        ohos:height="match_content"
        ohos:text_size="40vp"
        ohos:multiple_lines="true"
        ohos:max_text_lines="2"
        />

</DirectionalLayout>
```

图9-13展示了预览器显示的设置文本换行和最大显示行数之后的效果。由于文本长度超过了限制，因此文本内容无法显示完整。

图9-13　设置文本换行和最大显示行数之后的效果

9.5.6　设置自动调节字体大小

为了演示设置自动调节字体大小的过程，通过DevEco Studio创建名为“AutoFontSizeAbility”的Page，则会自动创建以下4个文件：AutoFontSizeAbility、AutoFontSizeAbilitySlice、ability_auto_font_size.xml和background_ability_auto_font_size.xml。

修改ability_auto_font_size.xml内容如下：

```
<?xml version="1.0" encoding="utf-8"?>
<DirectionalLayout
    xmlns:ohos="http://schemas.huawei.com/res/ohos"
    ohos:height="match_parent"
    ohos:width="match_parent"
    ohos:alignment="center"
    ohos:orientation="vertical">

    <Text
        ohos:id="$+id:text_auto_font_size"
        ohos:background_element="$graphic:background_ability_auto_font_size"
        ohos:layout_alignment="horizontal_center"
        ohos:text="Hello World"
        ohos:width="300"
        ohos:height="match_content"
        ohos:text_size="40vp"
        ohos:multiple_lines="true"
        ohos:max_text_lines="2"
```

```
        ohos:auto_font_size="true"
        />

</DirectionalLayout>
```

图 9-14 展示了预览器显示的设置自动调节字体大小之后的效果。

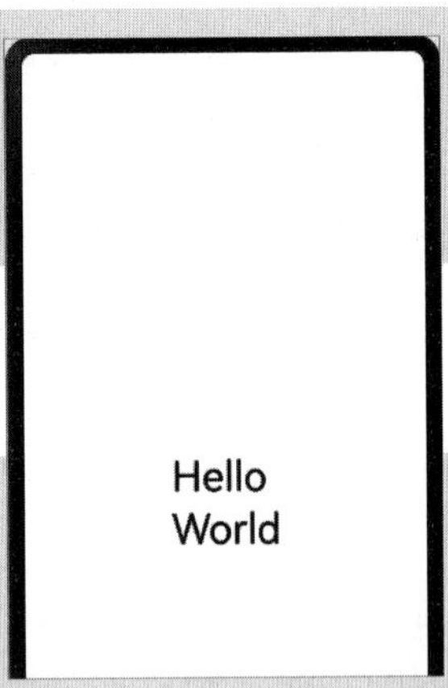

图 9-14　设置自动调节字体大小之后的效果

修改 AutoFontSizeAbilitySlice，代码如下：

```
package com.waylau.hmos.text.slice;

import com.waylau.hmos.text.ResourceTable;
import ohos.aafwk.ability.AbilitySlice;
import ohos.aafwk.content.Intent;
import ohos.agp.components.Component;
import ohos.agp.components.Text;

public class AutoFontSizeAbilitySlice extends AbilitySlice {
    @Override
    public void onStart(Intent intent) {
        super.onStart(intent);
        super.setUIContent(ResourceTable.Layout_ability_auto_font_size);

        Text textAutoFontSize = (Text) findComponentById(ResourceTable.Id_
text_auto_font_size);

        // 设置自动调整规则
        textAutoFontSize.setAutoFontSizeRule(30, 100, 1);

        // 设置点击一次增多一个"!"
        textAutoFontSize.setClickedListener(listener ->
                textAutoFontSize.setText(textAutoFontSize.getText() + "!"));
    }

    @Override
    public void onActive() {
        super.onActive();
    }

    @Override
    public void onForeground(Intent intent) {
        super.onForeground(intent);
    }
}
```

上述代码设置了点击事件。当点击Text内容之后，每点击一次，文本内容就增多一个“i”，同时可以看到字体也跟着缩小了。

图9-15显示了点击多次之后，字体缩小之后的效果。

图9-15　字体缩小之后的效果

9.5.7　实现跑马灯效果

当文本过长时，可以设置跑马灯效果，实现文本滚动显示。前提是文本换行关闭且最大显示行数为1，默认情况下即可满足前提要求。

为了演示实现跑马灯效果的过程，通过DevEco Studio创建名为“AutoScrollingAbility”的Page，则会自动创建以下4个文件：AutoScrollingAbility、AutoScrollingAbilitySlice、ability_auto_scrolling.xml和background_ability_auto_scrolling.xml。

修改ability_auto_scrolling.xml内容如下：

```
<?xml version="1.0" encoding="utf-8"?>
<DirectionalLayout
    xmlns:ohos="http://schemas.huawei.com/res/ohos"
    ohos:height="match_parent"
    ohos:width="match_parent"
    ohos:orientation="vertical">

    <Text
        ohos:id="$+id:text_auto_scrolling"
        ohos:background_element="$graphic:background_ability_auto_scrolling"
        ohos:layout_alignment="horizontal_center"
        ohos:text="Hello World"
        ohos:width="75vp"
        ohos:height="match_content"
        ohos:text_size="40vp"
        ohos:text_color="#0000FF"
        ohos:italic="true"
        ohos:text_weight="700"
        ohos:text_font="serif"
        />

</DirectionalLayout>
```

同时修改AutoScrollingAbilitySlice，代码如下：

```
package com.waylau.hmos.text.slice;
```

```
import com.waylau.hmos.text.ResourceTable;
import ohos.aafwk.ability.AbilitySlice;
import ohos.aafwk.content.Intent;
import ohos.agp.components.Text;

public class AutoScrollingAbilitySlice extends AbilitySlice {
    @Override
    public void onStart(Intent intent) {
        super.onStart(intent);
        super.setUIContent(ResourceTable.Layout_ability_auto_scrolling);

        Text textAutoScrolling =
            (Text) findComponentById(ResourceTable.Id_text_auto_scrolling);

        // 跑马灯效果
        textAutoScrolling.setTruncationMode(Text.TruncationMode.AUTO_
SCROLLING);

        // 始终处于自动滚动状态
        textAutoScrolling.setAutoScrollingCount(Text.AUTO_SCROLLING_FOREVER);

        // 启动跑马灯效果
        textAutoScrolling.startAutoScrolling();
    }

    @Override
    public void onActive() {
        super.onActive();
    }

    @Override
    public void onForeground(Intent intent) {
        super.onForeground(intent);
    }
}
```

上述代码启动了跑马灯效果，如图9-16所示。

图9-16　跑马灯效果

9.5.8　场景示例

接下来是一个场景示例，利用文本组件实现一个包含标题栏、详细内容及提交按钮的界面。

为了演示该示例，通过DevEco Studio创建一个名为“TitleDetailAbility”的Page，则会自动创建以下4个文件：TitleDetailAbility、Title

DetailAbilitySlice、ability_title_detail.xml和background_ability_title_detail.xml。

修改ability_title_detail.xml内容如下：

```
<?xml version="1.0" encoding="utf-8"?>
<DirectionalLayout
    xmlns:ohos="http://schemas.huawei.com/res/ohos"
    ohos:height="match_parent"
    ohos:width="match_parent"
    ohos:alignment="center"
    ohos:orientation="vertical">

    <Text
        ohos:id="$+id:text_title"
        ohos:height="match_content"
        ohos:width="match_parent"
        ohos:background_element="$graphic:background_ability_title_detail"
        ohos:left_margin="15vp"
        ohos:right_margin="15vp"
        ohos:text="Title"
        ohos:text_alignment="horizontal_center"
        ohos:text_size="40vp"
        ohos:text_weight="1000"
        ohos:top_margin="15vp"/>

    <Text
        ohos:id="$+id:text_content"
        ohos:height="100vp"
        ohos:width="match_parent"
        ohos:background_element="$graphic:background_ability_title_detail"
        ohos:below="$id:text_title"
        ohos:bottom_margin="15vp"
        ohos:left_margin="15vp"
        ohos:right_margin="15vp"
        ohos:text="Content"
        ohos:text_alignment="center"
        ohos:text_font="serif"
        ohos:text_size="40vp"
        ohos:top_margin="15vp"/>

    <Text
        ohos:id="$+id:text_submit"
        ohos:height="match_content"
        ohos:width="500"
        ohos:align_parent_end="true"
        ohos:background_element="$graphic:background_ability_title_detail"
        ohos:below="$id:text_content"
```

```
        ohos:bottom_margin="15vp"
        ohos:left_padding="5vp"
        ohos:right_margin="15vp"
        ohos:right_padding="5vp"
        ohos:text="Submit"
        ohos:text_font="serif"
        ohos:text_size="40vp"/>
</DirectionalLayout>
```

同时修改background_ability_title_detail.xml文件，内容如下：

```
<?xml version="1.0" encoding="UTF-8" ?>
<shape xmlns:ohos="http://schemas.huawei.com/res/ohos"
       ohos:shape="rectangle">
    <corners
        ohos:radius="20"/>
    <solid
        ohos:color="#878787"/>
</shape>
```

上述代码实现效果如图9-17所示。

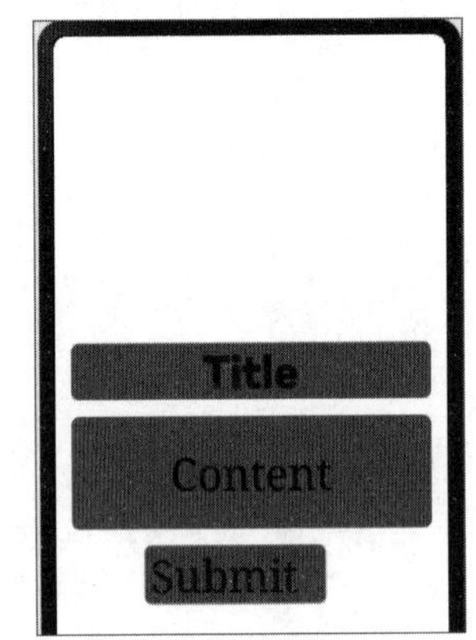

图9-17　最终效果

9.6 实战：常用显示类组件——Image

Image是用来显示图片的组件。创建一个名为“Image”的应用，来作为演示。

9.6.1 创建Image

在“Project”窗口，打开“entry > src > main > resources > base > media”，添加一个图片至media文件夹下，以“waylau_616_616.jpg”为例。

创建Image主要分为两种方式，既可以在XML中创建Image，也可以在代码中创建Image。

1. 在XML中创建Image

在XML中创建Image方式如下：

```
<Image
    ohos:id="$+id:image"
    ohos:width="match_content"
    ohos:height="match_content"
    ohos:layout_alignment="center"
    ohos:image_src="$media:waylau_616_616"/>
```

2. 在代码中创建 Image

在代码中创建Image方式如下：

```
Image image = new Image(getContext());
image.setPixelMap(ResourceTable.Media_plant);
```

3. 修改 ability_main.xml

在本例中采用在XML中创建Image方式，修改ability_main.xml文件，内容如下：

```
<?xml version="1.0" encoding="utf-8"?>
<DirectionalLayout
    xmlns:ohos="http://schemas.huawei.com/res/ohos"
    ohos:height="match_parent"
    ohos:width="match_parent"
    ohos:orientation="vertical">

    <Image
        ohos:id="$+id:image"
        ohos:width="match_content"
        ohos:height="match_content"
        ohos:layout_alignment="center"
        ohos:image_src="$media:waylau_616_616"/>

</DirectionalLayout>
```

上述文件中，ohos:image_src用来配置图片的位置。

最终效果如图9-18所示。

图9-18　界面显示效果

9.6.2　设置透明度

为了演示设置透明度的过程，通过DevEco Studio创建名为“AlphaAbility”的Page，则会自动创建AlphaAbility、AlphaAbilitySlice、ability_alpha.xml和background_ability_alpha.xml这4个文件。

修改ability_alpha.xml内容如下：

```
<?xml version="1.0" encoding="utf-8"?>
<DirectionalLayout
    xmlns:ohos="http://schemas.huawei.com/res/ohos"
    ohos:height="match_parent"
    ohos:width="match_parent"
    ohos:orientation="vertical">

    <Image
```

```
        ohos:id="$+id:image_alpha"
        ohos:width="match_content"
        ohos:height="match_content"
        ohos:layout_alignment="center"
        ohos:image_src="$media:waylau_616_616"
        ohos:alpha="0.3"/>

</DirectionalLayout>
```

在上述代码中，ohos:alpha用于配置透明度。图9-19展示了预览器显示的设置透明度之后的效果。

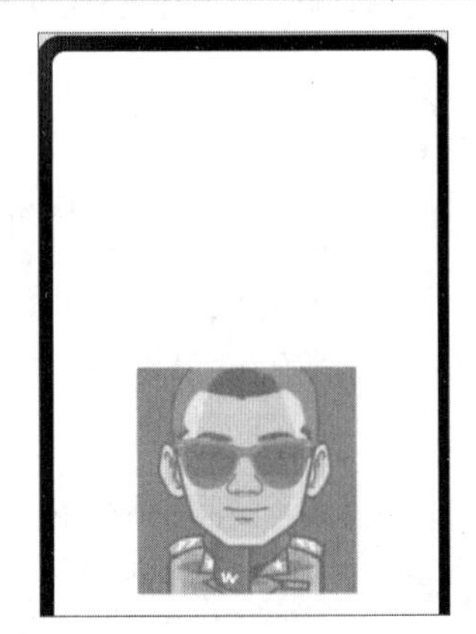

图9-19　设置透明度之后的效果

9.6.3　设置缩放系数

为了演示设置缩放系数的过程，通过DevEco Studio创建名为“ScaleAbility”的Page，则会自动创建以下4个文件：ScaleAbility、ScaleAbilitySlice、ability_scale.xml和background_ability_scale.xml。

修改ability_scale.xml内容如下：

```
<?xml version="1.0" encoding="utf-8"?>
<DirectionalLayout
    xmlns:ohos="http://schemas.huawei.com/res/ohos"
    ohos:height="match_parent"
    ohos:width="match_parent"
    ohos:orientation="vertical">

    <Image
        ohos:id="$+id:image_scale"
        ohos:width="match_content"
        ohos:height="match_content"
        ohos:layout_alignment="center"
        ohos:image_src="$media:waylau_616_616"
        ohos:scale_x="0.5"
        ohos:scale_y="0.5"/>

</DirectionalLayout>
```

图9-20展示了预览器显示的设置缩放系数之后的效果。

图9-20　设置缩放系数之后的效果

9.7 实战：常用显示类组件——ProgressBar

ProgressBar用于显示内容或操作的进度。创建一个名为“ProgressBar”的应用，来作为演示。

9.7.1 创建ProgressBar

在本例中采用在XML中创建ProgressBar方式，修改ability_main.xml文件，内容如下：

```
<?xml version="1.0" encoding="utf-8"?>
<DirectionalLayout
    xmlns:ohos="http://schemas.huawei.com/res/ohos"
    ohos:height="match_parent"
    ohos:width="match_parent"
    ohos:alignment="center"
    ohos:orientation="vertical">

    <ProgressBar
        ohos:id="$+id:progressbar"
        ohos:progress_width="20vp"
        ohos:height="60vp"
        ohos:width="260vp"
        ohos:max="100"
        ohos:min="0"
        ohos:progress="60"/>

</DirectionalLayout>
```

在上述文件中，ProgressBar标签用来创建一个ProgressBar对象。其中ohos:progress用于设置当前进度，ohos:max用于设置最大值，ohos:min用于设置最小值。

最终效果如图9-21所示。

9.7.2 设置方向

默认情况下，ProgressBar方向是水平，也可以设置为垂直。

为了演示设置方向的过程，通过DevEco Studio创建名为“OrientationAbility”的Page，则会自动创建OrientationAbility、OrientationAbilitySlice、ability_orientation.xml和background_ability_orientation.xml这4个文件。

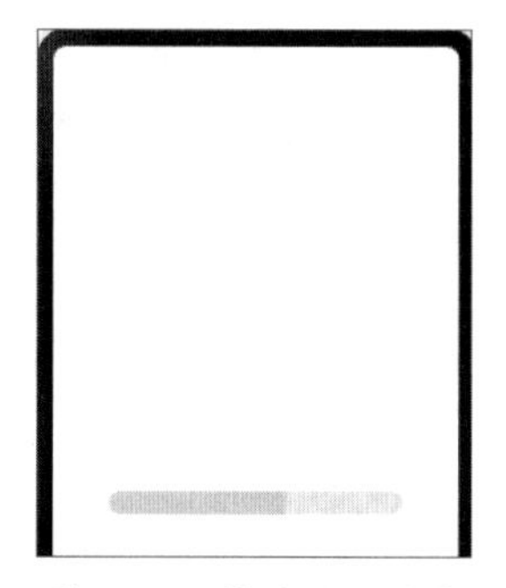
图9-21　界面显示效果

修改ability_orientation.xml内容如下：

```
<?xml version="1.0" encoding="utf-8"?>
<DirectionalLayout
    xmlns:ohos="http://schemas.huawei.com/res/ohos"
    ohos:height="match_parent"
    ohos:width="match_parent"
    ohos:alignment="center"
    ohos:orientation="vertical">

    <ProgressBar
        ohos:id="$+id:progressbar_orientation"
        ohos:progress_width="20vp"
        ohos:height="120vp"
        ohos:width="260vp"
        ohos:max="100"
        ohos:min="0"
        ohos:progress="60"
        ohos:orientation="vertical" />

</DirectionalLayout>
```

在上述代码中，ohos:orientation用于配置方向，本例配置的是垂直。图9-22显示了预览器设置方向之后的效果。

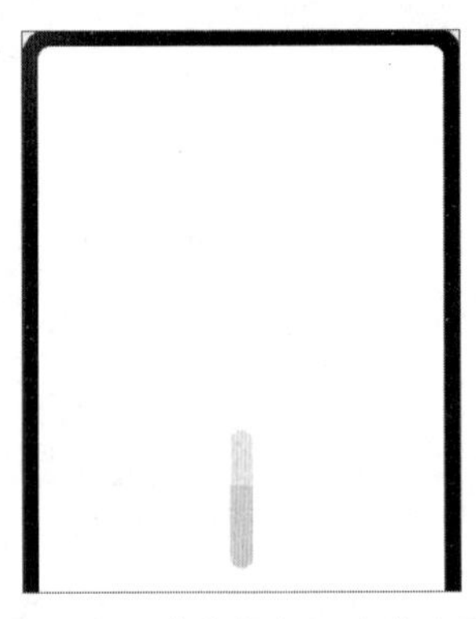
图9-22　设置方向之后的效果

9.7.3　设置颜色

为了演示设置颜色的过程，通过DevEco Studio创建名为“ElementAbility”的Page，则会自动创建以下4个文件：ElementAbility、ElementAbilitySlice、ability_element.xml和background_ability_element.xml。

修改ability_element.xml内容如下：

```
<?xml version="1.0" encoding="utf-8"?>
<DirectionalLayout
    xmlns:ohos="http://schemas.huawei.com/res/ohos"
    ohos:height="match_parent"
    ohos:width="match_parent"
    ohos:alignment="center"
    ohos:orientation="vertical">

    <ProgressBar
        ohos:id="$+id:progressbar_element"
        ohos:progress_width="20vp"
        ohos:height="60vp"
        ohos:width="260vp"
```

```
        ohos:max="100"
        ohos:min="0"
        ohos:progress="60"
        ohos:progress_element="#FF9900"
        ohos:background_instruct_element="#009900" />

</DirectionalLayout>
```

在上述代码中，ohos:progress_element用于配置进度条颜色，而ohos:background_instruct_element用于设置ProgressBar底色。

图9-23展示了预览器显示的设置颜色之后的效果。

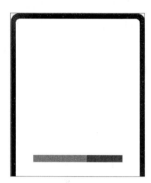

图9-23　设置颜色之后的效果

9.7.4　设置提示文字

为了演示设置提示文字的过程，通过DevEco Studio创建名为“HintAbility”的Page，则会自动创建以下4个文件：HintAbility、HintAbilitySlice、ability_hint.xml和background_ability_hint.xml。

修改ability_hint.xml内容如下：

```
<?xml version="1.0" encoding="utf-8"?>
<DirectionalLayout
    xmlns:ohos="http://schemas.huawei.com/res/ohos"
    ohos:height="match_parent"
    ohos:width="match_parent"
    ohos:alignment="center"
    ohos:orientation="vertical">

    <ProgressBar
        ohos:id="$+id:progressbar_hint"
        ohos:progress_width="20vp"
        ohos:height="60vp"
        ohos:width="260vp"
        ohos:max="100"
        ohos:min="0"
        ohos:progress="60"
        ohos:progress_hint_text="60%"
        ohos:progress_hint_text_color="#FFFC9F"
        />

</DirectionalLayout>
```

在上述代码中，ohos:progress_hint_text用于配置提示的文字内容，而ohos:progress_hint_text_color用于设置提示文字的颜色。

图9-24展示了预览器显示的设置提示文字之后的效果。

图9-24　设置提示文字之后的效果

9.8 实战：常用交互类组件——Button

常用交互类组件包括Button、TextField、Checkbox、RadioButton/RadioContainer、Switch、ScrollView、Tab/TabList、ListContainer、Picker、TimePicker、DatePicker、RoundProgressBar等，这些类提供了具体场景下与用户交互响应的功能，例如，Button提供了点击响应功能，Picker提供了滑动选择功能等。

Button应该是在UI界面设计中使用最为广泛的组件了，因为不管是提交表单，还是执行下一页，都少不了Button组件，点击Button可以触发对应的操作。

Button通常由文本或图标组成，也可以由图标和文本共同组成。

创建一个名为“Button”的应用，来作为演示。

9.8.1 创建Button

以下是创建Button应用时产生的ability_main.xml文件。在该文件中增加了创建Button的描述内容：

```
<?xml version="1.0" encoding="utf-8"?>
<DirectionalLayout
    xmlns:ohos="http://schemas.huawei.com/res/ohos"
    ohos:height="match_parent"
    ohos:width="match_parent"
    ohos:alignment="center"
    ohos:orientation="vertical">

    <Button
        ohos:id="$+id:button"
        ohos:width="match_content"
        ohos:height="match_content"
        ohos:text_size="40vp"
        ohos:text="I am Button"
        ohos:left_margin="15vp"
        ohos:bottom_margin="15vp"
        ohos:right_padding="8vp"
        ohos:left_padding="8vp"
        ohos:background_element="$graphic:background_button"
        />

    <Button
        ohos:id="$+id:button_icon"
        ohos:width="match_content"
```

```
        ohos:height="match_content"
        ohos:text_size="40vp"
        ohos:left_margin="15vp"
        ohos:bottom_margin="15vp"
        ohos:right_padding="8vp"
        ohos:left_padding="8vp"
        ohos:element_left="$graphic:ic_btn_reload"
        ohos:background_element="$graphic:background_button"
        />

    <Button
        ohos:id="$+id:button_icon_text"
        ohos:width="match_content"
        ohos:height="match_content"
        ohos:text_size="40vp"
        ohos:text="I am Button"
        ohos:left_margin="15vp"
        ohos:bottom_margin="15vp"
        ohos:right_padding="8vp"
        ohos:left_padding="8vp"
        ohos:element_left="$graphic:ic_btn_reload"
        ohos:background_element="$graphic:background_button"
        />

</DirectionalLayout>
```

在上述文件中，定义了三个Button组件。ohos:background_element可以用来配置Button的背景。上述配置引用了background_button.xml文件里面的内容，该文件放置在graphic目录下。

新增的background_button.xml文件内容如下：

```
<?xml version="1.0" encoding="UTF-8" ?>
<shape xmlns:ohos="http://schemas.huawei.com/res/ohos"
       ohos:shape="rectangle">
    <corners
        ohos:radius="10"/>
    <solid
        ohos:color="#007CFD"/>
</shape>
```

在定义的三个Button组件中，第一个Button是纯文本的按钮；第二个Button是纯图标的按钮；第三个Button是图标加文本的按钮。图标是通过ohos:element_left配置的。图标的定义配置在ic_btn_reload.xml文件中：

```
<?xml version="1.0" encoding="UTF-8"?>
<vector xmlns:ohos="http://schemas.huawei.com/res/ohos" ohos:width="64vp"
ohos:height="64vp" ohos:viewportWidth="1024" ohos:viewportHeight="1024">
```

```
    <path ohos:fillColor="#FF000000" ohos:pathData="M810.67,512 L952.32,512
741.12,723.2 529.92,512 724.05,512C725.33,446.29 700.59,381.01 650.24,330.67
550.4,230.83 388.27,230.83 288.43,330.67 188.59,430.51 188.59,593.07
288.43,692.91 366.93,771.41 484.69,788.05 579.41,742.83L642.13,805.
55C512,882.77 341.33,865.71 227.84,753.07 94.72,619.95 95.15,404.05
228.27,270.93 362.67,137.39 577.28,136.96 710.83,270.51 777.39,337.07
810.67,424.53 810.67,512Z"></path>
</vector>
```

最终效果如图9-25所示。

I am Button

I am Button

图9-25　界面显示效果

9.8.2　设置点击事件

按钮的重要作用是当用户单击按钮时，会执行相应的操作或界面出现相应的变化。实际上用户点击按钮时，Button对象将收到一个点击事件。开发者可以自定义响应点击事件的方法。例如，通过创建一个Component.ClickedListener对象，然后通过调用setClickedListener将其分配给按钮。这个设置与Text的点击事件类似。

为了演示点击事件的设置过程，通过DevEco Studio创建名为“ClickedListenerAbility”的Page，则会自动创建以下4个文件：ClickedListenerAbility、ClickedListenerAbilitySlice、ability_clicked_listener.xml和background_ability_clicked_listener.xml。

修改ability_clicked_listener.xml内容如下：

```
<?xml version="1.0" encoding="utf-8"?>
<DirectionalLayout
    xmlns:ohos="http://schemas.huawei.com/res/ohos"
    ohos:height="match_parent"
    ohos:width="match_parent"
    ohos:alignment="center"
    ohos:orientation="vertical">

    <Button
        ohos:id="$+id:button_clicked_listener"
        ohos:width="match_content"
        ohos:height="match_content"
        ohos:text_size="38vp"
        ohos:text="I am Button"
        ohos:left_margin="15vp"
        ohos:bottom_margin="15vp"
        ohos:right_padding="8vp"
        ohos:left_padding="8vp"
        ohos:background_element="$graphic:background_button"
        />
```

```
</DirectionalLayout>
```

图9-26展示了预览器显示的创建Button之后的效果。

修改ClickedListenerAbilitySlice，代码如下：

图9-26　创建Button之后的效果

```
package com.waylau.hmos.button.slice;

import com.waylau.hmos.button.ResourceTable;
import ohos.aafwk.ability.AbilitySlice;
import ohos.aafwk.content.Intent;
import ohos.agp.components.Button;

public class ClickedListenerAbilitySlice extends
    AbilitySlice {
    @Override
    public void onStart(Intent intent) {
        super.onStart(intent);
        super.setUIContent(ResourceTable.Layout_ability_clicked_listener);

        Button button =
                (Button) findComponentById(ResourceTable.Id_button_clicked_
                                                        listener);

        // 为按钮设置点击事件回调
        button.setClickedListener(listener ->
                button.setText("Button was clicked!"));
    }

    @Override
    public void onActive() {
        super.onActive();
    }

    @Override
    public void onForeground(Intent intent) {
        super.onForeground(intent);
    }
}
```

上述代码解释如下。

- 获取了Button组件。
- 在Button组件上设置了点击事件。

当点击按钮后，会将按钮上的文本修改为“Button was clicked!”。图9-27所显示的就是点击Button之后的效果。

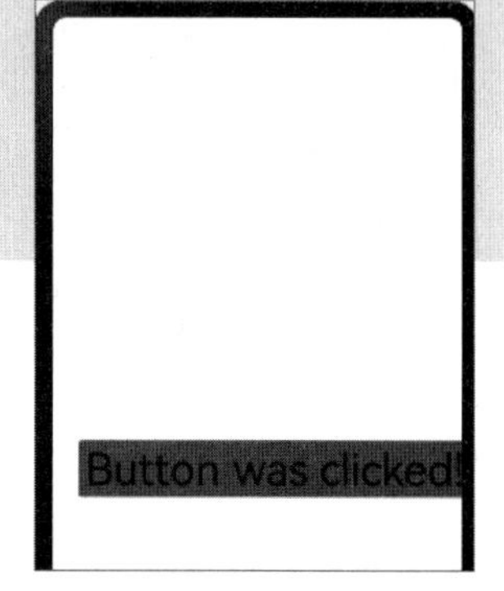

图9-27　点击Button之后的效果

9.8.3 设置椭圆按钮

为了演示设置椭圆按钮的过程，通过DevEco Studio创建名为“OvalAbility”的Page，则会自动创建以下4个文件：OvalAbility、OvalAbilitySlice、ability_oval.xml和background_ability_oval.xml。

椭圆按钮是通过设置background_element来实现的，background_element的shape设置为椭圆（oval）。

修改ability_oval.xml内容如下：

```
<?xml version="1.0" encoding="utf-8"?>
<DirectionalLayout
    xmlns:ohos="http://schemas.huawei.com/res/ohos"
    ohos:height="match_parent"
    ohos:width="match_parent"
    ohos:alignment="center"
    ohos:orientation="vertical">

    <Button
        ohos:id="$+id:button_oval"
        ohos:width="match_content"
        ohos:height="match_content"
        ohos:text_size="40vp"
        ohos:text="I am Button"
        ohos:left_margin="15vp"
        ohos:bottom_margin="15vp"
        ohos:right_padding="8vp"
        ohos:left_padding="8vp"
        ohos:background_element="$graphic:background_ability_oval"
        />

</DirectionalLayout>
```

同时修改background_ability_oval.xml文件，内容如下：

```
<?xml version="1.0" encoding="UTF-8" ?>
<shape xmlns:ohos="http://schemas.huawei.com/res/ohos"
       ohos:shape="oval">
    <solid
        ohos:color="#007CFD"/>
</shape>
```

上述配置将shape设置为椭圆（oval）。图9-28所显示的就是设置椭圆之后的效果。

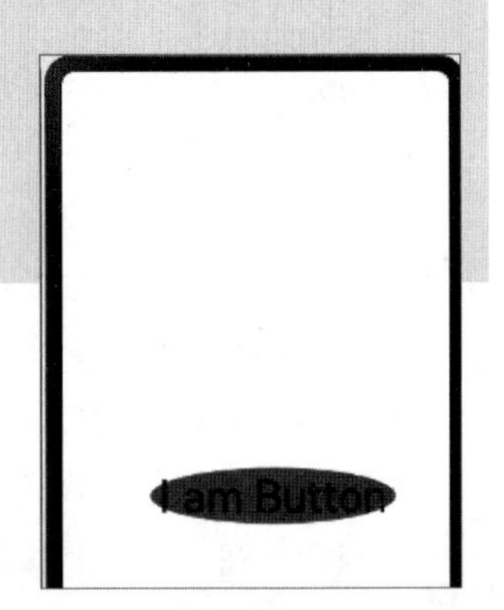

图9-28　设置椭圆之后的效果

9.8.4 设置圆形按钮

为了演示设置圆形按钮的过程，通过DevEco Studio创建名为“CircleAbility”的Page，则会自动创建以下4个文件：CircleAbility、CircleAbilitySlice、ability_circle.xml和background_ability_circle.xml。

圆形按钮和椭圆按钮的区别在于，组件本身的宽度和高度需要相同。

修改ability_circle.xml内容如下：

```
<?xml version="1.0" encoding="utf-8"?>
<DirectionalLayout
    xmlns:ohos="http://schemas.huawei.com/res/ohos"
    ohos:height="match_parent"
    ohos:width="match_parent"
    ohos:alignment="center"
    ohos:orientation="vertical"
    ohos:background_element="$graphic:color_light_gray_element">

    <Text
        ohos:width="match_content"
        ohos:height="match_content"
        ohos:text_size="20fp"
        ohos:text="778907484"
        ohos:background_element="$graphic:green_text_element"
        ohos:text_alignment="center"
        ohos:layout_alignment="horizontal_center"
        />
    <DirectionalLayout
        ohos:width="match_parent"
        ohos:height="match_content"
        ohos:alignment="horizontal_center"
        ohos:orientation="horizontal"
        ohos:top_margin="5vp"
        ohos:bottom_margin="5vp">
        <Button
            ohos:width="40vp"
            ohos:height="40vp"
            ohos:text_size="15fp"
            ohos:background_element="$graphic:green_circle_button_element"
            ohos:text="1"
            ohos:text_alignment="center"
            />
        <Button
            ohos:width="40vp"
            ohos:height="40vp"
```

```
            ohos:text_size="15fp"
            ohos:background_element="$graphic:green_circle_button_element"
            ohos:text="2"
            ohos:left_margin="5vp"
            ohos:right_margin="5vp"
            ohos:text_alignment="center"
            />
        <Button
            ohos:width="40vp"
            ohos:height="40vp"
            ohos:text_size="15fp"
            ohos:background_element="$graphic:green_circle_button_element"
            ohos:text="3"
            ohos:text_alignment="center"
            />
    </DirectionalLayout>
    <DirectionalLayout
        ohos:width="match_parent"
        ohos:height="match_content"
        ohos:alignment="horizontal_center"
        ohos:orientation="horizontal"
        ohos:bottom_margin="5vp">
        <Button
            ohos:width="40vp"
            ohos:height="40vp"
            ohos:text_size="15fp"
            ohos:background_element="$graphic:green_circle_button_element"
            ohos:text="4"
            ohos:text_alignment="center"
            />
        <Button
            ohos:width="40vp"
            ohos:height="40vp"
            ohos:text_size="15fp"
            ohos:left_margin="5vp"
            ohos:right_margin="5vp"
            ohos:background_element="$graphic:green_circle_button_element"
            ohos:text="5"
            ohos:text_alignment="center"
            />
        <Button
            ohos:width="40vp"
            ohos:height="40vp"
            ohos:text_size="15fp"
            ohos:background_element="$graphic:green_circle_button_element"
            ohos:text="6"
```

```
            ohos:text_alignment="center"
            />
    </DirectionalLayout>
    <DirectionalLayout
        ohos:width="match_parent"
        ohos:height="match_content"
        ohos:alignment="horizontal_center"
        ohos:orientation="horizontal"
        ohos:bottom_margin="5vp">
        <Button
            ohos:width="40vp"
            ohos:height="40vp"
            ohos:text_size="15fp"
            ohos:background_element="$graphic:green_circle_button_element"
            ohos:text="7"
            ohos:text_alignment="center"
            />
        <Button
            ohos:width="40vp"
            ohos:height="40vp"
            ohos:text_size="15fp"
            ohos:left_margin="5vp"
            ohos:right_margin="5vp"
            ohos:background_element="$graphic:green_circle_button_element"
            ohos:text="8"
            ohos:text_alignment="center"
            />
        <Button
            ohos:width="40vp"
            ohos:height="40vp"
            ohos:text_size="15fp"
            ohos:background_element="$graphic:green_circle_button_element"
            ohos:text="9"
            ohos:text_alignment="center"
            />
    </DirectionalLayout>
    <DirectionalLayout
        ohos:width="match_parent"
        ohos:height="match_content"
        ohos:alignment="horizontal_center"
        ohos:orientation="horizontal"
        ohos:bottom_margin="5vp">
        <Button
            ohos:width="40vp"
            ohos:height="40vp"
            ohos:text_size="15fp"
```

```
            ohos:background_element="$graphic:green_circle_button_element"
            ohos:text="*"
            ohos:text_alignment="center"
            />
        <Button
            ohos:width="40vp"
            ohos:height="40vp"
            ohos:text_size="15fp"
            ohos:left_margin="5vp"
            ohos:right_margin="5vp"
            ohos:background_element="$graphic:green_circle_button_element"
            ohos:text="0"
            ohos:text_alignment="center"
            />
        <Button
            ohos:width="40vp"
            ohos:height="40vp"
            ohos:text_size="15fp"
            ohos:background_element="$graphic:green_circle_button_element"
            ohos:text="#"
            ohos:text_alignment="center"
            />
    </DirectionalLayout>
    <Button
        ohos:width="match_content"
        ohos:height="match_content"
        ohos:text_size="15fp"
        ohos:text="CALL"
        ohos:background_element="$graphic:green_capsule_button_element"
        ohos:bottom_margin="5vp"
        ohos:text_alignment="center"
        ohos:layout_alignment="horizontal_center"
        ohos:left_padding="10vp"
        ohos:right_padding="10vp"
        ohos:top_padding="2vp"
        ohos:bottom_padding="2vp"
        />
</DirectionalLayout>
```

上述代码采用多个DirectionalLayout进行了嵌套组合。

同时新增了如下样式代码。

1. color_light_gray_element.xml

新增color_light_gray_element.xml代码如下：

```
<?xml version="1.0" encoding="utf-8"?>
```

```
<shape xmlns:ohos="http://schemas.huawei.com/res/ohos"
    ohos:shape="rectangle">
    <solid
        ohos:color="#EDEDED"/>
</shape>
```

2. green_text_element.xml

新增green_text_element.xml，代码如下：

```
<?xml version="1.0" encoding="utf-8"?>
<shape xmlns:ohos="http://schemas.huawei.com/res/ohos"
    ohos:shape="rectangle">
    <corners
        ohos:radius="20"/>
    <stroke
        ohos:width="2"
        ohos:color="#006E00"/>
    <solid
        ohos:color="#EDEDED"/>
</shape>
```

3. green_circle_button_element.xml

新增green_circle_button_element.xml，代码如下：

```
<?xml version="1.0" encoding="utf-8"?>
<shape xmlns:ohos="http://schemas.huawei.com/res/ohos"
    ohos:shape="oval">
    <stroke
        ohos:width="5"
        ohos:color="#006E00"/>
    <solid
        ohos:color="#EDEDED"/>
</shape>
```

4. green_capsule_button_element.xml

新增green_capsule_button_element.xml，代码如下：

```
<?xml version="1.0" encoding="utf-8"?>
<shape xmlns:ohos="http://schemas.huawei.com/res/ohos"
    ohos:shape="rectangle">
    <corners
        ohos:radius="100"/>
    <solid
        ohos:color="#006E00"/>
</shape>
```

上述代码实现效果如图9-29所示。

图9-29　最终效果

9.9 实战：常用交互类组件——TextField

TextField在UI界面设计中，提供了一种文本输入框。

创建一个名为“TextField”的应用，来作为演示。

9.9.1 创建TextField

以下是创建TextField应用时产生的ability_main.xml文件。在该文件中增加了创建TextField组件的描述内容：

```
<?xml version="1.0" encoding="utf-8"?>
<DirectionalLayout
    xmlns:ohos="http://schemas.huawei.com/res/ohos"
    ohos:height="match_parent"
    ohos:width="match_parent"
    ohos:alignment="center"
    ohos:orientation="vertical">


    <TextField
        ohos:id="$+id:textfiled"
        ohos:height="50vp"
        ohos:width="320vp"
        ohos:left_padding="40vp"
        ohos:hint="Enter your name"
        ohos:text_size="30vp"
        ohos:text_alignment="vertical_center"
        ohos:background_element="$graphic:background_ability_main"
        />

</DirectionalLayout>
```

在上述文件中，定义了TextField组件。

- ohos:hint用于设置提示文字。
- ohos:text_alignment设置文字对齐方式是垂直居中。
- ohos:background_element可以用来配置TextField的背景。上述配置引用了background_ability_main.xml文件里面的内容，该文件放置在graphic目录下。

新增的background_ability_main.xml文件内容如下：

```
<?xml version="1.0" encoding="UTF-8" ?>
<shape xmlns:ohos="http://schemas.huawei.com/res/ohos"
```

```
        ohos:shape="rectangle">
    <corners
        ohos:radius="40"/>
    <solid
        ohos:color="#FFFF00"/>
</shape>
```

最终效果如图9-30所示。

图9-30　最终效果

9.9.2　设置多行显示

TextField可以设置为多行显示。

为了演示多行显示的设置过程，通过DevEco Studio创建名为“MultipleLinesAbility”的Page，则会自动创建以下4个文件：MultipleLinesAbility、MultipleLinesAbilitySlice、ability_multiple_lines.xml和background_ability_multiple_lines.xml。

修改ability_multiple_lines.xml，内容如下：

```
<?xml version="1.0" encoding="utf-8"?>
<DirectionalLayout
    xmlns:ohos="http://schemas.huawei.com/res/ohos"
    ohos:height="match_parent"
    ohos:width="match_parent"
    ohos:alignment="center"
    ohos:orientation="vertical">

    <TextField
        ohos:id="$+id:textfiled_multiple_lines"
        ohos:height="120vp"
        ohos:width="320vp"
        ohos:text_size="30vp"
        ohos:left_padding="40vp"
        ohos:hint="Enter your name, your age and your country!"
        ohos:text_alignment="vertical_center"
        ohos:multiple_lines="true"
        ohos:background_element="$graphic:background_ability_multiple_lines"
        />

</DirectionalLayout>
```

在上述代码中，ohos:multiple_lines即为是否启用多行显示的开关。

同时修改background_ability_multiple_lines.xml内容如下：

```
<?xml version="1.0" encoding="UTF-8" ?>
```

```
<shape xmlns:ohos="http://schemas.huawei.com/res/ohos"
       ohos:shape="rectangle">
    <corners
        ohos:radius="40"/>
    <solid
        ohos:color="#FFFF00"/>
</shape>
```

图9-31显示了TextField多行显示的效果。

图9-31　TextField多行显示的效果

9.9.3　场景示例

接下来是一个场景示例，展示一个常见的登录界面。

为了演示该示例，通过DevEco Studio创建名为“LoginAbility”的Page，则会自动创建以下4个文件：LoginAbility、LoginAbilitySlice、ability_login.xml和background_ability_login.xml。

修改ability_login.xml内容如下：

```
<?xml version="1.0" encoding="utf-8"?>
<DirectionalLayout
    xmlns:ohos="http://schemas.huawei.com/res/ohos"
    ohos:height="match_parent"
    ohos:width="match_parent"
    ohos:alignment="center"
    ohos:orientation="vertical">

    <TextField
        ohos:id="$+id:textfield_name"
        ohos:height="match_content"
        ohos:width="320vp"
        ohos:background_element="$graphic:background_ability_login"
        ohos:bottom_padding="8vp"
        ohos:hint="Enter phone number"
        ohos:layout_alignment="center"
        ohos:min_height="44vp"
        ohos:multiple_lines="false"
        ohos:text_alignment="vertical_center"
        ohos:text_size="30vp"
        ohos:top_margin="10vp"
        ohos:top_padding="8vp"/>

    <TextField
        ohos:id="$+id:textfield_password"
        ohos:height="match_content"
        ohos:width="320vp"
```

```
        ohos:background_element="$graphic:background_ability_login"
        ohos:bottom_padding="8vp"
        ohos:hint="Enter password"
        ohos:layout_alignment="center"
        ohos:min_height="44vp"
        ohos:multiple_lines="false"
        ohos:text_alignment="vertical_center"
        ohos:text_size="30vp"
        ohos:top_margin="30vp"
        ohos:top_padding="8vp"/>

    <Button
        ohos:id="$+id:ensure_button"
        ohos:height="40vp"
        ohos:width="120vp"
        ohos:background_element="$graphic:background_ability_login"
        ohos:layout_alignment="horizontal_center"
        ohos:text="Log in"
        ohos:text_size="30vp"
        ohos:top_margin="40vp"/>

</DirectionalLayout>
```

同时修改background_ability_login.xml，代码如下：

```
<?xml version="1.0" encoding="UTF-8" ?>
<shape xmlns:ohos="http://schemas.huawei.com/res/ohos"
      ohos:shape="rectangle">
    <corners
        ohos:radius="40"/>
    <solid
        ohos:color="#FFFF00"/>
    <stroke
        ohos:width="6"
    />
</shape>
```

上述代码实现效果如图9-32所示。

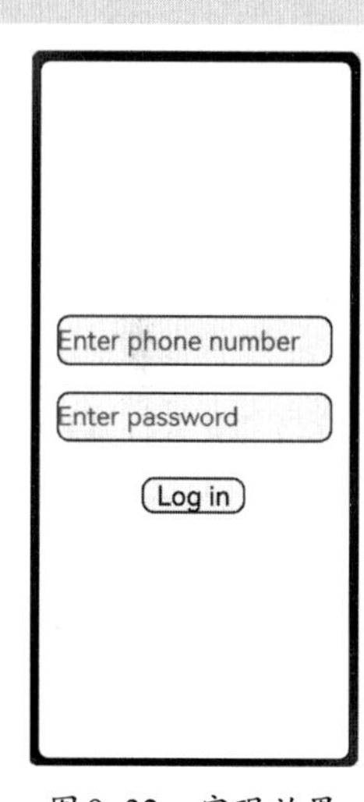

图9-32　实现效果

9.10 实战：常用交互类组件——Checkbox

Checkbox在UI界面设计中，提供了一种实现选中和取消选中的功能。

创建一个名为“Checkbox”的应用，来作为演示。

9.10.1 创建Checkbox

以下是创建Checkbox应用时产生的ability_main.xml文件。在该文件中增加了创建Checkbox组件的描述内容：

```
<?xml version="1.0" encoding="utf-8"?>
<DirectionalLayout
    xmlns:ohos="http://schemas.huawei.com/res/ohos"
    ohos:height="match_parent"
    ohos:width="match_parent"
    ohos:alignment="center"
    ohos:orientation="vertical"
    ohos:background_element="#FFFCCCCC">

    <Checkbox
        ohos:id="$+id:checkbox"
        ohos:height="match_content"
        ohos:width="match_content"
        ohos:text="I am Checkbox"
        ohos:text_size="40vp"/>
</DirectionalLayout>
```

在上述文件中，定义了Checkbox组件。

在初始化时，Checkbox组件如图9-33所示。

点击Checkbox组件之后，显示出了选中状态，如图9-34所示。

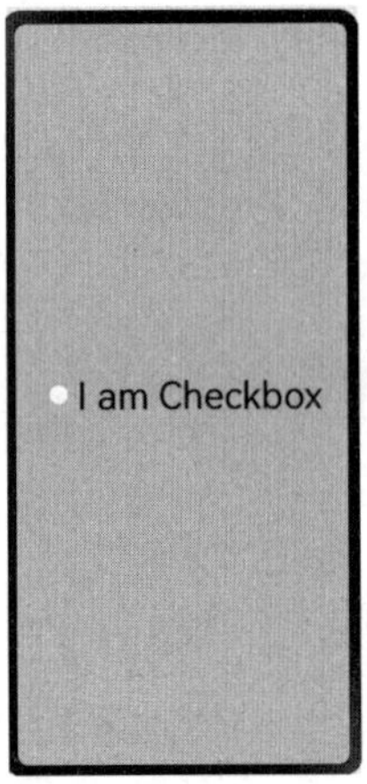

图9-33 初始化显示效果

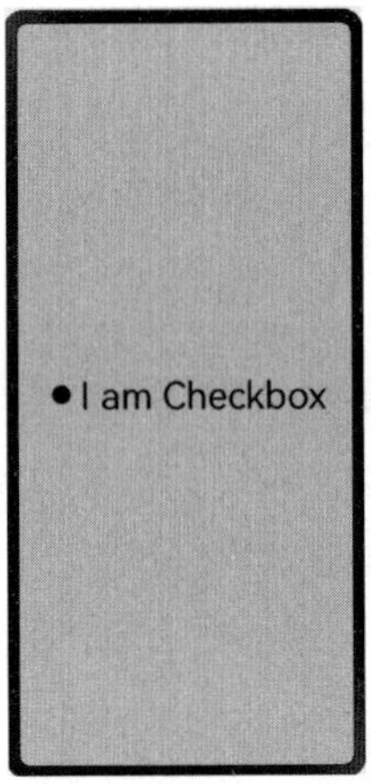

图9-34 选中显示效果

9.10.2 设置选中和取消选中时的颜色

Checkbox可以设置选中和取消选中时的颜色。

为了演示选中和取消选中时的颜色的设置过程，通过DevEco Studio创建名为“OnOffAbility”的Page，则会自动创建以下4个文件：OnOffAbility、OnOffAbilitySlice、ability_on_off.xml和background_ability_on_off.xml。

修改ability_on_off.xml内容如下：

```
<?xml version="1.0" encoding="utf-8"?>
<DirectionalLayout
    xmlns:ohos="http://schemas.huawei.com/res/ohos"
    ohos:height="match_parent"
    ohos:width="match_parent"
```

```
    ohos:orientation="vertical"
    ohos:background_element="#FFFCCCCC">

    <Checkbox
        ohos:id="$+id:checkbox_on_off"
        ohos:height="match_content"
        ohos:width="match_content"
        ohos:text="I am Checkbox"
        ohos:text_size="28fp"
        ohos:text_color_on="#00AAEE"
        ohos:text_color_off="#000000"/>

</DirectionalLayout>
```

在上述代码中，ohos:text_color_on设置为选中时的颜色，而ohos:text_color_off设置为取消时的颜色。

图9-35显示了Checkbox选中时的颜色效果。

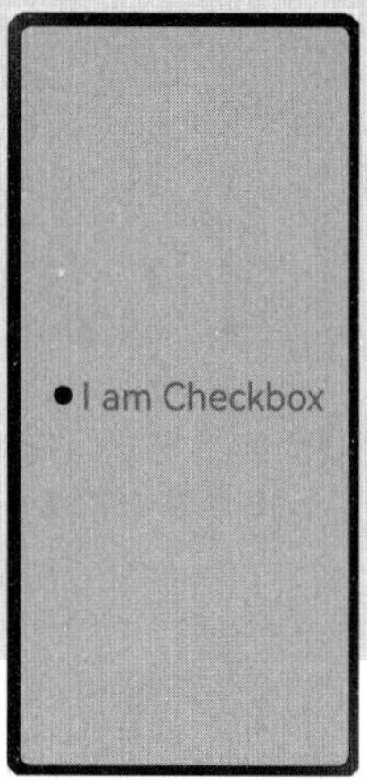

图9-35　Checkbox选中时的颜色效果

9.11 实战：常用交互类组件——RadioButton/RadioContainer

RadioButton用于多选一的操作，需要搭配RadioContainer使用，实现单选效果。

创建一个名为“RadioButtonRadioContainer”的应用，来作为演示。

9.11.1 创建RadioButton/RadioContainer

以下是创建RadioButtonRadioContainer应用时产生的ability_main.xml文件。在该文件中增加了创建RadioButton/RadioContainer组件的描述内容：

```
<?xml version="1.0" encoding="utf-8"?>
<DirectionalLayout
    xmlns:ohos="http://schemas.huawei.com/res/ohos"
    ohos:height="match_parent"
    ohos:width="match_parent"
    ohos:alignment="center"
    ohos:orientation="vertical"
    ohos:background_element="#FFFCCCCC">

    <Text
        ohos:height="match_content"
        ohos:width="match_content"
        ohos:text="最喜欢老卫哪部作品？"
```

```
        ohos:text_size="20fp"
        ohos:layout_alignment="left"
        ohos:multiple_lines="true"/>

    <RadioContainer
        ohos:id="$+id:radio_container"
        ohos:height="match_content"
        ohos:width="match_content">
        <RadioButton
            ohos:id="$+id:radio_button_1"
            ohos:height="30vp"
            ohos:width="match_content"
            ohos:text="1.《Spring Boot 企业级应用开发实战》"
            ohos:text_size="14fp"/>
        <RadioButton
            ohos:id="$+id:radio_button_2"
            ohos:height="30vp"
            ohos:width="match_content"
            ohos:text="2.《Spring Cloud 微服务架构开发实战》"
            ohos:text_size="14fp"/>
        <RadioButton
            ohos:id="$+id:radio_button_3"
            ohos:height="30vp"
            ohos:width="match_content"
            ohos:text="3.《Spring 5 开发大全》"
            ohos:text_size="14fp"/>
        <RadioButton
            ohos:id="$+id:radio_button_4"
            ohos:height="30vp"
            ohos:width="match_content"
            ohos:text="4.《Cloud Native 分布式架构原理与实践》"
            ohos:text_size="14fp"/>
        <RadioButton
            ohos:id="$+id:radio_button_5"
            ohos:height="30vp"
            ohos:width="match_content"
            ohos:text="5.《大型互联网应用轻量级架构实战》"
            ohos:text_size="14fp"/>
        <RadioButton
            ohos:id="$+id:radio_button_6"
            ohos:height="30vp"
            ohos:width="match_content"
            ohos:text="6.《Node.js 企业级应用开发实战》"
            ohos:text_size="14fp"/>
        <RadioButton
            ohos:id="$+id:radio_button_7"
            ohos:height="30vp"
            ohos:width="match_content"
```

```
            ohos:text="7.《Netty原理解析与开发实战》"
            ohos:text_size="14fp"/>
    </RadioContainer>
</DirectionalLayout>
```

在上述文件中，定义了RadioButton/RadioContainer组件。

在初始化时，RadioButton/RadioContainer组件如图9-36所示。

点击RadioButton/RadioContainer组件之后，显示出了选中状态，如图9-37所示。

9.11.2 设置显示单选结果

可以设置响应RadioContainer状态改变的事件，显示单选结果。

为了演示设置显示单选结果的过程，通过DevEco Studio创建了名为“MarkChangedAbility”的Page，则会自动创建以下4个文件：MarkChangedAbility、MarkChangedAbilitySlice、ability_mark_changed.xml和background_ability_mark_changed.xml。

修改ability_mark_changed.xml内容如下：

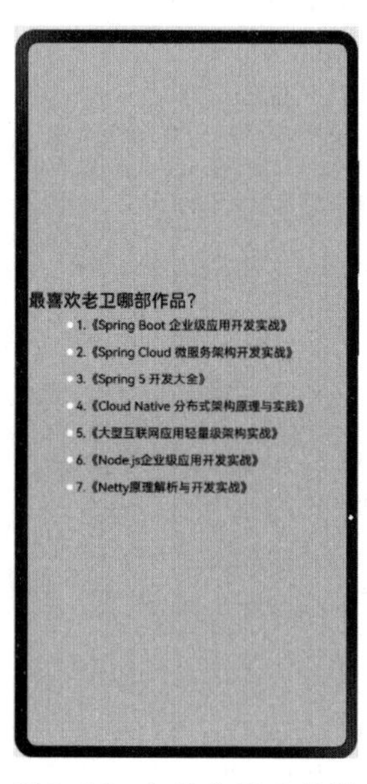

图9-36 初始化显示效果

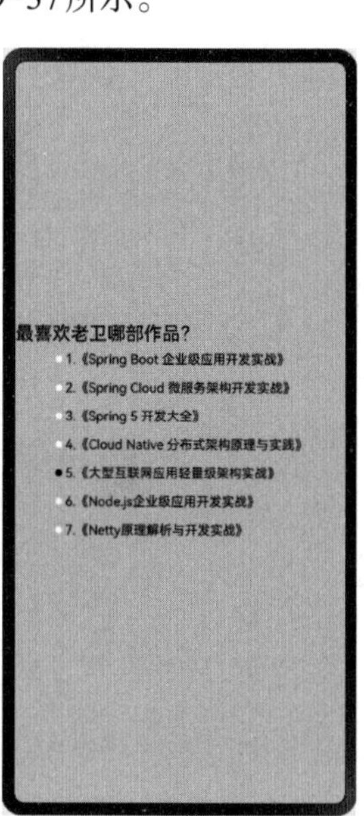

图9-37 选中显示效果

```
<?xml version="1.0" encoding="utf-8"?>
<DirectionalLayout
    xmlns:ohos="http://schemas.huawei.com/res/ohos"
    ohos:height="match_parent"
    ohos:width="match_parent"
    ohos:alignment="center"
    ohos:orientation="vertical"
    ohos:background_element="#FFFCCCCC">

    <DirectionalLayout
        ohos:width="match_parent"
        ohos:height="match_content"
        ohos:orientation="horizontal">
        <Text
            ohos:id="$+id:text_question"
            ohos:height="match_content"
            ohos:width="match_content"
            ohos:text="最喜欢老卫哪部作品？答案是："
```

```
            ohos:text_size="20fp"
            ohos:layout_alignment="left"
            ohos:multiple_lines="true"/>
        <Text
            ohos:id="$+id:text_answer"
            ohos:height="match_content"
            ohos:width="match_content"
            ohos:text=""
            ohos:text_size="20fp"
            ohos:layout_alignment="left"
            ohos:multiple_lines="true"/>
    </DirectionalLayout>

    <RadioContainer
        ohos:id="$+id:radio_container"
        ohos:height="match_content"
        ohos:width="match_content">
        <RadioButton
            ohos:id="$+id:radio_button_1"
            ohos:height="30vp"
            ohos:width="match_content"
            ohos:text="1.《Spring Boot 企业级应用开发实战》"
            ohos:text_size="14fp"/>
        <RadioButton
            ohos:id="$+id:radio_button_2"
            ohos:height="30vp"
            ohos:width="match_content"
            ohos:text="2.《Spring Cloud 微服务架构开发实战》"
            ohos:text_size="14fp"/>
        <RadioButton
            ohos:id="$+id:radio_button_3"
            ohos:height="30vp"
            ohos:width="match_content"
            ohos:text="3.《Spring 5 开发大全》"
            ohos:text_size="14fp"/>
        <RadioButton
            ohos:id="$+id:radio_button_4"
            ohos:height="30vp"
            ohos:width="match_content"
            ohos:text="4.《Cloud Native 分布式架构原理与实践》"
            ohos:text_size="14fp"/>
        <RadioButton
            ohos:id="$+id:radio_button_5"
            ohos:height="30vp"
            ohos:width="match_content"
            ohos:text="5.《大型互联网应用轻量级架构实战》"
```

```
            ohos:text_size="14fp"/>
        <RadioButton
            ohos:id="$+id:radio_button_6"
            ohos:height="30vp"
            ohos:width="match_content"
            ohos:text="6.《Node.js 企业级应用开发实战》"
            ohos:text_size="14fp"/>
        <RadioButton
            ohos:id="$+id:radio_button_7"
            ohos:height="30vp"
            ohos:width="match_content"
            ohos:text="7.《Netty 原理解析与开发实战》"
            ohos:text_size="14fp"/>
    </RadioContainer>

</DirectionalLayout>
```

图9-38 初始化时的效果

图9-38显示了RadioButton/RadioContainer初始化时的效果。

在RadioContainer上设置响应状态改变的事件。修改MarkChangedAbilitySlice代码如下：

```
package com.waylau.hmos.radiobuttonradiocontainer.slice;

import com.waylau.hmos.radiobuttonradiocontainer.ResourceTable;
import ohos.aafwk.ability.AbilitySlice;
import ohos.aafwk.content.Intent;
import ohos.agp.components.RadioContainer;
import ohos.agp.components.Text;

public class MarkChangedAbilitySlice extends AbilitySlice {
    @Override
    public void onStart(Intent intent) {
        super.onStart(intent);
        super.setUIContent(ResourceTable.Layout_ability_mark_changed);

        // 获取 Text
        Text answer = (Text) findComponentById(ResourceTable.Id_text_answer);

        // 获取 RadioContainer
        RadioContainer radioContainer =
            (RadioContainer) findComponentById(ResourceTable.Id_radio_
                                                container);

        // 设置状态监听
        radioContainer.setMarkChangedListener((radioContainer1, index) -> {
            answer.setText((++index) + "");
        });
    }
```

```
    @Override
    public void onActive() {
        super.onActive();
    }

    @Override
    public void onForeground(Intent intent) {
        super.onForeground(intent);
    }
}
```

上述代码，在RadioContainer被选中时，将会选取到RadioButton的索引。由于索引是从0开始的，因此需要增加1。

图9-39显示了RadioButton/RadioContainer被选中时的效果。

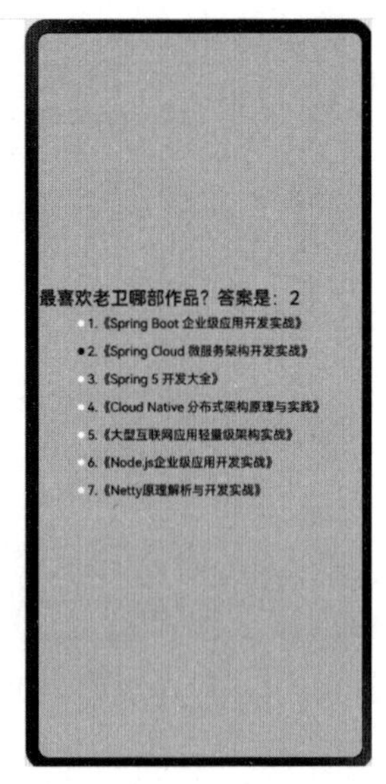

图9-39 被选中时的效果

9.12 实战：常用交互类组件——Switch

Switch是切换单个设置开/关两种状态的组件。

创建一个名为“Switch”的应用，来作为演示。

9.12.1 创建Switch

以下是创建Switch应用时产生的ability_main.xml文件。在该文件中增加了创建Switch组件的描述内容：

```
<?xml version="1.0" encoding="utf-8"?>
<DirectionalLayout
    xmlns:ohos="http://schemas.huawei.com/res/ohos"
    ohos:height="match_parent"
    ohos:width="match_parent"
    ohos:alignment="center"
    ohos:orientation="vertical">

    <Switch
        ohos:id="$+id:btn_switch"
        ohos:height="30vp"
        ohos:width="60vp"/>

</DirectionalLayout>
```

在上述文件中，定义了Switch组件。

在初始化时，Switch组件处于关闭状态，如图9-40所示。

点击Switch组件之后，Switch组件处于开启状态，如图9-41所示。

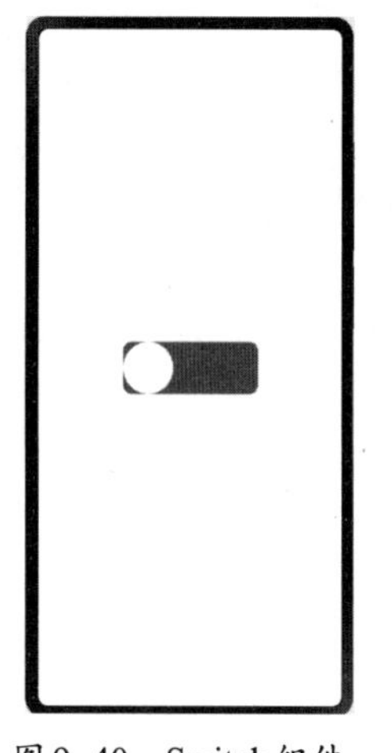

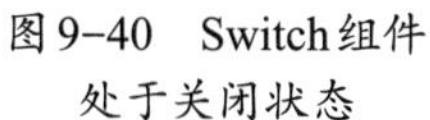

图9-40 Switch组件处于关闭状态

图9-41 Switch组件处于开启状态

9.12.2 设置文本

可以设置Switch在开启和关闭时的文本。

为了演示设置文本的过程，通过DevEco Studio创建名为“TextStateAbility”的Page，则会自动创建以下4个文件：TextStateAbility、TextStateAbilitySlice、ability_text_state.xml和background_ability_text_state.xml。

修改ability_text_state.xml内容如下：

```
<?xml version="1.0" encoding="utf-8"?>
<DirectionalLayout
    xmlns:ohos="http://schemas.huawei.com/res/ohos"
    ohos:height="match_parent"
    ohos:width="match_parent"
    ohos:alignment="center"
    ohos:orientation="vertical">

    <Switch
        ohos:id="$+id:btn_switch"
        ohos:height="60vp"
        ohos:width="160vp"
        ohos:text_size="40vp"
        ohos:text_state_off="OFF"
        ohos:text_state_on="ON"
        />
</DirectionalLayout>
```

图9-42显示了Switch设置文本之后的效果。

图9-42 Switch设置文本之后的效果

9.13 实战：常用交互类组件——ScrollView

ScrollView是一种带滚动功能的组件，它采用滑动的方式在有限的区域内显示更多的内容。这在显示面积非常受限的移动智能终端上非常实用。

下面创建一个名为“ScrollView”的应用，来作为演示。

9.13.1 创建ScrollView

以下是创建ScrollView应用时产生的ability_main.xml文件。在该文件中增加了创建ScrollView组件的描述内容：

```
<?xml version="1.0" encoding="utf-8"?>
<ScrollView
    xmlns:ohos="http://schemas.huawei.com/res/ohos"
    ohos:height="match_parent"
    ohos:width="match_parent"
    ohos:alignment="center"
    ohos:orientation="vertical">

    <Text
        ohos:id="$+id:text_lines"
        ohos:height="match_content"
        ohos:width="match_parent"
        ohos:italic="true"
        ohos:multiple_lines="true"
        ohos:text="$string:text_lines_content"
        ohos:text_color="#0000FF"
        ohos:text_font="serif"
        ohos:text_size="40vp"
        ohos:text_weight="700"/>
</ScrollView>
```

在上述文件中，定义了ScrollView组件。由于ScrollView本身就继承自StackLayout，因此，上述ability_main.xml文件无须再添加布局。

9.13.2 配置Text显示的内容

同时，在element的string.json文件中，新增了text_lines_content的配置，以配置Text显示的内容。配置如下：

```
{
  "string": [
    {
      "name": "entry_MainAbility",
      "value": "entry_MainAbility"
    },
```

```
    {
      "name": "mainability_description",
      "value": "Java_Empty Ability"
    },
    {
      "name": "mainability_HelloWorld",
      "value": "Hello World"
    },
    {
      "name": "text_lines_content",
      "value": "“鸿蒙操作系统”特指华为智能终端操作系统。“鸿蒙操作系统”具有以下三大特征：\n\n一是一套操作系统可以满足大大小小设备需求，实现统一OS，弹性部署；\n二是搭载该操作系统的设备在系统层面融为一体、形成超级终端，让设备的硬件能力可以弹性扩展，实现设备之间硬件互助，资源共享；\n三是面向开发者，实现一次开发，多端部署。"}
  ]
}
```

在初始化时，ScrollView组件显示如图9-43所示。

上述界面无法显示完整的文本内容，因此，往上进行拖动，可以将文本的后续内容进行显示，如图9-44所示。

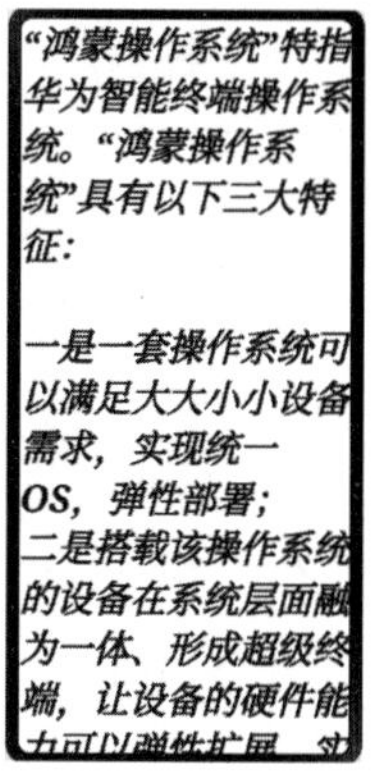

图9-43　无法显示完整的文本内容

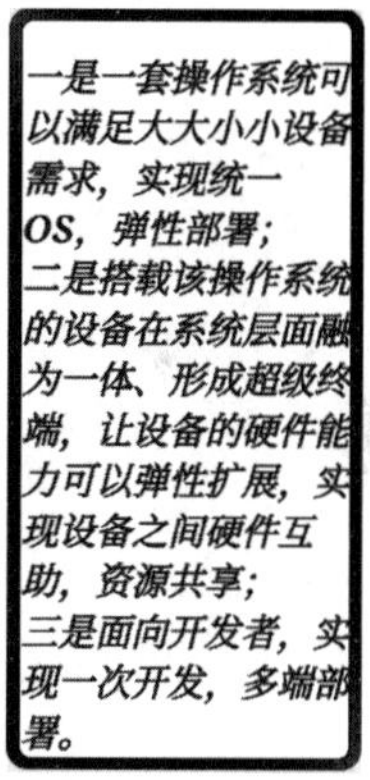

图9-44　拖动显示

9.14 实战：常用交互类组件——Tab/TabList

Tablist可以实现多个页签栏的切换，Tab为某个页签。子页签通常放在内容区上方，展示不同的分类。页签名称应该简洁明了，清晰描述分类的内容。

创建一个名为“TabList”的应用，来作为演示。

9.14.1 创建TabList

以下是创建TabList应用时产生的ability_main.xml文件。在该文件中增加了创建TabList组件的描述内容：

```
<?xml version="1.0" encoding="utf-8"?>
<DirectionalLayout
    xmlns:ohos="http://schemas.huawei.com/res/ohos"
    ohos:height="match_parent"
    ohos:width="match_parent"
    ohos:orientation="vertical">

    <TabList
        ohos:id="$+id:tab_list"
        ohos:height="36vp"
        ohos:width="match_parent"
        ohos:layout_alignment="center"
        ohos:orientation="horizontal"
        ohos:normal_text_color="#999999"
        ohos:selected_text_color="#FF0000"
        ohos:selected_tab_indicator_color="#FF0000"
        ohos:selected_tab_indicator_height="2vp"
        ohos:tab_length="140vp"
        ohos:tab_margin="24vp"
        ohos:text_alignment="center"
        ohos:text_size="30vp"
        ohos:top_margin="10vp"/>

</DirectionalLayout>
```

在上述文件中，定义了TabList组件。

- ohos:normal_text_color定义了默认文本颜色。
- ohos:selected_text_color定义了选中的文本颜色。
- ohos:selected_tab_indicator_color定义了指示器（文本下面的横线）颜色。
- ohos:selected_text_color定义了指示器的高度。

同时，修改MainAbilitySlice代码如下：

```
package com.waylau.hmos.tablist.slice;

import com.waylau.hmos.tablist.ResourceTable;
import ohos.aafwk.ability.AbilitySlice;
import ohos.aafwk.content.Intent;
import ohos.agp.components.TabList;
import ohos.hiviewdfx.HiLog;
```

```
import ohos.hiviewdfx.HiLogLabel;

public class MainAbilitySlice extends AbilitySlice {
    private static final String TAG = MainAbilitySlice.class.getSimpleName();
    private static final HiLogLabel LABEL_LOG =
            new HiLogLabel(HiLog.LOG_APP, 0x00001, TAG);

    @Override
    public void onStart(Intent intent) {
        super.onStart(intent);
        super.setUIContent(ResourceTable.Layout_ability_main);

        // 获取 TabList
        TabList tabList = (TabList) findComponentById(ResourceTable.Id_tab_list);

        // TabList 中添加 Tab
        TabList.Tab tab1 = tabList.new Tab(getContext());
        tab1.setText("Tab1");
        tabList.addTab(tab1);

        TabList.Tab tab2 = tabList.new Tab(getContext());
        tab2.setText("Tab2");
        tabList.addTab(tab2);

        TabList.Tab tab3 = tabList.new Tab(getContext());
        tab3.setText("Tab3");
        tabList.addTab(tab3);

        TabList.Tab tab4 = tabList.new Tab(getContext());
        tab4.setText("Tab4");
        tabList.addTab(tab4);

        // 设置 FixedMode
        tabList.setFixedMode(true);

        // 初始化选中的 Tab
        tabList.selectTab(tab1);



    }

    @Override
    public void onActive() {
        super.onActive();
    }
```

```
    @Override
    public void onForeground(Intent intent) {
        super.onForeground(intent);
    }
}
```

上述代码解释如下。

- 获取 TabList。
- 在 TabList 中添加了 4 个 Tab。
- 设置 FixedMode 为 true。默认为 false，该模式下 TabList 的总宽度是各 Tab 宽度的总和，若固定了 TabList 的宽度，当超出可视区域时，则可以通过滑动 TabList 来显示。如果设置为 true，TabList 的总宽度将与可视区域相同，各个 Tab 的宽度也会根据 TabList 的宽度而平均分配，该模式适用于 Tab 较少的情况。
- 初始化选中的 Tab 为 tab1。

在初始化时，TabList 组件显示如图 9-45 所示。

图 9-45　TabList 组件显示效果

9.14.2　响应焦点变化

可以在 TabList 设置响应焦点变化。代码如下：

```
// 设置响应焦点变化
tabList.addTabSelectedListener(new TabList.TabSelectedListener() {
    @Override
    public void onSelected(TabList.Tab tab) {
        // 当某个 Tab 从未选中状态变为选中状态时的回调
        HiLog.info(LABEL_LOG, "%{public}s, onSelected", tab.getText());
    }

    @Override
    public void onUnselected(TabList.Tab tab) {
        // 当某个 Tab 从选中状态变为未选中状态时的回调
        HiLog.info(LABEL_LOG, "%{public}s, onUnselected", tab.getText());
    }

    @Override
    public void onReselected(TabList.Tab tab) {
        // 当某个 Tab 已处于选中状态，再次被点击时的状态回调
        HiLog.info(LABEL_LOG, "%{public}s, onReselected", tab.getText());
    }
});
```

上述代码在 TabList 上设置了焦点变化的事件，其中，需要重写 TabSelectedListener 的方法。

- onSelected：当某个Tab从未选中状态变为选中状态时的回调。
- onUnselected：当某个Tab从选中状态变为未选中状态时的回调。
- onReselected：当某个Tab已处于选中状态，再次被点击时的状态回调。

运行应用，以观察回调的执行情况。初始化时，默认选中是Tab1。

先点击Tab2，则控制台日志输出如下：

```
 22:55:23.144 6980-6980/com.waylau.hmos.tablist I 00001/MainAbilitySlice:
Tab1, onUnselected
04-26 22:55:23.144 6980-6980/com.waylau.hmos.tablist I 00001/MainAbilitySlice:
Tab2, onSelected
```

再点击Tab3，则控制台日志输出如下：

```
04-26 22:55:32.321 6980-6980/com.waylau.hmos.tablist I 00001/MainAbilitySlice:
Tab2, onUnselected
04-26 22:55:32.321 6980-6980/com.waylau.hmos.tablist I 00001/MainAbilitySlice:
Tab3, onSelected
```

再次点击Tab3，则控制台日志输出如下：

```
04-26 22:55:33.800 6980-6980/com.waylau.hmos.tablist I 00001/MainAbilitySlice:
Tab3, onReselected
```

9.15 实战：常用交互类组件——Picker

Picker提供了滑动选择器，允许用户从预定义范围中进行选择。Picker还有两个特例，分别是日期滑动选择器DatePicker和时间滑动选择器TimePicker。

创建一个名为“Picker”的应用，来作为演示。

9.15.1 创建Picker

以下是创建Picker应用时产生的ability_main.xml文件。在该文件中增加了创建Picker组件的描述内容：

```
<?xml version="1.0" encoding="utf-8"?>
<DirectionalLayout
    xmlns:ohos="http://schemas.huawei.com/res/ohos"
    ohos:height="match_parent"
    ohos:width="match_parent"
    ohos:alignment="center"
```

```
    ohos:orientation="vertical">

    <Picker
        ohos:id="$+id:test_picker"
        ohos:height="match_content"
        ohos:width="300vp"
        ohos:background_element="#E1FFFF"
        ohos:layout_alignment="horizontal_center"
        ohos:normal_text_size="38vp"
        ohos:selected_text_size="40vp"/>

</DirectionalLayout>
```

在上述文件中，定义了Picker组件，可以选择的数字范围是从0到9。Picker组件显示如图9-46所示。

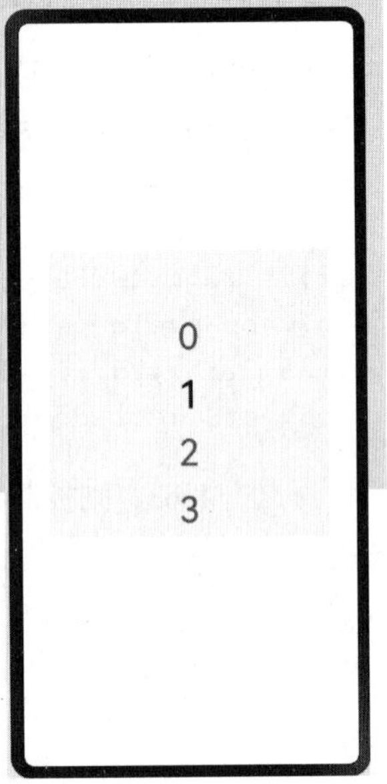

0
1
2
3

图9-46 Picker组件显示效果

9.15.2 格式化Picker的显示

可以在Picker设置格式化的显示。代码如下：

```
package com.waylau.hmos.picker.slice;

import com.waylau.hmos.picker.ResourceTable;
import ohos.aafwk.ability.AbilitySlice;
import ohos.aafwk.content.Intent;
import ohos.agp.components.Picker;

public class MainAbilitySlice extends AbilitySlice {
    @Override
    public void onStart(Intent intent) {
        super.onStart(intent);
        super.setUIContent(ResourceTable.Layout_ability_main);

        // 获取Picker
        Picker picker = (Picker) findComponentById(ResourceTable.Id_test_picker);

        // 设置格式化
        picker.setFormatter(new Picker.Formatter() {
            @Override
            public String format(int i) {
                String value = "";
                switch (i) {
                    case 0:
                        value = "零";
                        break;
```

```
                case 1:
                    value = "一";
                    break;
                case 2:
                    value = "二";
                    break;
                case 3:
                    value = "三";
                    break;
                case 4:
                    value = "四";
                    break;
                case 5:
                    value = "无";
                    break;
                case 6:
                    value = "六";
                    break;
                case 7:
                    value = "七";
                    break;
                case 8:
                    value = "八";
                    break;
                case 9:
                    value = "九";
                    break;
            }

            return value;
        }
    });

}

// 为节约篇幅，省略部分代码
}
```

上述代码在Picker上设置了格式化，将数字转为了中文，Picker组件格式化后显示如图9-47所示。

图9-47　Picker组件格式化后显示效果

9.15.3　日期滑动选择器DatePicker

为了演示日期滑动选择器DatePicker的显示效果，通过DevEco Studio创建名为“DatePickerAbility”的Page，则会自动创建以下4个文件：DatePickerAbility、DatePickerAbilitySlice、ability_date_picker.xml和background_ability_date_picker.xml。

修改ability_date_picker.xml内容如下：

```
<?xml version="1.0" encoding="utf-8"?>
<DirectionalLayout
    xmlns:ohos="http://schemas.huawei.com/res/ohos"
    ohos:height="match_parent"
    ohos:width="match_parent"
    ohos:alignment="center"
    ohos:orientation="vertical">

    <DatePicker
        ohos:id="$+id:date_pick"
        ohos:height="match_content"
        ohos:width="match_parent"
        ohos:normal_text_size="38vp"
        ohos:selected_text_size="40vp">
    </DatePicker>

</DirectionalLayout>
```

DatePicker的显示效果如图9-48所示。

图9-48　DatePicker的显示效果

9.15.4　时间滑动选择器TimePicker

为了演示日期滑动选择器TimePicker的显示效果，通过DevEco Studio创建名为“TimePickerAbility”的Page，则会自动创建以下4个文件：TimePickerAbility、TimePickerAbilitySlice、ability_time_picker.xml和background_ability_time_picker.xml。

修改ability_time_picker.xml内容如下：

```
<?xml version="1.0" encoding="utf-8"?>
<DirectionalLayout
    xmlns:ohos="http://schemas.huawei.com/res/ohos"
    ohos:height="match_parent"
    ohos:width="match_parent"
    ohos:alignment="center"
    ohos:orientation="vertical">

    <TimePicker
        ohos:id="$+id:time_picker"
        ohos:height="match_content"
        ohos:width="match_parent"
```

```
        ohos:normal_text_size="38vp"
        ohos:selected_text_size="40vp"/>

</DirectionalLayout>
```

TimePicker的显示效果如图9-49所示。

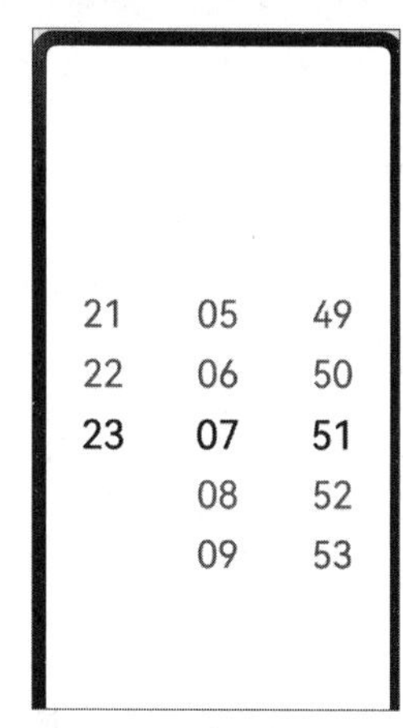

图9-49　TimePicker的显示效果

9.16 实战：常用交互类组件——ListContainer

ListContainer是用来呈现连续、多行数据的组件，包含一系列相同类型的列表项。

创建一个名为“ListContainer”的应用，来作为演示。

9.16.1　创建ListContainer

以下是创建ListContainer应用时产生的ability_main.xml文件。在该文件中增加了创建ListContainer组件的描述内容：

```
<?xml version="1.0" encoding="utf-8"?>
<DirectionalLayout
    xmlns:ohos="http://schemas.huawei.com/res/ohos"
    ohos:height="match_parent"
    ohos:width="match_parent"
    ohos:alignment="center"
    ohos:orientation="vertical">

    <ListContainer
        ohos:id="$+id:list_container"
        ohos:height="match_content"
        ohos:width="match_content"
        ohos:layout_alignment="horizontal_center"/>
</DirectionalLayout>
```

在上述文件中，定义了ListContainer组件。

9.16.2　创建ListContainer子布局

在layout目录下新建xml文件（本例为my_item.xml），作为ListContainer的子布局。内容如下：

```
<?xml version="1.0" encoding="utf-8"?>
<DirectionalLayout
    xmlns:ohos="http://schemas.huawei.com/res/ohos"
    ohos:height="match_parent"
    ohos:width="match_parent"
    ohos:orientation="vertical">

    <Text
        ohos:id="$+id:item_index"
        ohos:height="40vp"
        ohos:width="match_content"
        ohos:text="Item0"
        ohos:text_size="40vp"
        ohos:layout_alignment="center"/>
</DirectionalLayout>
```

9.16.3 创建ListContainer数据包装类

创建MyItem.java，作为ListContainer的数据包装类。代码如下：

```
package com.waylau.hmos.listcontainer;

public class MyItem {
    private String name;

    public MyItem(String name) {
        this.name = name;
    }

    public String getName() {
        return name;
    }

    public void setName(String name) {
        this.name = name;
    }
}
```

9.16.4 创建ListContainer数据提供者

ListContainer每一行可以为不同的数据，因此需要适配不同的数据结构，使其都能添加到ListContainer上。创建MyItemProvider.java，继承自RecycleItemProvider。代码如下：

```
package com.waylau.hmos.listcontainer;
```

```
import com.waylau.hmos.listcontainer.slice.MainAbilitySlice;
import ohos.aafwk.ability.AbilitySlice;
import ohos.agp.components.*;

import java.util.List;

public class MyItemProvider extends RecycleItemProvider {
    private List<MyItem> list;
    private AbilitySlice slice;

    public MyItemProvider(List<MyItem> list, MainAbilitySlice slice) {
        this.list = list;
        this.slice = slice;
    }

    @Override
    public int getCount() {
        return list.size();
    }

    @Override
    public Object getItem(int position) {
        return list.get(position);
    }

    @Override
    public long getItemId(int position) {
        return position;
    }

    @Override
    public Component getComponent(int position,
        Component convertComponent, ComponentContainer componentContainer) {
        Component cpt = convertComponent;
        if (cpt == null) {
            cpt = LayoutScatter.getInstance(slice)
                .parse(ResourceTable.Layout_my_item, null, false);
        }
        MyItem sampleItem = list.get(position);
        Text text = (Text) cpt.findComponentById(ResourceTable.Id_item_
index);
        text.setText(sampleItem.getName());
        return cpt;
    }
}
```

9.16.5 修改MainAbilitySlice

修改MainAbilitySlice，在Java代码中添加ListContainer的数据，并适配其数据结构。代码修改如下：

```
package com.waylau.hmos.listcontainer.slice;

import com.waylau.hmos.listcontainer.MyItem;
import com.waylau.hmos.listcontainer.MyItemProvider;
import com.waylau.hmos.listcontainer.ResourceTable;
import ohos.aafwk.ability.AbilitySlice;
import ohos.aafwk.content.Intent;
import ohos.agp.components.ListContainer;

import java.util.ArrayList;
import java.util.List;

public class MainAbilitySlice extends AbilitySlice {
    @Override
    public void onStart(Intent intent) {
        super.onStart(intent);
        super.setUIContent(ResourceTable.Layout_ability_main);

        // 获取 ListContainer
        ListContainer listContainer = (ListContainer)
findComponentById(ResourceTable.Id_list_container);

        // 提供 ListContainer 的数据
        List<MyItem> list = getData();
        MyItemProvider sampleItemProvider = new MyItemProvider(list, this);
        listContainer.setItemProvider(sampleItemProvider);
    }

    private ArrayList<MyItem> getData() {
        ArrayList<MyItem> list = new ArrayList<>();
        for (int i = 0; i <= 8; i++) {
            list.add(new MyItem("Item" + i));
        }
        return list;
    }

    // 为节约篇幅，省略部分代码
}
```


图9-50　最终显示效果

最终显示效果如图9-50所示。

9.17 实战：常用交互类组件——RoundProgressBar

RoundProgressBar继承自ProgressBar，拥有ProgressBar的属性，在设置同样的属性时用法和ProgressBar一致，用于显示环形进度。

创建一个名为“RoundProgressBar”的应用，来作为演示。

9.17.1 创建RoundProgressBar

以下是创建RoundProgressBar应用时产生的ability_main.xml文件。在该文件中增加了创建RoundProgressBar的描述内容：

```
<?xml version="1.0" encoding="utf-8"?>
<DirectionalLayout
    xmlns:ohos="http://schemas.huawei.com/res/ohos"
    ohos:height="match_parent"
    ohos:width="match_parent"
    ohos:alignment="center"
    ohos:orientation="vertical">

    <RoundProgressBar
        ohos:id="$+id:round_progress_bar"
        ohos:height="300vp"
        ohos:width="300vp"
        ohos:progress_width="30vp"
        ohos:progress="20"
        ohos:progress_color="#47CC47"/>

</DirectionalLayout>
```

在上述文件中，定义了一个RoundProgressBar组件。ohos:progress_width可以用来配置RoundProgressBar的环的宽度；ohos:progress用来配置RoundProgressBar的进度；而ohos:progress_color用来配置环的颜色。

最终效果如图9-51所示。

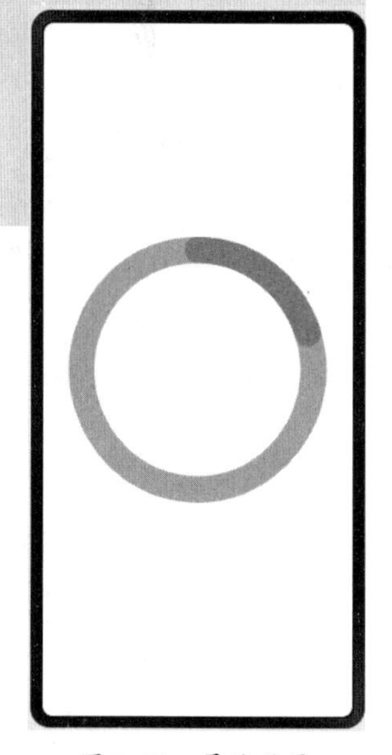

图9-51　最终效果

9.17.2 设置开始和结束角度

可以设置环形进度条的开始和结束位置所在的角度。

为了演示设置开始和结束角度的过程，通过DevEco Studio创建名为“StartMaxAngleAbility”的Page，则会自动创建以下4个文件：

StartMaxAngleAbility、StartMaxAngleAbilitySlice、ability_start_max_angle.xml和background_ability_start_max_angle.xml。

修改ability_start_max_angle.xml内容如下：

```
<?xml version="1.0" encoding="utf-8"?>
<DirectionalLayout
    xmlns:ohos="http://schemas.huawei.com/res/ohos"
    ohos:height="match_parent"
    ohos:width="match_parent"
    ohos:alignment="center"
    ohos:orientation="vertical">

    <RoundProgressBar
        ohos:id="$+id:round_progress_bar_start_max_angle"
        ohos:height="300vp"
        ohos:width="300vp"
        ohos:progress_width="30vp"
        ohos:progress="20"
        ohos:progress_color="#47CC47"
        ohos:start_angle="45"
        ohos:max_angle="270"
        ohos:progress_hint_text="RoundProgressBar"
        ohos:progress_hint_text_color="#007DFF"
        />

</DirectionalLayout>
```

图9-52展示了预览器显示的设置开始和结束角度之后的效果。

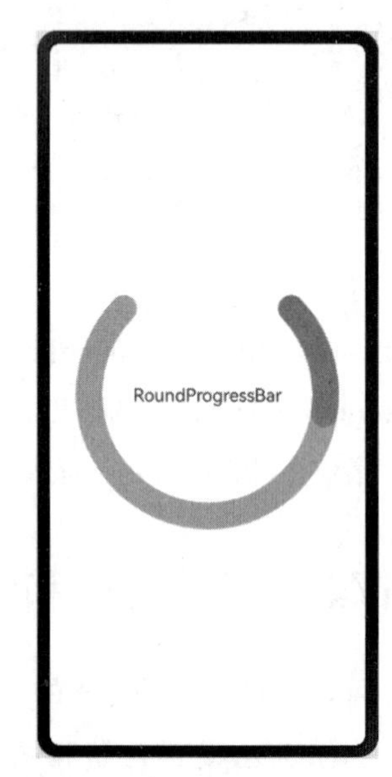

图9-52　预览器显示的设置开始和结束角度之后的效果

上述代码解释如下。

- ohos:start_angle设置了开始角度。
- ohos:max_angle设置了结束角度。
- ohos:progress_hint_text设置了提示文本内容。
- ohos:progress_hint_text_color设置了提示文本颜色。

9.18 实战：常用布局——DirectionalLayout

在Java UI中，常用的布局包括DirectionalLayout、StackLayout、

DependentLayout、TableLayout等。

DirectionalLayout是Java UI中一种非常重要的组件布局，在前面的章节中也频繁出现过。DirectionalLayout用于将一组组件按照水平或垂直方向排布，能够方便地对齐布局内的组件。该布局和其他布局的组合，可以实现更加丰富的布局方式。

创建一个名为“DirectionalLayout”的应用，来作为演示。

9.18.1 创建DirectionalLayout

以下是创建DirectionalLayout应用时产生的ability_main.xml文件。在该文件中增加了创建DirectionalLayout的描述内容：

```
<?xml version="1.0" encoding="utf-8"?>
<DirectionalLayout
    xmlns:ohos="http://schemas.huawei.com/res/ohos"
    ohos:height="match_parent"
    ohos:width="match_parent"
    ohos:alignment="center"
    ohos:orientation="vertical">

    <Button
        ohos:width="320vp"
        ohos:height="100vp"
        ohos:bottom_margin="3vp"
        ohos:left_margin="13vp"
        ohos:text_size="40vp"
        ohos:background_element="$graphic:background_ability_main"
        ohos:text="I am Button1"/>
    <Button
        ohos:width="320vp"
        ohos:height="100vp"
        ohos:bottom_margin="3vp"
        ohos:left_margin="13vp"
        ohos:text_size="40vp"
        ohos:background_element="$graphic:background_ability_main"
        ohos:text="I am Button2"/>
    <Button
        ohos:width="320vp"
        ohos:height="100vp"
        ohos:bottom_margin="3vp"
        ohos:left_margin="13vp"
        ohos:text_size="40vp"
        ohos:background_element="$graphic:background_ability_main"
        ohos:text="I am Button3"/>
```

```
</DirectionalLayout>
```

在上述文件中，定义了1个DirectionalLayout布局及3个Button。

DirectionalLayout布局中的ohos:orientation属性，可以用来指定排列方向是水平（horizontal）还是垂直（vertical）方向。默认为垂直排列。

Button的ohos:background_element定义在background_ability_main.xml文件中，内容如下：

```
<?xml version="1.0" encoding="UTF-8" ?>
<shape xmlns:ohos="http://schemas.huawei.com/res/ohos"
       ohos:shape="rectangle">
    <solid
        ohos:color="#00FFFD"/>
</shape>
```

最终效果如图9-53所示。

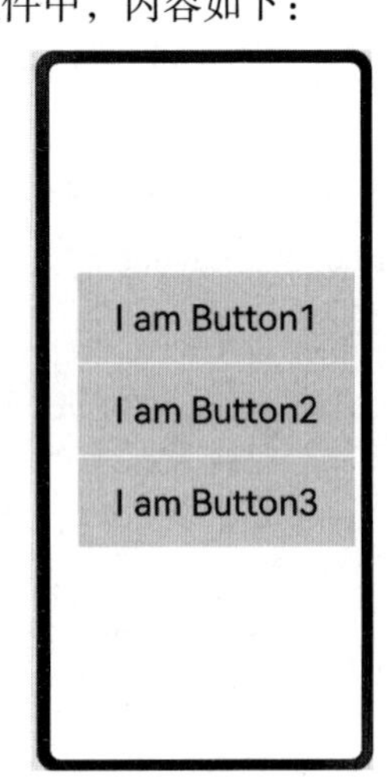

图9-53 DirectionalLayout的效果

9.18.2 设置水平排列

为了演示水平排列的设置过程，通过DevEco Studio创建名为“HorizontalAbility”的Page，则会自动创建以下4个文件：HorizontalAbility、HorizontalAbilitySlice、ability_horizontal.xml和background_ability_horizontal.xml。

修改ability_horizontal.xml内容如下：

```
<?xml version="1.0" encoding="utf-8"?>
<DirectionalLayout
    xmlns:ohos="http://schemas.huawei.com/res/ohos"
    ohos:height="match_parent"
    ohos:width="match_parent"
    ohos:alignment="center"
    ohos:orientation="horizontal">

    <Button
        ohos:width="100vp"
        ohos:height="60vp"
        ohos:bottom_margin="3vp"
        ohos:left_margin="13vp"
        ohos:text_size="16vp"
        ohos:background_element="$graphic:background_ability_main"
        ohos:text="I am Button1"/>
    <Button
        ohos:width="100vp"
```

```
        ohos:height="60vp"
        ohos:bottom_margin="3vp"
        ohos:left_margin="13vp"
        ohos:text_size="16vp"
        ohos:background_element="$graphic:background_ability_main"
        ohos:text="I am Button2"/>
    <Button
        ohos:width="100vp"
        ohos:height="60vp"
        ohos:bottom_margin="3vp"
        ohos:left_margin="13vp"
        ohos:text_size="16vp"
        ohos:background_element="$graphic:background_ability_main"
        ohos:text="I am Button3"/>

</DirectionalLayout>
```

上述代码将布局的ohos:orientatio属性改为了horizontal（水平）

图9-54展示了预览器显示的DirectionalLayout水平排列的效果。

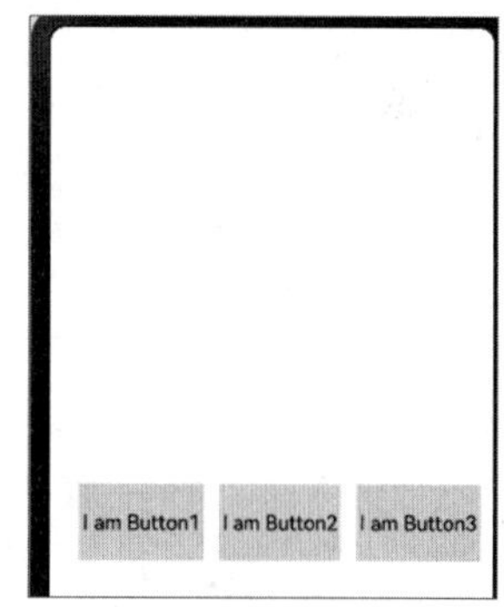

图9-54　DirectionalLayout水平排列的效果

9.18.3　设置权重

权重（weight）就是按比例来分配组件占用父组件的大小。在水平布局下，计算公式为：

```
父布局可分配宽度 = 父布局宽度 - 所有子组件 width 之和

组件宽度 = 组件 weight/ 所有组件 weight 之和 × 父布局可分配宽度
```

在实际使用过程中，建议使用width=0来按比例分配父布局的宽度。

为了演示权重的设置过程，通过DevEco Studio创建名为“WeightAbility”的Page，则会自动创建以下4个文件：WeightAbility、WeightAbilitySlice、ability_weight.xml和background_ability_weight.xml。

修改ability_weight.xml内容如下：

```
<?xml version="1.0" encoding="utf-8"?>
<DirectionalLayout
    xmlns:ohos="http://schemas.huawei.com/res/ohos"
    ohos:height="match_parent"
    ohos:width="match_parent"
    ohos:alignment="center"
```

```
    ohos:orientation="horizontal">

    <Button
        ohos:width="0vp"
        ohos:height="60vp"
        ohos:bottom_margin="3vp"
        ohos:left_margin="13vp"
        ohos:text_size="12vp"
        ohos:weight="1"
        ohos:background_element="$graphic:background_ability_main"
        ohos:text="I am Button1"/>
    <Button
        ohos:width="0vp"
        ohos:height="60vp"
        ohos:bottom_margin="3vp"
        ohos:left_margin="13vp"
        ohos:text_size="12vp"
        ohos:weight="1"
        ohos:background_element="$graphic:background_ability_main"
        ohos:text="I am Button2"/>
    <Button
        ohos:width="0vp"
        ohos:height="60vp"
        ohos:bottom_margin="3vp"
        ohos:left_margin="13vp"
        ohos:text_size="12vp"
        ohos:weight="2"
        ohos:background_element="$graphic:background_ability_main"
        ohos:text="I am Button3"/>

</DirectionalLayout>
```

上述代码按1:1:2的比例来分配父布局的宽度，效果如图9-55所示。

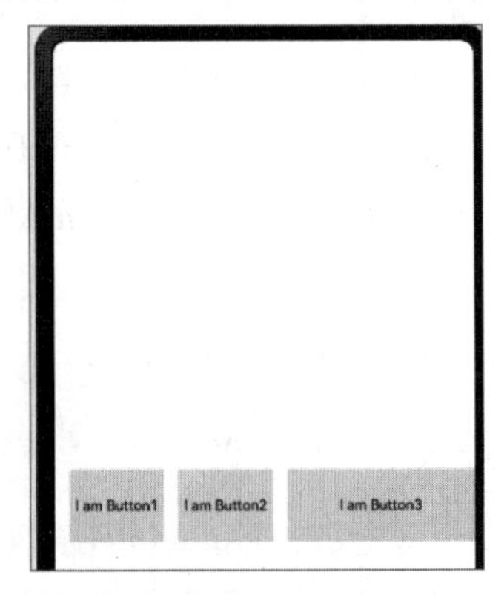

图 9-55　DirectionalLayout 的设置权重效果

9.19 实战：常用交互类组件——DependentLayout

DependentLayout是Java UI系统里的一种常见布局，与DirectionalLayout相比，其拥有更多的排布方式，每个组件可以指定相对于其他同级元素的位置，或者指定相对于父组件的位置。

下面创建一个名为“DependentLayout”的应用，来作为演示。

9.19.1 创建DependentLayout

以下是创建DependentLayout应用时，产生的ability_main.xml文件。在该文件中增加了创建DependentLayout的描述内容：

```
<?xml version="1.0" encoding="utf-8"?>
<DependentLayout
    xmlns:ohos="http://schemas.huawei.com/res/ohos"
    ohos:height="match_parent"
    ohos:width="match_parent"
    ohos:background_element="#ADEDED">
    <Text
        ohos:id="$+id:text1"
        ohos:width="match_content"
        ohos:height="match_content"
        ohos:left_margin="20vp"
        ohos:top_margin="20vp"
        ohos:bottom_margin="20vp"
        ohos:text="text1"
        ohos:text_size="20fp"
        ohos:background_element="#DDEDED"/>
    <Text
        ohos:id="$+id:text2"
        ohos:width="match_content"
        ohos:height="match_content"
        ohos:left_margin="20vp"
        ohos:top_margin="20vp"
        ohos:right_margin="20vp"
        ohos:bottom_margin="20vp"
        ohos:text="end_of text1"
        ohos:text_size="20fp"
        ohos:background_element="#FDEDED"
        ohos:end_of="$id:text1"/>

</DependentLayout>
```

在上述文件中，定义了一个DependentLayout布局及两个Text。两个Text使用了“相对于同级组件”的布局方式，即text2在处于同级组件的text1的结束侧。

最终效果如图9-56所示。

text1 end_of text1

图9-56 最终效果

9.19.2 相对于同级组件

上述示例使用了“相对于同级组件”布局方式。相对于同级组件的位置布局可配置项如下。

- above：处于同级组件的上侧。
- below：处于同级组件的下侧。
- start_of：处于同级组件的起始侧。
- end_of：处于同级组件的结束侧。
- left_of：处于同级组件的左侧。
- right_of：处于同级组件的右侧。

9.19.3 相对于父组件

相对于父组件的位置布局可配置项如下。

- align_parent_left：处于父组件的左侧。
- align_parent_right：处于父组件的右侧。
- align_parent_start：处于父组件的起始侧。
- align_parent_end：处于父组件的结束侧。
- align_parent_top：处于父组件的上侧。
- align_parent_bottom：处于父组件的下侧。
- center_in_parent：处于父组件的中间。

以上位置布局可以组合，形成处于左上角、左下角、右上角、右下角的布局。

9.19.4 场景示例

使用DependentLayout可以轻松实现内容丰富的布局。

为了演示该示例，通过DevEco Studio创建名为“FullAbility”的Page，则会自动创建以下4个文件：FullAbility、FullAbilitySlice、ability_full.xml和background_ability_full.xml。

修改ability_full.xml内容如下：

```
<?xml version="1.0" encoding="utf-8"?>
<DependentLayout
    xmlns:ohos="http://schemas.huawei.com/res/ohos"
    ohos:width="match_parent"
    ohos:height="match_content"
    ohos:background_element="#ADEDED">
    <Text
        ohos:id="$+id:text1"
        ohos:width="match_parent"
```

```
        ohos:height="match_content"
        ohos:text_size="25fp"
        ohos:top_margin="15vp"
        ohos:left_margin="15vp"
        ohos:right_margin="15vp"
        ohos:background_element="#DDEDED"
        ohos:text="Title"
        ohos:text_weight="1000"
        ohos:text_alignment="horizontal_center"
        />
    <Text
        ohos:id="$+id:text2"
        ohos:width="match_content"
        ohos:height="680vp"
        ohos:text_size="10vp"
        ohos:background_element="#FDEDED"
        ohos:text="Catalog"
        ohos:top_margin="15vp"
        ohos:left_margin="15vp"
        ohos:right_margin="15vp"
        ohos:bottom_margin="15vp"
        ohos:align_parent_left="true"
        ohos:text_alignment="center"
        ohos:multiple_lines="true"
        ohos:below="$id:text1"
        ohos:text_font="serif"/>
    <Text
        ohos:id="$+id:text3"
        ohos:width="match_parent"
        ohos:height="680vp"
        ohos:text_size="25fp"
        ohos:background_element="#FDEDED"
        ohos:text="Content"
        ohos:top_margin="15vp"
        ohos:right_margin="15vp"
        ohos:bottom_margin="15vp"
        ohos:text_alignment="center"
        ohos:below="$id:text1"
        ohos:end_of="$id:text2"
        ohos:text_font="serif"/>
    <Button
        ohos:id="$+id:button1"
        ohos:width="70vp"
        ohos:height="match_content"
        ohos:text_size="15fp"
        ohos:background_element="#FDEDED"
```

```
        ohos:text="Previous"
        ohos:right_margin="15vp"
        ohos:bottom_margin="15vp"
        ohos:below="$id:text3"
        ohos:left_of="$id:button2"
        ohos:italic="false"
        ohos:text_weight="5"
        ohos:text_font="serif"/>
    <Button
        ohos:id="$+id:button2"
        ohos:width="70vp"
        ohos:height="match_content"
        ohos:text_size="15fp"
        ohos:background_element="#FDEDED"
        ohos:text="Next"
        ohos:right_margin="15vp"
        ohos:bottom_margin="15vp"
        ohos:align_parent_end="true"
        ohos:below="$id:text3"
        ohos:italic="false"
        ohos:text_weight="5"
        ohos:text_font="serif"/>
</DependentLayout>
```

上述代码效果如图9-57所示。

图9-57　场景示例效果

9.20 实战：常用交互类组件——StackLayout

StackLayout直接在屏幕上开辟出一块空白的区域，添加到这个布局中的视图都是以层叠的方式显示，而它会把这些视图默认放到这块区域的左上角，第一个添加到布局中的视图显示在最底层，最后一个被放在最顶层。上一层的视图会覆盖下一层的视图。

创建一个名为“StackLayout”的应用，来作为演示。

以下是创建StackLayout应用时产生的ability_main.xml文件。在该文件中增加了创建StackLayout的描述内容：

```
<?xml version="1.0" encoding="utf-8"?>
<StackLayout
    xmlns:ohos="http://schemas.huawei.com/res/ohos"
```

```
    ohos:height="match_parent"
    ohos:width="match_parent">

    <Text
        ohos:text_alignment="bottom|horizontal_center"
        ohos:text_size="24fp"
        ohos:text="Layer 1"
        ohos:height="match_parent"
        ohos:width="match_parent"
        ohos:background_element="#3F56EA" />

    <Text
        ohos:text_alignment="bottom|horizontal_center"
        ohos:text_size="24fp"
        ohos:text="Layer 2"
        ohos:height="100vp"
        ohos:width="300vp"
        ohos:background_element="#00AAEE" />

    <Text
        ohos:text_alignment="center"
        ohos:text_size="24fp"
        ohos:text="Layer 3"
        ohos:height="80vp"
        ohos:width="80vp"
        ohos:background_element="#00BFC9" />

</StackLayout>
```

在上述文件中，定义了1个StackLayout布局及3个Text。StackLayout中组件的布局默认在区域的左上角，并且以后创建的组件会在上层，即text1处于最下层，而text3处于最上层。

最终效果如图9-58所示。

图 9-58　StackLayout的效果

9.21 实战：常用交互类组件——TableLayout

顾名思义，TableLayout使用表格的方式划分子组件。

创建一个名为“TableLayout”的应用，来作为演示。

以下是创建TableLayout应用时产生的ability_main.xml文件。在该文件中增加了创建TableLayout的描述内容：

```
<?xml version="1.0" encoding="utf-8"?>
<TableLayout
    xmlns:ohos="http://schemas.huawei.com/res/ohos"
    ohos:height="match_parent"
    ohos:width="match_parent"
    ohos:background_element="#87CEEB"
    ohos:layout_alignment="horizontal_center"
    ohos:padding="8vp"
    ohos:row_count="2"
    ohos:column_count="2">

    <Text
        ohos:height="100vp"
        ohos:width="100vp"
        ohos:background_element="$graphic:background_ability_main"
        ohos:margin="8vp"
        ohos:text_alignment="center"
        ohos:text_size="40vp"
        ohos:text="1"/>

    <Text
        ohos:height="100vp"
        ohos:width="100vp"
        ohos:background_element="$graphic:background_ability_main"
        ohos:margin="8vp"
        ohos:text_alignment="center"
        ohos:text_size="40vp"
        ohos:text="2"/>

    <Text
        ohos:height="100vp"
        ohos:width="100vp"
        ohos:background_element="$graphic:background_ability_main"
        ohos:margin="8vp"
        ohos:text_alignment="center"
        ohos:text_size="40vp"
        ohos:text="3"/>

    <Text
        ohos:height="100vp"
        ohos:width="100vp"
        ohos:background_element="$graphic:background_ability_main"
        ohos:margin="8vp"
        ohos:text_alignment="center"
```

```
        ohos:text_size="40vp"
        ohos:text="4"/>
</TableLayout>
```

在上述文件中，定义了1个TableLayout布局及4个Text。其中，ohos:row_count设置了行数，而ohos:column_count设置了列数。

Text引用的样式定义在background_ability_main.xml文件中，内容如下：

```
<?xml version="1.0" encoding="utf-8"?>
<shape xmlns:ohos="http://schemas.huawei.com/res/ohos"
       ohos:shape="rectangle">
    <corners
        ohos:radius="5vp"/>
    <stroke
        ohos:width="1vp"
        ohos:color="gray"/>
    <solid
        ohos:color="#00BFFF"/>
</shape>
```

最终效果如图9-59所示。

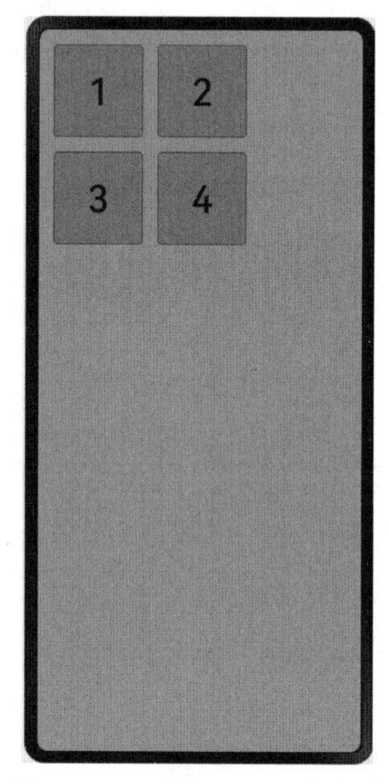

图9-59　TableLayout的效果

第10章 设备管理

HarmonyOS 设备是底层硬件的一种设备抽象概念。HarmonyOS 提供了一系列 API 针对不同的设备进行管理。

10.1 设备管理概述

HarmonyOS设备是对底层硬件的一种设备抽象概念。

HarmonyOS提供了一系列API针对不同的设备进行管理，这些设备包括以下几种。

- 传感器。传感器分为六大类：运动类传感器、环境类传感器、方向类传感器、光线类传感器、健康类传感器、其他类传感器（如霍尔传感器）。
- 控制类小器件。控制类小器件指的是设备上的LED灯和振动器。其中，LED灯主要用作指示（如充电状态）、闪烁功能（如三色灯）等；振动器主要用于闹钟、开关机振动、来电振动等场景。
- 位置。位置能力用于确定用户设备在哪里，系统使用位置坐标标示设备的位置，并用多种定位技术提供服务，如GNSS定位、基站定位、WLAN/蓝牙定位（基站定位、WLAN/蓝牙定位后续统称为“网络定位技术”）。通过这些定位技术，无论用户设备在室内或是户外，都可以准确地确定设备位置。
- 设置。应用程序可以对系统各类设置项进行查询。例如，三方应用提前注册飞行模式设置项的回调，当用户通过系统设置修改终端的飞行模式状态时，三方应用会检测到此设置项发生变化并进行适配。如检测到飞行模式开启，将进入离线状态；如检测到飞行模式关闭，其将重新获取在线数据。

10.1.1 传感器

HarmonyOS传感器包含如下4个模块：Sensor API、Sensor Framework、Sensor Service、HD_IDL层。图10-1展示的是HarmonyOS传感器的架构。

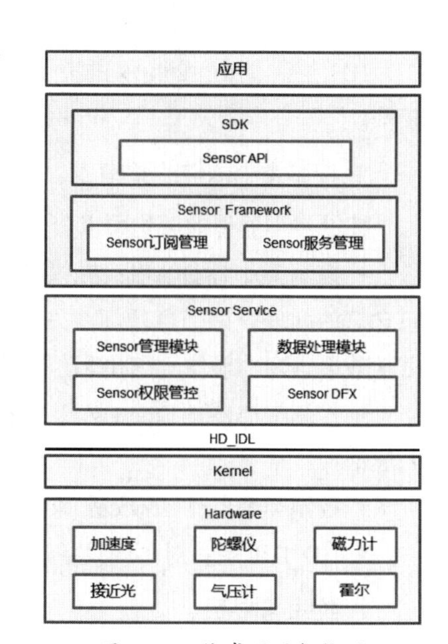

图10-1 传感器的架构图

1. 传感器的架构

传感器的架构可以分为以下几层。

- Sensor API：提供传感器的基础API，主要包含查询传感器的列表、订阅/取消传感器的数据、执行控制命令等，简化应用开发。
- Sensor Framework：主要实现传感器的订阅管理，数据通道的创建、销毁、订阅与取消订阅，实现与SensorService的通信。
- Sensor Service：主要实现HD_IDL层数据接收、解析、分发，前后台的策略管控，对该设备Sensor的管理，Sensor权限管控等。
- HD_IDL层：对不同的FIFO、频率进行策略选择，以及对不同设备进行适配。

2. 约束与限制

针对某些传感器，开发者需要请求相应的权限，才能获取到相应传感器的数据。表10-1展示的是传感器权限列表。

表10-1 传感器权限列表

传感器	权限名	敏感级别	权限描述
加速度传感器、加速度未校准传感器、线性加速度传感器	ohos.permission.ACCELEROMETER	system_grant	允许订阅Motion组对应的加速度传感器的数据
陀螺仪传感器、陀螺仪未校准传感器	ohos.permission.GYROSCOPE	system_grant	允许订阅Motion组对应的陀螺仪传感器的数据
计步器	ohos.permission.ACTIVITY_MOTION	user_grant	允许订阅运动状态
心率	ohos.permission.READ_HEALTH_DATA	user_grant	允许读取健康数据

传感器数据订阅和取消订阅接口成对调用，当不再需要订阅传感器数据时，开发者需要调用取消订阅接口进行资源释放。

10.1.2 控制类小器件

HarmonyOS控制类小器件主要包含以下4个模块：控制类小器件API、控制类小器件Framework、控制类小器件Service、HD_IDL层。图10-2展示的是HarmonyOS控制类小器件的架构。

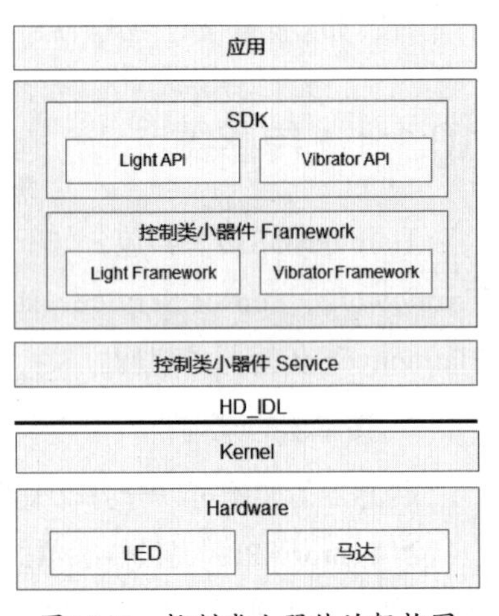

图10-2 控制类小器件的架构图

1. 控制类小器件的架构

控制类小器件的架构可以分为以下几层。

- 控制类小器件API：提供灯和振动器基础的API，主要包含灯的列表查询，打开灯、关闭灯等接口，振动器的列表和振动效果查询，触发/关闭振动器等接口。
- 控制类小器件Framework：主要实现灯和振动器的框架层管理，实现与控制类小器件Service的通信。
- 控制类小器件Service：实现灯和振动器的服务管理。
- HD_IDL层：对不同设备进行适配。

2. 约束与限制

使用控制类小器件时需要注意以下约束与限制。

- 在调用Light API时，请先通过getLightIdList接口查询设备所支持的灯的ID列表，以免调用打开接口异常。

• 在调用Vibrator API时，请先通过getVibratorIdList接口查询设备所支持的振动器的ID列表，以免调用振动接口异常。

• 在使用振动器时，开发者需要配置请求振动器的权限ohos.permission.VIBRATE，才能控制振动器振动。

10.1.3 位置

移动终端设备已经深入人们日常生活的方方面面，如查看所在城市的天气、新闻轶事、出行打车、旅行导航、运动记录。这些习以为常的活动，都离不开定位用户终端设备的位置。

当用户处于这些丰富的使用场景中时，系统的位置能力可以提供实时准确的位置数据。对于开发者，设计基于位置体验的服务，也可以使应用的使用体验更贴近每个用户。

当应用在实现基于设备位置的功能时，如驾车导航、记录运动轨迹等，可以调用该模块的API接口，完成位置信息的获取。

1. 基本概念

位置能力包含以下基本概念。

• 坐标：系统以1984年世界大地坐标系统为参考，使用经度、纬度数据描述地球上的一个位置。

• GNSS定位：全球导航卫星系统，包含GPS、GLONASS、北斗、Galileo等。通过导航卫星、设备芯片提供的定位算法，来确定设备准确位置。定位过程具体使用哪些定位系统，取决于用户设备的硬件能力。

• 基站定位：根据设备当前驻网基站和相邻基站的位置，估算设备当前位置。此定位方式的定位结果精度相对较低，并且需要设备可以访问蜂窝网络。

• WLAN、蓝牙定位：根据设备可搜索到的周围WLAN、蓝牙设备位置，估算设备当前位置。此定位方式的定位结果精度依赖设备周围可见的固定WLAN、蓝牙设备的分布，密度较高时，精度也比基站定位方式更高，同时也需要设备可以访问网络。

2. 运作机制

位置能力作为系统为应用提供的一种基础服务，需要应用在所使用的业务场景中，向系统主动发起请求，并在业务场景结束时，主动结束此请求，在此过程中系统会将实时的定位结果上报给应用。

3. 约束与限制

使用位置时需要注意以下约束与限制。

• 使用设备的位置能力，需要用户进行确认并主动开启位置开关。如果位置开关没有开启，系统不会向任何应用提供位置服务。

• 设备位置信息属于用户敏感数据，所以即使用户已经开启位置开关，应用在获取设备位置前仍需向用户申请位置访问权限。在用户确认允许后，系统才会向应用提供位置服务。

10.2 实战：传感器示例

本节演示如何获取方向传感器的数据。

HarmonyOS 传感器提供的功能有查询传感器的列表、订阅 / 取消订阅传感器数据、查询传感器的最小采样时间间隔、执行控制命令。

10.2.1 接口说明

方向类别的传感器 CategoryOrientationAgent 的主要接口有以下几种。

- getAllSensors()：获取属于方向类别的传感器列表。
- getAllSensors(int)：获取属于方向类别中特定类型的传感器列表。
- getSingleSensor(int)：查询方向类别中特定类型的默认 sensor（如果存在多个则返回第一个）。
- setSensorDataCallback(ICategoryOrientationDataCallback, CategoryOrientation, long)：以设定的采样间隔订阅给定传感器的数据。
- setSensorDataCallback(ICategoryOrientationDataCallback, CategoryOrientation, long, long)：以设定的采样间隔和时延订阅给定传感器的数据。
- releaseSensorDataCallback(ICategoryOrientationDataCallback, CategoryOrientation)：取消订阅指定传感器的数据。
- releaseSensorDataCallback(ICategoryOrientationDataCallback)：取消订阅所有的传感器数据。

10.2.2 创建应用

为了演示获取方向传感器的数据的功能，创建一个名为“CategoryOrientationAgent”的应用。

在应用的界面上，通过点击按钮，来触发获取方向传感器的数据的操作。

10.2.3 修改 ability_main.xml

修改 ability_main.xml 内容如下：

```
<?xml version="1.0" encoding="utf-8"?>
<DirectionalLayout
    xmlns:ohos="http://schemas.huawei.com/res/ohos"
    ohos:height="match_parent"
    ohos:width="match_parent"
    ohos:alignment="center"
    ohos:orientation="vertical">
```

```
    <Button
        ohos:id="$+id:button_add"
        ohos:height="match_content"
        ohos:width="match_parent"
        ohos:background_element="#F76543"
        ohos:layout_alignment="horizontal_center"
        ohos:margin="10vp"
        ohos:padding="10vp"
        ohos:text="Add"
        ohos:text_size="30fp"
        />

    <Button
        ohos:id="$+id:button_remove"
        ohos:height="match_content"
        ohos:width="match_parent"
        ohos:background_element="#F76543"
        ohos:layout_alignment="horizontal_center"
        ohos:margin="10vp"
        ohos:padding="10vp"
        ohos:text="Remove"
        ohos:text_size="30fp"
        />

</DirectionalLayout>
```

界面预览效果如图10-3所示。

图10-3　界面预览效果

上述代码，设置了“Add”按钮和“Remove”按钮，以备设置点击事件，以触发获取方向传感器数据的相关操作。

10.2.4　修改MainAbilitySlice

修改MainAbilitySlice内容如下：

```
package com.waylau.hmos.categoryorientationagent.slice;

import com.waylau.hmos.categoryorientationagent.ResourceTable;
import ohos.aafwk.ability.AbilitySlice;
import ohos.aafwk.content.Intent;
import ohos.agp.components.Button;
import ohos.hiviewdfx.HiLog;
import ohos.hiviewdfx.HiLogLabel;
import ohos.sensor.agent.CategoryOrientationAgent;
import ohos.sensor.bean.CategoryOrientation;
import ohos.sensor.data.CategoryOrientationData;
```

```
import ohos.sensor.listener.ICategoryOrientationDataCallback;
import ohos.utils.zson.ZSONObject;

public class MainAbilitySlice extends AbilitySlice {
    private static final String TAG = MainAbilitySlice.class.getSimpleName();
    private static final HiLogLabel LABEL_LOG =
            new HiLogLabel(HiLog.LOG_APP, 0x00001, TAG);
    private static final int MATRIX_LENGTH = 9;
    private static final long INTERVAL = 100000000L;

    private ICategoryOrientationDataCallback orientationDataCallback;

    private CategoryOrientationAgent categoryOrientationAgent;

    private CategoryOrientation orientationSensor;

    @Override
    public void onStart(Intent intent) {
        super.onStart(intent);
        super.setUIContent(ResourceTable.Layout_ability_main);

        // 初始化对象
        initData();

        // 为按钮设置点击事件回调
        Button buttonAdd =
                (Button) findComponentById(ResourceTable.Id_button_add);
        buttonAdd.setClickedListener(listener -> add());

        Button buttonRemove =
                (Button) findComponentById(ResourceTable.Id_button_remove);
        buttonRemove.setClickedListener(listener -> remove());
    }

    private void initData() {
        orientationDataCallback = new ICategoryOrientationDataCallback() {
            @Override
            public void onSensorDataModified(
                CategoryOrientationData categoryOrientationData) {
                // 对接收的 categoryOrientationData 传感器数据对象解析和使用
                //获取传感器的维度信息
                int dim = categoryOrientationData.getSensorDataDim();

                // 获取方向类传感器的第一维数据
                float degree = categoryOrientationData.getValues()[0];
```

```
            // 根据旋转矢量传感器的数据获得旋转矩阵
            float[] rotationMatrix = new float[MATRIX_LENGTH];
            CategoryOrientationData.getDeviceRotationMatrix(rotationMatrix,
                    categoryOrientationData.values);

            // 根据计算出来的旋转矩阵获取设备的方向
            float[] rotationAngle = new float[MATRIX_LENGTH];
            float[] rotationAngleResult =
                    CategoryOrientationData.getDeviceOrientation(rotation
                        Matrix, rotationAngle);

            HiLog.info(LABEL_LOG, "dim:%{public}s, degree: %{public}s, " +
                "rotationMatrix: %{public}s, rotationAngle: %{public}s",
                dim, degree, ZSONObject.toZSONString(rotationMatrix),
                ZSONObject.toZSONString(rotationAngleResult));
        }

        @Override
        public void onAccuracyDataModified(
            CategoryOrientation categoryOrientation, int i) {
            // 使用变化的精度
            HiLog.info(LABEL_LOG, "onAccuracyDataModified");
        }

        @Override
        public void onCommandCompleted(CategoryOrientation
                                        categoryOrientation) {
            // 传感器执行命令回调
            HiLog.info(LABEL_LOG, "onCommandCompleted");
        }
    };

    categoryOrientationAgent = new CategoryOrientationAgent();
}

private void remove() {
    HiLog.info(LABEL_LOG, "before remove");

    if (orientationSensor != null) {
        // 取消订阅传感器数据
        categoryOrientationAgent.releaseSensorDataCallback(
                orientationDataCallback, orientationSensor);
    }

    HiLog.info(LABEL_LOG, "end remove");
}
```

```
    private void add() {
        HiLog.info(LABEL_LOG, "before add");

        // 获取传感器对象，并订阅传感器数据
        orientationSensor = categoryOrientationAgent.getSingleSensor(
                CategoryOrientation.SENSOR_TYPE_ORIENTATION);
        if (orientationSensor != null) {
            categoryOrientationAgent.setSensorDataCallback(
                    orientationDataCallback, orientationSensor, INTERVAL);
        }

        HiLog.info(LABEL_LOG, "end add");
    }

    @Override
    public void onActive() {
        super.onActive();
    }

    @Override
    public void onForeground(Intent intent) {
        super.onForeground(intent);
    }
}
```

上述代码解释如下。

- 在Button上设置了点击事件。
- initData用于初始化ICategoryOrientationDataCallback和CategoryOrientationAgent对象。
- add方法用于订阅传感器数据。
- remove方法用于取消订阅传感器数据。

10.2.5 运行

运行应用后，点击界面按钮“Add”以触发订阅传感器数据的操作执行。此时，控制台输出内容如下：

```
04-27 17:58:18.660 4782-4782/com.waylau.hmos.categoryorientationagent I
00001/MainAbilitySlice: before add
04-27 17:58:18.707 4782-4782/com.waylau.hmos.categoryorientationagent I
00001/MainAbilitySlice: end add
04-27 17:58:18.775 4782-5986/com.waylau.hmos.categoryorientationagent I
00001/MainAbilitySlice: dim:3, degree: 304.9, rotationMatrix: [-14.3298,-
```

```
195.13599,-1676.95,-195.13599,-185942.14,1.76,-1676.95,1.76,-185927.22],
rotationAngle: [-3.1405432,-9.465305E-6,3.1325736,0.0,0.0,0.0,0.0,0.0,0.0]
04-27 17:58:19.364 4782-5986/com.waylau.hmos.categoryorientationagent
I 00001/MainAbilitySlice: dim:3, degree: 304.77, rotationMatrix:
[-15.023399,-207.24359,-1712.8073,-207.24359,-185784.3,1.9108,-
1712.8073,1.9108,-185768.73], rotationAngle: [-3.1404772,-1.0285039E-
5,3.1323729,0.0,0.0,0.0,0.0,0.0,0.0]
04-27 17:58:19.956 4782-5986/com.waylau.hmos.categoryorientationagent I
00001/MainAbilitySlice: dim:3, degree: 304.9, rotationMatrix: [-14.3298,-
195.13599,-1676.95,-195.13599,-185942.14,1.76,-1676.95,1.76,-185927.22],
rotationAngle: [-3.1405432,-9.465305E-6,3.1325736,0.0,0.0,0.0,0.0,0.0,0.0]
04-27 17:58:20.552 4782-5986/com.waylau.hmos.categoryorientationagent
I 00001/MainAbilitySlice: dim:3, degree: 304.77, rotationMatrix:
[-15.023399,-207.24359,-1712.8073,-207.24359,-185784.3,1.9108,-
1712.8073,1.9108,-185768.73], rotationAngle: [-3.1404772,-1.0285039E-
5,3.1323729,0.0,0.0,0.0,0.0,0.0,0.0]
04-27 17:58:21.144 4782-5986/com.waylau.hmos.categoryorientationagent I
00001/MainAbilitySlice: dim:3, degree: 304.9, rotationMatrix: [-14.3298,-
195.13599,-1676.95,-195.13599,-185942.14,1.76,-1676.95,1.76,-185927.22],
rotationAngle: [-3.1405432,-9.465305E-6,3.1325736,0.0,0.0,0.0,0.0,0.0,0.0]
04-27 17:58:21.732 4782-5986/com.waylau.hmos.categoryorientationagent
I 00001/MainAbilitySlice: dim:3, degree: 304.77, rotationMatrix:
[-15.023399,-207.24359,-1712.8073,-207.24359,-185784.3,1.9108,-
1712.8073,1.9108,-185768.73], rotationAngle: [-3.1404772,-1.0285039E-
5,3.1323729,0.0,0.0,0.0,0.0,0.0,0.0]
04-27 17:58:22.328 4782-5986/com.waylau.hmos.categoryorientationagent I
00001/MainAbilitySlice: dim:3, degree: 304.9, rotationMatrix: [-14.3298,-
195.13599,-1676.95,-195.13599,-185942.14,1.76,-1676.95,1.76,-185927.22],
rotationAngle: [-3.1405432,-9.465305E-6,3.1325736,0.0,0.0,0.0,0.0,0.0,0.0]
04-27 17:58:22.924 4782-5986/com.waylau.hmos.categoryorientationagent
I 00001/MainAbilitySlice: dim:3, degree: 304.77, rotationMatrix:
[-15.023399,-207.24359,-1712.8073,-207.24359,-185784.3,1.9108,-
1712.8073,1.9108,-185768.73], rotationAngle: [-3.1404772,-1.0285039E-
5,3.1323729,0.0,0.0,0.0,0.0,0.0,0.0]
04-27 17:58:23.520 4782-5986/com.waylau.hmos.categoryorientationagent I
00001/MainAbilitySlice: dim:3, degree: 304.9, rotationMatrix: [-14.3298,-
195.13599,-1676.95,-195.13599,-185942.14,1.76,-1676.95,1.76,-185927.22],
rotationAngle: [-3.1405432,-9.465305E-6,3.1325736,0.0,0.0,0.0,0.0,0.0,0.0]
04-27 17:58:24.116 4782-5986/com.waylau.hmos.categoryorientationagent
I 00001/MainAbilitySlice: dim:3, degree: 304.77, rotationMatrix:
[-15.023399,-207.24359,-1712.8073,-207.24359,-185784.3,1.9108,-
1712.8073,1.9108,-185768.73], rotationAngle: [-3.1404772,-1.0285039E-
5,3.1323729,0.0,0.0,0.0,0.0,0.0,0.0]
```

从上述日志可以看到，订阅的数据以规定的频率发送过来。

点击界面按钮“Remove”以触发取消订阅传感器数据的操作执行。此时，控制台输出内容如下：

```
04-27 17:58:10.640 4782-4782/com.waylau.hmos.categoryorientationagent I
00001/MainAbilitySlice: before remove
04-27 17:58:10.643 4782-4782/com.waylau.hmos.categoryorientationagent I
00001/MainAbilitySlice: end remove
```

10.3 实战：Light示例

本节演示如何使用Light模块。当设备需要设置不同的闪烁效果时，可以调用Light模块，例如，LED灯能够设置灯颜色、灯亮和灯灭时长的闪烁效果。

10.3.1 接口说明

灯模块主要提供的功能有：查询设备上灯的列表、查询某个灯设备支持的效果、打开和关闭灯设备。LightAgent类开放能力主要接口有以下几种。

- getLightIdList()：获取硬件设备上的灯列表。
- isSupport(int)：根据指定灯ID，查询硬件设备是否有该灯。
- isEffectSupport(int, String)：查询指定的灯是否支持指定的闪烁效果。
- turnOn(int, String)：对指定的灯创建指定效果的一次性闪烁。
- turnOn(int, LightEffect)：对指定的灯创建自定义效果的一次性闪烁。
- turnOn(String)：对指定的灯创建指定效果的一次性闪烁。
- turnOn(LightEffect)：对指定的灯创建自定义效果的一次性闪烁。
- turnOff(int)：关闭指定的灯。
- turnOff()：关闭灯。

10.3.2 创建应用

为了演示使用Light模块的功能，创建一个名为“LightAgent”的应用。

在应用的界面上，通过点击按钮，来触发使用Light模块的操作。

10.3.3 修改ability_main.xml

修改ability_main.xml内容如下：

```
<?xml version="1.0" encoding="utf-8"?>
```

```
<DirectionalLayout
    xmlns:ohos="http://schemas.huawei.com/res/ohos"
    ohos:height="match_parent"
    ohos:width="match_parent"
    ohos:alignment="center"
    ohos:orientation="vertical">

    <Button
        ohos:id="$+id:button_on"
        ohos:height="match_content"
        ohos:width="match_parent"
        ohos:background_element="#F76543"
        ohos:layout_alignment="horizontal_center"
        ohos:margin="10vp"
        ohos:padding="10vp"
        ohos:text="On"
        ohos:text_size="40vp"
        />

    <Button
        ohos:id="$+id:button_off"
        ohos:height="match_content"
        ohos:width="match_parent"
        ohos:background_element="#F76543"
        ohos:layout_alignment="horizontal_center"
        ohos:margin="10vp"
        ohos:padding="10vp"
        ohos:text="Off"
        ohos:text_size="40vp"
        />

</DirectionalLayout>
```

界面预览效果如图10-4所示。

上述代码，设置了“On”按钮和“Off”按钮，以备设置点击事件，以触发使用Light模块的相关操作。

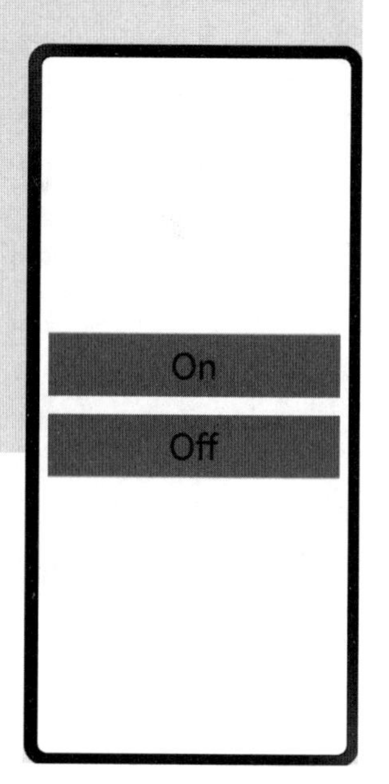

图10-4　界面预览效果

10.3.4　修改MainAbilitySlice

修改MainAbilitySlice内容如下：

```
package com.waylau.hmos.lightagent.slice;

import com.waylau.hmos.lightagent.ResourceTable;
import ohos.aafwk.ability.AbilitySlice;
```

```
import ohos.aafwk.content.Intent;
import ohos.agp.components.Button;
import ohos.hiviewdfx.HiLog;
import ohos.hiviewdfx.HiLogLabel;
import ohos.light.agent.LightAgent;
import ohos.light.bean.LightBrightness;
import ohos.light.bean.LightEffect;

import java.util.List;

public class MainAbilitySlice extends AbilitySlice {
    private static final String TAG = MainAbilitySlice.class.getSimpleName();
    private static final HiLogLabel LABEL_LOG =
            new HiLogLabel(HiLog.LOG_APP, 0x00001, TAG);

    private LightAgent lightAgent;
    private List<Integer> isEffectSupportLightIdList;

    @Override
    public void onStart(Intent intent) {
        super.onStart(intent);
        super.setUIContent(ResourceTable.Layout_ability_main);

        // 初始化对象
        initData();

        // 为按钮设置点击事件回调
        Button buttonAdd =
                (Button) findComponentById(ResourceTable.Id_button_on);
        buttonAdd.setClickedListener(listener -> on());

        Button buttonRemove =
                (Button) findComponentById(ResourceTable.Id_button_off);
        buttonRemove.setClickedListener(listener -> off());
    }

    private void initData() {
        HiLog.info(LABEL_LOG, "before initData");

        lightAgent = new LightAgent();

        // 查询硬件设备上的灯列表
        isEffectSupportLightIdList = lightAgent.getLightIdList();
        if (isEffectSupportLightIdList.isEmpty()) {
            HiLog.info(LABEL_LOG, "lightIdList is empty");
        }
```

```
        HiLog.info(LABEL_LOG, "end initData, size: %{public}s",
                isEffectSupportLightIdList.size());
    }

    private void off() {
        HiLog.info(LABEL_LOG, "before off");

        for (Integer lightId : isEffectSupportLightIdList) {
            // 关灯
            boolean turnOffResult = lightAgent.turnOff(lightId);

            HiLog.info(LABEL_LOG, "%{public}s turnOffResult : %{public}s ",
                    lightId, turnOffResult);
        }

        HiLog.info(LABEL_LOG, "end off");
    }

    private void on() {
        HiLog.info(LABEL_LOG, "before on");

        for (Integer lightId : isEffectSupportLightIdList) {
            // 创建自定义效果的一次性闪烁
            LightBrightness lightBrightness =
                    new LightBrightness(255, 255, 255);
            LightEffect lightEffect =
                    new LightEffect(lightBrightness, 1000, 1000);

            // 开灯
            boolean turnOnResult = lightAgent.turnOn(lightId, lightEffect);

            HiLog.info(LABEL_LOG, "%{public}s turnOnResult: %{public}s ",
                    lightId, turnOnResult);
        }

        HiLog.info(LABEL_LOG, "end on");
    }

    @Override
    public void onActive() {
        super.onActive();
    }

    @Override
    public void onForeground(Intent intent) {
        super.onForeground(intent);
    }
```

```
}
```

上述代码解释如下。

- 在Button上设置了点击事件。
- initData用于初始化硬件设备上的灯列表。
- on方法用于打开灯列表上所有的灯。
- off方法用于关闭灯列表上所有的灯。
- 通过LightBrightness和LightEffect可以创建自定义效果的一次性闪烁。

10.3.5 运行

运行应用后，控制台输出内容如下：

```
02-14 20:07:33.780 21003-21003/com.waylau.hmos.lightagent I 00001/
MainAbilitySlice: before initData
02-14 20:07:33.805 21003-21003/com.waylau.hmos.lightagent I 00001/
MainAbilitySlice: end initData, size: 1
```

点击界面按钮“On”以触发打开灯的操作执行。此时，控制台输出内容如下：

```
02-14 20:08:05.508 21003-21003/com.waylau.hmos.lightagent I 00001/
MainAbilitySlice: before on
02-14 20:08:05.522 21003-21003/com.waylau.hmos.lightagent I 00001/
MainAbilitySlice: 0 turnOnResult: true
02-14 20:08:05.522 21003-21003/com.waylau.hmos.lightagent I 00001/
MainAbilitySlice: end on
```

从上述日志可以看到，第0号灯已经打开了。

点击界面按钮“Off”以触发关闭灯的操作执行。此时，控制台输出内容如下：

```
02-14 20:08:51.508 21003-21003/com.waylau.hmos.lightagent I 00001/
MainAbilitySlice: before off
02-14 20:08:51.511 21003-21003/com.waylau.hmos.lightagent I 00001/
MainAbilitySlice: 0 turnOffResult : true
02-14 20:08:51.511 21003-21003/com.waylau.hmos.lightagent I 00001/
MainAbilitySlice: end off
```

从上述日志可以看到，第0号灯已经关闭了。

实战：获取设备的位置

开发者可以调用HarmonyOS位置相关接口，获取设备实时位置或者最近的历史位置。本节演示如何获取设备的位置。

对于位置敏感的应用业务，建议获取设备实时位置信息。如果不需要设备实时位置信息，并且希望尽可能节省耗电，开发者可以考虑获取最近的历史位置。

10.4.1 接口说明

获取设备的位置信息，所使用的接口有以下几种。

- Locator(Context context)：创建Locator实例对象。
- RequestParam(int scenario)：根据定位场景类型创建定位请求的RequestParam对象。
- onLocationReport(Location location)：获取定位结果。
- startLocating(RequestParam request, LocatorCallback callback)：向系统发起定位请求。
- requestOnce(RequestParam request, LocatorCallback callback)：向系统发起单次定位请求。
- stopLocating(LocatorCallback callback)：结束定位。
- getCachedLocation()：获取系统缓存的位置信息。

10.4.2 创建应用

为了演示获取设备的位置的功能，创建一个名为"Locator"的应用。

在应用的界面上，通过点击按钮，来触发获取设备的位置的操作。

10.4.3 声明权限

修改配置文件，声明使用位置的权限如下：

```
// 声明权限
"reqPermissions": [
    {
    "name": "ohos.permission.LOCATION"
    }
]
```

由于上述权限是敏感权限，因此，需要在代码中进行显式声明，代码如下：

```
package com.waylau.hmos.locator;

import com.waylau.hmos.locator.slice.MainAbilitySlice;
import ohos.aafwk.ability.Ability;
import ohos.aafwk.content.Intent;

import java.util.ArrayList;
import java.util.List;
```

```
public class MainAbility extends Ability {
    @Override
    public void onStart(Intent intent) {
        super.onStart(intent);
        super.setMainRoute(MainAbilitySlice.class.getName());

        // 显式声明需要使用的权限
        requestPermission();
    }

    // 显式声明需要使用的权限
    private void requestPermission() {
        String[] permission = {
                "ohos.permission.LOCATION"};
        List<String> applyPermissions = new ArrayList<>();
        for (String element : permission) {
            if (verifySelfPermission(element) != 0) {
                if (canRequestPermission(element)) {
                    applyPermissions.add(element);
                }
            }
        }
        requestPermissionsFromUser(applyPermissions.toArray(new String[0]), 0);
    }
}
```

10.4.4 修改ability_main.xml

修改ability_main.xml内容如下：

```
<?xml version="1.0" encoding="utf-8"?>
<DirectionalLayout
    xmlns:ohos="http://schemas.huawei.com/res/ohos"
    ohos:height="match_parent"
    ohos:width="match_parent"
    ohos:alignment="center"
    ohos:orientation="vertical">

    <Button
        ohos:id="$+id:button_on"
        ohos:height="match_content"
        ohos:width="match_parent"
        ohos:background_element="#F76543"
        ohos:layout_alignment="horizontal_center"
```

```
        ohos:margin="10vp"
        ohos:padding="10vp"
        ohos:text="On"
        ohos:text_size="40vp"
        />

    <Button
        ohos:id="$+id:button_off"
        ohos:height="match_content"
        ohos:width="match_parent"
        ohos:background_element="#F76543"
        ohos:layout_alignment="horizontal_center"
        ohos:margin="10vp"
        ohos:padding="10vp"
        ohos:text="Off"
        ohos:text_size="40vp"
        />

</DirectionalLayout>
```

界面预览效果如图10-5所示。

上述代码，设置了“On”按钮和“Off”按钮，以备设置点击事件，以触发使用位置的相关操作。

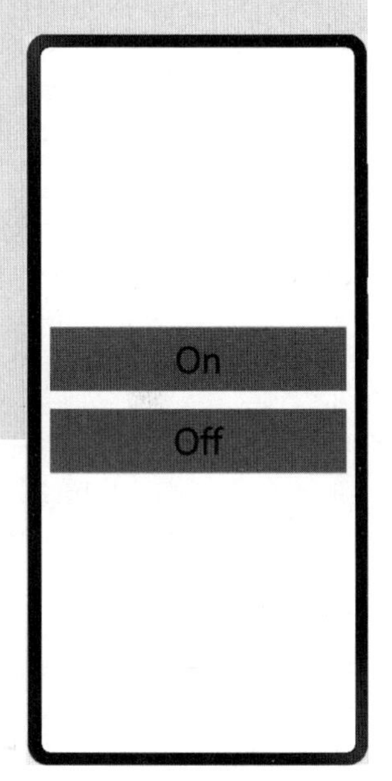

图10-5 界面预览效果

10.4.5 修改MainAbilitySlice

修改MainAbilitySlice内容如下：

```
package com.waylau.hmos.locator.slice;

import com.waylau.hmos.locator.ResourceTable;
import ohos.aafwk.ability.AbilitySlice;
import ohos.aafwk.content.Intent;
import ohos.agp.components.Button;
import ohos.hiviewdfx.HiLog;
import ohos.hiviewdfx.HiLogLabel;
import ohos.location.Location;
import ohos.location.Locator;
import ohos.location.LocatorCallback;
import ohos.location.RequestParam;
import ohos.utils.zson.ZSONObject;


public class MainAbilitySlice extends AbilitySlice {
    private static final String TAG = MainAbilitySlice.class.getSimpleName();
```

```
    private static final HiLogLabel LABEL_LOG =
            new HiLogLabel(HiLog.LOG_APP, 0x00001, TAG);

    private Locator locator;
    private MyLocatorCallback locatorCallback;

    @Override
    public void onStart(Intent intent) {
        super.onStart(intent);
        super.setUIContent(ResourceTable.Layout_ability_main);

        // 初始化对象
        initData();

        // 为按钮设置点击事件回调
        Button buttonOn =
                (Button) findComponentById(ResourceTable.Id_button_on);
        buttonOn.setClickedListener(listener -> on());

        Button buttonOff =
                (Button) findComponentById(ResourceTable.Id_button_off);
        buttonOff.setClickedListener(listener -> off());
    }

    private void initData() {
        HiLog.info(LABEL_LOG, "before initData");

        locator = new Locator(this.getContext());

        HiLog.info(LABEL_LOG, "end initData");
    }

    private void off() {
        HiLog.info(LABEL_LOG, "before off");

        locator.stopLocating(locatorCallback);

        HiLog.info(LABEL_LOG, "end off");
    }


    private void on() {
        HiLog.info(LABEL_LOG, "before on");

        locatorCallback = new MyLocatorCallback();
```

```
        // 定位精度优先策略
        RequestParam requestParam =
                new RequestParam(RequestParam.PRIORITY_ACCURACY, 0, 0);

        locator.startLocating(requestParam, locatorCallback);

        HiLog.info(LABEL_LOG, "end on");
    }

    private class MyLocatorCallback implements LocatorCallback {
        @Override
        public void onLocationReport(Location location) {
            HiLog.info(LABEL_LOG, "onLocationReport, location: %{public}s",
                    ZSONObject.toZSONString(location));
        }

        @Override
        public void onStatusChanged(int type) {
            HiLog.info(LABEL_LOG, "onStatusChanged, type: %{public}s", type);
        }

        @Override
        public void onErrorReport(int type) {
            HiLog.info(LABEL_LOG, "onErrorReport, type: %{public}s", type);
        }
    }

    @Override
    public void onActive() {
        super.onActive();
    }

    @Override
    public void onForeground(Intent intent) {
        super.onForeground(intent);
    }
}
```

上述代码解释如下。

- 在Button上设置了点击事件。
- initData用于初始化Locator对象。
- on方法用于启动定位。
- off方法用于关闭定位。
- 实例化LocatorCallback对象，用于向系统提供位置上报的途径。系统在定位成功确定设备的实时位置结果时，会通过onLocationReport接口上报给应用。应用程序可以在onLocationReport接口的实现中完成自己的业务逻辑。

10.4.6 运行

初次运行应用后，点击界面按钮“On”以触发操作的执行。此时，控制台输出内容如下：

```
04-27 22:18:03.572 29978-29978/com.waylau.hmos.locator
I 00001/MainAbilitySlice: before initData
04-27 22:18:03.609 29978-29978/com.waylau.hmos.locator
I 00001/MainAbilitySlice: end initData
04-27 22:18:22.653 29978-29978/com.waylau.hmos.locator
I 00001/MainAbilitySlice: before on
04-27 22:18:22.656 29978-15077/com.waylau.hmos.locator
I 00001/MainAbilitySlice: onErrorReport, type: 257
04-27 22:18:23.876 29978-29978/com.waylau.hmos.locator
I 00001/MainAbilitySlice: end on
```

从上述日志可以看出，并未获取到位置信息，这是因为位置信息服务并未开启。按图10-6所示的方式开启位置信息服务。

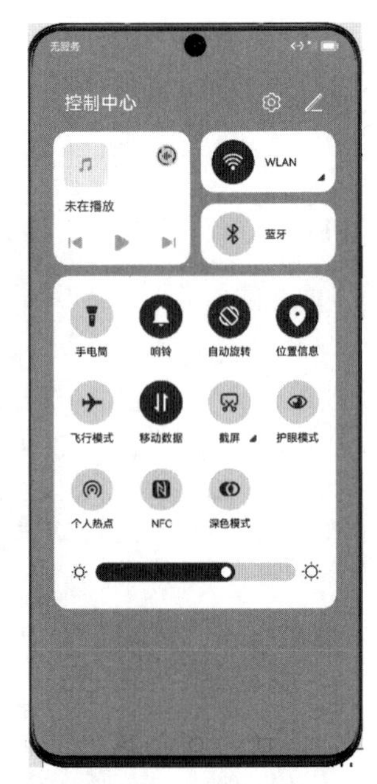

图 10-6　位置信息

再次点击界面按钮“On”以触发操作的执行。此时，控制台输出内容如下：

```
04-27 22:18:23.882 29978-15077/com.waylau.hmos.locator I 00001/
MainAbilitySlice: onStatusChanged, type: 2
04-27 22:18:36.083 29978-15077/com.waylau.hmos.locator I 00001/
MainAbilitySlice: onLocationReport, location: {"accuracy":10,"altitude":51,"d
irection":0,"latitude":30.495864,"longitude":114.535703,"speed":0,"timeSinceB
oot":22964904123500,"timeStamp":1489717530000}
04-27 22:18:41.566 29978-15077/com.waylau.hmos.locator I 00001/
MainAbilitySlice: onLocationReport, location: {"accuracy":10,"altitude":51,"d
irection":0,"latitude":30.495864,"longitude":114.535703,"speed":0,"timeSinceB
oot":22970390523580,"timeStamp":1489717535000}
04-27 22:18:42.416 29978-15077/com.waylau.hmos.locator I 00001/
MainAbilitySlice: onLocationReport, location: {"accuracy":10,"altitude":51,"d
irection":0,"latitude":30.495864,"longitude":114.535703,"speed":0,"timeSinceB
oot":22971291623680,"timeStamp":1489717536000}
04-27 22:18:43.422 29978-15077/com.waylau.hmos.locator I 00001/
MainAbilitySlice: onLocationReport, location: {"accuracy":10,"altitude":51,"d
irection":0,"latitude":30.495864,"longitude":114.535703,"speed":0,"timeSinceB
oot":22972294828940,"timeStamp":1489717537000}
04-27 22:18:54.423 29978-15077/com.waylau.hmos.locator I 00001/
MainAbilitySlice: onLocationReport, location: {"accuracy":10,"altitude":51,"d
```

```
irection":0,"latitude":30.495864,"longitude":114.535703,"speed":0,"timeSinceB
oot":22983298966500,"timeStamp":1489717548000}
```

从上述日志可以看出，已经能够成功获取到位置信息了。

点击界面按钮“Off”以关闭位置服务。此时，控制台输出内容如下：

```
04-27 22:19:01.411 29978-29978/com.waylau.hmos.locator I 00001/
MainAbilitySlice: before off
04-27 22:19:01.430 29978-29978/com.waylau.hmos.locator I 00001/
MainAbilitySlice: end off
04-27 22:19:01.456 29978-15077/com.waylau.hmos.locator I 00001/
MainAbilitySlice: onStatusChanged, type: 3
```

10.5 实战：（逆）地理编码转化

本节演示如何使用（逆）地理编码转化。

使用坐标描述一个位置，非常准确，但是并不直观，面向用户表达并不友好。

HarmonyOS向开发者提供了地理编码转化能力（将坐标转化为地理编码信息），以及逆地理编码转化能力（将地理描述转化为具体坐标）。其中地理编码包含多个属性来描述位置，包括国家、行政区划、街道、门牌号、地址描述等，这样的信息更便于用户理解。

10.5.1 接口说明

GeoConvert用于坐标和地理编码信息的相互转化，所使用的接口有以下几种。

- GeoConvert()：创建GeoConvert实例对象。
- GeoConvert(Locale locale)：根据自定义参数创建GeoConvert实例对象。
- getAddressFromLocation(double latitude, double longitude, int maxItems)：根据指定的经纬度坐标获取地理位置信息。
- getAddressFromLocationName(String description, int maxItems)：根据地理位置信息获取相匹配的包含坐标数据的地址列表。
- getAddressFromLocationName(String description, double minLatitude, double minLongitude, double maxLatitude, double maxLongitude, int maxItems)：根据指定的位置信息和地理区域获取相匹配的包含坐标数据的地址列表。

10.5.2 创建应用

为了演示使用（逆）地理编码转化的功能，创建一个名为“GeoConvert”的应用。

在应用的界面上，通过点击按钮，来触发使用（逆）地理编码转化的操作。

10.5.3 修改ability_main.xml

修改ability_main.xml内容如下：

```
<?xml version="1.0" encoding="utf-8"?>
<DirectionalLayout
    xmlns:ohos="http://schemas.huawei.com/res/ohos"
    ohos:height="match_parent"
    ohos:width="match_parent"
    ohos:orientation="vertical">

    <Button
        ohos:id="$+id:button_get"
        ohos:height="match_content"
        ohos:width="match_parent"
        ohos:background_element="#F76543"
        ohos:layout_alignment="horizontal_center"
        ohos:margin="10vp"
        ohos:padding="10vp"
        ohos:text="Get"
        ohos:text_size="30fp"
        />

    <Text
        ohos:id="$+id:text"
        ohos:height="match_content"
        ohos:width="match_content"
        ohos:background_element="$graphic:background_ability_main"
        ohos:layout_alignment="horizontal_center"
        ohos:multiple_lines="true"
        ohos:text=""
        ohos:text_size="20fp"
        />
</DirectionalLayout>
```

界面预览效果如图10-7所示。

上述代码，设置了“Get”按钮，以备设置点击事件，以触发使用（逆）地理编码转化的相关操作。

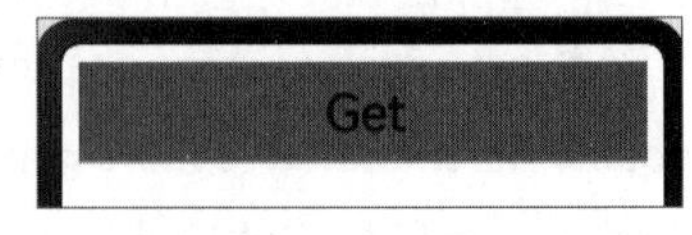

图10-7　界面预览效果

10.5.4 修改MainAbilitySlice

修改MainAbilitySlice内容如下：

```
package com.waylau.hmos.geoconvert.slice;

import com.waylau.hmos.geoconvert.ResourceTable;
import ohos.aafwk.ability.AbilitySlice;
import ohos.aafwk.content.Intent;
import ohos.agp.components.Button;
import ohos.agp.components.Text;
import ohos.hiviewdfx.HiLog;
import ohos.hiviewdfx.HiLogLabel;
import ohos.location.*;
import ohos.utils.zson.ZSONObject;

import java.io.IOException;
import java.util.List;


public class MainAbilitySlice extends AbilitySlice {
    private static final String TAG = MainAbilitySlice.class.getSimpleName();
    private static final HiLogLabel LABEL_LOG =
            new HiLogLabel(HiLog.LOG_APP, 0x00001, TAG);

    private Text text;

    @Override
    public void onStart(Intent intent) {
        super.onStart(intent);
        super.setUIContent(ResourceTable.Layout_ability_main);


        // 为按钮设置点击事件回调
        Button buttonOn =
                (Button) findComponentById(ResourceTable.Id_button_get);
        buttonOn.setClickedListener(listener -> {
            try {
                getInfo();
            } catch (IOException e) {
                e.printStackTrace();
            }
        });

        text = (Text) findComponentById(ResourceTable.Id_text);
    }

    private void getInfo() throws IOException {
        HiLog.info(LABEL_LOG, "before getInfo");

        // 实例化GeoConvert对象
```

```
        // 所有与（逆）地理编码转化能力相关的功能 API，都是通过 GeoConvert 提供的
        GeoConvert geoConvert = new GeoConvert();

        // 坐标转化地理位置信息
        int maxItems = 3;
        double latitude = 23.048062D;
        double longitude = 114.211902D;
        List<GeoAddress> addressList = geoConvert.getAddressFromLocation(
                                        latitude, longitude, maxItems);

        // 显示地理位置信息
        showAddress(addressList);

        // 位置描述转化坐标
        addressList = geoConvert.getAddressFromLocationName("华为", maxItems);

        // 显示地理位置信息
        showAddress(addressList);

        HiLog.info(LABEL_LOG, "end getInfo");
    }

    private void showAddress(List<GeoAddress> addressList) {
        HiLog.info(LABEL_LOG, "showAddress");

        for (GeoAddress address : addressList) {
            String addressInfo = ZSONObject.toZSONString(address);

            text.append(addressInfo + "\n");

            HiLog.info(LABEL_LOG, "addressInfo:%{public}s", addressInfo);
        }
    }

    @Override
    public void onActive() {
        super.onActive();
    }

    @Override
    public void onForeground(Intent intent) {
        super.onForeground(intent);
    }
}
```

上述代码解释如下。

- 在Button上设置了点击事件。

- getInfo方法实例化GeoConvert对象，并执行坐标转化地理位置信息和位置描述转化坐标等操作。
- showAddress方法用于在界面上显示获取到的地理位置信息。

10.5.5 运行

运行应用后，点击界面按钮“Get”以触发（逆）地理编码转化的操作执行。此时，控制台输出内容如下：

```
04-27 23:01:23.043 19665-19665/com.waylau.hmos.geoconvert I 00001/
MainAbilitySlice: before get
04-27 23:01:23.051 19665-19665/com.waylau.hmos.geoconvert I 00001/
MainAbilitySlice: showAddress
04-27 23:01:23.059 19665-19665/com.waylau.hmos.geoconvert I 00001/
MainAbilitySlice: addressInfo:{"administrativeArea":" 广东省 ","countryCode":
"CN","countryName":" 中国 ","descriptionsSize":0,"latitude":23.048069,"locale
":"zh_CN","locality":" 惠州市 ","longitude":114.211912,"placeName":" 潼湖镇西湖
","subLocality":" 惠城区 "}
04-27 23:01:31.310 19665-19665/com.waylau.hmos.geoconvert I 00001/
MainAbilitySlice: showAddress
04-27 23:01:31.327 19665-19665/com.waylau.hmos.geoconvert I 00001/
MainAbilitySlice: addressInfo:{"administrativeArea":" 广东省 ","countryCode
":"CN","countryName":" 中国 ","descriptionsSize":0,"latitude":22.547246,"lo
cale":"zh_CN","locality":" 深圳市 ","longitude":114.082036,"placeName":" 华为
","subLocality":" 福田区 "}
04-27 23:01:31.327 19665-19665/com.waylau.hmos.geoconvert I 00001/
MainAbilitySlice: end get
```

界面也能正常显示获取到的地理位置信息，如图10-8所示。

图10-8　界面效果

第11章 数据管理

HarmonyOS 的数据管理基于分布式软总线，实现了应用程序数据和用户数据的分布式管理。

11.1 数据管理概述

在全场景新时代，每个人拥有的设备越来越多，单一设备的数据往往无法满足用户的诉求，数据在设备间的流转变得越来越频繁。以一组照片数据在手机、平板、智慧屏和PC之间相互浏览和编辑为例，需要考虑到照片数据在多设备间是怎么存储、怎么共享和怎么访问的。

HarmonyOS的数据管理基于分布式软总线，实现了应用程序数据和用户数据的分布式管理。用户数据不再与单一物理设备绑定，业务逻辑与数据存储分离，跨设备的数据处理如同本地数据处理一样方便快捷，让开发者能够轻松实现全场景、多设备下的数据存储、共享和访问，为打造一致、流畅的用户体验创造了基础条件。

HarmonyOS分布式数据管理对开发者提供分布式数据库、分布式文件系统和分布式检索能力，开发者在多设备上开发应用时，对数据的操作、共享、检索可以跟使用本地数据一样进行处理，为开发者提供便捷、高效和安全的数据管理能力，大大降低了应用开发者实现数据分布式访问的门槛。同时，由于在系统层面实现了这样的功能，可以结合系统资源调度，大大提升跨设备数据远程访问和检索性能，让更多的开发者可以快速地上手实现流畅分布式应用!

HarmonyOS分布式数据管理的典型应用场景有以下几种。

- 协同办公场景：将手机上的文档投屏到智慧屏，在智慧屏上对文档执行翻页、缩放、涂鸦等操作，文档的最新状态可以在手机上同步显示。
- 家庭出游场景：一家人出游时，妈妈用手机拍了很多照片。通过家庭照片共享，爸爸可以在自己的手机上浏览、收藏和保存这些照片，家中的爷爷奶奶也可以通过智慧屏浏览这些照片。

HarmonyOS支持多样数据管理方式，包括：

- 关系型数据库；
- 对象关系映射数据库；
- 轻量级偏好数据库；
- 分布式数据服务；
- 分布式文件服务；
- 融合搜索；
- 数据存储管理。

11.2 关系型数据库

关系型数据库（Relational Database，RDB）是一种基于关系模型来管理数据的数据库。HarmonyOS关系型数据库基于SQLite组件，提供了一套完整的对本地数据库进行管理的机制，对外提供了一系列的增、删、改、查接口，也可以直接运行用户输入的SQL语句来满足复杂的场景需要。HarmonyOS提供的关系型数据库功能更加完善，查询效率更高。

11.2.1 基本概念

使用关系型数据库时，主要涉及以下概念。

• 关系型数据库：创建在关系模型基础上的数据库，以行和列的形式存储数据。

• 谓词：数据库中用来代表数据实体的性质、特征或数据实体之间关系的词项，主要用来定义数据库的操作条件。

• 结果集：指用户查询之后的结果集合，可以对数据进行访问。结果集提供了灵活的数据访问方式，可以更方便地获取到用户想要的数据。

• SQLite数据库：一款轻型的数据库，是遵守ACID的关系型数据库管理系统。它是一个开源的项目。

11.2.2 运作机制

HarmonyOS关系型数据库对外提供通用的操作接口，底层使用SQLite作为持久化存储引擎，支持SQLite具有的所有数据库特性，包括但不限于事务、索引、视图、触发器、外键、参数化查询和预编译SQL语句。

图11-1展示了关系型数据库的运作机制架构。

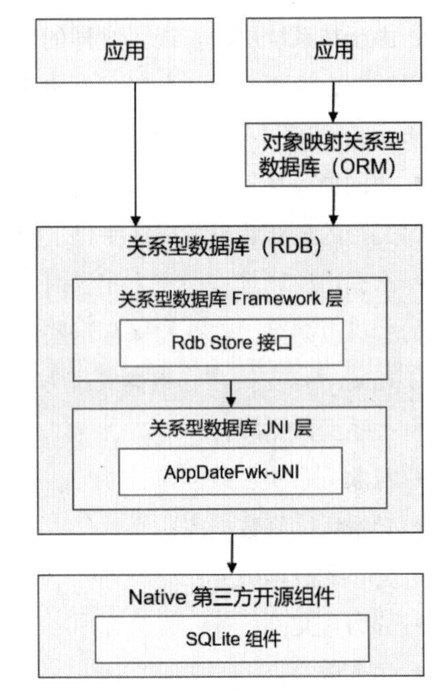

图11-1　关系型数据库的运作机制架构

11.2.3 默认配置

HarmonyOS关系型数据库的默认配置如下。

• 如果不指定数据库的日志模式，那么系统默认日志方式是WAL（Write Ahead Log，预写日志）模式。

• 如果不指定数据库的落盘模式，那么系统默认落盘方式是FULL（英文full的大写）模式。

• HarmonyOS数据库使用的共享内存默认大小是2MB。

11.2.4 约束与限制

HarmonyOS关系型数据库的约束与限制如下。

• 数据库中连接池的最大数量是4个，用以管理用户的读写操作。

• 为保证数据的准确性，数据库同一时间只能支持一个写操作。

11.2.5 接口说明

HarmonyOS关系型数据库的接口如下。

1. 数据库的创建和删除

关系型数据库提供了数据库创建方式，以及对应的删除接口方式，涉及的API如下所示。

- StoreConfig.Builder的builder()：对数据库进行配置，包括设置数据库名、存储模式、日志模式、同步模式，是否为只读及对数据库加密。
- RdbOpenCallback的onCreate(RdbStore store)：数据库创建时被回调，开发者可以在该方法中初始化表结构，并添加一些应用使用到的初始化数据。
- RdbOpenCallback的onUpgrade(RdbStore store, int currentVersion, int targetVersion)：数据库升级时被回调。
- DatabaseHelper的getRdbStore(StoreConfig config, int version, RdbOpenCallback openCallback, ResultSetHook resultSetHook)：根据配置创建或打开数据库。
- DatabaseHelper的deleteRdbStore(String name)：删除指定的数据库。

2. 数据库的加密

关系型数据库提供数据库加密的能力，创建数据库时传入指定密钥、创建加密数据库，后续打开加密数据库时，需要传入正确密钥。

- StoreConfig.Builder的setEncryptKey(byte[] encryptKey)：为数据库配置类设置数据库加密密钥，创建或打开数据库时传入包含数据库加密密钥的配置类，即可创建或打开加密数据库。

3. 数据库的增删改查

关系型数据库提供本地数据增删改查操作的能力，相关RdbStore的API如下所示。

- insert(String table, ValuesBucket initialValues)：向数据库插入数据。
- update(ValuesBucket values, AbsRdbPredicates predicates)：更新数据库表中符合谓词指定条件的数据。
- delete(AbsRdbPredicates predicates)：删除数据。
- query(AbsRdbPredicates predicates, String[] columns)：查询数据。
- querySql(String sql, String[] sqlArgs)：执行原生的用于查询操作的SQL语句。

4. 数据库谓词的使用

关系型数据库提供了用于设置数据库操作条件的谓词AbsRdbPredicates，其中包括两个实现子类RdbPredicates和RawRdbPredicates。

- RdbPredicates：开发者无须编写复杂的SQL语句，仅通过调用该类中条件相关的方法，如equalTo、notEqualTo、groupBy、orderByAsc、beginsWith等，就可自动完成SQL语句拼接，方便用户聚焦业务操作。

RdbPredicates包含的API如下。

- equalTo(String field, String value)：设置谓词条件，满足filed字段与value值相等。
- notEqualTo(String field, String value)：设置谓词条件，满足filed字段与value值不相等。
- beginsWith(String field, String value)：设置谓词条件，满足field字段以value值开头。
- between(String field, int low, int high)：设置谓词条件，满足field字段在最小值low和最大值high之间。
- orderByAsc(String field)：设置谓词条件，根据field字段升序排列。
- RawRdbPredicates：可满足复杂SQL语句的场景，支持开发者自己设置where条件子句和whereArgs参数。不支持equalTo等条件接口的使用。

RawRdbPredicates包含的API如下。

- setWhereClause(String whereClause)：设置where条件子句。
- setWhereArgs(List<String> whereArgs)：设置whereArgs参数，该值表示where子句中占位符的值。

5. 查询结果集的使用

关系型数据库提供了查询返回的结果集ResultSet，它指向查询结果中的一行数据，供用户对查询结果进行遍历和访问。ReusltSet的对外API如下。

- goTo(int offset)：从结果集当前位置移动指定偏移量。
- goToRow(int position)：将结果集移动到指定位置。
- goToNextRow()：将结果集向后移动一行。
- goToPreviousRow()：将结果集向前移动一行。
- isStarted()：判断结果集是否被移动过。
- isEnded()：判断结果集当前位置是否在最后一行之后。
- isAtFirstRow()：判断结果集当前位置是否在第一行。
- isAtLastRow()：判断结果集当前位置是否在最后一行。
- getRowCount()：获取当前结果集中的记录条数。
- getColumnCount()：获取结果集中的列数。
- getString(int columnIndex)：获取当前行指定列的值，以String类型返回。
- getBlob(int columnIndex)：获取当前行指定列的值，以字节数组形式返回。
- getDouble(int columnIndex)：获取当前行指定列的值，以double类型返回。

6. 事务

关系型数据库提供事务机制，来保证用户操作的原子性。对单条数据进行数据库操作时，无须开启事务；插入大量数据时，开启事务可以保证数据的准确性。如果中途操作出现失败，会执行回滚操作。

RdbStore事务API如下。

- beginTransaction()：开启事务。
- markAsCommit()：设置事务的标记为成功。
- endTransaction()：结束事务。

7. 事务和结果集观察者

关系型数据库提供了事务和结果集观察者能力，当对应的事件被触发时，观察者会收到通知。

- RdbStore 的 beginTransactionWithObserver(TransactionObserver transactionObserver)：开启事务，并观察事务的启动、提交和回滚。
- ResultSet 的 registerObserver(DataObserver observer)：注册结果集的观察者。
- unregisterObserver(DataObserver observer)：注销结果集的观察者。

8. 数据库的备份和恢复

用户可以将当前数据库的数据进行保存备份，还可以在需要的时候进行数据恢复。

RdbStore 提供的数据库备份和恢复 API 如下。

- restore(String srcName)：数据库恢复接口，从指定的非加密数据库文件中恢复数据。
- restore(String srcName, byte[] srcEncryptKey, byte[] destEncryptKey)：数据库恢复接口，从指定的数据库文件（加密和非加密均可）中恢复数据。
- backup(String destName)：数据库备份接口，备份的数据库文件是非加密的。
- backup(String destName, byte[] destEncryptKey)：数据库备份接口，此方法经常用在备份加密数据库场景。

11.2.6 开发过程

HarmonyOS 关系型数据库的开发过程总结如下。

1. 创建数据库

- 配置数据库相关信息，包括数据库的名称、存储模式、是否为只读模式等。
- 初始化数据库表结构和相关数据。
- 创建数据库。

示例代码如下：

```
StoreConfig config = StoreConfig.newDefaultConfig("RdbStoreTest.db");
private static final RdbOpenCallback callback = new RdbOpenCallback() {
    @Override
    public void onCreate(RdbStore store) {
        store.executeSql("CREATE TABLE IF NOT EXISTS test "
            +"(id INTEGER PRIMARY KEY AUTOINCREMENT, name TEXT NOT NULL, "
            +"age INTEGER, salary REAL, blobType BLOB)");
    }
    @Override
```

```
    public void onUpgrade(RdbStore store, int oldVersion, int newVersion) {
    }
};


DatabaseHelper helper = new DatabaseHelper(context);
RdbStore store = helper.getRdbStore(config, 1, callback, null);
```

2. 插入数据

- 构造要插入的数据，以 ValuesBucket 形式存储。
- 调用关系型数据库提供的插入接口。

示例代码如下：

```
ValuesBucket values = new ValuesBucket();
values.putInteger("id", 1);
values.putString("name", "zhangsan");
values.putInteger("age", 18);
values.putDouble("salary", 100.5);
values.putByteArray("blobType", new byte[] {1, 2, 3});
long id = store.insert("test", values);
```

3. 查询数据

- 构造用于查询的谓词对象，设置查询条件。
- 指定查询返回的数据列。
- 调用查询接口查询数据。
- 调用结果集接口，遍历返回结果。

示例代码如下：

```
String[] columns = new String[] {"id", "name", "age", "salary"};
RdbPredicates rdbPredicates =
    new RdbPredicates("test").equalTo("age", 25).orderByAsc("salary");
ResultSet resultSet = store.query(rdbPredicates, columns);
resultSet.goToNextRow();
```

有关关系型数据库的详细开发示例可以参见“5.11 实战：通过 DataAbilityHelper 访问数据库”。

11.3 对象关系映射数据库

HarmonyOS 对象关系映射（Object Relational Mapping，ORM）数据库是一款基于 SQLite 的数据库框架，屏蔽了底层 SQLite 数据库的 SQL 操作，针对实体和关系提供了增删改查等一系列的面向对象接口。应用开发者不必再去编写复杂的 SQL 语句，以操作对象的形式来操作数据库，在提升效率的同时也能聚焦于业务开发。

HarmonyOS对象关系映射数据库是建立在HarmonyOS关系型数据库的基础之上的，所以关系型数据库的一些约束与限制、默认配置，请参考关系型数据库的约束、限制及默认配置。

11.3.1 基本概念

使用对象关系映射数据库时，主要涉及以下概念。

- 对象关系映射数据库的三个主要组件如下。

– 数据库：被开发者用@Database注解，且继承了OrmDatabase的类，对应关系型数据库。

– 实体对象：被开发者用@Entity注解，且继承了OrmObject的类，对应关系型数据库中的表。

– 对象数据操作接口：包括数据库操作的入口OrmContext类和谓词接口（OrmPredicate）等。

- 谓词：数据库中用来代表数据实体的性质、特征或数据实体之间关系的词项，主要用来定义数据库的操作条件。对象关系映射数据库将SQLite数据库中的谓词封装成了接口方法供开发者调用。开发者通过对象数据操作接口，可以访问到应用持久化的关系型数据。
- 对象关系映射数据库：通过将实例对象映射到关系上，实现使用操作实例对象的语法，来操作关系型数据库。它是在SQLite数据库的基础上提供的一个抽象层。
- SQLite数据库：一款轻型的数据库，是遵守ACID的关系型数据库管理系统。

11.3.2 运作机制

对象关系映射数据库操作是基于关系型数据库操作接口完成的，实际是在关系型数据库操作的基础上又实现了对象关系映射等特性。因此，对象关系映射数据库跟关系型数据库一样，都使用SQLite作为持久化引擎，底层使用的是同一套数据库连接池和数据库连接机制。

使用对象关系映射数据库的开发者需要先配置实体模型与关系映射文件。应用数据管理框架提供的类生成工具会解析这些文件，生成数据库帮助类，这样应用数据管理框架就能在运行时，根据开发者的配置创建好数据库，并在存储过程中自动完成对象关系映射。开发者再通过对象数据操作接口，如OrmContext接口和谓词接口等操作持久化数据库。

对象数据操作接口提供一组基于对象映射的数据操作接口，实现了基于SQL的关系模型数据到对象的映射，让用户不需要再和复杂的SQL语句打交道，只需简单地操作实体对象的属性和方法。对象数据操作接口支持对象的增删改查操作，同时支持事务操作等。

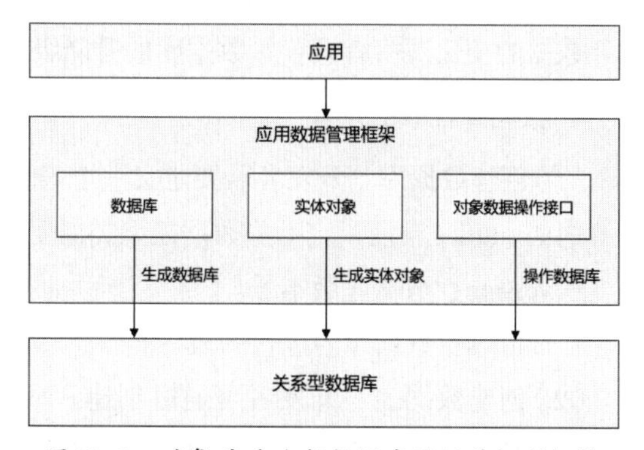

图11-2　对象关系映射数据库的运作机制架构

图11-2展示了对象关系映射数据库的运作

机制架构。

11.3.3 实体对象属性支持的类型

开发者建立实体对象类时，对象属性的类型可以在表11-1的类型中选择，不支持使用自定义类型。

表11-1 实体对象属性支持的类型

类型名称	描述	初始值	类型名称	描述	初始值
Integer	封装整型	null	boolean	布尔型	0
int	整型	0	Byte	封装字节型	null
Long	封装长整型	null	byte	字节型	0
long	长整型	0L	Character	封装字符型	null
Double	封装双精度浮点型	null	char	字符型	''
double	双精度浮点型	0	Date	日期类	null
Float	封装单精度浮点型	null	Time	时间类	null
float	单精度浮点型	0	Timestamp	时间戳类	null
Short	封装短整型	null	Calendar	日历类	null
short	短整型	0	Blob	二进制大对象	null
String	字符串型	null	Clob	字符大对象	null
Boolean	封装布尔型	null			

11.3.4 接口说明

对象关系映射数据库目前可以支持数据库和表的创建、数据库的加密、对象数据的增删改查、对象数据的变化观察者设置、数据库的升降级和数据库备份与恢复接口等功能。

1. 数据库和表的创建

（1）创建数据库。开发者需要定义一个表示数据库的类，继承OrmDatabase，再通过@Database注解内的entities属性指定哪些数据模型类属于这个数据库。属性包括以下两种。

- version：数据库版本号。
- entities：数据库内包含的表。

（2）创建数据表。开发者可通过创建一个继承了OrmObject并用@Entity注解的类，获取数据库实体对象，也就是表的对象。属性包括以下几种。

- tableName：表名。
- primaryKeys：主键名，一个表里只能有一个主键，一个主键可以由多个字段组成。
- foreignKeys：外键列表。
- indices：索引列表。

主要涉及如下注解。

- @Database：被@Database注解且继承了OrmDatabase的类，对应数据库类。
- @Entity：被@Entity注解且继承了OrmObject的类，对应数据表类。
- @Column：被@Column注解的变量，对应数据表的字段。
- @PrimaryKey：被@PrimaryKey注解的变量，对应数据表的主键。
- @ForeignKey：被@ForeignKey注解的变量，对应数据表的外键。
- @Index：被@Index注解的内容，对应数据表索引的属性。

2. 数据库的加密

对象关系映射数据库提供数据库加密的能力，创建数据库时传入指定密钥、创建加密数据库，后续打开加密数据库时，需要传入正确密钥。

OrmConfig.Builder传入密钥接口如下。

- setEncryptKey(byte[] encryptKey)：为数据库配置类设置数据库加密密钥。创建或打开数据库时传入包含数据库加密密钥的配置类，即可创建或打开加密数据库。

3. 对象数据的增删改查

通过对象数据操作接口，开发者可以对对象数据进行增删改查操作。

OrmContext对象数据操作接口如下。

- insert(Tobject)：添加方法。
- update(Tobject)：更新方法。
- query(OrmPredicates predicates)：查询方法。
- delete(Tobject)：删除方法。
- where(Class<T> clz)：设置谓词方法。

4. 对象数据的变化观察者设置

通过使用对象数据操作接口，开发者可以在某些数据上设置观察者，接收数据变化的通知。

OrmContext数据变化观察者接口如下。

- registerStoreObserver(String alias, OrmObjectObserver observer)：注册数据库变化回调。
- registerContextObserver(OrmContext watchedContext, OrmObjectObserver observer)：注册上下文变化回调。
- registerEntityObserver(String entityName, OrmObjectObserver observer)：注册数据库实体变化回调。

- registerObjectObserver(OrmObject ormObject, OrmObjectObserver observer)：注册对象变化回调。

5. 数据库的升降级

通过调用数据库升降级接口，开发者可以将数据库切换到不同的版本。

OrmMigration数据库升降级接口如下。

- onMigrate(int beginVersion, int endVersion)：数据库版本升降级接口。

6. 数据库备份与恢复接口

开发者可以将当前数据库的数据进行备份，在必要的时候进行数据恢复。

OrmContext数据库备份与恢复接口如下。

- backup(String destPath)：数据库备份接口。
- restore(String srcPath)：数据库恢复备份接口。

11.4 实战：使用对象关系映射数据库

“5.11 实战：通过DataAbilityHelper访问数据库”一节演示了如何通过DataAbilityHelper类来访问关系型数据库数据。本节将演示如何访问对象关系映射数据库。

创建一个名为“DataAbilityHelperAccessORM”的应用作为演示。

11.4.1 修改build.gradle

修改build.gradle配置文件，在其中增加compileOptions相关的配置，否则无法识别ORM相关的注解。

完整的build.gradle文件如下：

```
apply plugin: 'com.huawei.ohos.hap'
apply plugin: 'com.huawei.ohos.decctest'

ohos {
    compileSdkVersion 5
    defaultConfig {
        compatibleSdkVersion 5
    }
    buildTypes {
        release {
            proguardOpt {
                proguardEnabled false
```

```
                rulesFiles 'proguard-rules.pro'
            }
        }
    }

    // 添加 compileOptions 相关的配置
    compileOptions {
        annotationEnabled true
    }

}

dependencies {
    implementation fileTree(dir: 'libs', include: ['*.jar', '*.har'])
    testImplementation 'junit:junit:4.13'
    ohosTestImplementation 'com.huawei.ohos.testkit:runner:1.0.0.200'
}
decc {
    supportType = ['html','xml']
}
```

11.4.2 新增User

新增User类，用于表示用户实体。代码如下：

```
package com.waylau.hmos.dataabilityhelperaccessorm;

import ohos.data.orm.OrmObject;
import ohos.data.orm.annotation.Entity;
import ohos.data.orm.annotation.Index;
import ohos.data.orm.annotation.PrimaryKey;

@Entity(tableName = "user",
        indices = {@Index(value = {"userName"}, name = "name_index", unique =
true)})
public class User extends OrmObject {
    // 此处将 userId 设为了自增的主键。注意只有在数据类型为包装类型时，自增主键才能生效
    @PrimaryKey(autoGenerate = true)
    private Integer userId;
    private String userName;
    private int age;

    public Integer getUserId() {
        return userId;
    }
```

```
    public void setUserId(Integer userId) {
        this.userId = userId;
    }

    public String getUserName() {
        return userName;
    }

    public void setUserName(String userName) {
        this.userName = userName;
    }

    public int getAge() {
        return age;
    }

    public void setAge(int age) {
        this.age = age;
    }
}
```

上述代码构造数据表，即创建数据库实体类并配置对应的属性（如对应表的主键等）。数据表必须与其所在的数据库在同一个模块中。

上述代码定义了一个实体类User.java，对应数据库内的表名为“user”。其中indices用于设置索引，上述例子为“userName”这个字段建立的索引为“name_index”，并且这个索引值是唯一的。

其中，将userId设为了自增的主键。注意，只有在数据类型为包装类型时，自增主键才能生效。

11.4.3 新增UserStore

新增UserStore类，用于表示数据库。代码如下：

```
package com.waylau.hmos.dataabilityhelperaccessorm;

import ohos.data.orm.OrmDatabase;
import ohos.data.orm.annotation.Database;


@Database(entities = {User.class}, version = 1)
public abstract class UserStore extends OrmDatabase {
}
```

上述代码定义了类UserStore.java，包含的实体是“User”类，版本号为“1”。数据库类的getVersion方法和getHelper方法不需要实现，直接将数据库类设为虚类即可。

11.4.4 创建DataAbility

在DevEco Studio中，创建一个名为“UserDataAbility”的Data。UserDataAbility初始化时代码如下：

```
package com.waylau.hmos.dataabilityhelperaccessorm;

import ohos.aafwk.ability.Ability;
import ohos.aafwk.content.Intent;
import ohos.data.DatabaseHelper;
import ohos.data.orm.OrmContext;
import ohos.data.orm.OrmPredicates;
import ohos.data.resultset.ResultSet;
import ohos.data.rdb.ValuesBucket;
import ohos.data.dataability.DataAbilityPredicates;
import ohos.hiviewdfx.HiLog;
import ohos.hiviewdfx.HiLogLabel;
import ohos.utils.net.Uri;
import ohos.utils.PacMap;

import java.io.FileDescriptor;
import java.util.List;

public class UserDataAbility extends Ability {
    private static final String TAG = UserDataAbility.class.getSimpleName();
    private static final HiLogLabel LABEL_LOG =
            new HiLogLabel(HiLog.LOG_APP, 0x00001, TAG);

    private static final String DATABASE_NAME = "RdbStoreTest.db";
    private static final String DATABASE_NAME_ALIAS = TAG;

    private DatabaseHelper manager;
    private OrmContext ormContext = null;

    @Override
    public void onStart(Intent intent) {
        super.onStart(intent);
        HiLog.info(LABEL_LOG, "UserDataAbility onStart");

        // 初始化DatabaseHelper、OrmContext
        manager = new DatabaseHelper(this);
        ormContext = manager.getOrmContext(DATABASE_NAME_ALIAS,
                                           DATABASE_NAME, UserStore.class);
    }
```

```
    @Override
    public void onStop() {
        super.onStop();
        HiLog.info(LABEL_LOG, "UserDataAbility onStop");

        // 删除数据库
        manager.deleteRdbStore(DATABASE_NAME);
    }

    public List<User> queryAll() {
        OrmPredicates predicates = ormContext.where(User.class);
        List<User> users = ormContext.query(predicates);
        users.forEach(user -> {
            HiLog.info(LABEL_LOG, "query user: %{public}s", user);
        });

        return users;
    }

    public int insert(User user) {
        HiLog.info(LABEL_LOG, "before insert");

        // 插入数据库
        ormContext.insert(user);
        boolean isSuccessed = ormContext.flush();

        // 获取 UserId
        int userId = user.getUserId();

        HiLog.info(LABEL_LOG, "end insert: %{public}s, isSuccessed: %{public}s",
                userId, isSuccessed);
        return userId;
    }

    public int deleteAll() {
        HiLog.info(LABEL_LOG, "before delete");
        OrmPredicates predicates = ormContext.where(User.class);
        List<User> users = ormContext.query(predicates);

        users.forEach(user -> {
            boolean isSuccessed = ormContext.delete(user);
            HiLog.info(LABEL_LOG, "delete user: %{public}s,
                        isSuccessed: %{public}s",
                    user.getUserId(), isSuccessed);
        });
```

```
        boolean isSuccessed = ormContext.flush();

        HiLog.info(LABEL_LOG, "end delete, isSuccessed: %{public}s",
                isSuccessed);

        return users.size();
    }

    public int update(User user) {
        HiLog.info(LABEL_LOG, "before update");

        ormContext.update(user);

        boolean isSuccessed = ormContext.flush();

        HiLog.info(LABEL_LOG, "end update, isSuccessed: %{public}s",
                isSuccessed);
        return 1;
    }
}
```

UserDataAbility自动在配置文件中添加了相应的配置，内容如下：

```
"abilities": [
    {
    "skills": [
        {
        "entities": [
            "entity.system.home"
        ],
        "actions": [
            "action.system.home"
        ]
        }
    ],
    "orientation": "unspecified",
    "name": "com.waylau.hmos.dataabilityhelperaccessorm.MainAbility",
    "icon": "$media:icon",
    "description": "$string:mainability_description",
    "label": "DataAbilityHelperAccessORM",
    "type": "page",
    "launchType": "standard"
    },
    // 新增UserDataAbility配置
    {
    "permissions": [
```

```
            "com.waylau.hmos.dataabilityhelperaccessorm.DataAbilityShellProvider.
                PROVIDER"
        ],
        "name": "com.waylau.hmos.dataabilityhelperaccessorm.UserDataAbility",
        "icon": "$media:icon",
        "description": "$string:userdataability_description",
        "type": "data",
        "uri": "dataability://com.waylau.hmos.dataabilityhelperaccessorm.
            UserDataAbility"
    }
]
```

从上述配置可以看出以下几点。

- type: 类型设置为 data。
- uri: 对外提供的访问路径，全局唯一。
- permissions: 访问该 Data Ability 时需要申请的访问权限。

11.4.5 初始化数据库

修改 UserDataAbility 代码如下：

```
public class UserDataAbility extends Ability {
    private static final String TAG = UserDataAbility.class.getSimpleName();
    private static final HiLogLabel LABEL_LOG =
            new HiLogLabel(HiLog.LOG_APP, 0x00001, TAG);

    private static final String DATABASE_NAME = "RdbStoreTest.db";
    private static final String DATABASE_NAME_ALIAS = TAG;

    private DatabaseHelper manager;
    private OrmContext ormContext = null;

    @Override
    public void onStart(Intent intent) {
        super.onStart(intent);
        HiLog.info(LABEL_LOG, "UserDataAbility onStart");

        // 初始化 DatabaseHelper、OrmContext
        manager = new DatabaseHelper(this);
        ormContext = manager.getOrmContext(DATABASE_NAME_ALIAS,
                                          DATABASE_NAME, UserStore.class);
    }

    @Override
```

```
    public void onStop() {
        super.onStop();
        HiLog.info(LABEL_LOG, "UserDataAbility onStop");

        // 删除数据库
        manager.deleteRdbStore(DATABASE_NAME);
    }
}
```

上述代码解释如下。

- 在onStart方法中初始化了一个名为“RdbStoreTest.db”的数据库。
- 实例化了DatabaseHelper、OrmContext对象。
- 重写onStop方法，以删除数据库。

11.4.6 新增queryAll方法

在UserDataAbility中新增queryAll方法，代码如下：

```
public List<User> queryAll() {
    OrmPredicates predicates = ormContext.where(User.class);
    List<User> users = ormContext.query(predicates);
    users.forEach(user -> {
        HiLog.info(LABEL_LOG, "query user: %{public}s", user);
    });

    return users;
}
```

上述代码中，将会查询数据库中的所有用户数据。

11.4.7 新增insert方法

在UserDataAbility中新增insert方法，代码如下：

```
public int insert(User user) {
    HiLog.info(LABEL_LOG, "before insert");

    // 插入数据库
    ormContext.insert(user);
    boolean isSuccessed = ormContext.flush();

    // 获取UserId
    int userId = user.getUserId();
```

```
    HiLog.info(LABEL_LOG, "end insert: %{public}s, isSuccessed: %{public}s",
            userId, isSuccessed);
    return userId;
}
```

上述代码中，将在数据库中插入用户信息。

11.4.8 新增update方法

在UserDataAbility中新增update方法，代码如下：

```
public int update(User user) {
    HiLog.info(LABEL_LOG, "before update");

    ormContext.update(user);

    boolean isSuccessed = ormContext.flush();

    HiLog.info(LABEL_LOG, "end update, isSuccessed: %{public}s",
            isSuccessed);
    return 1;
}
```

上述代码中，将会更新指定的用户信息。

11.4.9 新增deleteAll方法

在UserDataAbility中新增deleteAll方法，代码如下：

```
public int deleteAll() {
    HiLog.info(LABEL_LOG, "before delete");
    OrmPredicates predicates = ormContext.where(User.class);
    List<User> users = ormContext.query(predicates);

    users.forEach(user -> {
        boolean isSuccessed = ormContext.delete(user);
        HiLog.info(LABEL_LOG, "delete user: %{public}s, isSuccessed: %{public}s",
                user.getUserId(), isSuccessed);
    });

    boolean isSuccessed = ormContext.flush();

    HiLog.info(LABEL_LOG, "end delete, isSuccessed: %{public}s",isSuccessed);

    return users.size();
}
```

上述代码中，将会删除所有用户信息。

11.4.10 修改ability_main.xml

修改ability_main.xml内容如下：

```
<?xml version="1.0" encoding="utf-8"?>
<DirectionalLayout
    xmlns:ohos="http://schemas.huawei.com/res/ohos"
    ohos:height="match_parent"
    ohos:width="match_parent"
    ohos:orientation="vertical">

    <DirectionalLayout
        ohos:height="60vp"
        ohos:width="match_parent"
        ohos:orientation="horizontal">
        <Button
            ohos:id="$+id:button_query"
            ohos:height="40vp"
            ohos:width="0vp"
            ohos:background_element="#F76543"
            ohos:layout_alignment="horizontal_center"
            ohos:margin="10vp"
            ohos:padding="10vp"
            ohos:text="Query"
            ohos:text_size="16fp"
            ohos:weight="1"
            />

        <Button
            ohos:id="$+id:button_insert"
            ohos:height="40vp"
            ohos:width="0vp"
            ohos:background_element="#F76543"
            ohos:layout_alignment="horizontal_center"
            ohos:margin="10vp"
            ohos:padding="10vp"
            ohos:text="Insert"
            ohos:text_size="16fp"
            ohos:weight="1"
            />

        <Button
            ohos:id="$+id:button_update"
```

```
            ohos:height="40vp"
            ohos:width="0vp"
            ohos:background_element="#F76543"
            ohos:layout_alignment="horizontal_center"
            ohos:margin="10vp"
            ohos:padding="10vp"
            ohos:text="Update"
            ohos:text_size="16fp"
            ohos:weight="1"
            />

        <Button
            ohos:id="$+id:button_delete"
            ohos:height="40vp"
            ohos:width="0vp"
            ohos:background_element="#F76543"
            ohos:layout_alignment="horizontal_center"
            ohos:margin="10vp"
            ohos:padding="10vp"
            ohos:text="Delete"
            ohos:text_size="16fp"
            ohos:weight="1"
            />
    </DirectionalLayout>

    <Text
        ohos:id="$+id:text"
        ohos:height="match_content"
        ohos:width="match_content"
        ohos:background_element="$graphic:background_ability_main"
        ohos:multiple_lines="true"
        ohos:text=""
        ohos:text_size="20fp"
        />
</DirectionalLayout>
```

界面预览效果如图 11-3 所示。

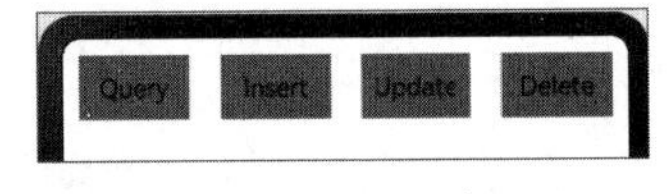

图 11-3　界面预览效果

上述代码：

- 设置了 4 个按钮，用于触发操作数据库的相关操作。
- Text 组件，用于展示查询到的用户信息。

11.4.11　修改 MainAbilitySlice

修改 MainAbilitySlice，代码如下：

```
package com.waylau.hmos.dataabilityhelperaccessorm.slice;

import com.waylau.hmos.dataabilityhelperaccessorm.ResourceTable;
import com.waylau.hmos.dataabilityhelperaccessorm.User;
import com.waylau.hmos.dataabilityhelperaccessorm.UserDataAbility;
import ohos.aafwk.ability.AbilitySlice;
import ohos.aafwk.content.Intent;
import ohos.agp.components.Button;
import ohos.agp.components.Text;
import ohos.hiviewdfx.HiLog;
import ohos.hiviewdfx.HiLogLabel;
import ohos.utils.zson.ZSONObject;

import java.util.List;
import java.util.Random;

public class MainAbilitySlice extends AbilitySlice {
    private static final String TAG = MainAbilitySlice.class.getSimpleName();
    private static final HiLogLabel LABEL_LOG =
            new HiLogLabel(HiLog.LOG_APP, 0x00001, TAG);

    private UserDataAbility userDataAbility = new UserDataAbility();

    private Text text;

    @Override
    public void onStart(Intent intent) {
        super.onStart(intent);
        super.setUIContent(ResourceTable.Layout_ability_main);

        userDataAbility.onStart(intent);

        // 添加点击事件来触发访问数据
        Button buttonQuery = (Button) findComponentById(ResourceTable.Id_
                                                        button_query);
        buttonQuery.setClickedListener(listener -> this.doQuery());

        Button buttonInsert = (Button) findComponentById(ResourceTable.Id_
                                                         button_insert);
        buttonInsert.setClickedListener(listener -> this.doInsert());

        Button buttonUpdate = (Button) findComponentById(ResourceTable.Id_
                                                         button_update);
        buttonUpdate.setClickedListener(listener -> this.doUpdate());

        Button buttonDelete = (Button) findComponentById(ResourceTable.Id_
```

```
                                                    button_delete);
    buttonDelete.setClickedListener(listener -> this.doDelete());

    text = (Text) findComponentById(ResourceTable.Id_text);
}

@Override
public void onStop() {
    super.onStop();
    userDataAbility.onStop();
}

private void doQuery() {
    // 查询所有用户
    List<User> users = userDataAbility.queryAll();

    // 用户的信息显示在界面
    text.setText(ZSONObject.toZSONString(users) + "\n");
}

private void doInsert() {
    // 生成随机数据
    Random random = new Random();
    int age = random.nextInt();
    String name = "n" + System.currentTimeMillis();

    //  生成用户
    User user = new User();
    user.setUserName(name);
    user.setAge(age);

    // 插入用户
    userDataAbility.insert(user);
}

private void doUpdate() {
    // 查询所有用户
    List<User> users = userDataAbility.queryAll();

    // 更新所有用户
    users.forEach(user -> {
        user.setAge(43);
        userDataAbility.update(user);
    });
}
```

```
    private void doDelete() {
        // 删除所有用户
        userDataAbility.deleteAll();
    }

    @Override
    public void onActive() {
        super.onActive();
    }

    @Override
    public void onForeground(Intent intent) {
        super.onForeground(intent);
    }
}
```

在上述方法中：

- 初始化了UserDataAbility，并通过UserDataAbility来实现用户的查询、插入、更新、删除操作；
- Text组件，用于展示查询到的用户信息。

11.4.12 运行

运行应用后，点击3次“Insert”按钮，以触发插入用户的操作，可以看到控制台输出内容如下：

```
04-27 13:28:29.253 19890-19890/com.waylau.hmos.dataabilityhelperaccessorm I
00001/UserDataAbility: before insert
04-27 13:28:29.269 19890-19890/com.waylau.hmos.dataabilityhelperaccessorm I
00001/UserDataAbility: end insert: 1, isSuccessed: true
04-27 13:28:29.673 19890-19890/com.waylau.hmos.dataabilityhelperaccessorm I
00001/UserDataAbility: before insert
04-27 13:28:29.676 19890-19890/com.waylau.hmos.dataabilityhelperaccessorm I
00001/UserDataAbility: end insert: 2, isSuccessed: true
04-27 13:28:30.042 19890-19890/com.waylau.hmos.dataabilityhelperaccessorm I
00001/UserDataAbility: before insert
04-27 13:28:30.045 19890-19890/com.waylau.hmos.dataabilityhelperaccessorm I
00001/UserDataAbility: end insert: 3, isSuccessed: true
```

此时，点击“Query”按钮，以触发查询用户的操作，可以看到控制台输出内容如下：

```
04-27 13:29:30.355 19890-19890/com.waylau.hmos.dataabilityhelperaccessorm I
00001/UserDataAbility: query user: com.waylau.hmos.dataabilityhelperaccessorm.
User@52c68dca
04-27 13:29:30.356 19890-19890/com.waylau.hmos.dataabilityhelperaccessorm I
00001/UserDataAbility: query user: com.waylau.hmos.dataabilityhelperaccessorm.
```

```
User@52c68dea
04-27 13:29:30.356 19890-19890/com.waylau.hmos.dataabilityhelperaccessorm I
00001/UserDataAbility: query user: com.waylau.hmos.dataabilityhelperaccessorm.
User@52c68e0a
```

界面效果如图 11-4 所示。

此时，点击“Update”按钮，以触发更新用户的操作，可以看到控制台输出内容如下：

```
04-27 13:30:51.934 19890-19890/com.waylau.hmos.dataabilityhelperaccessorm I
00001/UserDataAbility: before update
04-27 13:30:51.940 19890-19890/com.waylau.hmos.dataabilityhelperaccessorm I
00001/UserDataAbility: end update, isSuccessed: true
04-27 13:30:51.940 19890-19890/com.waylau.hmos.dataabilityhelperaccessorm I
00001/UserDataAbility: before update
04-27 13:30:51.943 19890-19890/com.waylau.hmos.dataabilityhelperaccessorm I
00001/UserDataAbility: end update, isSuccessed: true
04-27 13:30:51.943 19890-19890/com.waylau.hmos.dataabilityhelperaccessorm I
00001/UserDataAbility: before update
04-27 13:30:51.944 19890-19890/com.waylau.hmos.dataabilityhelperaccessorm I
00001/UserDataAbility: end update, isSuccessed: true
```

点击“Query”按钮，界面效果如图 11-5 所示。

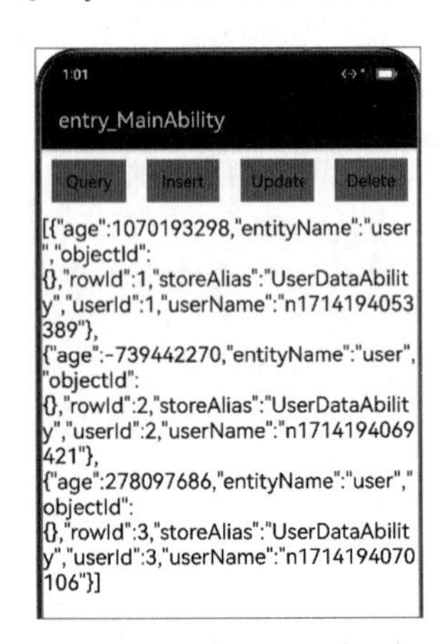

图 11-4　界面效果（一）

图 11-5　界面效果（二）

可以看到，“age”已经被更新为“43”。

此时，点击“Delete”按钮，以触发删除用户的操作，可以看到控制台输出内容如下：

```
04-27 13:33:01.119 19890-19890/com.waylau.hmos.dataabilityhelperaccessorm I
00001/UserDataAbility: before delete
04-27 13:33:01.120 19890-19890/com.waylau.hmos.dataabilityhelperaccessorm I
00001/UserDataAbility: delete user: 1, isSuccessed: true
04-27 13:33:01.120 19890-19890/com.waylau.hmos.dataabilityhelperaccessorm I
00001/UserDataAbility: delete user: 2, isSuccessed: true
04-27 13:33:01.120 19890-19890/com.waylau.hmos.dataabilityhelperaccessorm I
```

```
00001/UserDataAbility: delete user: 3, isSuccessed: true
04-27 13:33:01.125 19890-19890/com.waylau.hmos.dataabilityhelperaccessorm I
00001/UserDataAbility: end delete, isSuccessed: true
```

点击“Query”按钮，界面效果如图 11-6 所示。

可以看到，用户信息都已经被删除了。

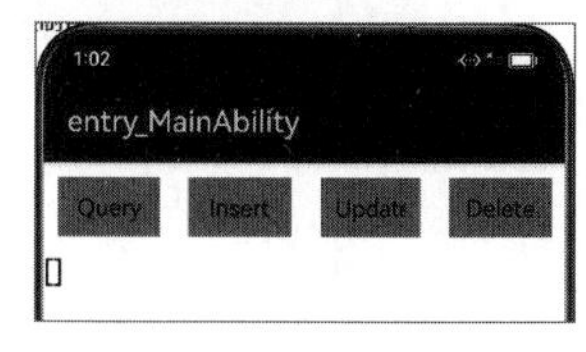

图 11-6　界面效果（三）

11.5 轻量级偏好数据库

轻量级偏好数据库（也称为“用户首选项”）主要提供轻量级Key-Value操作，支持本地应用存储少量数据，数据存储在本地文件中，同时也加载在内存中，所以访问速度更快，效率更高。轻量级偏好数据库属于非关系型数据库，不宜存储大量数据，经常用于操作键值对形式数据的场景。

11.5.1　基本概念

轻量级偏好数据库主要涉及以下概念。

- Key-Value数据库：一种以键值对存储数据的数据库，类似Java中的Map。Key是关键字，Value是值。常见的Key-Value数据库产品有Redis、Berkley DB等。
- 非关系型数据库：区别于关系数据库，不保证遵循ACID（Atomic、Consistency、Isolation及Durability）特性，不采用关系模型来组织数据，数据之间无关系，扩展性好。除了上面几款Key-Value数据库产品外，还有Cassandra、MongoDB、HBase等。
- 偏好数据：用户经常访问和使用的数据。

有关Key-Value数据库、非关系型数据库的更多内容可以详见笔者所著的《分布式系统常用技术及案例分析》。

11.5.2　运作机制

HarmonyOS提供偏好型数据库的操作类，应用通过这些操作类完成数据库操作。

- 借助DatabaseHelper的API，应用可以将指定文件的内容加载到Preferences实例，每个文件最多有一个Preferences实例，系统会通过静态容器将该实例存储在内存中，直到应用主动从内存中移除该实例或删除该文件。
- 获取到文件对应的Preferences实例后，应用可以借助Preferences的API，从Preferences实例中读取数据或将数据写入Preferences实例，通过flush或flushSync将Preferences实例持久化。

图 11-7 所示是轻量级偏好数据库的架构。

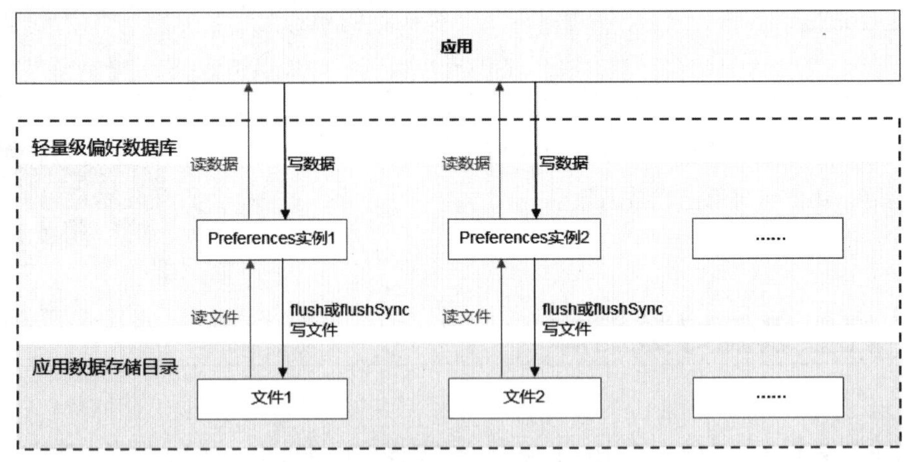

图 11-7　轻量级偏好数据库的架构

11.5.3　约束与限制

使用偏好型数据库时，需要注意以下约束与限制。

- Key键为String类型，要求非空且大小不超过80个字符。
- 如果Value值为String类型，可以为空但是长度不超过8192个字符。
- 存储的数据量应该是轻量级的，建议存储的数据不超过一万条，否则会在内存方面产生较大的开销。
- 轻量级偏好数据库主要用于保存应用的一些常用配置，并不适合频繁改变数据的场景。

11.5.4　接口说明

轻量级偏好数据库向本地应用提供了操作偏好型数据库的API，支持本地应用读写少量数据及观察数据变化。数据存储形式为键值对，键的类型为字符串型，值的存储数据类型包括整型、字符串型、布尔型、浮点型、长整型、字符串型Set集合。

1. 创建数据库

通过数据库操作的辅助类可以获取到要操作的Preferences实例，用于进行数据库的操作。

DatabaseHelper轻量级偏好数据库创建接口如下。

- getPreferences(String name)：获取文件对应的Preferences单实例，用于数据操作。

2. 查询数据

通过调用Get系列的方法，可以查询不同类型的数据。

Preferences轻量级偏好数据库查询接口如下。

- getInt(String key, int defValue)：获取键对应的int类型的值。
- getFloat(String key, float defValue)：获取键对应的float类型的值。

3. 插入数据

通过Put系列的方法可以修改Preferences实例中的数据，通过flush或flushSync将Preferences实例持久化。

Preferences轻量级偏好数据库插入接口如下。

- putInt(String key, int value)：设置Preferences实例中键对应的int类型的值。
- putString(String key, String value)：设置Preferences实例中键对应的String类型的值。
- flush()：将Preferences实例异步写入文件。
- flushSync()：将Preferences实例同步写入文件。

4. 观察数据变化

轻量级偏好数据库还提供了一系列的接口变化回调，用于观察数据的变化。开发者可以通过重写onChange方法来定义观察者的行为。

Preferences轻量级偏好数据库接口变化回调如下。

- registerObserver(PreferencesObserver preferencesObserver)：注册观察者，用于观察数据变化。
- unRegisterObserver(PreferencesObserver preferencesObserver)：注销观察者。
- onChange(Preferences preferences, String key)：观察者的回调方法，任意数据变化都会回调该方法。

5. 删除数据文件

DatabaseHelper轻量级偏好数据库删除接口如下。

- deletePreferences(String name)：删除文件和文件对应的Preferences单实例。
- removePreferencesFromCache(String name)：删除文件对应的Preferences单实例。

6. 移动数据库文件

DatabaseHelper轻量级偏好数据库移动接口如下。

- movePreferences(Context sourceContext, String sourceName, String targetName)：移动数据库文件。

11.6 实战：使用轻量级偏好数据库

本节将演示如何使用轻量级偏好数据库。

创建一个名为“Preferences”的应用作为演示。

11.6.1 修改ability_main.xml

修改ability_main.xml内容如下：

```
<?xml version="1.0" encoding="utf-8"?>
<DirectionalLayout
    xmlns:ohos="http://schemas.huawei.com/res/ohos"
    ohos:height="match_parent"
    ohos:width="match_parent"
    ohos:orientation="vertical">

    <DirectionalLayout
        ohos:height="60vp"
        ohos:width="match_parent"
        ohos:orientation="horizontal">
        <Button
            ohos:id="$+id:button_query"
            ohos:height="40vp"
            ohos:width="0vp"
            ohos:background_element="#F76543"
            ohos:layout_alignment="horizontal_center"
            ohos:margin="10vp"
            ohos:padding="10vp"
            ohos:text="Query"
            ohos:text_size="16fp"
            ohos:weight="1"
            />

        <Button
            ohos:id="$+id:button_insert"
            ohos:height="40vp"
            ohos:width="0vp"
            ohos:background_element="#F76543"
            ohos:layout_alignment="horizontal_center"
            ohos:margin="10vp"
            ohos:padding="10vp"
            ohos:text="Insert"
            ohos:text_size="16fp"
            ohos:weight="1"
            />

        <Button
            ohos:id="$+id:button_update"
            ohos:height="40vp"
            ohos:width="0vp"
            ohos:background_element="#F76543"
            ohos:layout_alignment="horizontal_center"
            ohos:margin="10vp"
            ohos:padding="10vp"
            ohos:text="Update"
```

```
            ohos:text_size="16fp"
            ohos:weight="1"
            />

        <Button
            ohos:id="$+id:button_delete"
            ohos:height="40vp"
            ohos:width="0vp"
            ohos:background_element="#F76543"
            ohos:layout_alignment="horizontal_center"
            ohos:margin="10vp"
            ohos:padding="10vp"
            ohos:text="Delete"
            ohos:text_size="16fp"
            ohos:weight="1"
            />
    </DirectionalLayout>

    <Text
        ohos:id="$+id:text"
        ohos:height="match_content"
        ohos:width="match_content"
        ohos:background_element="$graphic:background_ability_main"
        ohos:multiple_lines="true"
        ohos:text=""
        ohos:text_size="20fp"
        />
</DirectionalLayout>
```

界面预览效果如图11-8所示。

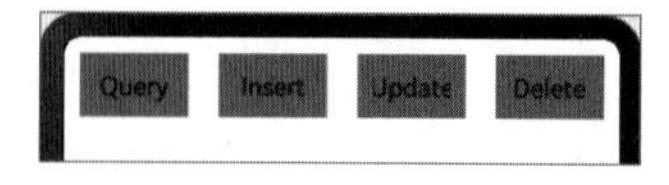

图11-8　界面预览效果

上述代码解释如下。

- 设置了4个按钮，用于触发操作数据库的相关操作。
- Text组件，用于展示查询到的数据信息。

11.6.2　修改MainAbilitySlice

修改MainAbilitySlice，代码如下：

```
package com.waylau.hmos.preferences.slice;
```

```
import com.waylau.hmos.preferences.ResourceTable;
import ohos.aafwk.ability.AbilitySlice;
import ohos.aafwk.content.Intent;
import ohos.agp.components.Button;
import ohos.agp.components.Text;
import ohos.data.DatabaseHelper;
import ohos.data.preferences.Preferences;
import ohos.hiviewdfx.HiLog;
import ohos.hiviewdfx.HiLogLabel;

public class MainAbilitySlice extends AbilitySlice {
    private static final String TAG = MainAbilitySlice.class.getSimpleName();
    private static final HiLogLabel LABEL_LOG =
            new HiLogLabel(HiLog.LOG_APP, 0x00001, TAG);

    private static final String PREFERENCES_FILE = "preferences-file";
    private static final String PREFERENCES_KEY = "preferences-key";

    private DatabaseHelper databaseHelper;
    private Preferences preferences;
    private Preferences.PreferencesObserver observer;
    private Text text;

    @Override
    public void onStart(Intent intent) {
        super.onStart(intent);
        super.setUIContent(ResourceTable.Layout_ability_main);

        // 初始化数据
        initData();

        // 添加点击事件来触发访问数据
        Button buttonQuery = (Button) findComponentById(ResourceTable.Id_
                                                            button_query);
        buttonQuery.setClickedListener(listener -> this.doQuery());

        Button buttonInsert = (Button) findComponentById(ResourceTable.Id_
                                                            button_insert);
        buttonInsert.setClickedListener(listener -> this.doInsert());

        Button buttonUpdate = (Button) findComponentById(ResourceTable.Id_
                                                            button_update);
        buttonUpdate.setClickedListener(listener -> this.doUpdate());

        Button buttonDelete = (Button) findComponentById(ResourceTable.Id_
                                                            button_delete);
```

```
        buttonDelete.setClickedListener(listener -> this.doDelete());

        text = (Text) findComponentById(ResourceTable.Id_text);
    }

    private void initData() {
        HiLog.info(LABEL_LOG, "before initData");

        databaseHelper = new DatabaseHelper(this.getContext());

        // fileName 表示文件名，其取值不能为空，也不能包含路径
        // 默认存储目录可以通过 context.getPreferencesDir() 获取
        preferences = databaseHelper.getPreferences(PREFERENCES_FILE);

        // 观察者
        observer = new MyPreferencesObserver();

        HiLog.info(LABEL_LOG, "end initData");
    }

    @Override
    public void onStop() {
        super.onStop();

        // 从内存中移除指定文件对应的 Preferences 单实例
        // 并删除指定文件及其备份文件、损坏文件
        boolean result = databaseHelper.deletePreferences(PREFERENCES_FILE);
    }

    private void doQuery() {
        HiLog.info(LABEL_LOG, "before doQuery");

        // 查询
        String result = preferences.getString(PREFERENCES_KEY, "");

        // 将查询结果显示在界面
        text.setText(result + "\n");

        HiLog.info(LABEL_LOG, "end doQuery, result: %{public}s", result);
    }

    private void doInsert() {
        HiLog.info(LABEL_LOG, "before doInsert");

        // 生成随机数据
        String data = "d" + System.currentTimeMillis();
```

```
        // 将数据写入 Preferences 实例
        preferences.putString(PREFERENCES_KEY, data);

        // 注册观察者
        preferences.registerObserver(observer);

        // 通过 flush 或 flushSync 将 Preferences 实例持久化
        preferences.flush(); // 异步
        // preferences.flushSync(); // 同步

        HiLog.info(LABEL_LOG, "end doInsert, data: %{public}s", data);
    }

    private void doUpdate() {
        HiLog.info(LABEL_LOG, "before doUpdate");

        // 更新就是重新做一次插入
        doInsert();

        HiLog.info(LABEL_LOG, "end doUpdate");
    }

    private void doDelete() {
        HiLog.info(LABEL_LOG, "before doDelete");

        // 删除
        preferences.delete(PREFERENCES_KEY);

        HiLog.info(LABEL_LOG, "end doDelete");
    }

    private class MyPreferencesObserver implements Preferences.
PreferencesObserver {

        @Override
        public void onChange(Preferences preferences, String key) {
            HiLog.info(LABEL_LOG, "onChange, key: %{public}s", key);
        }
    }

    @Override
    public void onActive() {
        super.onActive();
    }
```

```
    @Override
    public void onForeground(Intent intent) {
        super.onForeground(intent);
    }
}
```

在上述方法中：

- initData方法初始化了DatabaseHelper、Preferences、PreferencesObserver对象；
- 通过Preferences对象来实现数据的查询、插入、更新、删除操作；
- onStop方法从内存中移除指定文件对应的Preferences单实例，并删除指定文件及其备份文件、损坏文件；
- Text组件，用于展示查询到的数据信息。

11.6.3 运行

运行应用后，点击“Query”按钮，以触发查询的操作，可以看到控制台输出内容如下：

```
04-27 16:31:05.641 29080-29080/com.waylau.hmos.preferences I 00001/
MainAbilitySlice: before doQuery
04-27 16:31:05.642 29080-29080/com.waylau.hmos.preferences I 00001/
MainAbilitySlice: end doQuery, result:
```

从日志可以看出，并没有查到任何数据，界面效果如图11-9所示。

此时，点击“Insert”按钮，以触发插入的操作，可以看到控制台输出内容如下：

```
04-27 16:32:58.167 29080-29080/com.waylau.hmos.preferences I 00001/
MainAbilitySlice: before doInsert
04-27 16:32:58.168 29080-29080/com.waylau.hmos.preferences I 00001/
MainAbilitySlice: onChange, key: preferences-key
04-27 16:32:58.168 29080-29080/com.waylau.hmos.preferences I 00001/
MainAbilitySlice: end doInsert, data: d1613377978167
```

此时，再次点击“Query”按钮，界面就能查到刚插入的数据，效果如图11-10所示。

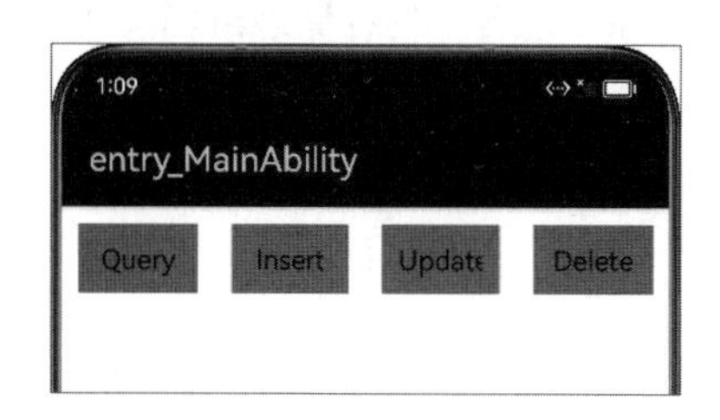

图11-9　界面效果（一）

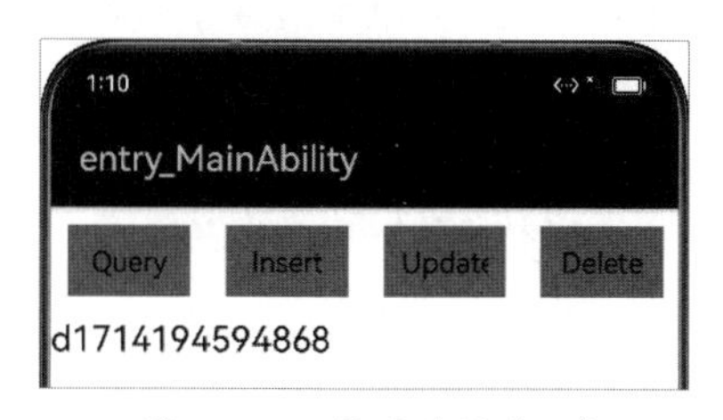

图11-10　界面效果（二）

点击“Update”按钮，以触发更新的操作，可以看到控制台输出内容如下：

```
04-27 16:34:23.407 29080-29080/com.waylau.hmos.preferences I 00001/
```

```
MainAbilitySlice: before doUpdate
04-27 16:34:23.407 29080-29080/com.waylau.hmos.preferences I 00001/
MainAbilitySlice: before doInsert
04-27 16:34:23.408 29080-29080/com.waylau.hmos.preferences I 00001/
MainAbilitySlice: onChange, key: preferences-key
04-27 16:34:23.408 29080-29080/com.waylau.hmos.preferences I 00001/
MainAbilitySlice: end doInsert, data: d1613378063407
04-27 16:34:23.408 29080-29080/com.waylau.hmos.preferences I 00001/
MainAbilitySlice: end doUpdate
```

此时，点击“Query”按钮，界面就能查到刚更新的数据，效果如图 11-11 所示。

点击“Delete”按钮，以触发删除的操作，可以看到控制台输出内容如下：

```
04-27 16:35:56.404 29080-29080/com.waylau.hmos.preferences I 00001/
MainAbilitySlice: before doDelete
04-27 16:35:56.404 29080-29080/com.waylau.hmos.preferences I 00001/
MainAbilitySlice: end doDelete
```

此时，点击“Query”按钮，界面效果如图 11-12 所示。

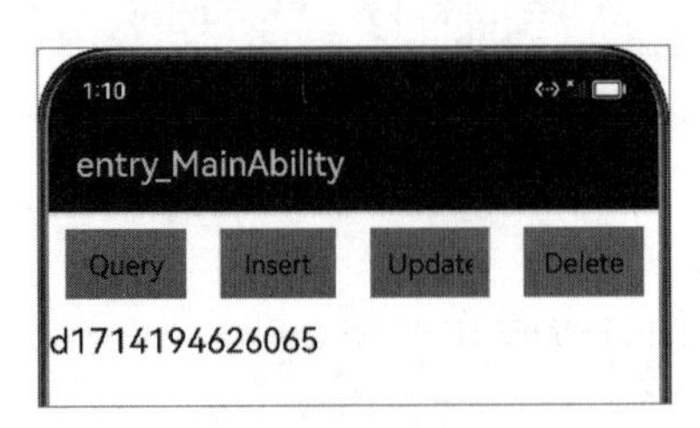

图 11-11　界面效果（三）

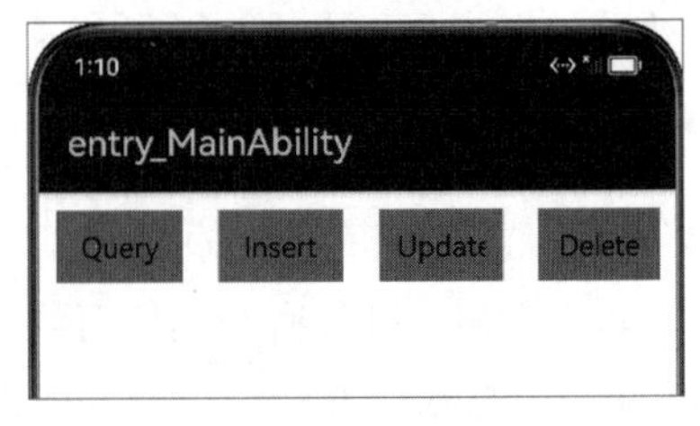

图 11-12　界面效果（四）

可以看到，信息都已经被删除了。

11.7 数据存储管理

数据存储管理指导开发者基于 HarmonyOS，进行存储设备（包含本地存储、SD 卡、U 盘等）的数据存储管理能力的开发，包括获取存储设备列表、获取存储设备视图等。

11.7.1 基本概念

数据存储管理涉及以下基本概念。

- 数据存储管理：包括获取存储设备列表、获取存储设备视图，同时也可以按照条件获取对应的存储设备视图信息。
- 设备存储视图：提供了存储设备的抽象及访问存储设备自身信息的接口。

11.7.2 运作机制

用统一的视图结构可以表示各种存储设备，该视图结构的内部属性会因为设备的不同而不同。每个存储设备可以抽象成两部分，一部分是存储设备自身信息区域，一部分是用来真正存放数据的区域。

图11-13展示了存储设备视图。

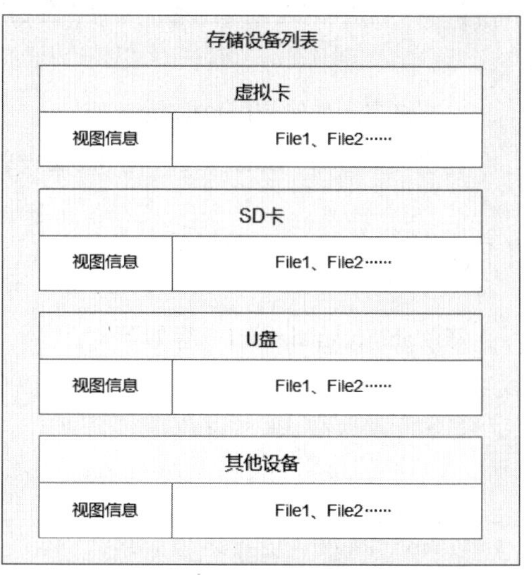

图11-13　存储设备视图

11.7.3 接口说明

为了给用户展示存储设备信息，开发者可以使用数据存储管理接口获取存储设备视图信息，也可以根据用户提供的文件名获取对应存储设备的视图信息。

数据存储管理为开发者提供下面几种功能，主要涉及DataUsage和Volume的API。

1. DataUsage

DataUsage的API有以下几种。

- getVolumes()：获取当前用户可用的设备列表视图。
- getVolume(File file)：获取存储该文件的存储设备视图。
- getVolume(Context context, Uri uri)：获取该URI对应文件所在的存储设备视图。
- getDiskMountedStatus()：获取默认存储设备的挂载状态。
- getDiskMountedStatus(File path)：获取存储该文件设备的挂载状态。
- isDiskPluggable()：默认存储设备是否为可插拔设备。
- isDiskPluggable(File path)：存储该文件的设备是否为可插拔设备。
- isDiskEmulated()：默认存储设备是否为虚拟设备。
- isDiskEmulated(File path)：存储该文件的设备是否为虚拟设备。

2. Volume

Volume的API有以下几种。

- isEmulated()：该设备是否为虚拟存储设备。
- isPluggable()：该设备是否支持插拔。
- getDescription()：获取设备描述信息。
- getState()：获取设备挂载状态。
- getVolUuid()：获取设备唯一标识符。

11.8 实战：使用数据存储管理

本节将演示如何使用数据存储管理。

创建一个名为“DataUsage”的Phone设备类型应用作为演示。

11.8.1 修改ability_main.xml

修改ability_main.xml内容如下：

```
<?xml version="1.0" encoding="utf-8"?>
<DirectionalLayout
    xmlns:ohos="http://schemas.huawei.com/res/ohos"
    ohos:height="match_parent"
    ohos:width="match_parent"
    ohos:orientation="vertical">

    <Button
        ohos:id="$+id:button_query"
        ohos:height="match_content"
        ohos:width="match_parent"
        ohos:background_element="#F76543"
        ohos:layout_alignment="horizontal_center"
        ohos:margin="10vp"
        ohos:padding="10vp"
        ohos:text="Query"
        ohos:text_size="30fp"
        />

    <Text
        ohos:id="$+id:text"
        ohos:height="match_content"
        ohos:width="match_content"
        ohos:background_element="$graphic:background_ability_main"
        ohos:multiple_lines="true"
        ohos:text=""
        ohos:text_size="20fp"
        />
</DirectionalLayout>
```

界面预览效果如图11-14所示。

图11-14 界面预览效果

上述代码解释如下。

- 设置了“Query”按钮，以备设置点击事件，以触发查询数据的相关的操作。

- Text组件，用于展示查询到的数据信息。

11.8.2 修改MainAbilitySlice

修改MainAbilitySlice，代码如下：

```
package com.waylau.hmos.datausage.slice;

import com.waylau.hmos.datausage.ResourceTable;
import ohos.aafwk.ability.AbilitySlice;
import ohos.aafwk.content.Intent;
import ohos.agp.components.Button;
import ohos.agp.components.Text;
import ohos.data.usage.DataUsage;
import ohos.data.usage.MountState;
import ohos.data.usage.Volume;
import ohos.hiviewdfx.HiLog;
import ohos.hiviewdfx.HiLogLabel;
import ohos.utils.zson.ZSONObject;

import java.util.List;
import java.util.Optional;

public class MainAbilitySlice extends AbilitySlice {
    private static final String TAG = MainAbilitySlice.class.getSimpleName();
    private static final HiLogLabel LABEL_LOG =
            new HiLogLabel(HiLog.LOG_APP, 0x00001, TAG);

    private Text text;

    @Override
    public void onStart(Intent intent) {
        super.onStart(intent);
        super.setUIContent(ResourceTable.Layout_ability_main);

        // 添加点击事件来触发访问数据
        Button buttonQuery = (Button) findComponentById(ResourceTable.Id_
                                                    button_query);
        buttonQuery.setClickedListener(listener -> this.doQuery());

        text = (Text) findComponentById(ResourceTable.Id_text);
    }

    private void doQuery() {
```

```
        HiLog.info(LABEL_LOG, "before doQuery");

        // 查询
        // 获取默认存储设备挂载状态
        MountState status = DataUsage.getDiskMountedStatus();
        String statusString = ZSONObject.toZSONString(status);
        text.append("MountState: " + statusString + "\n");

        // 默认存储设备是否为可插拔设备
        boolean isDiskPluggable = DataUsage.isDiskPluggable();
        text.append("isDiskPluggable: " + isDiskPluggable + "\n");

        // 默认存储设备是否为虚拟设备
        boolean isDiskEmulated = DataUsage.isDiskEmulated();
        text.append("isDiskEmulated: " + isDiskEmulated + "\n");

        // 获取存储设备列表
        Optional<List<Volume>> listOptional = DataUsage.getVolumes();
        if (listOptional.isPresent()) {
            text.append("Volume:\n");

            listOptional.get().forEach(volume -> {
                // 查询 Volume 的信息
                String volUuid = volume.getVolUuid();
                String description = volume.getDescription();
                boolean isEmulated = volume.isEmulated();
                boolean isPluggable = volume.isPluggable();

                String volumeString = ZSONObject.toZSONString(volume);
                text.append(volumeString + "\n");

                HiLog.info(LABEL_LOG, "volUuid: %{public}s,
                           description: %{public}s, " +
                           "isEmulated: %{public}s, isPluggable: %{public}s",
                           volUuid, description, isEmulated, isPluggable);
            });
        }

        HiLog.info(LABEL_LOG, "end doQuery");
    }

    @Override
    public void onActive() {
        super.onActive();
    }
```

```
    @Override
    public void onForeground(Intent intent) {
        super.onForeground(intent);
    }
}
```

在上述方法中：

- doQuery方法通过DataUsage对象来查询默认存储设备的信息，包括挂载状态、是否为可插拔设备、是否为虚拟设备等；
- DataUsage.getVolumes可以获得所有的存储设备列表，遍历该列表，则可以获取每个Volume的信息；
- Text组件，用于展示查询到的数据信息。

11.8.3 运行

运行应用后，点击“Query”按钮，以触发查询的操作，可以看到控制台输出内容如下：

```
04-27 17:42:46.873 21732-21732/com.waylau.hmos.datausage I 00001/
MainAbilitySlice: before doQuery
04-27 17:42:46.889 21732-21732/com.waylau.hmos.datausage I 00001/
MainAbilitySlice: volUuid: , description: 内部存储, isEmulated: true,
isPluggable: false
04-27 17:42:46.889 21732-21732/com.waylau.hmos.datausage I 00001/
MainAbilitySlice: end doQuery
```

从日志可以看出，已经查到了设备的信息。界面效果如图11-15所示。

图11-15 界面效果

11.9 实战：Stage模型的关系型数据库开发

本节以一个“账本”为例，使用关系型数据库的相关接口实现对账单的增、删、改、查操作。

为了演示该功能，创建一个名为“ArkTSRdb”的应用。

11.9.1 操作RdbStore

首先要获取一个RdbStore来操作关系型数据库。

在src/main/ets目录下创建名为“common”的目录，用于存放常用的工具类。在该common目录创建工具类RdbUtil，代码如下：

```
// 导入 rdb 模块。
import data_rdb from '@ohos.data.rdb'
// 导入 common
import common from '@ohos.app.ability.common';

type Context = common.UIAbilityContext;
const STORE_CONFIG = { name: "rdbstore.db" }

export default class RdbUtil {
  private rdbStore: any = null;
  private tableName: string;
  private sqlCreateTable: string;
  private columns: Array<string>;

  constructor(tableName: string, sqlCreateTable: string, columns: Array<string>) {
    this.tableName = tableName;
    this.sqlCreateTable = sqlCreateTable;
    this.columns = columns;
  }

  getRdbStore(callback) {
    // 如果已经获取到 RdbStore 则不做操作
    if (this.rdbStore != null) {
      console.info('The rdbStore exists.');
      callback();
      return;
    }

    // 应用上下文，本例子是使用 API9 Stage 模型的 Context
    let context: Context = getContext(this) as Context;
    data_rdb.getRdbStore(context, STORE_CONFIG, 1, (err, rdb) => {
      if (err) {
        console.error('gerRdbStore() failed, err: ' + err);
        return;
      }
      this.rdbStore = rdb;

      // 获取到 RdbStore 后，需使用 executeSql 接口初始化数据库表结构和相关数据
      this.rdbStore.executeSql(this.sqlCreateTable);
      console.info('getRdbStore() finished.');
      callback();
    });
  }
}
```

为了对数据进行增、删、改、查操作，我们要封装对应接口。关系型数据库接口提供的增、删、改、查方法均有callback和Promise两种异步回调方式，本例使用callback异步回调。代码如下：

```
// 插入数据
insertData(data, callback) {
    let resFlag: boolean = false;   // 用于记录插入是否成功的 flag
    const valueBucket = data;  // 存储键值对的类型，表示要插入表中的数据行
    this.rdbStore.insert(this.tableName, valueBucket, function (err, ret) {
        if (err) {
        console.error('Rdb', 'insertData() failed, err: ' + err);
        callback(resFlag);
        return;
        }
        callback(!resFlag);
    });
}

// 删除数据
deleteData(predicates, callback) {
    let resFlag: boolean = false;
    // predicates 表示待删除数据的操作条件
    this.rdbStore.delete(predicates, function (err, ret) {
        if (err) {
        console.error('Rdb', 'deleteData() failed, err: ' + err);
        callback(resFlag);
        return;
        }
        callback(!resFlag);
    });
}

// 更新数据
updateData(predicates, data, callback) {
    let resFlag: boolean = false;
    const valueBucket = data;
    this.rdbStore.update(valueBucket, predicates, function (err, ret) {
        if (err) {
        console.error('Rdb', 'updateData() failed, err: ' + err);
        callback(resFlag);
        return;
        }
        callback(!resFlag);
    });
}

// 查找数据
```

```
query(predicates, callback){
    // columns 表示要查询的列，如果为空则表示查询所有列
    this.rdbStore.query(predicates, this.columns, function (err, resultSet) {
        if (err) {
        console.error('Rdb', 'query() failed, err: ' + err);
        return;
        }
        callback(resultSet);  // 如果查找成功则返回 resultSet 结果集
        resultSet.close();    // 操作完成后关闭结果集
    });
}
```

11.9.2 账目信息的表示

由于需要记录账目的类型（收入/支出）、具体类别和金额，因此我们需要创建一张存储账目信息的表，SQL脚本如下：

```
CREATE TABLE IF NOT EXISTS accountTable(
    id INTEGER PRIMARY KEY AUTOINCREMENT,
    accountType INTEGER,
    typeText TEXT,
    amount INTEGER
)
```

accountTable表的各字段含义如下。

- id：主键。
- accountType：账目类型。0表示支出；1表示收入。
- typeText：账目的具体类别。
- amount：账目金额。

在src/main/ets目录下创建名为“database”的目录，并在该database目录下创建与上述脚本对应的类AccountData，代码如下：

```
export default interface AccountData {
  id: number;
  accountType: number;
  typeText: string;
  amount: number;
}
```

11.9.3 操作账目信息表

在database目录下创建针对账目信息表的操作类AccountTable。AccountTable类封装了增、删、

改、查接口。代码如下：

```
import data_rdb from '@ohos.data.rdb';
import RdbUtil from '../common/RdbUtil';
import AccountData from './AccountData';

const ACCOUNT_TABLE = {
  tableName: 'accountTable',
  sqlCreate: 'CREATE TABLE IF NOT EXISTS accountTable(' +
  'id INTEGER PRIMARY KEY AUTOINCREMENT, accountType INTEGER, ' +
  'typeText TEXT, amount INTEGER)',
  columns: ['id', 'accountType', 'typeText', 'amount']
};

export default class AccountTable {

  private accountTable = new RdbUtil(ACCOUNT_TABLE.tableName, ACCOUNT_TABLE.
                                    sqlCreate, ACCOUNT_TABLE.columns);

  constructor(callback: Function = () => {}) {
    this.accountTable.getRdbStore(callback);
  }

  getRdbStore(callback: Function = () => {}) {
    this.accountTable.getRdbStore(callback);
  }

  // 插入数据
  insertData(account: AccountData, callback) {
    // 根据输入数据创建待插入的数据行
    const valueBucket = generateBucket(account);
    this.accountTable.insertData(valueBucket, callback);
  }

  // 删除数据
  deleteData(account: AccountData, callback) {
    let predicates = new data_rdb.RdbPredicates(ACCOUNT_TABLE.tableName);

    // 根据 id 匹配待删除的数据行
    predicates.equalTo('id', account.id);
    this.accountTable.deleteData(predicates, callback);
  }

  // 修改数据
  updateData(account: AccountData, callback) {
    const valueBucket = generateBucket(account);
    let predicates = new data_rdb.RdbPredicates(ACCOUNT_TABLE.tableName);
```

```
    // 根据 id 匹配待修改的数据行
    predicates.equalTo('id', account.id);
    this.accountTable.updateData(predicates, valueBucket, callback);
  }

  // 查找数据
  query(amount: number, callback, isAll: boolean = true){
    let predicates = new data_rdb.RdbPredicates(ACCOUNT_TABLE.tableName);

    // 是否查找全部数据
    if (!isAll) {
      predicates.equalTo('amount', amount);  // 根据金额匹配要查找的数据行
    }
    this.accountTable.query(predicates, function(resultSet) {
      let count = resultSet.rowCount;

      // 查找结果为空则返回空数组，否则返回查找结果数组
      if (count === 0 || typeof count === 'string') {
        console.log('Query no results!');
        callback([]);
      } else {
        resultSet.goToFirstRow();
        const result = [];
        for (let i = 0; i < count; i++) {
          let tmp: AccountData = {id: 0, accountType: 0, typeText: '', amount: 0 };
          tmp.id = resultSet.getDouble(resultSet.getColumnIndex('id'));
          tmp.accountType = resultSet.getDouble(resultSet.
                            getColumnIndex('accountType'));
          tmp.typeText = resultSet.getString(resultSet.
                         getColumnIndex('typeText'));
          tmp.amount = resultSet.getDouble(resultSet.
                       getColumnIndex('amount'));
          result[i] = tmp;
          resultSet.goToNextRow();
        }
        callback(result);
      }
    });
  }
}

function generateBucket(account: AccountData) {
  let obj = {};
  ACCOUNT_TABLE.columns.forEach((item) => {
    if (item != 'id') {
      obj[item] = account[item];
```

```
    }
  });
  return obj;
}
```

11.9.4 设计界面

为了简化程序，突出核心逻辑，我们的界面设计得非常简单，只是1个Text组件和4个Button组件。4个Button组件用于触发增、删、改、查操作，而Text组件用于展示每次操作后的结果。修改Index代码如下：

```
// 导入AccountData
import AccountData from '../database/AccountData';
// 导入AccountTable
import AccountTable from '../database/AccountTable';

@Entry
@Component
struct Index {
  @State message: string = 'Hello World'
  private accountTable = new AccountTable();

  aboutToAppear() {
    // 初始化数据库
    this.accountTable.getRdbStore(() => {
      this.accountTable.query(0, (result) => {
        this.message = result;
      }, true);
    });
  }

  build() {
    Row() {
      Column() {
        Text(this.message)
          .fontSize(50)
          .fontWeight(FontWeight.Bold)

        // 增加
        Button(('增加'), { type: ButtonType.Capsule })
          .width(140)
          .fontSize(40)
          .fontWeight(FontWeight.Medium)
          .margin({ top: 20, bottom: 20 })
          .onClick(() => {
```

```
        let newAccount: AccountData = { id: 0, accountType: 0,
                                        typeText: '苹果', amount: 0 };
        this.accountTable.insertData(newAccount, () => {
        })
      })

    // 查询
    Button(('查询'), { type: ButtonType.Capsule })
      .width(140)
      .fontSize(40)
      .fontWeight(FontWeight.Medium)
      .margin({ top: 20, bottom: 20 })
      .onClick(() => {
        this.accountTable.query(0, (result) => {
          this.message = JSON.stringify(result);
        }, true);
      })

    // 修改
    Button(('修改'), { type: ButtonType.Capsule })
      .width(140)
      .fontSize(40)
      .fontWeight(FontWeight.Medium)
      .margin({ top: 20, bottom: 20 })
      .onClick(() => {
        let newAccount: AccountData = { id: 1, accountType: 1,
                                        typeText: '栗子', amount: 1 };
        this.accountTable.updateData(newAccount, () => {
        })
      })

    // 删除
    Button(('删除'), { type: ButtonType.Capsule })
      .width(140)
      .fontSize(40)
      .fontWeight(FontWeight.Medium)
      .margin({ top: 20, bottom: 20 })
      .onClick(() => {
        let newAccount: AccountData = { id: 2, accountType: 1,
                                        typeText: '栗子', amount: 1 };
        this.accountTable.deleteData(newAccount, () => {
        })
      })
  }
  .width('100%')
}
.height('100%')
```

```
    }
}
```

上述代码，在aboutToAppear生命周期阶段，初始化了数据库。点击“新增”会将预设好的数据“{ id: 0, accountType: 0, typeText: '苹果', amount: 0 }”写入数据库；点击“修改”会将预设好的数据“{ id: 1, accountType: 1, typeText: '栗子', amount: 1 }”更新到数据库；点击“删除”则会将预设好的“{ id: 2, accountType: 1, typeText: '栗子', amount: 1 }”数据从数据库删除。

11.9.5 运行

运行应用，显示的界面效果如图11-16所示。

当用户点击“增加”后再点击“查询”时，界面如图11-17所示，证明数据已经成功写入数据库。

再次点击“增加”后再点击“查询”时，界面如图11-18所示，证明数据又一次成功写入数据库。

当用户点击“修改”后再点击“查询”时，界面如图11-19所示，证明数据已经被修改并更新回数据库。

当用户点击“删除”后再点击“查询”时，界面如图11-20所示，证明数据已经从数据库删除。

图 11-16 界面效果

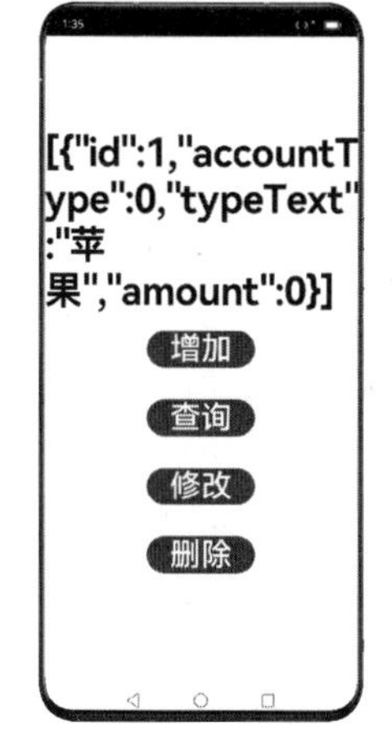

图 11-17 数据已经成功写入数据库

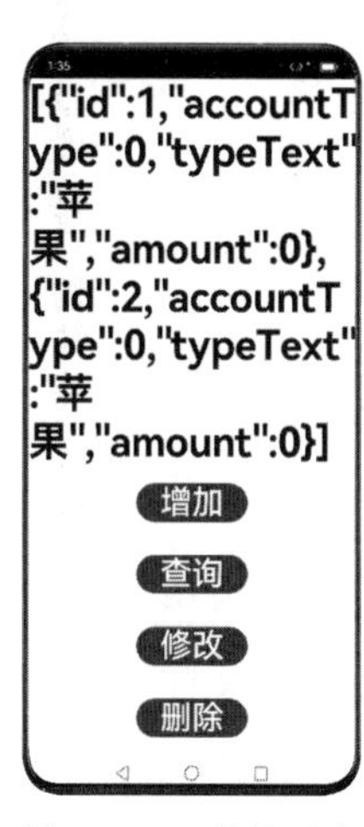

图 11-18 数据再次写入数据库

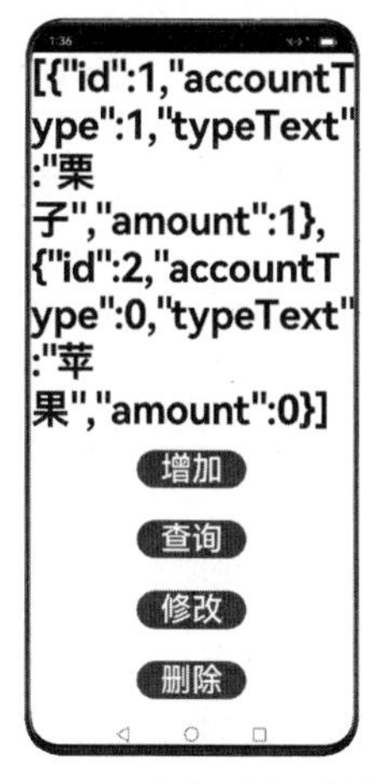

图 11-19 数据已经被修改并更新回数据库

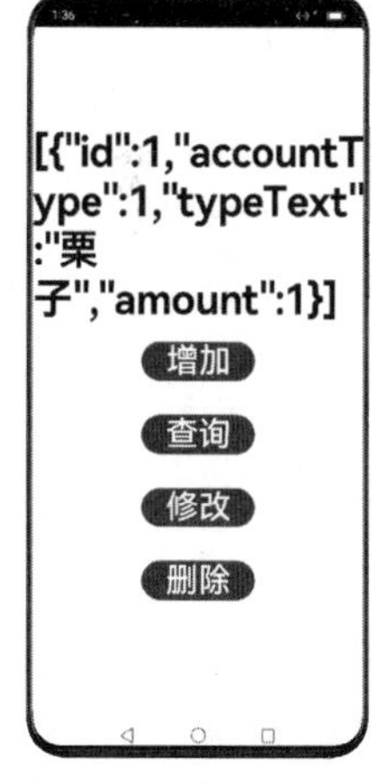

图 11-20 数据已经从数据库删除

11.10 实战：Stage模型的首选项开发

本节以一个“账本”为例，使用首选项的相关接口实现对账单的增、删、改、查操作。

为了演示该功能，创建一个名为“ArkTSPreferences”的应用。

11.10.1 操作Preferences

首先要获取一个Preferences来操作首选项。

在src/main/ets目录下创建名为“common”的目录，用于存放常用的工具类。在该common目录创建工具类PreferencesUtil，代码如下：

```
// 导入 preferences 模块
import dataPreferences from '@ohos.data.preferences';
// 导入 ctx 模块。
import ctx from '@ohos.application.context';

let context = getContext(this) as ctx.AbilityContext;
const PREFERENCES_NAME = 'fruit.db';

export default class PreferencesUtil {
  private preferences;

  // 调用 getPreferences 方法读取指定首选项持久化文件
  // 将数据加载到 Preferences 实例，用于数据操作
  async getPreferencesFromStorage() {
    await dataPreferences.getPreferences(context, PREFERENCES_NAME).
                                        then((data) => {
      this.preferences = data;
      console.info(`Succeeded in getting preferences`);
    }).catch((err) => {
      console.error(`Failed to get preferences, Cause:` + err);
    });
  }
}
```

为了对数据进行保存、查询、删除操作，我们要封装对应接口。首选项接口提供的保存、查询、删除方法均有callback和Promise两种异步回调方式，本例使用Promise异步回调。代码如下：

```
// 将用户输入的数据，保存到缓存的 Preference 实例中
async putPreference(key: string, data: string) {
    if (this.preferences === null) {
        await this.getPreferencesFromStorage();
    }

    await this.preferences.put(key, data).then(() => {
        console.info(`Succeeded in putting value`);
    }).catch((err) => {
        console.error(`Failed to get preferences, Cause:` + err);
```

```
    });

    // 将 Preference 实例存储到首选项持久化文件中
    await this.preferences.flush();
}

// 使用 Preferences 的 get 方法读取数据
async getPreference(key: string) {
    let result = '';
    if (this.preferences === null) {
        await this.getPreferencesFromStorage();
    }
    await this.preferences.get(key, '').then((data) => {
        result = data;
        console.info(`Succeeded in getting value`);
    }).catch((err) => {
        console.error(`Failed to get preferences, Cause:` + err);
    });

    return result;
}

// 从内存中移除指定文件对应的 Preferences 单实例
// 移除 Preferences 单实例时，应用不允许再使用该实例进行数据操作，否则会出现数据一致性问题
async deletePreferences() {
    await dataPreferences.deletePreferences(context, PREFERENCES_NAME).
                                                    then(() => {
        console.info(`Succeeded in delete preferences`);
    }).catch((err) => {
        console.error(`Failed to get preferences, Cause:` + err);
    });
    this.preferences = null;
}
```

11.10.2 账目信息的表示

在src/main/ets目录下创建名为“database”的目录，并在该database目录下创建类AccountData，代码如下：

```
export default interface AccountData {
  id: number;
  accountType: number;
  typeText: string;
  amount: number;
}
```

AccountData各属性含义如下。

- id：主键。
- accountType：账目类型。0表示支出；1表示收入。
- typeText：账目的具体类别。
- amount：账目金额。

11.10.3 设计界面

为了简化程序，突出核心逻辑，我们的界面设计得非常简单，只是1个Text组件和4个Button组件。4个Button组件用于触发增、删、改、查操作，而Text组件用于展示每次操作后的结果。修改Index代码如下：

```
// 导入 PreferencesUtil
import PreferencesUtil from '../common/PreferencesUtil';
// 导入 AccountData
import AccountData from '../database/AccountData';

const PREFERENCES_KEY = 'fruit';

@Entry
@Component
struct Index {
  @State message: string = 'Hello World'
  private preferencesUtil = new PreferencesUtil();

  async aboutToAppear() {
    // 初始化首选项
    await this.preferencesUtil.getPreferencesFromStorage();

    // 获取结果
    this.preferencesUtil.getPreference(PREFERENCES_KEY).then(resultData => {
      this.message = resultData;
    });
  }

  build() {
    Row() {
      Column() {
        Text(this.message)
          .fontSize(50)
          .fontWeight(FontWeight.Bold)

        // 增加
```

```
Button(('增加'), { type: ButtonType.Capsule })
  .width(140)
  .fontSize(40)
  .fontWeight(FontWeight.Medium)
  .margin({ top: 20, bottom: 20 })
  .onClick(() => {
    // 保存数据
    let newAccount: AccountData = { id: 0, accountType: 0, typeText:
                                      '苹果', amount: 0 };
    this.preferencesUtil.putPreference(PREFERENCES_KEY, JSON.
                                         stringify(newAccount));
  })

// 查询
Button(('查询'), { type: ButtonType.Capsule })
  .width(140)
  .fontSize(40)
  .fontWeight(FontWeight.Medium)
  .margin({ top: 20, bottom: 20 })
  .onClick(() => {
    // 获取结果
    this.preferencesUtil.getPreference(PREFERENCES_KEY).
                                         then(resultData => {
      this.message = resultData;
    });
  })

// 修改
Button(('修改'), { type: ButtonType.Capsule })
  .width(140)
  .fontSize(40)
  .fontWeight(FontWeight.Medium)
  .margin({ top: 20, bottom: 20 })
  .onClick(() => {
    // 修改数据
    let newAccount: AccountData = { id: 1, accountType: 1, typeText:
                                      '栗子', amount: 1 };
    this.preferencesUtil.putPreference(PREFERENCES_KEY, JSON.
                                         stringify(newAccount));
  })

// 删除
Button(('删除'), { type: ButtonType.Capsule })
  .width(140)
  .fontSize(40)
  .fontWeight(FontWeight.Medium)
```

```
            .margin({ top: 20, bottom: 20 })
            .onClick(() => {
              this.preferencesUtil.deletePreferences();
            })
        }
        .width('100%')
      }
      .height('100%')
    }
  }
```

上述代码，在aboutToAppear生命周期阶段，初始化了Preferences。点击“新增”会将预设好的数据“{ id: 0, accountType: 0, typeText: '苹果', amount: 0 }”写入Preferences；点击“修改”会将预设好的数据“{ id: 1, accountType: 1, typeText: '栗子', amount: 1 }”更新到Preferences；点击“删除”则会从内存中移除指定文件对应的Preferences单实例。

11.10.4 运行

运行应用，显示的界面效果如图11-21所示。

当用户点击“增加”后再点击“查询”时，界面如图11-22所示，证明数据已经成功写入Preferences。

当用户点击“修改”后再点击“查询”时，界面如图11-23所示，证明数据已经被修改并更新回Preferences。

当用户点击“删除”后再点击“查询”时，界面如图11-24所示，证明数据已经从Preferences删除。

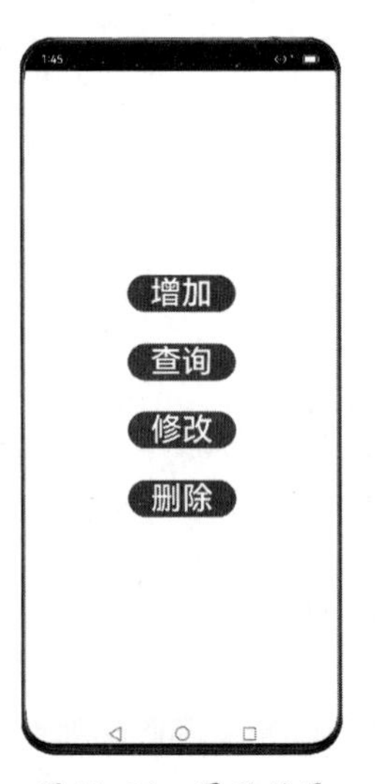

图11-21　界面效果

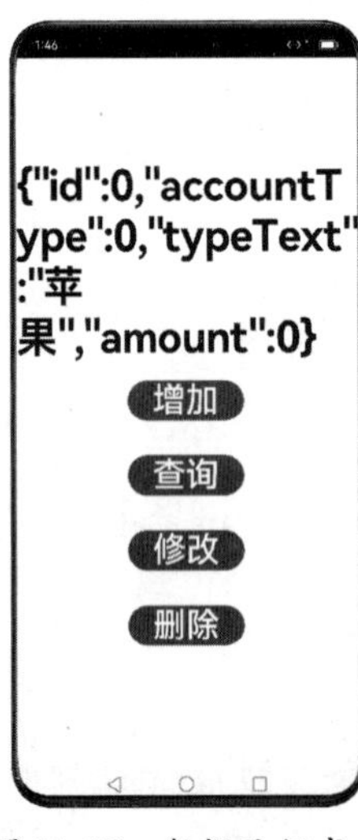

图11-22　数据已经成功写入Preferences

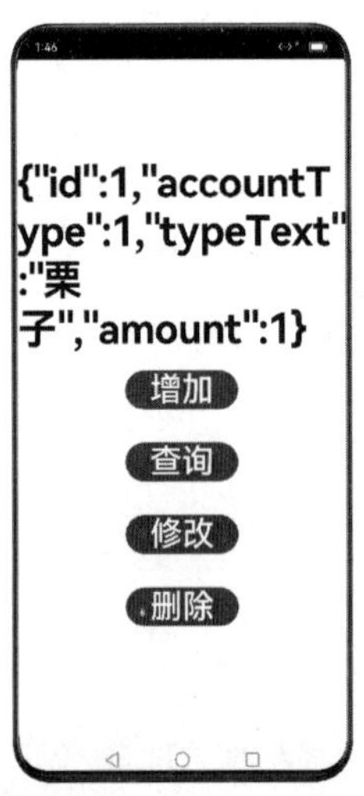

图11-23　数据已经被修改并更新回Preferences

图11-24　数据已经从Preferences删除

第12章 线程管理

当今的计算机，在硬件上往往能够在同一时间执行多个线程任务，因此，当多个任务并发执行时，整体处理性能可以得到提升。

本章介绍 HarmonyOS 关于线程的管理。

12.1 线程管理概述

不同应用在各自独立的进程中运行。当应用以任何形式启动时，系统为其创建进程，该进程将持续运行。当进程完成当前任务处于等待状态，且系统资源不足时，系统自动回收。

在启动应用时，系统会为该应用创建一个名为“主线程”的执行线程。该线程随着应用创建或消失，是应用的核心线程。UI界面的显示和更新等操作，都是在主线程上进行的。主线程又称UI线程，默认情况下，所有的操作都在主线程上执行。如果需要执行比较耗时的任务（如下载文件、查询数据库），可创建其他线程来处理。

12.2 场景介绍

如果应用的业务逻辑比较复杂，可能需要创建多个线程来执行多个任务。Java语言本身是支持创建多线程的。

12.2.1 传统Java多线程管理

下面看一个多线程的示例。SimpleThreads示例有两个线程，第一个线程是每个Java应用程序都有的主线程，主线程创建Runnable对象的MessageLoop（也就是第二个线程），并等待它完成。如果MessageLoop需要很长时间才能完成，主线程就中断它。

该MessageLoop线程输出一系列消息。如果中断之前就已经输出了所有消息，则MessageLoop线程输出一条消息并退出。

```
class SimpleThreads {

    // 显示当前执行线程的名称、信息
    static void threadMessage(String message) {
        String threadName =
            Thread.currentThread().getName();
        System.out.format("%s: %s%n",
                          threadName,
                          message);
    }

    private static class MessageLoop
        implements Runnable {
        public void run() {
```

```
        String importantInfo[] = {
            "Mares eat oats",
            "Does eat oats",
            "Little lambs eat ivy",
            "A kid will eat ivy too"
        };
        try {
            for (int i = 0; i < importantInfo.length; i++) {

                // 暂停 4 秒
                Thread.sleep(4000);

                // 打印消息
                threadMessage(importantInfo[i]);
            }
        } catch (InterruptedException e) {
            threadMessage("I wasn't done!");
        }
    }
}

public static void main(String args[])
    throws InterruptedException {

    // 在中断 MessageLoop 线程（默认为 1 小时）前先延迟一段时间（单位是毫秒）
    long patience = 1000 * 60 * 60;

    // 如果命令行参数出现
    // 设置 patience 的时间值
    // 单位是秒
    if (args.length > 0) {
        try {
            patience = Long.parseLong(args[0]) * 1000;
        } catch (NumberFormatException e) {
            System.err.println("Argument must be an integer.");
            System.exit(1);
        }
    }

    threadMessage("Starting MessageLoop thread");
    long startTime = System.currentTimeMillis();
    Thread t = new Thread(new MessageLoop());
    t.start();

    threadMessage("Waiting for MessageLoop thread to finish");
```

```
        // 循环直到 MessageLoop 线程退出
        while (t.isAlive()) {
            threadMessage("Still waiting...");

            // 最长等待 1 秒
            // 给 MessageLoop 线程来完成
            t.join(1000);
            if (((System.currentTimeMillis() - startTime) > patience)
                  && t.isAlive()) {
                threadMessage("Tired of waiting!");
                t.interrupt();

                // 等待
                t.join();
            }
        }
        threadMessage("Finally!");
    }
}
```

如果线程的数量再增多的话，则需要引入Executor框架。Executor框架最核心的类是ThreadPoolExecutor，它是线程池的实现类。通过线程池，可以使线程得到复用，避免创建过多的线程对象。

有关Java线程池的示例不再赘述，有兴趣的读者可以自行参阅相关资料。

12.2.2 HarmonyOS多线程管理

通过Java语言自行实现线程管理是困难的，代码复杂且难以维护，任务与线程的交互也会更加繁杂。为解决此问题，HarmonyOS提供了TaskDispatcher（任务分发器），开发者通过TaskDispatcher可以分发不同的任务。

接下来将详细介绍TaskDispatcher接口。

12.3 接口说明

TaskDispatcher是Ability分发任务的基本接口，隐藏任务所在线程的实现细节。

为保证应用有更好的响应性，需要设计任务的优先级。在UI线程上运行的任务默认以高优先级运行，如果某个任务无须等待结果，则可以用低优先级。

HarmonyOS线程优先级分类如下。

- HIGH：最高任务优先级，比默认优先级、低优先级的任务有更高的概率得到执行。
- DEFAULT：默认任务优先级，比低优先级的任务有更高的概率得到执行。
- LOW：低任务优先级，比高优先级、默认优先级的任务有更低的概率得到执行。

TaskDispatcher具有多种实现，每种实现对应不同的任务分发器。在分发任务时可以指定任务的优先级，由同一个任务分发器分发出的任务具有相同的优先级。系统提供的任务分发器有GlobalTaskDispatcher、ParallelTaskDispatcher、SerialTaskDispatcher 、SpecTaskDispatcher。

12.3.1 GlobalTaskDispatcher

全局并发任务分发器，由Ability执行getGlobalTaskDispatcher()获取。适用于任务之间没有联系的情况。一个应用只有一个GlobalTaskDispatcher，它在程序结束时才被销毁。

代码示例如下：

```
TaskDispatcher globalTaskDispatcher = getGlobalTaskDispatcher(TaskPriority.
DEFAULT);
```

12.3.2 ParallelTaskDispatcher

并发任务分发器，由Ability执行createParallelTaskDispatcher()创建并返回。与GlobalTaskDispatcher不同的是，ParallelTaskDispatcher不具有全局唯一性，可以创建多个。开发者在创建或销毁分发器时，需要持有对应的对象引用。

代码示例如下：

```
String dispatcherName = "parallelTaskDispatcher";
TaskDispatcher parallelTaskDispatcher =
    createParallelTaskDispatcher(dispatcherName, TaskPriority.DEFAULT);
```

12.3.3 SerialTaskDispatcher

串行任务分发器，由Ability执行createSerialTaskDispatcher()创建并返回。由该分发器分发的所有的任务都是按顺序执行的，但是执行这些任务的线程并不是固定的。如果要执行并行任务，应使用ParallelTaskDispatcher或GlobalTaskDispatcher，而不是创建多个SerialTaskDispatcher。如果任务之间没有依赖，应使用GlobalTaskDispatcher来实现。它的创建和销毁由开发者自己管理，开发者在使用期间需要持有该对象引用。

代码示例如下：

```
String dispatcherName = "serialTaskDispatcher";
TaskDispatcher serialTaskDispatcher =
```

```
createSerialTaskDispatcher(dispatcherName, TaskPriority.DEFAULT);
```

12.3.4 SpecTaskDispatcher

专有任务分发器，即绑定到专有线程上的任务分发器，目前已有的专有线程是主线程。UITaskDispatcher和MainTaskDispatcher都属于SpecTaskDispatcher，建议使用UITaskDispatcher。

1. UITaskDispatcher

UITaskDispatcher是绑定到应用主线程的专有任务分发器，由Ability执行getUITaskDispatcher()创建并返回。由该分发器分发的所有的任务都是在主线程上按顺序执行的，它在应用程序结束时被销毁。

代码示例如下：

```
TaskDispatcher uiTaskDispatcher = getUITaskDispatcher();
```

2. MainTaskDispatcher

MainTaskDispatcher是由Ability执行getMainTaskDispatcher()创建并返回的。

代码示例如下：

```
TaskDispatcher mainTaskDispatcher= getMainTaskDispatcher()
```

12.4 实战：线程管理示例

创建一个名为“ParallelTaskDispatcher”的应用，用于演示使用ParallelTaskDispatcher任务分发器派发任务。

12.4.1 修改ability_main.xml

修改ability_main.xml内容如下：

```
<?xml version="1.0" encoding="utf-8"?>
<DirectionalLayout
    xmlns:ohos="http://schemas.huawei.com/res/ohos"
    ohos:height="match_parent"
    ohos:width="match_parent"
    ohos:alignment="center"
    ohos:orientation="vertical">
```

```
    <Text
        ohos:id="$+id:text_start_parallel_task_dispatcher"
        ohos:height="match_content"
        ohos:width="match_content"
        ohos:background_element="$graphic:background_ability_main"
        ohos:layout_alignment="horizontal_center"
        ohos:text="Start ParallelTaskDispatcher"
        ohos:text_size="28vp"
        />

</DirectionalLayout>
```

显示界面效果如图12-1所示。

图12-1　界面效果

12.4.2　自定义任务

MyTask是自定义的一个任务。该任务逻辑比较简单，只是模拟了一个耗时的操作。其代码如下：

```
package com.waylau.hmos.paralleltaskdispatcher;

import ohos.hiviewdfx.HiLog;
import ohos.hiviewdfx.HiLogLabel;

import java.util.concurrent.TimeUnit;

public class MyTask implements Runnable {
    private static final String TAG = MyTask.class.
getSimpleName();
    private static final HiLogLabel LABEL_LOG =
            new HiLogLabel(HiLog.LOG_APP, 0x00001, TAG);

    private String taskName;

    public MyTask(String taskName) {
        this.taskName = taskName;
    }

    @Override
    public void run() {
        HiLog.info(LABEL_LOG, "before %{public}s run", taskName);
        int task1Result = getRandomInt();
        try {
            // 模拟一个耗时的操作
            TimeUnit.MILLISECONDS.sleep(task1Result);
```

```
        } catch (InterruptedException e) {
            e.printStackTrace();
        }

        HiLog.info(LABEL_LOG, "after %{public}s run, result is: %{public}s",
taskName, task1Result);
    }

    // 返回随机整数
    private int getRandomInt() {
        // 获取 [0, 1000) 之间的 int 整数。方法如下：
        double a = Math.random();
        int result = (int) (a * 1000);
        return result;
    }

}
```

这个耗时操作是通过获取一个随机数，而后根据随机数执行线程sleep实现的。

12.4.3 执行任务派发

修改MainAbilitySlice，增加任务派发器相关的逻辑。其代码如下：

```
package com.waylau.hmos.paralleltaskdispatcher.slice;

import com.waylau.hmos.paralleltaskdispatcher.MyTask;
import com.waylau.hmos.paralleltaskdispatcher.ResourceTable;
import ohos.aafwk.ability.AbilitySlice;
import ohos.aafwk.content.Intent;
import ohos.agp.components.Text;
import ohos.app.dispatcher.Group;
import ohos.app.dispatcher.TaskDispatcher;
import ohos.app.dispatcher.task.TaskPriority;
import ohos.hiviewdfx.HiLog;
import ohos.hiviewdfx.HiLogLabel;

public class MainAbilitySlice extends AbilitySlice {
    private static final String TAG = MainAbilitySlice.class.getSimpleName();
    private static final HiLogLabel LABEL_LOG =
            new HiLogLabel(HiLog.LOG_APP, 0x00001, TAG);

    @Override
    public void onStart(Intent intent) {
        super.onStart(intent);
        super.setUIContent(ResourceTable.Layout_ability_main);
```

```
        // 添加点击事件来触发
        Text textStartDispatcher =
                (Text) findComponentById(ResourceTable.Id_text_start_parallel_
                                                task_dispatcher);
        textStartDispatcher.setClickedListener(listener -> startDispatcher());
    }

    // 指定任务派发
    private void startDispatcher() {
        String dispatcherName = "MyDispatcher";

        TaskDispatcher dispatcher =
                this.getContext().createParallelTaskDispatcher(
                    dispatcherName, TaskPriority.DEFAULT);

        // 创建任务组
        Group group = dispatcher.createDispatchGroup();

        // 将任务 1 加入任务组
        dispatcher.asyncGroupDispatch(group, new MyTask("task1"));

        // 将与任务 1 相关联的任务 2 加入任务组
        dispatcher.asyncGroupDispatch(group, new MyTask("task2"));

        // task3 必须要等任务组中的所有任务执行完成后才会执行
        dispatcher.groupDispatchNotify(group, new MyTask("task3"));
    }

    @Override
    public void onActive() {
        super.onActive();
    }

    @Override
    public void onForeground(Intent intent) {
        super.onForeground(intent);
    }
}
```

上述代码解释如下。

- Text增加了点击事件，以触发startDispatcher任务。
- startDispatcher方法中创建了ParallelTaskDispatcher任务派发器。
- 创建了3个MyTask任务实例，这些任务都是一个任务组。
- task1和task2通过asyncGroupDispatch方式异步派发。

- task3通过groupDispatchNotify方式派发。groupDispatchNotify方式需要等任务组中的所有任务执行完成后才会执行指定任务。

12.4.4 运行

运行应用后，点击2次界面文本“Start ParallelTaskDispatcher”以触发任务派发。此时，控制台输出如下：

```
04-27 11:29:18.180 15991-16685/com.waylau.hmos.paralleltaskdispatcher I
00001/MyTask: before task1 run
04-27 11:29:18.181 15991-16686/com.waylau.hmos.paralleltaskdispatcher I
00001/MyTask: before task2 run
04-27 11:29:18.338 15991-16686/com.waylau.hmos.paralleltaskdispatcher I
00001/MyTask: after task2 run, result is: 157
04-27 11:29:18.974 15991-16685/com.waylau.hmos.paralleltaskdispatcher I
00001/MyTask: after task1 run, result is: 793
04-27 11:29:18.976 15991-16744/com.waylau.hmos.paralleltaskdispatcher I
00001/MyTask: before task3 run
04-27 11:29:19.248 15991-16744/com.waylau.hmos.paralleltaskdispatcher I
00001/MyTask: after task3 run, result is: 269


04-27 11:29:22.499 15991-16946/com.waylau.hmos.paralleltaskdispatcher I
00001/MyTask: before task1 run
04-27 11:29:22.500 15991-16947/com.waylau.hmos.paralleltaskdispatcher I
00001/MyTask: before task2 run
04-27 11:29:22.666 15991-16947/com.waylau.hmos.paralleltaskdispatcher I
00001/MyTask: after task2 run, result is: 166
04-27 11:29:23.203 15991-16946/com.waylau.hmos.paralleltaskdispatcher I
00001/MyTask: after task1 run, result is: 704
04-27 11:29:23.204 15991-16995/com.waylau.hmos.paralleltaskdispatcher I
00001/MyTask: before task3 run
04-27 11:29:23.750 15991-16995/com.waylau.hmos.paralleltaskdispatcher I
00001/MyTask: after task3 run, result is: 545
```

分别执行了2次，可以看到task1和taks2的先后顺序是随机的，但task3一定是在task1和taks2完成之后，才会执行。

12.5 线程间通信概述

大家如果有开发过Netty或Node.js应用的话，那么对于事件循环器就不大陌生了，事件循环器是高并发非阻塞的“秘籍”。在HarmonyOS中，事件循环器实现方式就是EventHandler机制。当前线程中处理较为耗时的操作时，如果不希望当前的线程受到阻塞，此时，就可以使用EventHandler机制。EventHandler是HarmonyOS用于处理线程间通信的一种机制，可以通过EventRunner创建新线程，将耗时的操作放到新线程上执行。这样既不阻塞原来的线程，任务又可以得到合理的处理。比如：主线程使用EventHandler创建子线程，子线程做耗时的下载图片操作，下载完成后，子线程通过EventHandler通知主线程，主线程再更新UI。

Netty或Node.js方面的内容，可以参见本书最后“参考文献”部分。

12.5.1 基本概念

EventRunner是一种事件循环器，循环处理从该EventRunner创建的新线程的事件队列中获取InnerEvent事件或Runnable任务。InnerEvent是EventHandler投递的事件。

EventHandler是一种用户在当前线程上投递InnerEvent事件或Runnable任务到异步线程上处理的机制。每一个EventHandler和指定的EventRunner所创建的新线程绑定，并且该新线程内部有一个事件队列。EventHandler可以投递指定的InnerEvent事件或Runnable任务到这个事件队列。EventRunner从事件队列里循环地取出事件，如果取出的事件是InnerEvent事件，将在EventRunner所在线程执行processEvent回调；如果取出的事件是Runnable任务，将在EventRunner所在线程执行Runnable的run回调。一般，EventHandler有两个主要作用：

- 在不同线程间分发和处理InnerEvent事件或Runnable任务；
- 延迟处理InnerEvent事件或Runnable任务。

12.5.2 运作机制

EventHandler的运作机制如图12-2所示。

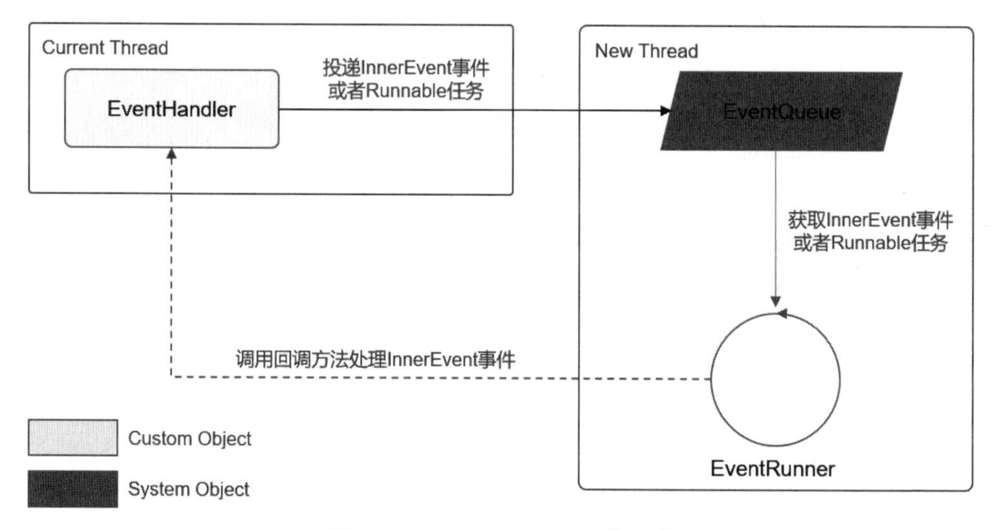

图12-2 EventHandler的运作机制

使用EventHandler实现线程间通信的主要流程如下。

- EventHandler投递具体的InnerEvent事件或Runnable任务到EventRunner所创建的线程的事件队列。
- EventRunner循环从事件队列中获取InnerEvent事件或Runnable任务：

①如果EventRunner取出的事件为InnerEvent事件，则触发EventHandler的回调方法并触发EventHandler的处理方法，在新线程上处理该事件。

②如果EventRunner取出的事件为Runnable任务，则EventRunner直接在新线程上处理Runnable任务。

12.5.3 约束限制

在进行线程间通信的时候，EventHandler只能和EventRunner所创建的线程进行绑定，EventRunner创建时需要判断是否创建成功，只有确保获取的EventRunner实例非空时，才可以使用EventHandler绑定EventRunner。

一个EventHandler只能同时与一个EventRunner绑定，一个EventRunner上可以创建多个EventHandler。

12.6 实战：线程间通信示例

创建一个名为“EventHandler”的应用，用于演示使用EventHandler处理线程间通信。

12.6.1 修改ability_main.xml

修改ability_main.xml内容如下：

```
<?xml version="1.0" encoding="utf-8"?>
<DirectionalLayout
    xmlns:ohos="http://schemas.huawei.com/res/ohos"
    ohos:height="match_parent"
    ohos:width="match_parent"
    ohos:alignment="center"
    ohos:orientation="vertical">

    <Text
        ohos:id="$+id:text_send_event"
        ohos:height="match_content"
        ohos:width="match_content"
```

```
        ohos:background_element="$graphic:background_ability_main"
        ohos:layout_alignment="horizontal_center"
        ohos:text="Send Event"
        ohos:text_size="40vp"
        />

</DirectionalLayout>
```

显示界面效果如图12-3所示。

Send Event

图12-3　界面效果

12.6.2　自定义事件处理器

MyEventHandler是一个自定义的事件处理。该事件处理器逻辑比较简单，只是模拟了一个耗时的操作。其代码如下：

```
package com.waylau.hmos.eventhandler;

import ohos.eventhandler.EventHandler;
import ohos.eventhandler.EventRunner;
import ohos.eventhandler.InnerEvent;
import ohos.hiviewdfx.HiLog;
import ohos.hiviewdfx.HiLogLabel;

import java.util.concurrent.TimeUnit;

public class MyEventHandler extends EventHandler {
    private static final String TAG = MyEventHandler.class.getSimpleName();
    private static final HiLogLabel LABEL_LOG =
            new HiLogLabel(HiLog.LOG_APP, 0x00001, TAG);

    public MyEventHandler(EventRunner runner) throws IllegalArgumentException
{
        super(runner);
    }

    @Override
    public void processEvent(InnerEvent event) {
        super.processEvent(event);
        if (event == null) {
            HiLog.info(LABEL_LOG, "before processEvent event is null");
            return;
        }
        int eventId = event.eventId;
        HiLog.info(LABEL_LOG, "before processEvent eventId: %{public}s", eventId);
```

```
        int task1Result = getRandomInt();
        try {
            // 模拟一个耗时的操作
            TimeUnit.MILLISECONDS.sleep(task1Result);
        } catch (InterruptedException e) {
            e.printStackTrace();
        }

        HiLog.info(LABEL_LOG, "after processEvent eventId %{public}s", eventId);
    }

    // 返回随机整数
    private int getRandomInt() {
        // 获取 [0, 1000) 之间的 int 整数。方法如下：
        double a = Math.random();
        int result = (int) (a * 1000);
        return result;
    }
}
```

这个耗时操作，是通过获取一个随机数，而后根据随机数执行线程sleep实现的。

12.6.3 执行事件发送

修改MainAbilitySlice，增加事件发送相关的逻辑。其代码如下：

```
package com.waylau.hmos.eventhandler.slice;

import com.waylau.hmos.eventhandler.MyEventHandler;
import com.waylau.hmos.eventhandler.ResourceTable;
import ohos.aafwk.ability.AbilitySlice;
import ohos.aafwk.content.Intent;
import ohos.agp.components.Text;
import ohos.eventhandler.EventRunner;
import ohos.hiviewdfx.HiLog;
import ohos.hiviewdfx.HiLogLabel;

public class MainAbilitySlice extends AbilitySlice {
    private static final String TAG = MainAbilitySlice.class.getSimpleName();
    private static final HiLogLabel LABEL_LOG =
            new HiLogLabel(HiLog.LOG_APP, 0x00001, TAG);

    // 创建 EventRunner
    private EventRunner eventRunner = EventRunner.create("MyEventRunner"); //
内部会新建一个线程
```

```
    private int eventId = 0; // 事件ID，递增的序列

    @Override
    public void onStart(Intent intent) {
        super.onStart(intent);
        super.setUIContent(ResourceTable.Layout_ability_main);

        // 添加点击事件来触发
        Text textSendEvent =
                (Text) findComponentById(ResourceTable.Id_text_send_event);
        textSendEvent.setClickedListener(listener -> sendEvent());
    }

    private void sendEvent() {
        HiLog.info(LABEL_LOG, "before sendEvent");

        // 创建MyEventHandler实例
        MyEventHandler handler = new MyEventHandler(eventRunner);

        eventId++;

        // 向EventRunner发送事件
        handler.sendEvent(eventId);

        HiLog.info(LABEL_LOG, "end sendEvent eventId: %{public}s", eventId);
    }

    @Override
    public void onActive() {
        super.onActive();
    }

    @Override
    public void onForeground(Intent intent) {
        super.onForeground(intent);
    }
}
```

上述代码解释如下。

- Text增加了点击事件，以触发sendEvent任务。
- sendEvent方法中创建了EventHandler事件处理器。
- 通过EventHandler的sendEvent方法发送了一个事件ID。事件ID是自增的序列。

12.6.4 运行

运行应用后，点击3次界面文本“Start Event”以触发事件派发。此时，控制台输出如下：

```
04-27 14:47:05.056 3943-3943/com.waylau.hmos.eventhandler I 00001/
MainAbilitySlice: before sendEvent
04-27 14:47:05.057 3943-3943/com.waylau.hmos.eventhandler I 00001/
MainAbilitySlice: end sendEvent eventId: 1
04-27 14:47:05.058 3943-4024/com.waylau.hmos.eventhandler I 00001/
MyEventHandler: before processEvent eventId: 1
04-27 14:47:05.196 3943-4024/com.waylau.hmos.eventhandler I 00001/
MyEventHandler: after processEvent eventId 1

04-27 14:47:07.731 3943-3943/com.waylau.hmos.eventhandler I 00001/
MainAbilitySlice: before sendEvent
04-27 14:47:07.732 3943-4024/com.waylau.hmos.eventhandler I 00001/
MyEventHandler: before processEvent eventId: 2
04-27 14:47:07.732 3943-3943/com.waylau.hmos.eventhandler I 00001/
MainAbilitySlice: end sendEvent eventId: 2
04-27 14:47:08.104 3943-4024/com.waylau.hmos.eventhandler I 00001/
MyEventHandler: after processEvent eventId 2

04-27 14:47:09.149 3943-3943/com.waylau.hmos.eventhandler I 00001/
MainAbilitySlice: before sendEvent
04-27 14:47:09.151 3943-3943/com.waylau.hmos.eventhandler I 00001/
MainAbilitySlice: end sendEvent eventId: 3
04-27 14:47:09.159 3943-4024/com.waylau.hmos.eventhandler I 00001/
MyEventHandler: before processEvent eventId: 3
04-27 14:47:09.300 3943-4024/com.waylau.hmos.eventhandler I 00001/
MyEventHandler: after processEvent eventId 3
```

从上述日志可以看出，MainAbilitySlice先是发送了3次事件，而后MyEventHandler就处理了这些事件。

第13章

视频

HarmonyOS 视频模块支持视频业务的开发和生态开放，开发者可以通过已开放的接口很容易地实现视频媒体的播放、操作和新功能开发。

13.1 视频概述

通过HarmonyOS视频模块，开发者可以通过已开放的接口很容易地实现视频媒体的播放、操作和新功能开发。视频媒体的常见操作有视频编解码、视频合成、视频提取、视频播放及视频录制等。

视频媒体常见的概念包括以下几种。

- 编码：编码是信息从一种形式或格式转换为另一种形式或格式的过程。用预先规定的方法将文字、数字或其他对象编成数码，或将信息、数据转换成规定的电脉冲信号。在本章中，编码是指编码器将原始的视频信息压缩为另一种格式的过程。
- 解码：解码是一种用特定方法，把数码还原成它所代表的内容，或将电脉冲信号、光信号、无线电波等转换成它所代表的信息、数据等的过程。在本章中，解码是指解码器将接收到的数据还原为视频信息的过程，与编码过程相对应。
- 帧率：帧率是以帧为单位的位图图像连续出现在显示器上的频率（速率），以赫兹（Hz）为单位。该术语同样适用于胶片、摄像机、计算机图形和动作捕捉系统。

13.2 实战：媒体编解码能力查询

媒体编解码能力查询主要指查询设备所支持的编解码器的MIME（Multipurpose Internet Mail Extensions，多用途互联网邮件扩展）列表，并判断设备是否支持指定MIME对应的编码器/解码器。

13.2.1 接口说明

媒体编解码能力查询类CodecDescriptionList的主要接口有以下几种。

- getSupportedMimes()：获取某设备所支持的编解码器的MIME列表。
- isDecodeSupportedByMime(String mime)：判断某设备是否支持指定MIME对应的解码器。
- isEncodeSupportedByMime(String mime)：判断某设备是否支持指定MIME对应的编码器。
- isDecoderSupportedByFormat(Format format)：判断某设备是否支持指定媒体格式对应的解码器。
- isEncoderSupportedByFormat(Format format)：判断某设备是否支持指定媒体格式对应的编码器。

13.2.2 创建应用

为了演示媒体编解码能力查询的功能，创建一个名为“CodecDescriptionList”的应用。

在应用的界面上，通过点击按钮来触发获取查询媒体编解码能力，并将能力文本信息输出到界面上。

13.2.3 修改ability_main.xml

修改ability_main.xml内容如下：

```
<?xml version="1.0" encoding="utf-8"?>
<DirectionalLayout
    xmlns:ohos="http://schemas.huawei.com/res/ohos"
    ohos:height="match_parent"
    ohos:width="match_parent"
    ohos:alignment="center"
    ohos:orientation="vertical">

    <Button
        ohos:id="$+id:button"
        ohos:height="match_content"
        ohos:width="match_content"
        ohos:background_element="#F76543"
        ohos:text="Get"
        ohos:text_size="40vp"
        />

    <Text
        ohos:id="$+id:text"
        ohos:height="match_content"
        ohos:width="match_content"
        ohos:background_element="$graphic:background_ability_main"
        ohos:layout_alignment="horizontal_center"
        ohos:multiple_lines="true"
        ohos:text="Hello World"
        ohos:text_size="30vp"
        />

</DirectionalLayout>
```

界面预览效果如图13-1所示。

图13-1　界面预览效果

13.2.4 修改MainAbilitySlice

修改MainAbilitySlice内容如下：

```
package com.waylau.hmos.codecdescriptionlist.slice;
```

```
import com.waylau.hmos.codecdescriptionlist.ResourceTable;
import ohos.aafwk.ability.AbilitySlice;
import ohos.aafwk.content.Intent;
import ohos.agp.components.Button;
import ohos.agp.components.Text;
import ohos.media.codec.CodecDescriptionList;
import ohos.media.common.Format;

import java.util.List;

public class MainAbilitySlice extends AbilitySlice {
    @Override
    public void onStart(Intent intent) {
        super.onStart(intent);
        super.setUIContent(ResourceTable.Layout_ability_main);

        Button button =
                (Button) findComponentById(ResourceTable.Id_button);

        Text text =
                (Text) findComponentById(ResourceTable.Id_text);

        // 为按钮设置点击事件回调
        button.setClickedListener(listener -> {
                    showInfo(text);
                }
        );
    }

    private void showInfo(Text text) {
        // 获取某设备所支持的编解码器的 MIME 列表
        List<String> mimes = CodecDescriptionList.getSupportedMimes();
        text.setText("mimes:" + mimes);

        // 判断某设备是否支持指定 MIME 对应的解码器，支持返回 true，否则返回 false
        boolean isDecodeSupportedByMime =
            CodecDescriptionList.isDecodeSupportedByMime(Format.VIDEO_VP9);
        text.insert("isDecodeSupportedByMime:" + isDecodeSupportedByMime);

        // 判断某设备是否支持指定 MIME 对应的编码器，支持返回 true，否则返回 false
        boolean isEncodeSupportedByMime =
            CodecDescriptionList.isEncodeSupportedByMime(Format.AUDIO_FLAC);
        text.insert("isEncodeSupportedByMime:" + isEncodeSupportedByMime);

        // 判断某设备是否支持指定 Format 的编解码器，支持返回 true，否则返回 false
        Format format = new Format();
```

```
        format.putStringValue(Format.MIME, Format.VIDEO_AVC);
        format.putIntValue(Format.WIDTH, 2560);
        format.putIntValue(Format.HEIGHT, 1440);
        format.putIntValue(Format.FRAME_RATE, 30);
        format.putIntValue(Format.FRAME_INTERVAL, 1);
        boolean isDecoderSupportedByFormat =
            CodecDescriptionList.isDecoderSupportedByFormat(format);
        text.insert("isDecoderSupportedByFormat:" +
                    isDecoderSupportedByFormat);
        boolean isEncoderSupportedByFormat =
            CodecDescriptionList.isEncoderSupportedByFormat(format);
        text.insert("isEncoderSupportedByFormat:" +
                    isEncoderSupportedByFormat);
    }

    @Override
    public void onActive() {
        super.onActive();
    }

    @Override
    public void onForeground(Intent intent) {
        super.onForeground(intent);
    }
}
```

上述代码解释如下。

- 在按钮上设置了点击事件，以触发获取查询媒体编解码能力。
- 将获取到媒体编解码能力以文本信息形式输出到界面上。

13.2.5 运行

运行应用后，界面效果如图13-2所示。

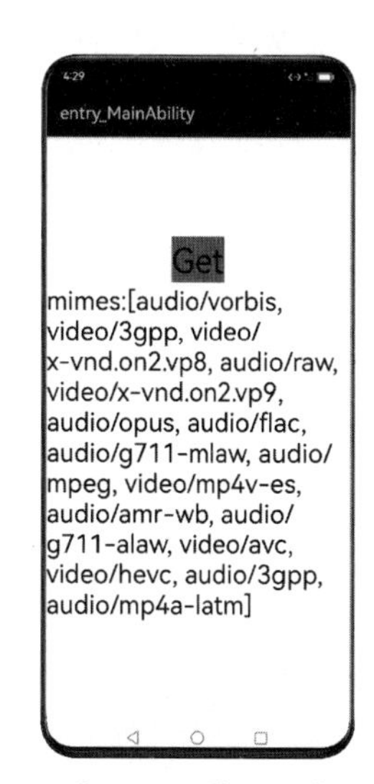

图13-2 界面效果

13.3 实战：视频编解码

视频编解码的主要工作是将视频进行编码和解码。

13.3.1 接口说明

视频编解码类Codec的主要接口有以下几种。

- createDecoder()：创建解码器 Codec 实例。
- createEncoder()：创建编码器 Codec 实例。
- registerCodecListener(ICodecListener listener)：注册侦听器用来异步接收编码或解码后的数据。
- setSource(Source source, TrackInfo trackInfo)：根据解码器的源轨道信息设置数据源，对于编码器 trackInfo 无效。
- setSourceFormat(Format format)：在编码器的管道模式下，设置编码器编码格式。
- setCodecFormat(Format format)：普通模式设置编 / 解码器参数。
- setVideoSurface(Surface surface)：设置解码器的 Surface。
- getAvailableBuffer(long timeout)：普通模式获取可用 ByteBuffer。
- writeBuffer(ByteBuffer buffer, BufferInfo info)：推送源数据给 Codec。
- getBufferFormat(ByteBuffer buffer)：获取输出 Buffer 数据格式。
- start()：启动编 / 解码。
- stop()：停止编 / 解码。
- release()：释放所有资源。

13.3.2 创建应用

为了演示视频编解码的功能，创建一个名为“Codec”的应用。

在应用的界面上，分别设置编码和解码按钮，通过触发按钮来执行相应的编码和解码操作。

同时，在应用的“resources/rawfile”目录下，增加 big_buck_bunny.mp4 视频文件。

13.3.3 修改 ability_main.xml

修改 ability_main.xml 内容如下：

```
<?xml version="1.0" encoding="utf-8"?>
<DirectionalLayout
    xmlns:ohos="http://schemas.huawei.com/res/ohos"
    ohos:height="match_parent"
    ohos:width="match_parent"
    ohos:alignment="center"
    ohos:orientation="vertical">

    <Button
        ohos:id="$+id:button_encode"
        ohos:height="match_content"
        ohos:width="match_content"
        ohos:background_element="#F76543"
        ohos:text="Encode"
```

```
            ohos:text_size="40vp"
            />

        <Button
            ohos:id="$+id:button_decode"
            ohos:height="match_content"
            ohos:width="match_content"
            ohos:background_element="#F76543"
            ohos:text="Decode"
            ohos:text_size="40vp"
            />

</DirectionalLayout>
```

界面预览效果如图13-3所示。

图13-3 界面预览效果

13.3.4 修改MainAbilitySlice

修改MainAbilitySlice内容如下：

```
package com.waylau.hmos.codec.slice;

import com.waylau.hmos.codec.ResourceTable;
import ohos.aafwk.ability.AbilitySlice;
import ohos.aafwk.content.Intent;
import ohos.agp.components.Button;
import ohos.global.resource.RawFileEntry;
import ohos.hiviewdfx.HiLog;
import ohos.hiviewdfx.HiLogLabel;
import ohos.media.codec.Codec;
import ohos.media.codec.TrackInfo;
import ohos.media.common.BufferInfo;
import ohos.media.common.Format;
import ohos.media.common.Source;

import java.io.FileDescriptor;
import java.io.IOException;
import java.nio.ByteBuffer;

public class MainAbilitySlice extends AbilitySlice {
    private static final String TAG = MainAbilitySlice.class.getSimpleName();
    private static final HiLogLabel LABEL_LOG =
            new HiLogLabel(HiLog.LOG_APP, 0x00001, TAG);

    @Override
```

```
    public void onStart(Intent intent) {
        super.onStart(intent);
        super.setUIContent(ResourceTable.Layout_ability_main);

        Button buttonEncode =
                (Button) findComponentById(ResourceTable.Id_button_encode);

        Button buttonDecode =
                (Button) findComponentById(ResourceTable.Id_button_decode);

        // 为按钮设置点击事件回调
        buttonEncode.setClickedListener(listener -> {
            try {
                encode();
            } catch (IOException e) {
                HiLog.error(LABEL_LOG, "exception : %{public}s", e);
            }
        });

        buttonDecode.setClickedListener(listener -> {
            try {
                decode();
            } catch (IOException e) {
                HiLog.error(LABEL_LOG, "exception : %{public}s", e);
            }
        });
    }

    private void decode() throws IOException {
        HiLog.info(LABEL_LOG, "before decode()");

        //1. 创建解码 Codec 实例，可调用 createDecoder() 创建
        final Codec decoder = Codec.createDecoder();

        //2. 调用 setSource() 设置数据源，支持设定文件路径或文件 File Descriptor
        RawFileEntry rawFileEntry = this.getResourceManager().
            getRawFileEntry("resources/rawfile/big_buck_bunny.mp4");
        FileDescriptor fd = rawFileEntry.openRawFileDescriptor().
            getFileDescriptor();
        decoder.setSource(new Source(fd), new TrackInfo());

        //3. 构造数据源格式或从 Extractor 中读取数据源格式，并设置给 Codec 实例
        // 调用 setSourceFormat()
        Format fmt = new Format();
        fmt.putStringValue(Format.MIME, Format.VIDEO_AVC);
        fmt.putIntValue(Format.WIDTH, 1920);
```

```
        fmt.putIntValue(Format.HEIGHT, 1080);
        fmt.putIntValue(Format.BIT_RATE, 392000);
        fmt.putIntValue(Format.FRAME_RATE, 30);
        fmt.putIntValue(Format.FRAME_INTERVAL, -1);
        decoder.setSourceFormat(fmt);

        //4. 设置监听器
        decoder.registerCodecListener(listener);

        //5. 开始编码
        decoder.start();

        //6. 停止编码
        decoder.stop();

        //7. 释放资源
        decoder.release();

        HiLog.info(LABEL_LOG, "end decode()");
    }

    private void encode() throws IOException {
        HiLog.info(LABEL_LOG, "before encode()");

        //1. 创建编码 Codec 实例，可调用 createEncoder() 创建
        final Codec encoder = Codec.createEncoder();

        //2. 调用 setSource() 设置数据源，支持设定文件路径或文件 File Descriptor
        // 获取源文件
        RawFileEntry rawFileEntry = this.getResourceManager().
getRawFileEntry("resources/rawfile/big_buck_bunny.mp4");
        FileDescriptor fd = rawFileEntry.openRawFileDescriptor().
getFileDescriptor();
        encoder.setSource(new Source(fd), new TrackInfo());

        //3. 构造数据源格式或从 Extractor 中读取数据源格式，并设置给 Codec 实例
        // 调用 setSourceFormat()
        Format fmt = new Format();
        fmt.putStringValue(Format.MIME, Format.VIDEO_AVC);
        fmt.putIntValue(Format.WIDTH, 1920);
        fmt.putIntValue(Format.HEIGHT, 1080);
        fmt.putIntValue(Format.BIT_RATE, 392000);
        fmt.putIntValue(Format.FRAME_RATE, 30);
        fmt.putIntValue(Format.FRAME_INTERVAL, -1);
        encoder.setSourceFormat(fmt);
```

```
        //4.设置监听器
        encoder.registerCodecListener(listener);

        //5.开始编码
        encoder.start();

        //6.停止编码
        encoder.stop();

        //7.释放资源
        encoder.release();

        HiLog.info(LABEL_LOG, "end encode()");
    }

    private Codec.ICodecListener listener = new Codec.ICodecListener() {
        @Override
        public void onReadBuffer(ByteBuffer byteBuffer, BufferInfo bufferInfo,
int trackId) {
            HiLog.info(LABEL_LOG, "onReadBuffer trackId: %{public}s",
trackId);
        }

        @Override
        public void onError(int errorCode, int act, int trackId) {
            throw new RuntimeException();
        }
    };

    @Override
    public void onActive() {
        super.onActive();
    }

    @Override
    public void onForeground(Intent intent) {
        super.onForeground(intent);
    }
}
```

上述代码解释如下。

- 在按钮上设置了点击事件，以触发编码和解码。
- 创建解码 Codec 实例，可调用 createDecoder() 创建。
- 创建编码 Codec 实例，可调用 createEncoder() 创建。

- 通过registerCodecListener方法分别给解码Codec实例、编码Codec实例注册监听器。该监听器是Codec.ICodecListener的实例。

13.4 实战：视频播放

视频播放包括播放控制、播放设置和播放查询，如播放的开始/停止、播放速度设置和是否循环播放等。

13.4.1 接口说明

视频播放类Player的主要接口有以下几种。

- Player(Context context)：创建Player实例。
- setSource(Source source)：设置媒体源。
- prepare()：准备播放。
- play()：开始播放。
- pause()：暂停播放。
- stop()：停止播放。
- rewindTo(long microseconds)：拖曳播放。
- setVolume(float volume)：调节播放音量。
- setVideoSurface(Surface surface)：设置视频播放的窗口。
- enableSingleLooping(boolean looping)：设置为单曲循环。
- isSingleLooping()：检查是否单曲循环播放。
- isNowPlaying()：检查是否播放。
- getCurrentTime()：获取当前播放位置。
- getDuration()：获取媒体文件总时长。
- getVideoWidth()：获取视频宽度。
- getVideoHeight()：获取视频高度。
- setPlaybackSpeed(float speed)：设置播放速度。
- getPlaybackSpeed()：获取播放速度。
- setAudioStreamType(int type)：设置音频类型。
- getAudioStreamType()：获取音频类型。
- setNextPlayer(Player next)：设置当前播放结束后的下一个播放器。
- reset()：重置播放器。

- release()：释放播放资源。
- setPlayerCallback(IPlayerCallback callback)：注册回调，接收播放器的事件通知或异常通知。

13.4.2 创建应用

为了演示视频播放的功能，创建一个名为"Player"的Tablet设备类型的应用。

在应用的界面上，通过点击按钮来触发播放视频。

在rawfile目录下，放置一个video_00.mp4视频文件，以备测试。

13.4.3 修改ability_main.xml

修改ability_main.xml内容如下：

```
<?xml version="1.0" encoding="utf-8"?>
<DirectionalLayout
    xmlns:ohos="http://schemas.huawei.com/res/ohos"
    ohos:height="match_parent"
    ohos:width="match_parent"
    ohos:alignment="center"
    ohos:orientation="vertical">

    <DirectionalLayout
        ohos:height="60vp"
        ohos:width="match_content"
        ohos:orientation="horizontal">

        <Button
            ohos:id="$+id:button_play"
            ohos:height="match_content"
            ohos:width="match_content"
            ohos:background_element="#F76543"
            ohos:margin="10vp"
            ohos:padding="10vp"
            ohos:text="Play"
            ohos:text_size="20vp"
            />

        <Button
            ohos:id="$+id:button_pause"
            ohos:height="match_content"
            ohos:width="match_content"
            ohos:background_element="#F76543"
            ohos:margin="10vp"
```

```
            ohos:padding="10vp"
            ohos:text="Pause"
            ohos:text_size="20vp"
            />

        <Button
            ohos:id="$+id:button_stop"
            ohos:height="match_content"
            ohos:width="match_content"
            ohos:background_element="#F76543"
            ohos:margin="10vp"
            ohos:padding="10vp"
            ohos:text="Stop"
            ohos:text_size="20vp"
            />
    </DirectionalLayout>

    <DependentLayout
        ohos:id="$+id:layout_surface_provider"
        ohos:height="match_parent"
        ohos:width="match_content"
        ohos:orientation="horizontal"/>

</DirectionalLayout>
```

界面预览效果如图13-4所示。

图13-4　界面预览效果

13.4.4　修改MainAbilitySlice

修改MainAbilitySlice内容如下：

```
package com.waylau.hmos.player.slice;

import com.waylau.hmos.player.ResourceTable;
import ohos.aafwk.ability.AbilitySlice;
import ohos.aafwk.content.Intent;
import ohos.agp.components.Button;
import ohos.agp.components.ComponentContainer;
import ohos.agp.components.DependentLayout;
import ohos.agp.components.surfaceprovider.SurfaceProvider;
import ohos.agp.graphics.Surface;
import ohos.agp.graphics.SurfaceOps;
import ohos.global.resource.RawFileDescriptor;
import ohos.hiviewdfx.HiLog;
import ohos.hiviewdfx.HiLogLabel;
```

```
import ohos.media.common.Source;
import ohos.media.player.Player;

public class MainAbilitySlice extends AbilitySlice {
    private static final String TAG = MainAbilitySlice.class.getSimpleName();
    private static final HiLogLabel LABEL_LOG =
            new HiLogLabel(HiLog.LOG_APP, 0x00001, TAG);

    private static Player player;

    private SurfaceProvider surfaceProvider;

    @Override
    public void onStart(Intent intent) {
        super.onStart(intent);
        super.setUIContent(ResourceTable.Layout_ability_main);

        Button buttonPlay =
                (Button) findComponentById(ResourceTable.Id_button_play);

        // 为按钮设置点击事件回调
        buttonPlay.setClickedListener(listener -> {
            player.play();

            HiLog.info(LABEL_LOG, "getCurrentTime: %{public}s", player.
                getCurrentTime() );
            HiLog.info(LABEL_LOG, "getDuration: %{public}s", player.
                getDuration() );
        });

        Button buttonPause =
                (Button) findComponentById(ResourceTable.Id_button_pause);

        // 为按钮设置点击事件回调
        buttonPause.setClickedListener(listener -> {
            player.pause();
            HiLog.info(LABEL_LOG, "getCurrentTime: %{public}s", player.
                getCurrentTime() );
            HiLog.info(LABEL_LOG, "getDuration: %{public}s", player.
                getDuration() );
        });

        Button buttonStop =
                (Button) findComponentById(ResourceTable.Id_button_stop);

        // 为按钮设置点击事件回调
```

```
        buttonStop.setClickedListener(listener -> player.stop());
    }

    @Override
    public void onForeground(Intent intent) {
        super.onForeground(intent);
    }


    @Override
    protected void onActive() {
        super.onActive();
        initSurfaceProvider();
    }

    private void initSurfaceProvider() {
        HiLog.info(LABEL_LOG, "before initSurfaceProvider");

        player = new Player(this);

        surfaceProvider = new SurfaceProvider(this);
        surfaceProvider.getSurfaceOps().get().addCallback(new
            VideoSurfaceCallback());
        surfaceProvider.pinToZTop(true);
        surfaceProvider.setWidth(ComponentContainer.LayoutConfig.MATCH_CONTENT);
        surfaceProvider.setHeight(ComponentContainer.LayoutConfig.MATCH_PARENT);

        DependentLayout layout
                = (DependentLayout) findComponentById(ResourceTable.Id_
layout_surface_provider);
        layout.addComponent(surfaceProvider);

        HiLog.info(LABEL_LOG, "end initSurfaceProvider");
    }

    class VideoSurfaceCallback implements SurfaceOps.Callback {
        @Override
        public void surfaceCreated(SurfaceOps surfaceOps) {
            if (surfaceProvider.getSurfaceOps().isPresent()) {
                Surface surface = surfaceProvider.getSurfaceOps().get().
                    getSurface();
                playLocalFile(surface);
            }

            HiLog.info(LABEL_LOG, "surfaceCreated");
        }
```

```
    @Override
    public void surfaceChanged(SurfaceOps surfaceOps, int i, int i1, int i2) {
        HiLog.info(LABEL_LOG, "surfaceChanged, %{public}s, %{public}s,
            %{public}s", i, i1, i2);
    }

    @Override
    public void surfaceDestroyed(SurfaceOps surfaceOps) {
        HiLog.info(LABEL_LOG, "surfaceDestroyed");
    }
}

private void playLocalFile(Surface surface) {
    HiLog.info(LABEL_LOG, "before playLocalFile");

    try {
        RawFileDescriptor filDescriptor =
                getResourceManager()
                        .getRawFileEntry("resources/rawfile/video_00.mp4")
                        .openRawFileDescriptor();

        Source source = new Source(filDescriptor.getFileDescriptor(),
                filDescriptor.getStartPosition(), filDescriptor.
                    getFileSize());
        player.setSource(source);
        player.setVideoSurface(surface);
        player.setPlayerCallback(new VideoPlayerCallback());
        player.prepare();

        surfaceProvider.setTop(0);
    } catch (Exception e) {
        e.printStackTrace();
    }

    HiLog.info(LABEL_LOG, "before playLocalFile");
}

@Override
protected void onStop() {
    super.onStop();
    if (player != null) {
        player.stop();
    }
    surfaceProvider.removeFromWindow();
}
```

```
private class VideoPlayerCallback implements Player.IPlayerCallback {
    @Override
    public void onPrepared() {
        HiLog.info(LABEL_LOG, "onPrepared");
    }

    @Override
    public void onMessage(int i, int i1) {
        HiLog.info(LABEL_LOG, "onMessage, %{public}s, %{public}s", i, i1);
    }

    @Override
    public void onError(int i, int i1) {
        HiLog.info(LABEL_LOG, "onError, %{public}s, %{public}s", i, i1);
    }

    @Override
    public void onResolutionChanged(int i, int i1) {
        HiLog.info(LABEL_LOG, "onResolutionChanged, %{public}s, %{public}
                s", i, i1);
    }

    @Override
    public void onPlayBackComplete() {
        HiLog.info(LABEL_LOG, "onPlayBackComplete");
    }

    @Override
    public void onRewindToComplete() {
        HiLog.info(LABEL_LOG, "onRewindToComplete");
    }

    @Override
    public void onBufferingChange(int i) {
        HiLog.info(LABEL_LOG, "onBufferingChange, %{public}s", i);
    }

    @Override
    public void onNewTimedMetaData(Player.MediaTimedMetaData
                                   mediaTimedMetaData) {
        HiLog.info(LABEL_LOG, "onNewTimedMetaData");
    }

    @Override
    public void onMediaTimeIncontinuity(Player.MediaTimeInfo
```

```
mediaTimeInfo) {
                HiLog.info(LABEL_LOG, "onMediaTimeIncontinuity");
            }
        }
}
```

上述代码解释如下。

- 在按钮上设置了点击事件，以触发Player的播放、暂停和停止操作。
- 重写了onActive方法，以初始化SurfaceProvider。
- SurfaceProvider会设置VideoSurfaceCallback的回调。
- 当VideoSurfaceCallback的surfaceCreated回调执行时，Player会准备好视频源以备后续操作。
- 重写了onStop方法，以关闭Player。

13.4.5 运行

运行应用后，界面效果如图13-5所示。

图13-5 界面效果

13.5 实战：视频录制

视频录制的主要工作是选择视频/音频来源后，录制并生成视频/音频文件。

13.5.1 接口说明

视频录制类Recorder的主要接口有以下几种。

- Recorder()：创建Recorder实例。
- setSource(Source source)：设置音视频源。
- setAudioProperty(AudioProperty property)：设置音频属性。
- setVideoProperty(VideoProperty property)：设置视频属性。
- setStorageProperty(StorageProperty property)：设置音视频存储属性。
- prepare()：准备录制资源。
- start()：开始录制。
- stop()：停止录制。
- pause()：暂停录制。
- resume()：恢复录制。

- reset()：重置录制。
- setRecorderLocation(float latitude, float longitude)：设置视频的经纬度。
- setOutputFormat(int outputFormat)：设置输出文件格式。
- getVideoSurface()：获取视频窗口。
- setRecorderProfile(RecorderProfile profile)：设置媒体录制配置信息。
- registerRecorderListener(IRecorderListener listener)：注册媒体录制回调。
- release()：释放媒体录制资源。

13.5.2 创建应用

为了演示视频播放的功能，创建一个名为“Recorder”的应用。

在应用的界面上，通过点击按钮，来触发录制视频。

在rawfile目录下，放置一个big_buck_bunny.mp4视频文件，以备测试。

13.5.3 修改ability_main.xml

修改ability_main.xml内容如下：

```
<?xml version="1.0" encoding="utf-8"?>
<DirectionalLayout
    xmlns:ohos="http://schemas.huawei.com/res/ohos"
    ohos:height="match_parent"
    ohos:width="match_parent"
    ohos:alignment="center"
    ohos:orientation="vertical">

    <DirectionalLayout
        ohos:height="60vp"
        ohos:width="match_content"
        ohos:orientation="horizontal">

        <Button
            ohos:id="$+id:button_start"
            ohos:height="match_content"
            ohos:width="match_content"
            ohos:background_element="#F76543"
            ohos:margin="10vp"
            ohos:padding="10vp"
            ohos:text="Start"
            ohos:text_size="27fp"
            />
```

```
        <Button
            ohos:id="$+id:button_pause"
            ohos:height="match_content"
            ohos:width="match_content"
            ohos:background_element="#F76543"
            ohos:margin="10vp"
            ohos:padding="10vp"
            ohos:text="Pause"
            ohos:text_size="27fp"
            />

        <Button
            ohos:id="$+id:button_stop"
            ohos:height="match_content"
            ohos:width="match_content"
            ohos:background_element="#F76543"
            ohos:margin="10vp"
            ohos:padding="10vp"
            ohos:text="Stop"
            ohos:text_size="27fp"
            />
    </DirectionalLayout>

    <Text
        ohos:id="$+id:text_helloworld"
        ohos:height="match_parent"
        ohos:width="match_content"
        ohos:background_element="$graphic:background_ability_main"
        ohos:layout_alignment="horizontal_center"
        ohos:text="Hello World"
        ohos:text_size="50"
        />

</DirectionalLayout>
```

界面预览效果如图 13-6 所示。

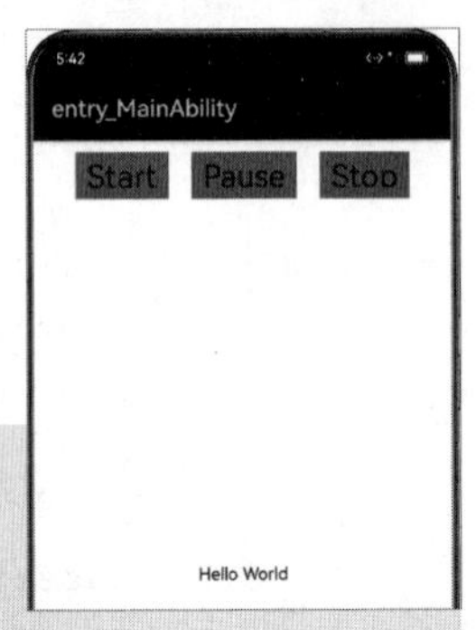

图 13-6　界面预览效果

13.5.4　修改 MainAbilitySlice

修改 MainAbilitySlice 内容如下：

```
package com.waylau.hmos.recorder.slice;

import com.waylau.hmos.recorder.ResourceTable;
import ohos.aafwk.ability.AbilitySlice;
```

```
import ohos.aafwk.content.Intent;
import ohos.agp.components.Button;
import ohos.agp.components.Text;
import ohos.global.resource.RawFileDescriptor;
import ohos.media.common.Source;
import ohos.media.common.StorageProperty;
import ohos.media.recorder.Recorder;

import java.io.IOException;

public class MainAbilitySlice extends AbilitySlice {
    private final Recorder recorder = new Recorder();

    @Override
    public void onStart(Intent intent) {
        super.onStart(intent);
        super.setUIContent(ResourceTable.Layout_ability_main);

        Text text =
                (Text) findComponentById(ResourceTable.Id_text_helloworld);

        Button buttonStart =
                (Button) findComponentById(ResourceTable.Id_button_start);

        // 为按钮设置点击事件回调
        buttonStart.setClickedListener(listener -> {
            recorder.start();
            text.setText("Start");
        });

        Button buttonPause =
                (Button) findComponentById(ResourceTable.Id_button_pause);

        // 为按钮设置点击事件回调
        buttonPause.setClickedListener(listener -> {
            recorder.pause();
            text.setText("Pause");
        });

        Button buttonStop =
                (Button) findComponentById(ResourceTable.Id_button_stop);

        // 为按钮设置点击事件回调
        buttonStop.setClickedListener(listener -> {
            recorder.stop();
```

```
            text.setText("Stop");
        });
    }

    @Override
    public void onActive() {
        super.onActive();

        RawFileDescriptor filDescriptor = null;
        try {
            filDescriptor = getResourceManager()
                    .getRawFileEntry("resources/rawfile/big_buck_bunny.mp4")
                    .openRawFileDescriptor();
        } catch (IOException e) {
            e.printStackTrace();
        }

        // 设置媒体源
        Source source = new Source(filDescriptor.getFileDescriptor(),
                filDescriptor.getStartPosition(), filDescriptor.getFileSize());
        source.setRecorderAudioSource(Recorder.AudioSource.DEFAULT);
        recorder.setSource(source);

        // 设置存储属性
        String path = "record.mp4";
        StorageProperty storageProperty = new StorageProperty.Builder()
                .setRecorderPath(path)
                .setRecorderMaxDurationMs(-1)
                .setRecorderMaxFileSizeBytes(-1)
                .build();
        recorder.setStorageProperty(storageProperty);

        // 准备
        recorder.prepare();
    }

    @Override
    protected void onStop() {
        super.onStop();
        if (recorder != null) {
            recorder.stop();
        }
    }

    @Override
```

```
    public void onForeground(Intent intent) {
        super.onForeground(intent);
    }
}
```

上述代码解释如下。

- 调用Recorder()方法，创建Recorder实例recorder。
- 在按钮上设置了点击事件，以触发Recorder的开始、暂停和停止操作，并将状态信息回写到界面的Text。
- 重写了onActive()方法，以设置Recorder实例的属性。这些属性包括媒体源、存储属性等。
- 重写了onStop()方法，以关闭Player。

13.5.5 运行

运行应用后，界面效果如图13-7所示。

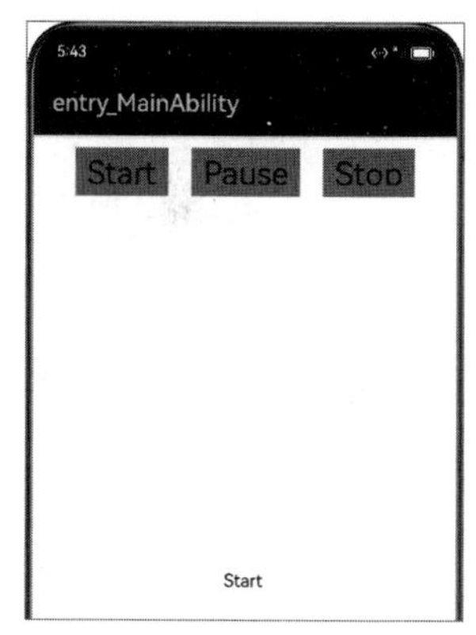

图13-7 界面效果

13.6 Stage模型的视频开发

在HarmonyOS系统中，提供以下两种视频播放开发的方案。

- AVPlayer：功能较完善的音视频播放ArkTS/JS API，集成了流媒体和本地资源解析、媒体资源解封装、视频解码和渲染功能，适用于对媒体资源进行端到端播放的场景，可直接播放MP4、MKV等格式的视频文件。
- Video组件：封装了视频播放的基础能力，需要设置数据源及基础信息即可播放视频，但扩展能力相对较弱。Video组件由ArkUI提供能力，相关指导请参考“8.6 媒体组件详解”。

本节重点介绍如何使用AVPlayer开发视频播放功能，以完整地播放一个视频作为示例，实现端到端播放原始媒体资源。

13.6.1 视频开发指导

播放的全流程包含：创建AVPlayer，设置播放资源和窗口，设置播放参数（音量/倍速/缩放模式），播放控制（播放/暂停/跳转/停止），重置，销毁资源。在进行应用开发的过程中，开发者可以通过AVPlayer的state属性主动获取当前状态，或用on方法监听stateChange事件，从而感知到状态的变化。如果应用在视频播放器处于错误状态时执行操作，系统可能会抛出异常或生成其他未定义的行为。

图13-8展示的是视频播放状态变化示意图。

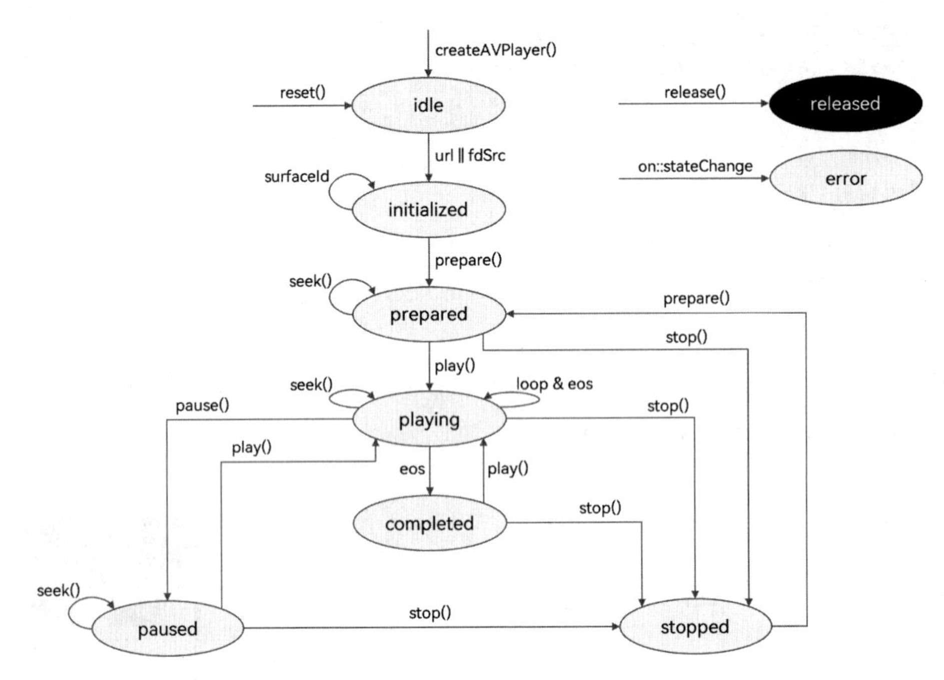

图13-8　视频播放状态变化示意图

状态的详细说明请参考AVPlayerState。当播放处于prepared/playing/paused/completed状态时，播放引擎处于工作状态，这需要占用系统较多的运行内存。当客户端暂时不使用播放器时，可调用reset()或release()回收内存资源，做好资源利用。

13.6.2　视频开发步骤

使用AVPlayer开发视频播放功能的步骤如下。

创建实例createAVPlayer()，AVPlayer初始化idle状态。

设置业务需要的监听事件，搭配全流程场景使用。支持的监听事件包括以下几种。

- stateChange：监听播放器的state属性改变。
- error：监听播放器的错误信息。
- durationUpdate：用于进度条，监听进度条长度，刷新资源时长。
- timeUpdate：监听进度条当前位置，刷新当前时间。
- seekDone：响应API调用，监听seek()请求完成情况。
- speedDone：响应API调用，监听setSpeed()请求完成情况。
- volumeChange：响应API调用，监听setVolume()请求完成情况。
- bitrateDone：响应API调用，用于HLS协议流，监听setBitrate()请求完成情况。
- availableBitrates：用于HLS协议流，监听HLS资源的可选bitrates，用于setBitrate()。
- bufferingUpdate：用于网络播放，监听网络播放缓冲信息。

- startRenderFrame：用于视频播放，监听视频播放首帧渲染时间。
- videoSizeChange：用于视频播放，监听视频播放的宽高信息，可用于调整窗口大小、比例。
- audioInterrupt：监听音频焦点切换信息，搭配属性audioInterruptMode使用。

设置资源：设置属性url，AVPlayer进入initialized状态。

设置窗口：获取并设置属性SurfaceID，用于设置显示画面。应用需要从XComponent组件获取surfaceID。

准备播放：调用prepare()，AVPlayer进入prepared状态，此时可以获取duration，设置缩放模式、音量等。

视频播控：实现播放play()、暂停pause()、跳转seek()、停止stop() 等操作。

（可选）更换资源：调用reset()重置资源，AVPlayer重新进入idle状态，允许更换资源url。

退出播放：调用release()销毁实例，AVPlayer进入released状态，退出播放。

完整示例如下：

```
import media from '@ohos.multimedia.media';
import fs from '@ohos.file.fs';
import common from '@ohos.app.ability.common';

export class AVPlayerDemo {
  private avPlayer;
  private count: number = 0;
  private surfaceID: string; // surfaceID用于播放画面显示，具体的值需要通过
Xcomponent接口获取

  // 注册avplayer回调函数
  setAVPlayerCallback() {
    // seek操作结果回调函数
    this.avPlayer.on('seekDone', (seekDoneTime) => {
      console.info(`AVPlayer seek succeeded, seek time is ${seekDoneTime}`);
    })
    // error回调监听函数，当avPlayer在操作过程中出现错误时，调用reset接口触发重置流程
    this.avPlayer.on('error', (err) => {
      console.error(`Invoke avPlayer failed, code is ${err.code}, message is
                    ${err.message}`);
      this.avPlayer.reset(); // 调用reset重置资源，触发idle状态
    })
    // 状态机变化回调函数
    this.avPlayer.on('stateChange', async (state, reason) => {
      switch (state) {
        case 'idle': // 成功调用reset接口后触发该状态机上报
          console.info('AVPlayer state idle called.');
          this.avPlayer.release(); // 调用release接口销毁实例对象
          break;
```

```
      case 'initialized': // avplayer 设置播放源后触发该状态上报
        console.info('AVPlayerstate initialized called.');
        this.avPlayer.surfaceId = this.surfaceID; // 设置显示画面，当播放的资源
                                                 // 为纯音频时无须设置
        this.avPlayer.prepare().then(() => {
          console.info('AVPlayer prepare succeeded.');
        }, (err) => {
          console.error(`Invoke prepare failed, code is ${err.code}, message
                     is ${err.message}`);
        });
        break;
      case 'prepared': // prepare 调用成功后上报该状态机
        console.info('AVPlayer state prepared called.');
        this.avPlayer.play(); // 调用播放接口开始播放
        break;
      case 'playing': // play 成功调用后触发该状态机上报
        console.info('AVPlayer state playing called.');
        if (this.count !== 0) {
          console.info('AVPlayer start to seek.');
          this.avPlayer.seek(this.avPlayer.duration); //seek 到视频末尾
        } else {
          this.avPlayer.pause(); // 调用暂停接口暂停播放
        }
        this.count++;
        break;
      case 'paused': // pause 成功调用后触发该状态机上报
        console.info('AVPlayer state paused called.');
        this.avPlayer.play(); // 再次播放接口开始播放
        break;
      case 'completed': // 播放结束后触发该状态机上报
        console.info('AVPlayer state completed called.');
        this.avPlayer.stop(); // 调用播放结束接口
        break;
      case 'stopped': // stop 接口成功调用后触发该状态机上报
        console.info('AVPlayer state stopped called.');
        this.avPlayer.reset(); // 调用 reset 接口初始化 avplayer 状态
        break;
      case 'released':
        console.info('AVPlayer state released called.');
        break;
      default:
        console.info('AVPlayer state unknown called.');
        break;
    }
  })
}
```

```
  // 以下 demo 为使用 fs 文件系统打开沙箱地址，获取媒体文件地址并通过 url 属性进行播放示例
  async avPlayerUrlDemo() {
    // 创建 avPlayer 实例对象
    this.avPlayer = await media.createAVPlayer();
    // 创建状态机变化回调函数
    this.setAVPlayerCallback();
    let fdPath = 'fd://';
    let context = getContext(this) as common.UIAbilityContext;
    // 通过 UIAbilityContext 获取沙箱地址 filesDir，以下为 Stage 模型获方式
    // 如需在 FA 模型上获取，请参考“访问应用沙箱”获取地址
    let pathDir = context.filesDir;
    let path = pathDir  + '/H264_AAC.mp4';
    // 打开相应的资源文件地址获取 fd，并为 url 赋值触发 initialized 状态机上报
    let file = await fs.open(path);
    fdPath = fdPath + '' + file.fd;
    this.avPlayer.url = fdPath;
  }

  // 以下 demo 为使用资源管理接口获取打包在 HAP 内的媒体资源文件，并通过 fdSrc 属性进行播放的示例
  async avPlayerFdSrcDemo() {
    // 创建 avPlayer 实例对象
    this.avPlayer = await media.createAVPlayer();
    // 创建状态机变化回调函数
    this.setAVPlayerCallback();
    // 通过 UIAbilityContext 的 resourceManager 成员的 getRawFd 接口获取媒体资源播放地址
    // 返回类型为 {fd,offset,length},fd 为 HAP 包 fd 地址，offset 为媒体资源偏移量，
    // length 为播放长度
    let context = getContext(this) as common.UIAbilityContext;
    let fileDescriptor = await context.resourceManager.getRawFd('H264_AAC.mp4');
    // 为 fdSrc 赋值触发 initialized 状态机上报
    this.avPlayer.fdSrc = fileDescriptor;
  }
}
```

13.7 实战：实现Stage模型的视频播放器

本节介绍使用ArkTS语言实现视频播放器，主要包括视频获取和视频播放功能：

- 获取本地视频；
- 通过AVPlayer进行视频播放。

为了演示该功能，创建一个名为“ArkTSVideoPlayer”的应用。

13.7.1 获取本地视频

1. 准备资源文件

在resources下面的rawfile文件夹里放置两个视频文件video1.mp4和video2.mp4，作为本例的视频素材。

2. 创建视频文件对象 VideoBean

在ets目录下创建一个新的目录“common/bean”，在该目录下创建视频文件对象VideoBean.ets，用来表示视频的信息，代码如下：

```
import image from '@ohos.multimedia.image';

@Observed export class VideoBean {
  name: string;
  src: string;
  pixelMap?: image.PixelMap;

  constructor(name: string, src: string, pixelMap?: image.PixelMap) {
    this.name = name;
    this.src = src;
    this.pixelMap = pixelMap;
  }
}
```

3. 获取本地文件

在ets目录下，创建一个新的目录“viewmodel”，在该目录下创建获取视频文件的对象HomeVideoListModel.ets。HomeVideoListModel可以通过resourceManager.ResourceManager对象获取rawfile文件夹中的本地视频文件，再通过“fd://${videoBean.fd}”组装视频地址。其代码如下：

```
import { VideoBean } from '../common/bean/VideoBean';
import { VIDEO_DATA } from '../common/constants/CommonConstants';

export class HomeVideoListModel {
  private videoLocalList: Array<VideoBean> = [];

  /**
   * 获取本地视频
   */
  async getLocalVideo() {
    this.videoLocalList = [];
    await this.assemblingVideoBean();
    globalThis.videoLocalList = this.videoLocalList;
    return this.videoLocalList;
```

```
  }

  /**
   * 组装本地视频对象
   */
  async assemblingVideoBean () {
    VIDEO_DATA.forEach(async (item: VideoBean) => {
      let videoBean =  await globalThis.resourceManager.getRawFd(item.src);
      let uri = `fd://${videoBean.fd}`;
      this.videoLocalList.push(new VideoBean(item.name, uri));
    });
  }
}

let homeVideoListModel = new HomeVideoListModel();
export default homeVideoListModel as HomeVideoListModel;
```

CommonConstants.ets用于存放常量，而VIDEO_DATA用于存储视频列表信息，代码如下：

```
import { VideoBean } from '../bean/VideoBean';

/**
 * 常量
 */
export class CommonConstants {
  /**
   * 比例
   */
  static readonly FULL_PERCENT: string = '100%';
  static readonly NINETY_PERCENT: string = '90%';
  static readonly FIFTY_PERCENT: string = '50%';

  /**
   * 播放其页面路径 .
   */
  static readonly PAGE: string = 'pages/Index';

  /**
   * 本地视频 ID.
   */
  static readonly TYPE_LOCAL: number = 0;

  /**
   * 开始 .
   */
  static readonly STATUS_START: number = 1;
```

```
/**
 * 暂停.
 */
static readonly STATUS_PAUSE: number = 2;

/**
 * 停止.
 */
static readonly STATUS_STOP: number = 3;

/**
 * 比例.
 */
static readonly ASPECT_RATIO: number = 1;

/**
 * 一百.
 */
static readonly ONE_HUNDRED: number = 100;

/**
 * 一千.
 */
static readonly A_THOUSAND: number = 1000;

/**
 * 设置速度.
 */
static readonly SPEED_ARRAY = [
  { text: '0.75X', value: 0 },
  { text: '1.0X', value: 1 },
  { text: '1.25X', value: 2 },
  { text: '1.75X', value: 3 },
  { text: '2.0X', value: 4 }
];

/**
 * 分转秒.
 */
static readonly TIME_UNIT: number = 60;

/**
 * 初始时间单位.
 */
static readonly INITIAL_TIME_UNIT: string = '00';
```

```
  /**
   * 填充间距 .
   */
  static readonly PADDING_LENGTH: number = 2;

  /**
   * 填充间距 .
   */
  static readonly PADDING_STR: string = '0';

  /**
   * 屏幕状态 .
   */
  static readonly SCREEN_OFF: string = 'usual.event.SCREEN_OFF';

  /**
   * 操作状态 .
   */
  static readonly OPERATE_STATE: Array<string> = ['prepared','playing',
'paused', 'completed'];
}

/**
 * 播放器组件状态 .
 */
export enum AvplayerStatus {
  IDLE = 'idle',
  INITIALIZED = 'initialized',
  PREPARED = 'prepared',
  PLAYING = 'playing',
  PAUSED = 'paused',
  COMPLETED = 'completed',
  STOPPED = 'stopped',
  RELEASED = 'released',
  ERROR = 'error'
}

/**
 * 绑定事件 .
 */
export enum Events {
  STATE_CHANGE = 'stateChange',
  TIME_UPDATE = 'timeUpdate',
  ERROR = 'error'
}
```

```
/**
 * 视频集合 .
 */
export const VIDEO_DATA: VideoBean[] = [
  {
    'name': 'video1',
    'src': 'video1.mp4'
  },
  {
    'name': 'video2',
    'src': 'video2.mp4'
  }
]
```

13.7.2 视频播放控制

在 ets 目录下创建一个新的目录“controller”，在该目录创建视频播放控制器 VideoController.ets。

1. 构建 AVPlayer 实例对象

使用 AVPlayer 前需要通过 createAVPlayer() 构建一个实例对象，并为 AVPlayer 实例绑定状态机状态 AVPlayerState。其代码如下：

```
import media from '@ohos.multimedia.media';
import prompt from '@ohos.promptAction';
import Logger from '../common/util/Logger';
import DateFormatUtil from '../common/util/DateFormatUtil';
import { CommonConstants, AvplayerStatus, Events } from '../common/constants/
CommonConstants';
import { PlayConstants } from '../common/constants/PlayConstants';

@Observed
export class VideoController {
  private avPlayer;
  private duration: number = 0;
  private status: number;
  private loop: boolean = false;
  private index: number;
  private url: string;
  private surfaceId: number;
  private playSpeed: number = PlayConstants.PLAY_PAGE.PLAY_SPEED;
  private seekTime: number = PlayConstants.PLAY_PROGRESS.SEEK_TIME;
  private progressThis;
  private playerThis;
```

```
private playPageThis;
private titleThis;

constructor() {
  this.createAVPlayer();
}

/**
 * 创建 videoPlayer 对象
 */
createAVPlayer() {
  media.createAVPlayer().then((avPlayer) => {
    if (avPlayer) {
      this.avPlayer = avPlayer;
      this.bindState();
    } else {
      Logger.error('[PlayVideoModel] createAvPlayer fail!');
    }
  });
}

/**
 * AVPlayer 绑定事件
 */
bindState() {
  this.avPlayer.on(Events.STATE_CHANGE, async (state) => {
    switch (state) {
      case AvplayerStatus.IDLE:
        this.resetProgress();
        this.avPlayer.url = this.url;
        break;
      case AvplayerStatus.INITIALIZED:
        this.avPlayer.surfaceId = this.surfaceId;
        this.avPlayer.prepare();
        break;
      case AvplayerStatus.PREPARED:
        this.avPlayer.videoScaleType = 0;
        this.setVideoSize();
        this.avPlayer.play();
        this.duration = this.avPlayer.duration;
        break;
      case AvplayerStatus.PLAYING:
        this.avPlayer.setVolume(this.playerThis.volume);
        this.setBright();
        this.status = CommonConstants.STATUS_START;
        this.watchStatus();
```

```
          break;
        case AvplayerStatus.PAUSED:
          this.status = CommonConstants.STATUS_PAUSE;
          this.watchStatus();
          break;
        case AvplayerStatus.COMPLETED:
          this.titleThis.playSpeed = 1;
          this.duration = PlayConstants.PLAY_PLAYER.DURATION;
          if (!this.loop) {
            let curIndex = this.index + PlayConstants.PLAY_PLAYER.NEXT;
            this.index = (curIndex === globalThis.videoList.length) ?
            PlayConstants.PLAY_PLAYER.FIRST : curIndex;
            this.url = globalThis.videoList[this.index].src;
          } else {
            this.url = this.avPlayer.url;
          }
          this.avPlayer.reset();
          break;
        case AvplayerStatus.RELEASED:
          this.avPlayer.release();
          this.status = CommonConstants.STATUS_STOP;
          this.watchStatus();
          Logger.info('[PlayVideoModel] state released called')
          break;
        default:
          Logger.info('[PlayVideoModel] unKnown state: ' + state);
          break;
      }
    });
    this.avPlayer.on(Events.TIME_UPDATE, (time: number) => {
      this.initProgress(time);
    });
    this.avPlayer.on(Events.ERROR, (error) => {
      this.playError();
    })
  }

  ...
}
```

AVPlayer实例需设置播放路径和XComponent中获取的surfaceID，设置播放路径之后AVPlayer状态机变为initialized状态，在此状态下调用prepare()，进入prepared状态。其代码如下：

```
firstPlay(index: number, url: string, surfaceId: number) {
  this.index = index;
  this.url = url;
```

```
  this.surfaceId = surfaceId;
  this.avPlayer.url = this.url;
}
```

2. 切换播放状态

视频播放后，变为playing状态，可通过“播放/暂停”按钮切换播放状态，当视频暂停时状态机变为paused状态。其代码如下：

```
switchPlayOrPause() {
  if (this.status === CommonConstants.STATUS_START) {
    this.avPlayer.pause();
  } else {
    this.avPlayer.play();
  }
}
```

3. 设置播放速度

可拖动进度条设置视频播放位置来设置播放速度。其代码如下：

```
// 设置当前播放位置
setSeekTime(value: number, mode: SliderChangeMode) {
  if (mode === SliderChangeMode.Moving) {
    this.progressThis.progressVal = value;
    this.progressThis.currentTime = DateFormatUtil.secondToTime(Math.
floor(value * this.duration /
      CommonConstants.ONE_HUNDRED / CommonConstants.A_THOUSAND));
  }
  if (mode === SliderChangeMode.End) {
    this.seekTime = value * this.duration / CommonConstants.ONE_HUNDRED;
    this.avPlayer.seek(this.seekTime, media.SeekMode.SEEK_PREV_SYNC);
  }
}

// 设置播放速度
setSpeed(playSpeed: number) {
  if (CommonConstants.OPERATE_STATE.indexOf(this.avPlayer.state) === -1) {
    return;
  }
  this.playSpeed = playSpeed;
  this.avPlayer.setSpeed(this.playSpeed);
}
```

13.7.3 创建播放器界面

在ets目录下，创建一个新的目录“view”，该目录用于存放界面相关的类。

1. 创建 PlayPlayer.ets 类

创建PlayPlayer.ets类，该类用于实现播放器画面。其代码如下：

```
import { VideoController } from '../controller/VideoController';
import { CommonConstants } from '../common/constants/CommonConstants';
import { PlayConstants } from '../common/constants/PlayConstants';

@Component
export struct PlayPlayer {
  private playVideoModel: VideoController;
  @Consume src: string;
  @Consume index: number;
  @State volume: number = PlayConstants.PLAY_PAGE.VOLUME;
  @State volumeShow: boolean = PlayConstants.PLAY_PAGE.VOLUME_SHOW;
  @State bright: number = PlayConstants.PLAY_PAGE.BRIGHT;
  @State brightShow: boolean = PlayConstants.PLAY_PAGE.BRIGHT_SHOW;
  private xComponentController;
  private surfaceID: number;

  aboutToAppear() {
    if (this.playVideoModel !== null) {
      this.playVideoModel.initPlayerThis(this);
    }
    this.xComponentController = new XComponentController();
  }

  build() {
    Stack() {
      XComponent({
        id: PlayConstants.PLAY_PLAYER.ID,
        type: PlayConstants.PLAY_PLAYER.TYPE,
        libraryname: PlayConstants.PLAY_PLAYER.LIBRARY_NAME,
        controller: this.xComponentController
      })
        .onLoad(async () => {
          this.xComponentController.setXComponentSurfaceSize({
            surfaceWidth: PlayConstants.PLAY_PLAYER.SURFACE_WIDTH,
            surfaceHeight: PlayConstants.PLAY_PLAYER.SURFACE_HEIGHT
          });
          this.surfaceID = this.xComponentController.getXComponentSurfaceId();
          this.playVideoModel.firstPlay(this.index, this.src, this.surfaceID);
```

```
        })
        .width(CommonConstants.FULL_PERCENT)
        .height(CommonConstants.FULL_PERCENT)

      Stack() {
        Progress({
          value: Math.floor(this.volume * CommonConstants.ONE_HUNDRED),
          type: ProgressType.Ring
        })
          .width(CommonConstants.FULL_PERCENT)
          .aspectRatio(CommonConstants.ASPECT_RATIO)
        Image($r('app.media.ic_volume'))
          .width(PlayConstants.PLAY_PLAYER.IMAGE_WIDTH)
          .aspectRatio(CommonConstants.ASPECT_RATIO)
      }
      .width(PlayConstants.PLAY_PLAYER.STACK_WIDTH)
      .aspectRatio(CommonConstants.ASPECT_RATIO)
      .visibility(this.volumeShow ? Visibility.Visible : Visibility.Hidden)

      Stack() {
        Progress({
          value: Math.floor(this.bright * CommonConstants.ONE_HUNDRED),
          type: ProgressType.Ring
        })
          .width(CommonConstants.FULL_PERCENT)
          .aspectRatio(CommonConstants.ASPECT_RATIO)
        Image($r('app.media.ic_brightness'))
          .width(PlayConstants.PLAY_PLAYER.IMAGE_WIDTH)
          .aspectRatio(CommonConstants.ASPECT_RATIO)
      }
      .width(PlayConstants.PLAY_PLAYER.STACK_WIDTH)
      .aspectRatio(CommonConstants.ASPECT_RATIO)
      .visibility(this.brightShow ? Visibility.Visible : Visibility.Hidden)
    }
    .width(CommonConstants.FULL_PERCENT)
    .height(CommonConstants.FULL_PERCENT)
  }
}
```

2. 创建 PlayProgress.ets 类

创建PlayProgress.ets类，该类用于实现播放进度条。其代码如下：

```
import { VideoController } from '../controller/VideoController';
import { CommonConstants } from '../common/constants/CommonConstants';
import { PlayConstants } from '../common/constants/PlayConstants';
```

```
@Component
export struct PlayProgress {
  private playVideoModel: VideoController;
  @State currentTime: string = PlayConstants.PLAY_PROGRESS.CURRENT_TIME;
  @State totalTime: string = PlayConstants.PLAY_PROGRESS.TOTAL_TIME;
  @State progressVal: number = PlayConstants.PLAY_PROGRESS.PROGRESS_VAL;

  aboutToAppear() {
    if (this.playVideoModel !== null) {
      this.playVideoModel.initProgressThis(this);
    }
  }

  build() {
    Column() {
      Row() {
        Text(this.currentTime)
          .fontSize($r('app.float.slider_font_size'))
          .fontColor(Color.White)
        Slider({
          value: this.progressVal,
          step: PlayConstants.PLAY_PROGRESS.STEP,
          style: SliderStyle.OutSet
        })
          .blockColor(Color.White)
          .trackColor($r('app.color.track_color'))
          .selectedColor(Color.White)
          .trackThickness(PlayConstants.PLAY_PROGRESS.TRACK_THICKNESS)
          .layoutWeight(1)
          .margin({ left: PlayConstants.PLAY_PROGRESS.MARGIN_LEFT })
          .onChange((value: number, mode: SliderChangeMode) => {
            this.playVideoModel.setSeekTime(value, mode);
          })
        Text(this.totalTime)
          .fontSize($r('app.float.slider_font_size'))
          .fontColor(Color.White)
          .margin({ left: PlayConstants.PLAY_PROGRESS.MARGIN_LEFT })
      }
      .width(PlayConstants.PLAY_PROGRESS.ROW_WIDTH)
    }
    .width(CommonConstants.FULL_PERCENT)
    .height(CommonConstants.FULL_PERCENT)
    .justifyContent(FlexAlign.Center)
  }
}
```

3. 创建 PlayControl.ets 类

创建PlayControl.ets类，该类用于实现播放控制按钮。其代码如下：

```
import { VideoController } from '../controller/VideoController';
import { CommonConstants } from '../common/constants/CommonConstants';
import { PlayConstants } from '../common/constants/PlayConstants';

@Component
export struct PlayControl {
  private playVideoModel: VideoController;
  @Consume status: number;

  build() {
    Column() {
      Row() {
        Image($r('app.media.ic_previous'))
          .width($r('app.float.control_image_width'))
          .aspectRatio(CommonConstants.ASPECT_RATIO)
          .onClick(async () => {
            this.playVideoModel.previousVideo();
            this.status = CommonConstants.STATUS_START;
          })
        Column() {
          Image(this.status === CommonConstants.STATUS_START ?
            $r('app.media.ic_pause') : $r('app.media.ic_play'))
            .width($r('app.float.control_image_width'))
            .aspectRatio(CommonConstants.ASPECT_RATIO)
            .onClick(async () => {
              let curStatus = (this.playVideoModel.getStatus() ===
                                CommonConstants.STATUS_START);
              this.status = curStatus ? CommonConstants.STATUS_PAUSE :
                             CommonConstants.STATUS_START;
              this.playVideoModel.switchPlayOrPause();
            })
        }
        .layoutWeight(1)
        Image($r('app.media.ic_next'))
          .width($r('app.float.control_image_width'))
          .aspectRatio(CommonConstants.ASPECT_RATIO)
          .onClick(() => {
            this.playVideoModel.nextVideo();
            this.status = CommonConstants.STATUS_START;
          })
      }
      .width(PlayConstants.PLAY_CONTROL.ROW_WIDTH)
```

```
    }
    .width(CommonConstants.FULL_PERCENT)
    .height(CommonConstants.FULL_PERCENT)
    .justifyContent(FlexAlign.Center)
  }
}
```

4. 创建 PlayPage.ets 类

创建PlayPage.ets类，该类用于实现播放器整体界面。其代码如下：

```
import router from '@ohos.router';
import { PlayTitle } from '../view/PlayTitle';
import { PlayPlayer } from '../view/PlayPlayer';
import { PlayControl } from '../view/PlayControl';
import { PlayProgress } from '../view/PlayProgress';
import { VideoController } from '../controller/VideoController';
import { CommonConstants } from '../common/constants/CommonConstants';
import { PlayConstants } from '../common/constants/PlayConstants';

@Entry
@Component
struct PlayPage {
  @State videoHeight: string = PlayConstants.PLAY_PAGE.PLAY_PLAYER_HEIGHT;
  @State videoWidth: string = CommonConstants.FULL_PERCENT;
  @State videoMargin: string = PlayConstants.PLAY_PAGE.MARGIN_ZERO;
  @State videoPosition: FlexAlign  = FlexAlign.Center;
  private playVideoModel: VideoController = new VideoController();
  @Provide src: string = router.getParams()['src'];
  @Provide index: number = router.getParams()['index'];
  @Provide type: number = router.getParams()['type'];
  @Provide status: number = CommonConstants.STATUS_START;
  private panOptionBright: PanGestureOptions = new PanGestureOptions({
direction: PanDirection.Vertical });
  private panOptionVolume: PanGestureOptions = new PanGestureOptions({
direction: PanDirection.Horizontal });

  aboutToAppear() {
    this.playVideoModel.initPlayPageThis(this);
  }

  aboutToDisappear() {
    this.playVideoModel.release();
  }

  onPageHide() {
```

```
    this.status = CommonConstants.STATUS_PAUSE;
    this.playVideoModel.pause();
  }

  build() {
    Stack() {
      Column () {
        Column(){
        }
        .height(this.videoMargin)
        PlayPlayer({ playVideoModel: this.playVideoModel })
          .width(this.videoWidth)
          .height(this.videoHeight)
      }
      .height(CommonConstants.FULL_PERCENT)
      .width(CommonConstants.FULL_PERCENT)
      .justifyContent(this.videoPosition)
      .zIndex(0)
      Column() {
        PlayTitle({ playVideoModel: this.playVideoModel })
          .width(CommonConstants.FULL_PERCENT)
          .height(PlayConstants.PLAY_PAGE.HEIGHT)
        Column()
          .width(CommonConstants.FULL_PERCENT)
          .height(PlayConstants.PLAY_PAGE.COLUMN_HEIGHT_ONE)
          .gesture(
            PanGesture(this.panOptionBright)
              .onActionStart((event: GestureEvent) => {
                this.playVideoModel.onBrightActionStart(event);
              })
              .onActionUpdate((event: GestureEvent) => {
                this.playVideoModel.onBrightActionUpdate(event);
              })
              .onActionEnd(() => {
                this.playVideoModel.onActionEnd();
              })
          )
        Column() {
        }
        .width(CommonConstants.FULL_PERCENT)
        .height(PlayConstants.PLAY_PAGE.PLAY_PLAYER_HEIGHT)
        Column()
          .width(CommonConstants.FULL_PERCENT)
          .height(PlayConstants.PLAY_PAGE.COLUMN_HEIGHT_TWO)
          .gesture(
            PanGesture(this.panOptionVolume)
```

```
            .onActionStart((event: GestureEvent) => {
              this.playVideoModel.onVolumeActionStart(event);
            })
            .onActionUpdate((event: GestureEvent) => {
              this.playVideoModel.onVolumeActionUpdate(event);
            })
            .onActionEnd(() => {
              this.playVideoModel.onActionEnd();
            })
        )
      PlayControl({ playVideoModel: this.playVideoModel })
        .width(CommonConstants.FULL_PERCENT)
        .height(PlayConstants.PLAY_PAGE.HEIGHT)
      PlayProgress({ playVideoModel: this.playVideoModel })
        .width(CommonConstants.FULL_PERCENT)
        .height(PlayConstants.PLAY_PAGE.PLAY_PROGRESS_HEIGHT)
    }
    .height(CommonConstants.FULL_PERCENT)
    .width(CommonConstants.FULL_PERCENT)
    .zIndex(1)
  }
  .height(CommonConstants.FULL_PERCENT)
  .width(CommonConstants.FULL_PERCENT)
  .backgroundColor(Color.Black)
 }
}
```

13.7.4 运行

运行应用，可以看到播放器运行效果如图 13-9 所示。

可以通过播放操作按钮来实现视频的快进、切换等。切换视频效果如图 13-10 所示。

图 13-9 播放器运行效果

图 13-10 切换视频效果

第14章 图像

HarmonyOS 图像模块支持图像业务的开发，常见功能如图像解码、图像编码、基本的位图操作、图像编辑等。

14.1 图像概述

HarmonyOS图像模块支持图像业务的开发，常见功能如图像解码、图像编码、基本的位图操作、图像编辑等。当然，也支持通过接口组合来实现更复杂的图像处理逻辑。

14.1.1 基本概念

图像业务包含以下基本概念。

- 图像解码：图像解码就是将不同的存档格式图片（如JPEG、PNG等）解码为无压缩的位图格式，以方便在应用或系统中进行相应的处理。
- PixelMap：PixelMap是图像解码后无压缩的位图格式，用于图像显示或进一步的处理。
- 渐进式解码：渐进式解码是在无法一次性提供完整图像文件数据的场景下，随着图像文件数据的逐步增加，通过多次增量解码逐步完成图像解码的模式。
- 预乘：预乘时，RGB各通道的值被替换为原始值乘以Alpha通道不透明的比例（0 ~ 1）后的值，方便后期直接合成叠加；不预乘指RGB各通道的数值是图像的原始值，与Alpha通道的值无关。
- 图像编码：图像编码就是将无压缩的位图格式，编码成不同格式的存档格式图片（JPEG、PNG等），以方便在应用或系统中进行相应的处理。

14.1.2 约束与限制

为及时释放本地资源，建议在图像解码的ImageSource对象、位图图像PixelMap对象或图像编码的ImagePacker对象使用完成后，主动调用release()方法。

14.2 实战：图像解码和编码

图像解码就是将所支持格式的存档图片解码成统一的PixelMap图像，用于后续图像显示或其他处理，比如旋转、缩放、裁剪等。当前支持格式包括JPEG、PNG、GIF、HEIF、WebP、BMP。

14.2.1 接口说明

ImageSource主要用于图像解码。ImageSource的主要接口有以下几种。

- create(String pathName, SourceOptions opts)：从图像文件路径创建图像数据源。
- create(InputStream is, SourceOptions opts)：从输入流创建图像数据源。

- create(byte[] data, SourceOptions opts)：从字节数组创建图像源。
- create(byte[] data, int offset, int length, SourceOptions opts)：从字节数组指定范围创建图像源。
- create(File file, SourceOptions opts)：从文件对象创建图像数据源。
- create(FileDescriptor fd, SourceOptions opts)：从文件描述符创建图像数据源。
- createIncrementalSource(SourceOptions opts)：创建渐进式图像数据源。
- createIncrementalSource(IncrementalSourceOptions opts)：创建渐进式图像数据源，支持设置渐进式数据更新模式。
- createPixelmap(DecodingOptions opts)：从图像数据源解码并创建PixelMap图像。
- createPixelmap(int index, DecodingOptions opts)：从图像数据源解码并创建PixelMap图像，如果图像数据源支持多张图片的话，支持指定图像索引。
- updateData(byte[] data, boolean isFinal)：更新渐进式图像源数据。
- updateData(byte[] data, int offset, int length, boolean isFinal)：更新渐进式图像源数据，支持设置输入数据的有效数据范围。
- getImageInfo()：获取图像基本信息。
- getImageInfo(int index)：根据特定的索引获取图像基本信息。
- getSourceInfo()：获取图像源信息。
- release()：释放对象关联的本地资源。

14.2.2 创建应用

为了演示图像编解码的功能，创建一个名为“ImageCodec”的应用。

在应用的界面上，通过点击按钮来触发解码或编码图片的操作。

在media目录下，放置一张用于测试的图片waylau_616_616.jpeg。

14.2.3 修改ability_main.xml

修改ability_main.xml内容如下：

```
<?xml version="1.0" encoding="utf-8"?>
<DirectionalLayout
    xmlns:ohos="http://schemas.huawei.com/res/ohos"
    ohos:height="match_parent"
    ohos:width="match_parent"
    ohos:orientation="horizontal">

    <DirectionalLayout
        ohos:height="match_content"
        ohos:width="0vp"
```

```
        ohos:orientation="vertical"
        ohos:weight="1">

        <Button
            ohos:id="$+id:button_decode"
            ohos:height="match_content"
            ohos:width="match_content"
            ohos:background_element="#F76543"
            ohos:left_padding="10vp"
            ohos:text="Decode"
            ohos:text_size="40vp"
            />

        <Image
            ohos:id="$+id:image_decode"
            ohos:height="100vp"
            ohos:width="100vp"/>

    </DirectionalLayout>

    <DirectionalLayout
        ohos:height="match_content"
        ohos:width="0vp"
        ohos:orientation="vertical"
        ohos:weight="1">

        <Button
            ohos:id="$+id:button_encode"
            ohos:height="match_content"
            ohos:width="match_content"
            ohos:background_element="#F76543"
            ohos:left_padding="10vp"
            ohos:text="Encode"
            ohos:text_size="40vp"
            />

        <Image
            ohos:id="$+id:image_encode"
            ohos:height="100vp"
            ohos:width="100vp"/>
    </DirectionalLayout>
</DirectionalLayout>
```

界面预览效果如图 14-1 所示。

图 14-1　界面预览效果

14.2.4 修改MainAbilitySlice

修改MainAbilitySlice内容如下：

```
package com.waylau.hmos.imagedecoder.slice;

import com.waylau.hmos.imagedecoder.ResourceTable;
import ohos.aafwk.ability.AbilitySlice;
import ohos.aafwk.content.Intent;
import ohos.agp.components.Button;
import ohos.agp.components.Image;
import ohos.global.resource.NotExistException;
import ohos.hiviewdfx.HiLog;
import ohos.hiviewdfx.HiLogLabel;
import ohos.media.image.ImagePacker;
import ohos.media.image.ImageSource;
import ohos.media.image.PixelMap;
import ohos.media.image.common.Rect;
import ohos.media.image.common.Size;

import java.io.*;
import java.nio.file.Paths;

public class MainAbilitySlice extends AbilitySlice {
    private static final String TAG = MainAbilitySlice.class.getSimpleName();
    private static final HiLogLabel LABEL_LOG =
            new HiLogLabel(HiLog.LOG_APP, 0x00001, TAG);

    private ImageSource imageSource;
    private PixelMap pixelMap;

    @Override
    public void onStart(Intent intent) {
        super.onStart(intent);
        super.setUIContent(ResourceTable.Layout_ability_main);

        Button buttonDecode =
                (Button) findComponentById(ResourceTable.Id_button_decode);

        // 为按钮设置点击事件回调
        buttonDecode.setClickedListener(listener -> {
            try {
                decode();
            } catch (IOException e) {
                e.printStackTrace();
```

```
        } catch (NotExistException e) {
            e.printStackTrace();
        }
    });

    Button buttonEncode =
            (Button) findComponentById(ResourceTable.Id_button_encode);

    // 为按钮设置点击事件回调
    buttonEncode.setClickedListener(listener -> {
        try {
            encode();
        } catch (IOException e) {
            e.printStackTrace();
        }
    });
}

private void decode() throws IOException, NotExistException {
    // 获取图片流
    InputStream drawableInputStream =  getResourceManager().
        getResource(ResourceTable.Media_waylau_616_616);

    // 创建图像数据源 ImageSource 对象
    imageSource = ImageSource.create(drawableInputStream, this.
        getSourceOptions());

    // 普通解码叠加旋转、缩放、裁剪
    pixelMap = imageSource.createPixelmap(this.getDecodingOptions());

    Image imageDecode =
            (Image) findComponentById(ResourceTable.Id_image_decode);

    imageDecode.setPixelMap(pixelMap);
}

private void encode() throws IOException {
    HiLog.info(LABEL_LOG, "before encode()");

    // 创建图像编码 ImagePacker 对象
    ImagePacker imagePacker = ImagePacker.create();

    // 获取数据目录
    File dataDir = new File(this.getDataDir().toString());
    if(!dataDir.exists()){
        dataDir.mkdirs();
```

```
    }

    // 文件路径
    String filePath = Paths.get(dataDir.toString(),"test.jpeg").
        toString();

    // 构建目标文件
    File targetFile = new File(filePath);

    // 设置编码输出流和编码参数
    FileOutputStream outputStream = new FileOutputStream(targetFile);

    // 初始化打包
    imagePacker.initializePacking(outputStream, this.getPackingOptions());

    // 添加需要编码的 PixelMap 对象，进行编码操作
    imagePacker.addImage(pixelMap);

    // 完成图像打包任务
    imagePacker.finalizePacking();

    Image imageEncode =
            (Image) findComponentById(ResourceTable.Id_image_encode);

    // 文件转成图像
    imageSource = ImageSource.create(targetFile,this.getSourceOptions());

    pixelMap =  imageSource.createPixelmap(this.getDecodingOptions());

    imageEncode.setPixelMap(pixelMap);

    HiLog.info(LABEL_LOG, "end encode()");
}


// 设置打包格式
private ImagePacker.PackingOptions getPackingOptions() {
    ImagePacker.PackingOptions packingOptions = new ImagePacker.
        PackingOptions();
    packingOptions.format = "image/jpeg"; // 设置 format 为编码的图像格式，当前
        支持 jpeg 格式
    packingOptions.quality = 90; // 设置 quality 为图像质量，范围为 0 ～ 100，100
        为最佳质量
    return packingOptions;
}
```

```
    // 设置解码格式
    private ImageSource.DecodingOptions getDecodingOptions() {
        ImageSource.DecodingOptions decodingOpts = new ImageSource.
            DecodingOptions();
        decodingOpts.desiredSize = new Size(600, 600);
        decodingOpts.desiredRegion = new Rect(0, 0, 600, 600);
        decodingOpts.rotateDegrees = 90;

        return decodingOpts;
    }

    // 设置数据源的格式信息
    private ImageSource.SourceOptions getSourceOptions() {
        ImageSource.SourceOptions sourceOptions = new ImageSource.
            SourceOptions();
        sourceOptions.formatHint = "image/jpeg";

        return sourceOptions;
    }

    @Override
    protected void onStop() {
        super.onStop();
        if (imageSource != null) {
            // 释放资源
            imageSource.release();
        }
    }

    @Override
    public void onForeground(Intent intent) {
        super.onForeground(intent);
    }
}
```

上述代码，在按钮上设置了点击事件，以触发解码和编码操作。

14.2.5 解码操作说明

解码的过程是这样的：

- 在decode()方法内部，通过getResourceManager().getResource方法获取测试图片，并转为输入流；
- 将上述输入流作为创建图像数据源ImageSource对象的参数之一；
- 通过imageSource.createPixelmap方法，将ImageSource转为PixelMap；

- 通过imageDecode.setPixelMap方法将图像信息设置到Image组件上，从而在界面上显示图片。

运行应用后，点击“Decode”按钮，界面效果如图14-2所示。

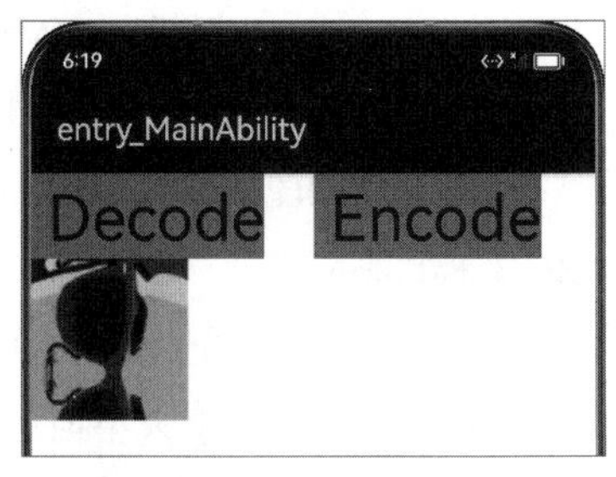

图14-2 界面效果（一）

14.2.6 编码操作说明

编码的过程是这样的：

- 在encode()方法内部，通过ImagePacker.create()方法来创建图像编码ImagePacker对象；
- 构建目标文件targetFile；
- 执行initializePacking、addImage、finalizePacking等系列操作后完成打包；
- 最后将图像信息设置到Image组件上，从而在界面上显示图片。

运行应用后，点击“Encode”按钮，界面效果如图14-3所示。

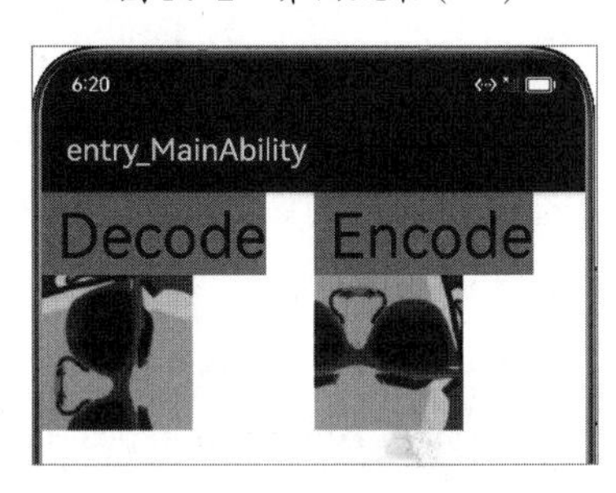

图14-3 界面效果（二）

14.3 实战：位图操作

位图操作是指对PixelMap图像进行相关的操作，比如创建、查询信息、读写像素数据等。

14.3.1 接口说明

位图操作类PixelMap的主要接口有以下几种。

- create(InitializationOptions opts)：根据图像大小、像素格式、Alpha类型等初始化选项创建PixelMap。
- create(int[] colors, InitializationOptions opts)：根据图像大小、像素格式、Alpha类型等初始化选项，以像素颜色数组为数据源创建PixelMap。
- create(int[] colors, int offset, int stride, InitializationOptions opts)：根据图像大小、像素格式、alpha类型等初始化选项，以像素颜色数组、起始偏移量、行像素大小描述的数据源创建PixelMap。
- create(PixelMap source, InitializationOptions opts)：根据图像大小、像素格式、Alpha类型等初始化选项，以源PixelMap为数据源创建PixelMap。
- create(PixelMap source, Rect srcRegion, InitializationOptions opts)：根据图像大小、像素格式、Alpha类型等初始化选项，以源PixelMap、源裁剪区域描述的数据源创建PixelMap。
- getBytesNumberPerRow()：获取每行像素数据占用的字节数。

- getPixelBytesCapacity()：获取存储Pixelmap像素数据的内存容量。
- isEditable()：判断PixelMap是否允许修改。
- isSameImage(PixelMap other)：判断两个图像是否相同，包括ImageInfo属性信息和像素数据。
- readPixel(Position pos)：读取指定位置像素的颜色值，返回的颜色格式为PixelFormat.ARGB_8888。
- readPixels(int[] pixels, int offset, int stride, Rect region)：读取指定区域像素的颜色值，输出到以起始偏移量、行像素大小描述的像素数组，返回的颜色格式为PixelFormat.ARGB_8888。
- readPixels(Buffer dst)：读取像素的颜色值到缓冲区，返回的数据是PixelMap中像素数据的原样拷贝，即返回的颜色数据格式与PixelMap中的像素格式一致。
- resetConfig(Size size, PixelFormat pixelFormat)：重置PixelMap的大小和像素格式配置，但不会改变原有的像素数据，也不会重新分配像素数据的内存，重置后图像数据的字节数不能超过PixelMap的内存容量。
- setAlphaType(AlphaType alphaType)：设置PixelMap的Alpha类型。
- writePixel(Position pos, int color)：向指定位置像素写入颜色值，写入颜色格式为PixelFormat.ARGB_8888。
- writePixels(int[] pixels, int offset, int stride, Rect region)：将像素颜色数组、起始偏移量、行像素的个数描述的源像素数据写入PixelMap的指定区域,写入颜色格式为PixelFormat.ARGB_8888。
- writePixels(Buffer src)：将缓冲区描述的源像素数据写入PixelMap，写入的数据将原样覆盖PixelMap中的像素数据，即写入数据的颜色格式应与PixelMap的配置兼容。
- writePixels(int color)：将所有像素都填充为指定的颜色值，写入颜色格式为 PixelFormat.ARGB_8888。
- getPixelBytesNumber()：获取全部像素数据包含的字节数。
- setBaseDensity(int baseDensity)：设置PixelMap的基础像素密度值。
- getBaseDensity()：获取PixelMap的基础像素密度值。
- setUseMipmap(boolean useMipmap)：设置PixelMap渲染是否使用mipmap。
- useMipmap()：获取PixelMap渲染是否使用mipmap。
- getNinePatchChunk()：获取图像的NinePatchChunk数据。
- getFitDensitySize(int targetDensity)：获取适应目标像素密度的图像缩放的尺寸。
- getImageInfo()：获取图像基本信息。
- release()：释放对象关联的本地资源。

14.3.2 创建应用

为了演示位图操作的功能，创建一个名为“PixelMap”的应用。

在应用的界面上，通过点击按钮，来触发位图操作。

14.3.3 修改ability_main.xml

修改ability_main.xml内容如下：

```
<?xml version="1.0" encoding="utf-8"?>
<DirectionalLayout
    xmlns:ohos="http://schemas.huawei.com/res/ohos"
    ohos:height="match_parent"
    ohos:width="match_parent"
    ohos:alignment="center"
    ohos:orientation="vertical">
    <Button
        ohos:id="$+id:button_create"
        ohos:height="match_content"
        ohos:width="match_content"
        ohos:background_element="#F76543"
        ohos:text="Create"
        ohos:text_size="40vp"
        />
    <Button
        ohos:id="$+id:button_get"
        ohos:height="match_content"
        ohos:width="match_content"
        ohos:background_element="#F76543"
        ohos:text="Get Info"
        ohos:text_size="40vp"
        />
    <Button
        ohos:id="$+id:button_read_write"
        ohos:height="match_content"
        ohos:width="match_content"
        ohos:background_element="#F76543"
        ohos:text="Read and write"
        ohos:text_size="40vp"
        />
</DirectionalLayout>
```

界面预览效果如图14-4所示。

图14-4　界面预览效果

14.3.4 修改MainAbilitySlice

修改MainAbilitySlice内容如下：

```
package com.waylau.hmos.pixelmap.slice;
```

```
import com.waylau.hmos.pixelmap.ResourceTable;
import ohos.aafwk.ability.AbilitySlice;
import ohos.aafwk.content.Intent;
import ohos.agp.components.Button;
import ohos.hiviewdfx.HiLog;
import ohos.hiviewdfx.HiLogLabel;
import ohos.media.image.PixelMap;
import ohos.media.image.common.PixelFormat;
import ohos.media.image.common.Position;
import ohos.media.image.common.Rect;
import ohos.media.image.common.Size;

import java.nio.IntBuffer;

public class MainAbilitySlice extends AbilitySlice {
    private static final String TAG = MainAbilitySlice.class.getSimpleName();
    private static final HiLogLabel LABEL_LOG =
            new HiLogLabel(HiLog.LOG_APP, 0x00001, TAG);

    private PixelMap pixelMap1;
    private PixelMap pixelMap2;

    @Override
    public void onStart(Intent intent) {
        super.onStart(intent);
        super.setUIContent(ResourceTable.Layout_ability_main);

        Button buttonCreate =
                (Button) findComponentById(ResourceTable.Id_button_create);
        buttonCreate.setClickedListener(listener -> createPixelMap());

        Button buttonGet =
                (Button) findComponentById(ResourceTable.Id_button_get);
        buttonGet.setClickedListener(listener -> getPixelMapInfo());

        Button buttonReadWrite =
                (Button) findComponentById(ResourceTable.Id_button_read_write);
        buttonReadWrite.setClickedListener(listener -> readWritePixels());
    }

    ...
}
```

上述代码在按钮上设置了点击事件，以触发位图操作。

14.3.5 创建PixelMap操作说明

createPixelMap方法用于创建PixelMap操作，其代码如下：

```
private void createPixelMap() {
    HiLog.info(LABEL_LOG, "before createPixelMap");

    // 像素颜色数组
    int[] defaultColors = new int[]{5, 5, 5, 5, 6, 6, 3, 3, 3, 0};

    // 初始化选项
    PixelMap.InitializationOptions initializationOptions = new PixelMap.
        InitializationOptions();
    initializationOptions.size = new Size(3, 2);
    initializationOptions.pixelFormat = PixelFormat.ARGB_8888;

    // 根据像素颜色数组、初始化选项创建位图对象 PixelMap
    pixelMap1 = PixelMap.create(defaultColors, initializationOptions);

    // 根据 PixelMap 作为数据源创建
    pixelMap2 = PixelMap.create(pixelMap1, initializationOptions);

    HiLog.info(LABEL_LOG, "end createPixelMap");
}
```

上述方法中：

- pixelMap1是根据像素颜色数组、初始化选项创建位图对象PixelMap的。
- pixelMap2是根据PixelMap作为数据源创建的。

运行应用后，点击“Create”按钮，控制台输出内容如下所示：

```
04-27 14:40:18.664 2717-2717/com.waylau.hmos.pixelmap I 00001/
MainAbilitySlice: before createPixelMap
04-27 14:40:18.669 2717-2717/com.waylau.hmos.pixelmap I 00001/
MainAbilitySlice: end createPixelMap
```

14.3.6 从位图对象中获取信息操作说明

getPixelMapInfo方法用于从位图对象中获取信息，其代码如下：

```
private void getPixelMapInfo() {
    // 从位图对象中获取信息
    long capacity = pixelMap1.getPixelBytesCapacity();
    long bytesNumber = pixelMap1.getPixelBytesNumber();
    int rowBytes = pixelMap1.getBytesNumberPerRow();
```

```
    byte[] ninePatchData = pixelMap1.getNinePatchChunk();

    HiLog.info(LABEL_LOG, "capacity: %{public}s", capacity);
    HiLog.info(LABEL_LOG, "bytesNumber: %{public}s", bytesNumber);
    HiLog.info(LABEL_LOG, "rowBytes: %{public}s", rowBytes);
    HiLog.info(LABEL_LOG, "ninePatchData: %{public}s", ninePatchData);
}
```

上述方法从 pixelMap1 中获取位图的信息。

运行应用后，点击“Get Info”按钮，控制台输出内容如下所示：

```
04-27 14:42:52.853 2717-2717/com.waylau.hmos.pixelmap I 00001/
MainAbilitySlice: capacity: 24
04-27 14:42:52.853 2717-2717/com.waylau.hmos.pixelmap I 00001/
MainAbilitySlice: bytesNumber: 24
04-27 14:42:52.853 2717-2717/com.waylau.hmos.pixelmap I 00001/
MainAbilitySlice: rowBytes: 12
04-27 14:42:52.853 2717-2717/com.waylau.hmos.pixelmap I 00001/
MainAbilitySlice: ninePatchData: null
```

14.3.7 读取和写入像素操作说明

readWritePixels 方法用于读取和写入像素，其代码如下：

```
private void readWritePixels() {
    // 读取指定位置像素
    int color = pixelMap1.readPixel(new Position(1, 1));
    HiLog.info(LABEL_LOG, "readPixel color: %{public}s", color);

    // 读取指定区域像素
    int[] pixelArray = new int[50];
    Rect region = new Rect(0, 0, 3, 2);
    pixelMap1.readPixels(pixelArray, 0, 10, region);
    HiLog.info(LABEL_LOG, "readPixel pixelArray: %{public}s", pixelArray);

    // 读取像素到 Buffer
    IntBuffer pixelBuf = IntBuffer.allocate(50);
    pixelMap1.readPixels(pixelBuf);
    HiLog.info(LABEL_LOG, "readPixel pixelBuf: %{public}s", pixelBuf);

    // 在指定位置写入像素
    pixelMap1.writePixel(new Position(1, 1), 0xFF112233);

    // 在指定区域写入像素
    pixelMap1.writePixels(pixelArray, 0, 10, region);
```

```
    // 写入 Buffer 中的像素
    pixelMap1.writePixels(pixelBuf);
}
```

上述方法分别执行readPixels和writePixels来进行像素的读取和写入。

运行应用后，点击“Read and write”按钮，控制台输出内容如下所示：

```
04-27 14:47:05.295 2717-2717/com.waylau.hmos.pixelmap I 00001/
MainAbilitySlice: readPixel color: 0
04-27 14:47:05.296 2717-2717/com.waylau.hmos.pixelmap I 00001/
MainAbilitySlice: readPixel pixelArray: [I@21e5496
04-27 14:47:05.296 2717-2717/com.waylau.hmos.pixelmap I 00001/
MainAbilitySlice: readPixel pixelBuf: java.nio.HeapIntBuffer[pos=6 lim=50
cap=50]
```

实战：图像属性解码

图像属性解码就是获取图像中包含的属性信息，比如EXIF属性。

14.4.1 接口说明

图像属性解码的功能主要由ImageSource和ExifUtils提供。

ImageSource的主要接口有以下几种。

- getThumbnailInfo()：获取嵌入图像文件的缩略图的基本信息。
- getImageThumbnailBytes()：获取嵌入图像文件缩略图的原始数据。
- getThumbnailFormat()：获取嵌入图像文件缩略图的格式。

ExifUtils的主要接口有以下几种。

- getLatLong(ImageSource imageSource)：获取嵌入图像文件的经纬度信息。
- getAltitude(ImageSource imageSource, double defaultValue)：获取嵌入图像文件的海拔信息。

14.4.2 创建应用

为了演示图像属性解码的功能，创建一个名为“ImageSourceExifUtils”的应用。

在应用的界面上，通过点击按钮，来触发图像属性解码的操作。

在media目录下，放置一张用于测试的照片IMG_20210219_175445.jpg。需要注意，测试照片需要包含EXIF属性信息。

14.4.3 修改ability_main.xml

修改ability_main.xml内容如下：

```
<?xml version="1.0" encoding="utf-8"?>
<DirectionalLayout
    xmlns:ohos="http://schemas.huawei.com/res/ohos"
    ohos:height="match_parent"
    ohos:width="match_parent"
    ohos:alignment="center"
    ohos:orientation="vertical">

    <Button
        ohos:id="$+id:button_get_info"
        ohos:height="match_content"
        ohos:width="match_content"
        ohos:background_element="#F76543"
        ohos:left_padding="10vp"
        ohos:text="Get Info"
        ohos:text_size="40vp"
        />

</DirectionalLayout>
```


图 14-5 界面预览效果

界面预览效果如图14-5所示。

14.4.4 修改MainAbilitySlice

修改MainAbilitySlice内容如下：

```
package com.waylau.hmos.imagesourceexifutils.slice;

import com.waylau.hmos.imagesourceexifutils.ResourceTable;
import ohos.aafwk.ability.AbilitySlice;
import ohos.aafwk.content.Intent;
import ohos.agp.components.Button;
import ohos.global.resource.RawFileEntry;
import ohos.hiviewdfx.HiLog;
import ohos.hiviewdfx.HiLogLabel;
import ohos.media.image.ExifUtils;
import ohos.media.image.ImageSource;
import ohos.media.image.common.ImageInfo;
import ohos.utils.Pair;

import java.io.*;
```

```
public class MainAbilitySlice extends AbilitySlice {
    private static final String TAG = MainAbilitySlice.class.getSimpleName();
    private static final HiLogLabel LABEL_LOG =
            new HiLogLabel(HiLog.LOG_APP, 0x00001, TAG);

    private ImageSource imageSource;

    @Override
    public void onStart(Intent intent) {
        super.onStart(intent);
        super.setUIContent(ResourceTable.Layout_ability_main);

        Button buttonGetInfo =
                (Button) findComponentById(ResourceTable.Id_button_get_info);

        // 为按钮设置点击事件回调
        buttonGetInfo.setClickedListener(listener -> {
            try {
                getInfo();
            } catch (IOException e) {
                e.printStackTrace();
            }
        });
    }

    private void getInfo() throws IOException {
        // 获取图片
        RawFileEntry fileEntry = getResourceManager().
                getRawFileEntry("resources/base/media/IMG_20210219_175445.jpg");

        //获取文件大小
        int fileSize = (int) fileEntry.openRawFileDescriptor().getFileSize();

        //定义读取文件的字节
        byte[] fileData = new byte[fileSize];

        //读取文件字节
        fileEntry.openRawFile().read(fileData);

        imageSource = ImageSource.create(fileData, this.getSourceOptions());

        // 获取嵌入图像文件的缩略图的基本信息
        ImageInfo imageInfo = imageSource.getThumbnailInfo();
        HiLog.info(LABEL_LOG, "imageInfo: %{public}s", imageInfo);
```

```
        // 获取嵌入图像文件缩略图的原始数据
        byte[] imageThumbnailBytes = imageSource.getImageThumbnailBytes();
        HiLog.info(LABEL_LOG, "imageThumbnailBytes: %{public}s",
            imageThumbnailBytes);

        // 获取嵌入图像文件缩略图的格式
        int thumbnailFormat = imageSource.getThumbnailFormat();
        HiLog.info(LABEL_LOG, "thumbnailFormat: %{public}s", thumbnailFormat);

        // 获取嵌入图像文件的经纬度信息。
        Pair<Float, Float> lat = ExifUtils.getLatLong(imageSource);
        HiLog.info(LABEL_LOG, "lat first: %{public}s", lat.f);
        HiLog.info(LABEL_LOG, "lat second: %{public}s", lat.s);

        // 获取嵌入图像文件的海拔信息
        double defaultValue = 100;
        double altitude = ExifUtils.getAltitude(imageSource, defaultValue);
        HiLog.info(LABEL_LOG, "altitude: %{public}s", altitude);
    }

    // 设置数据源的格式信息
    private ImageSource.SourceOptions getSourceOptions() {
        ImageSource.SourceOptions sourceOptions = new ImageSource.
            SourceOptions();
        sourceOptions.formatHint = "image/jpeg";

        return sourceOptions;
    }

    @Override
    protected void onStop() {
        super.onStop();
        if (imageSource != null) {
            // 释放资源
            imageSource.release();
        }
    }

    @Override
    public void onForeground(Intent intent) {
        super.onForeground(intent);
    }
}
```

上述代码解释如下。

- 在按钮上设置了点击事件，以触发获取图像属性信息的操作。
- 通过ImageSource的主要接口来获取嵌入图像文件的缩略图的基本信息、原始数据和格式。
- 通过ExifUtils的主要接口获取嵌入图像文件的经纬度信息、海拔信息。

14.4.5 运行

运行应用，点击“Get Info”按钮，可以看到控制台输出内容如下：

```
04-27 16:01:44.045 2947-2947/com.waylau.hmos.imagesourceexifutils I 00001/
MainAbilitySlice: imageInfo: ohos.media.image.common.ImageInfo@b982677
04-27 16:01:44.045 2947-2947/com.waylau.hmos.imagesourceexifutils I 00001/
MainAbilitySlice: imageThumbnailBytes: [B@28ad2e4
04-27 16:01:44.046 2947-2947/com.waylau.hmos.imagesourceexifutils I 00001/
MainAbilitySlice: thumbnailFormat: 3
04-27 16:01:44.047 2947-2947/com.waylau.hmos.imagesourceexifutils I 00001/
MainAbilitySlice: lat first: 23.081373
04-27 16:01:44.047 2947-2947/com.waylau.hmos.imagesourceexifutils I 00001/
MainAbilitySlice: lat second: 114.40377
04-27 16:01:44.047 2947-2947/com.waylau.hmos.imagesourceexifutils I 00001/
MainAbilitySlice: altitude: 100.0
```

14.5 实现Stage模型的图片开发

应用开发中的图片开发是对图片像素数据进行解析、处理、构造的过程，达到目标图片效果，主要涉及图片解码、图片处理、图片编码等。

14.5.1 图片开发基本概念

在学习图片开发前，需要熟悉以下基本概念。

- 图片解码：指将所支持格式的存档图片解码成统一的PixelMap，以便在应用或系统中进行图片显示或图片处理。当前支持的存档图片格式包括JPEG、PNG、GIF、RAW、WebP、BMP、SVG。
- PixelMap：指图片解码后无压缩的位图，用于图片显示或图片处理。
- 图片处理：指对PixelMap进行相关的操作，如旋转、缩放、设置透明度、获取图片信息、读写像素数据等。
- 图片编码：指将PixelMap编码成不同格式的存档图片（当前仅支持JPEG和WebP），用于后续处理，如保存、传输等。

除上述基本图片开发能力外，HarmonyOS还提供常用的图片工具，供开发者选择使用。

14.5.2 图片开发主要流程

图片开发的主要流程如图14-6所示。

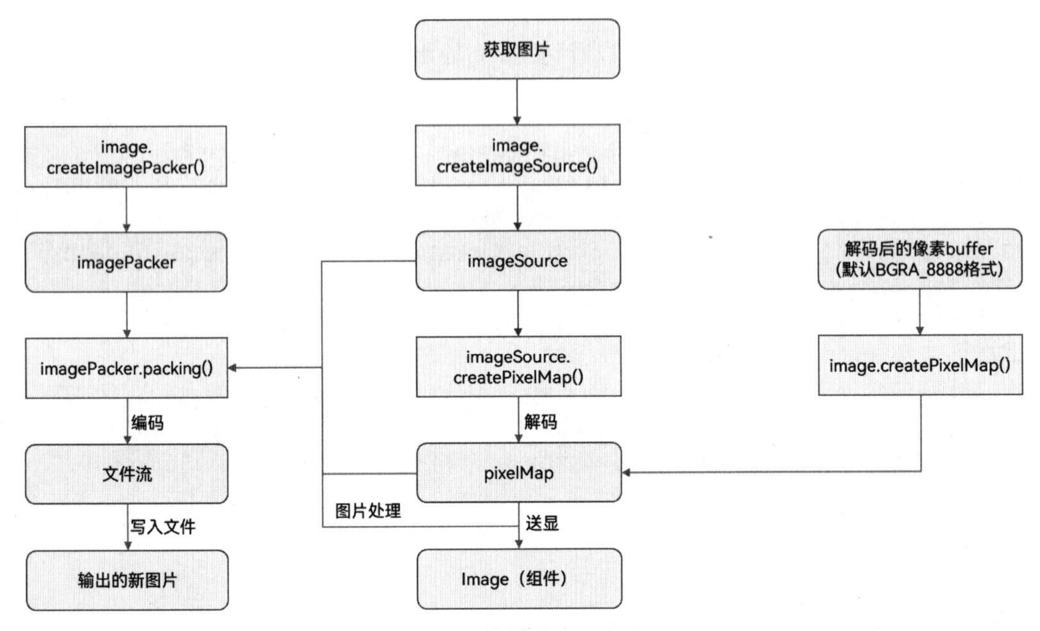

图 14-6　图片开发流程

- 获取图片：通过应用沙箱等方式获取原始图片。
- 创建ImageSource实例：ImageSource是图片解码出来的图片源类，用于获取或修改图片相关信息。
- 图片解码：通过ImageSource解码生成PixelMap。
- 图片处理：对PixelMap进行处理，更改图片属性实现图片的旋转、缩放、裁剪等效果，然后通过Image组件显示图片。
- 图片编码：使用图片打包器类ImagePacker，将PixelMap或ImageSource进行压缩编码，生成一张新的图片。

14.5.3 图片解码

图片解码指将所支持格式的存档图片解码成统一的PixelMap，以便在应用或系统中进行图片显示或图片处理。当前支持的存档图片格式包括JPEG、PNG、GIF、RAW、WebP、BMP、SVG。

图片解码的开发步骤如下。

第1步，全局导入Image模块。

```
import image from '@ohos.multimedia.image';
```

第2步，获取图片。方法有以下三种。

方法一：获取沙箱路径。具体请参考获取应用文件路径。应用沙箱的介绍及如何向应用沙箱推送文件，请参考文件管理。

```
// Stage 模型参考如下代码
const context = getContext(this);
const filePath = context.cacheDir + '/test.jpg';
```

方法二：通过沙箱路径获取图片的文件描述符。具体请参考file.fs API文档。 该方法需要先导入@ohos.file.fs模块。

```
import fs from '@ohos.file.fs';
```

然后调用fs.openSync()获取文件描述符。

```
// Stage 模型参考如下代码
const context = getContext(this);
const filePath = context.cacheDir + '/test.jpg';
const file = fs.openSync(filePath, fs.OpenMode.READ_WRITE);
const fd = file?.fd;
```

方法三：通过资源管理器获取资源文件的ArrayBuffer。具体请参考ResourceManager API文档。

```
// Stage 模型
const context = getContext(this);
// 获取 resourceManager 资源管理器
const resourceMgr = context.resourceManager;
```

不同模型获取资源管理器的方式不同，获取资源管理器后，再调用resourceMgr.getRawFileContent()获取资源文件的ArrayBuffer。

```
const fileData = await resourceMgr.getRawFileContent('test.jpg');
// 获取图片的 ArrayBuffer
const buffer = fileData.buffer;
```

第3步，创建ImageSource实例。方法有以下三种。

方法一：通过沙箱路径创建ImageSource。沙箱路径可以通过步骤2的方法一获取。

```
// path 为已获得的沙箱路径
const imageSource = image.createImageSource(filePath);
```

方法二：通过文件描述符fd创建ImageSource。文件描述符可以通过步骤2的方法二获取。

```
// fd 为已获得的文件描述符
const imageSource = image.createImageSource(fd);
```

方法三：通过缓冲区数组创建ImageSource。缓冲区数组可以通过步骤2的方法三获取。

```
const imageSource = image.createImageSource(buffer);
```

第4步，设置解码参数DecodingOptions，解码获取PixelMap图片对象。

```
let decodingOptions = {
    editable: true,
    desiredPixelFormat: 3,
}
// 创建 pixelMap 并进行简单的旋转和缩放
const pixelMap = await imageSource.createPixelMap(decodingOptions);
```

解码完成，获取到PixelMap对象后，可以进行后续图片处理。

14.5.4 图像变换

图像变换的开发步骤如下。

第1步，完成图片解码，获取PixelMap对象。

第2步，获取图片信息。

```
// 获取图片大小
pixelMap.getImageInfo().then( info => {
  console.info('info.width = ' + info.size.width);
  console.info('info.height = ' + info.size.height);
}).catch((err) => {
  console.error("Failed to obtain the image pixel map information.And the
error is: " + err);
});
```

第3步，进行图像变换操作。

图14-7展示的是图片的原图。

通过以下步骤对图片进行裁剪：

```
// x: 裁剪起始点横坐标 0
// y: 裁剪起始点纵坐标 0
// height: 裁剪高度 400，方向为从上往下（裁剪后的图片高度为 400）
// width: 裁剪宽度 400，方向为从左到右（裁剪后的图片宽度为 400）
pixelMap.crop({ x: 0, y: 0, size: { height: 400, width: 400 } });
```

图片裁剪后的效果如图14-8所示。

图 14-7　图片的原图

图 14-8　图片裁剪后的效果

通过以下步骤对图片进行缩放：

```
// 宽为原来的 0.5
// 高为原来的 0.5
pixelMap.scale(0.5, 0.5);
```

图片缩放后的效果如图 14-9 所示。

通过以下步骤对图片进行偏移：

```
// 向下偏移 100
// 向右偏移 100
pixelMap.translate(100, 100);
```

图片偏移后的效果如图 14-10 所示。

通过以下步骤对图片进行旋转：

```
// 顺时针旋转 90°
pixelMap.rotate(90);
```

图片旋转后的效果如图 14-11 所示。

图 14-9　图片缩放后的效果

图 14-10　图片偏移后的效果

图 14-11　图片旋转后的效果

通过以下步骤对图片进行垂直翻转：

```
// 垂直翻转
pixelMap.flip(false, true);
```

图片翻转后的效果如图 14-12 所示。

通过以下步骤对图片进行水平翻转：

```
// 水平翻转
pixelMap.flip(true, false);
```

图片翻转后的效果如图 14-13 所示。

通过以下步骤对图片进行透明度设置：

```
// 透明度 0.5
pixelMap.opacity(0.5);
```

图片设置透明度后的效果如图 14-14 所示。

图 14-12　图片垂直翻转后的效果

图 14-13　图片水平翻转后的效果

图 14-14　图片设置透明度后的效果

14.5.5　位图操作

当需要对目标图片中的部分区域进行处理时，可以使用位图操作功能。此功能常用于图片美化等操作。

如图 14-15 所示，一张图片中，将指定的矩形区域像素数据读取出来进行修改后，再写回原图片对应区域。

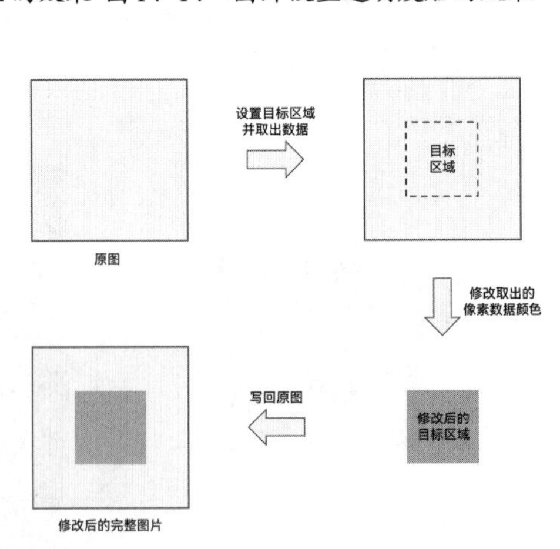

图 14-15　位图操作示意图

位图操作的开发步骤如下。

第 1 步，完成图片解码，获取 PixelMap 位图对象。

第 2 步，从 PixelMap 位图对象中获取信息。

```
// 获取图像像素的总字节数
let pixelBytesNumber = pixelMap.getPixelBytesNumber();
// 获取图像像素每行字节数
let rowCount = pixelMap.getBytesNumberPerRow();
// 获取当前图像像素密度。像素密度是指每英寸图片所拥有的像素数量。像素密度越大，图片越精细
let getDensity = pixelMap.getDensity();
```

第3步，读取并修改目标区域像素数据，写回原图。

```
// 场景一：将读取的整张图像像素数据结果写入 ArrayBuffer 中
const readBuffer = new ArrayBuffer(pixelBytesNumber);
pixelMap.readPixelsToBuffer(readBuffer).then(() => {
  console.info('Succeeded in reading image pixel data.');
}).catch(error => {
  console.error('Failed to read image pixel data. And the error is: ' +
error);
})

// 场景二：读取指定区域内的图片数据，结果写入 area.pixels 中
const area = {
  pixels: new ArrayBuffer(8),
  offset: 0,
  stride: 8,
  region: { size: { height: 1, width: 2 }, x: 0, y: 0 }
}
pixelMap.readPixels(area).then(() => {
  console.info('Succeeded in reading the image data in the area.');
}).catch(error => {
  console.error('Failed to read the image data in the area. And the error is:
' + error);
})

// 对于读取的图片数据，可以独立使用（创建新的 pixelMap），也可以对 area.pixels 进行修改
// 将图片数据 area.pixels 写入指定区域内
pixelMap.writePixels(area).then(() => {
  console.info('Succeeded to write pixelMap into the specified area.');
})

// 将图片数据结果写入 pixelMap 中
const writeColor = new ArrayBuffer(96);
pixelMap.writeBufferToPixels(writeColor, () => {});
```

14.5.6 图片编码

图片编码的开发步骤如下。

第1步，创建图像编码ImagePacker对象。

```
// 导入相关模块包
import image from '@ohos.multimedia.image';

const imagePackerApi = image.createImagePacker();
```

第2步，设置编码输出流和编码参数。format为图像的编码格式；quality为图像质量，范围为0～100，100为最佳质量。

```
let packOpts = { format:"image/jpeg", quality:98 };
```

第3步，创建PixelMap对象或创建ImageSource对象。

第4步，进行图片编码，并保存编码后的图片。有两种方法进行图片编码。

方法一：通过PixelMap进行编码。

```
imagePackerApi.packing(pixelMap, packOpts).then( data => {
  // data 为打包获取到的文件流，写入文件保存即可得到一张图片
}).catch(error => {
  console.error('Failed to pack the image. And the error is: ' + error);
})
```

方法二：通过imageSource进行编码。

```
imagePackerApi.packing(imageSource, packOpts).then( data => {
    // data 为打包获取到的文件流，写入文件保存即可得到一张图片
}).catch(error => {
  console.error('Failed to pack the image. And the error is: ' + error);
})
```

14.5.7 图片工具

图片工具当前主要提供图片EXIF信息的读取与编辑能力。EXIF（Exchangeable image file format）是专门为数码相机的照片设定的文件格式，可以记录数码照片的属性信息和拍摄数据，当前仅支持JPEG格式图片。在图库等应用中，需要查看或修改数码照片的EXIF信息。由于摄像机的手动镜头的参数无法自动写入EXIF信息中，或因为相机断电等原因经常会导致拍摄时间出错，这时候就需要手动修改错误的EXIF数据，即可使用本功能。

HarmonyOS目前仅支持对部分EXIF信息的查看和修改。

EXIF信息的读取与编辑的开发步骤如下。

第1步，获取图片，创建图片源ImageSource。

```
// 导入相关模块包
import image from '@ohos.multimedia.image';
```

```
// 获取沙箱路径创建 ImageSource
const fd = ...; // 获取需要被处理的图片的 fd
const imageSource = image.createImageSource(fd);
```

第2步，读取、编辑EXIF信息。

```
// 读取 EXIF 信息，BitsPerSample 为每个像素比特数
imageSource.getImageProperty('BitsPerSample', (error, data) => {
  if (error) {
    console.error('Failed to get the value of the specified attribute key of
                the image.And the error is: ' + error);
  } else {
    console.info('Succeeded in getting the value of the specified attribute
               key of the image ' + data);
  }
})

// 编辑 EXIF 信息
imageSource.modifyImageProperty('ImageWidth', '120').then(() => {
  const width = imageSource.getImageProperty("ImageWidth");
  console.info('The new imageWidth is ' + width);
})
```

第15章 网络管理

HarmonyOS 提供了网络管理模块以支持多场景的网络管理。

15.1 网络管理概述

HarmonyOS提供了网络管理模块以支持多场景的网络管理。

15.1.1 支持的场景

HarmonyOS网络管理模块主要提供以下功能。

- 数据连接管理：网卡绑定，打开URL，数据链路参数查询。
- 数据网络管理：指定数据网络传输，获取数据网络状态变更，数据网络状态查询。
- 流量统计：获取蜂窝网络、所有网卡、指定应用或指定网卡的数据流量统计值。
- HTTP缓存：有效管理HTTP缓存，减少数据流量。
- 创建本地套接字：实现本机不同进程间的通信，目前只支持流式套接字。

15.1.2 约束与限制

使用网络管理模块的相关功能时，需要请求相应的权限。

- ohos.permission.GET_NETWORK_INFO：获取网络连接信息。
- ohos.permission.SET_NETWORK_INFO：修改网络连接状态。
- ohos.permission.INTERNET：允许程序打开网络套接字，进行网络连接。

另外，请求网络的操作不应该放在主线程中，需要另外新启一个线程进行操作。

15.2 实战：使用当前网络打开一个URL链接

本节演示如何打开一个URL链接。

15.2.1 接口说明

应用使用当前网络打开一个URL链接，主要涉及NetManager和NetHandle两个类。

1. NetManager

NetManager的主要接口有以下几种。

- getInstance(Context context)：获取网络管理的实例对象。
- hasDefaultNet()：查询当前是否有默认可用的数据网络。

- getDefaultNet()：获取当前默认的数据网络句柄。
- addDefaultNetStatusCallback(NetStatusCallback callback)：获取当前默认的数据网络状态变化。
- setAppNet(NetHandle netHandle)：应用绑定该数据网络。

2. NetHandle

NetHandle 的主要接口如下。

- openConnection(URL url, Proxy proxy)：使用该网络打开一个 URL 链接。

15.2.2 创建应用

为了演示打开一个 URL 链接的功能，创建一个名为“NetManagerHandleURL”的应用。在应用的界面上，通过点击按钮，来触发打开一个 URL 链接的操作。

15.2.3 声明权限

修改配置文件，声明使用网络的权限如下：

```
// 声明权限
"reqPermissions": [
    {
    "name": "ohos.permission.GET_NETWORK_INFO"
    },
    {
    "name": "ohos.permission.SET_NETWORK_INFO"
    },
    {
    "name": "ohos.permission.INTERNET"
    }
]
```

由于上述权限不是敏感权限，因此，不需要在代码中进行显式声明。

15.2.4 修改 ability_main.xml

修改 ability_main.xml 内容如下：

```
<?xml version="1.0" encoding="utf-8"?>
<DirectionalLayout
    xmlns:ohos="http://schemas.huawei.com/res/ohos"
    ohos:height="match_parent"
    ohos:width="match_parent"
    ohos:orientation="vertical">
```

```
    <Button
        ohos:id="$+id:button_open"
        ohos:height="match_content"
        ohos:width="match_parent"
        ohos:background_element="#F76543"
        ohos:layout_alignment="horizontal_center"
        ohos:margin="10vp"
        ohos:padding="10vp"
        ohos:text="Open"
        ohos:text_size="40vp"
        />

    <Image
        ohos:id="$+id:image"
        ohos:height="match_content"
        ohos:width="match_parent"/>

</DirectionalLayout>
```

界面预览效果如图 15-1 所示。

图 15-1　界面预览效果

上述代码解释如下。

- 设置了 “Open” 按钮，以备设置点击事件，以触发打开链接的相关操作。
- Image组件，用于展示读取图片信息。

15.2.5　修改 MainAbilitySlice

修改 MainAbilitySlice 内容如下：

```
package com.waylau.hmos.netmanagerhandleurl.slice;

import com.waylau.hmos.netmanagerhandleurl.ResourceTable;
import ohos.aafwk.ability.AbilitySlice;
import ohos.aafwk.content.Intent;
import ohos.agp.components.Button;
import ohos.agp.components.Image;
import ohos.app.dispatcher.TaskDispatcher;
import ohos.app.dispatcher.task.TaskPriority;
import ohos.hiviewdfx.HiLog;
import ohos.hiviewdfx.HiLogLabel;
import ohos.media.image.ImageSource;
import ohos.media.image.PixelMap;
import ohos.media.image.common.PixelFormat;
```

```
import ohos.net.*;

import java.io.InputStream;
import java.net.*;

public class MainAbilitySlice extends AbilitySlice {
    private static final String TAG = MainAbilitySlice.class.getSimpleName();
    private static final HiLogLabel LABEL_LOG =
            new HiLogLabel(HiLog.LOG_APP, 0x00001, TAG);

    private Image image;

    @Override
    public void onStart(Intent intent) {
        super.onStart(intent);
        super.setUIContent(ResourceTable.Layout_ability_main);

        // 为按钮设置点击事件回调
        Button buttonOpen =
                (Button) findComponentById(ResourceTable.Id_button_open);
        buttonOpen.setClickedListener(listener -> open());

        image =
                (Image) findComponentById(ResourceTable.Id_image);
    }

    private void open() {
        HiLog.info(LABEL_LOG, "before open");

        // 启动线程任务
        getUITaskDispatcher().syncDispatch(() -> {
            NetManager netManager = NetManager.getInstance(getContext());

            if (!netManager.hasDefaultNet()) {
                return;
            }
            NetHandle netHandle = netManager.getDefaultNet();

            // 可以获取网络状态的变化
            netManager.addDefaultNetStatusCallback(callback);

            // 通过 openConnection 来获取 URLConnection
            HttpURLConnection connection = null;
            try {
                String urlString = "https://waylau.com/images/waylau_181_181.jpg";
                URL url = new URL(urlString);
```

```
            URLConnection urlConnection = netHandle.openConnection(url,
                    java.net.Proxy.NO_PROXY);
            if (urlConnection instanceof HttpURLConnection) {
                connection = (HttpURLConnection) urlConnection;
            }
            connection.setRequestMethod("GET");
            connection.setReadTimeout(10000);
            connection.setConnectTimeout(10000);
            connection.connect();

            // 之后可进行 url 的其他操作
            int code = connection.getResponseCode();
            HiLog.info(LABEL_LOG, "ResponseCode: %{public}s", code);

            if (code == HttpURLConnection.HTTP_OK) {
                // 得到服务器返回过来的图片流对象，在界面显示出来
                InputStream inputStream = urlConnection.getInputStream();
                ImageSource imageSource = ImageSource.create(inputStream,
                    new ImageSource.SourceOptions());
                ImageSource.DecodingOptions decodingOptions = new
                    ImageSource.DecodingOptions();
                decodingOptions.desiredPixelFormat = PixelFormat.
                    ARGB_8888;
                PixelMap pixelMap = imageSource.createPixelmap(decodingOp
                    tions);

                // 使用 UI 线程来更新 UI
                image.setPixelMap(pixelMap);
                pixelMap.release();
            }
        } catch (Exception e) {
            e.printStackTrace();
        } finally {
            connection.disconnect();
            HiLog.info(LABEL_LOG, "connection disconnect");
        }
    });

    HiLog.info(LABEL_LOG, "end open");
}

private NetStatusCallback callback = new NetStatusCallback() {
    public void onAvailable(NetHandle handle) {
        HiLog.info(LABEL_LOG, "onAvailable");
    }
```

```
        public void onBlockedStatusChanged(NetHandle handle, boolean blocked) {
            HiLog.info(LABEL_LOG, "onBlockedStatusChanged");
        }

        public void onLosing(NetHandle handle, long maxMsToLive) {
            HiLog.info(LABEL_LOG, "onLosing");
        }

        public void onLost(NetHandle handle) {
            HiLog.info(LABEL_LOG, "onLosing");
        }

        public void onUnavailable() {
            HiLog.info(LABEL_LOG, "onUnavailable");
        }

        public void onCapabilitiesChanged(NetHandle handle, NetCapabilities
                                          networkCapabilities) {
            HiLog.info(LABEL_LOG, "onCapabilitiesChanged");
        }

        public void onConnectionPropertiesChanged(NetHandle handle,
            ConnectionProperties connectionProperties) {
            HiLog.info(LABEL_LOG, "onConnectionPropertiesChanged");
        }
    };

    @Override
    public void onActive() {
        super.onActive();
    }

    @Override
    public void onForeground(Intent intent) {
        super.onForeground(intent);
    }
}
```

上述代码解释如下。

- 在 Button 上设置了点击事件。
- open 方法用于执行打开链接的操作。
- 调用 NetManager.getInstance(Context) 获取网络管理的实例对象。
- 调用 NetManager.getDefaultNet() 获取默认的数据网络。

- 调用NetHandle.openConnection()打开一个URL。
- 通过URL链接实例访问网站。
- 得到服务器返回过来的图片流对象在界面显示出来。
- 网络操作比较耗时，为了避免阻塞主线程，需要将Socket的操作都放置到独立的线程任务里面去执行。

15.2.6 运行

运行应用后，点击界面按钮“Open”以触发操作的执行。此时，控制台输出内容如下：

```
04-27 10:29:35.982 10071-10071/com.waylau.hmos.netmanagerhandleurl I 00001/
MainAbilitySlice: before open
04-27 10:29:36.004 10071-12307/com.waylau.hmos.netmanagerhandleurl I 00001/
MainAbilitySlice: onAvailable
04-27 10:29:36.004 10071-12307/com.waylau.hmos.netmanagerhandleurl I 00001/
MainAbilitySlice: onCapabilitiesChanged
04-27 10:29:36.004 10071-12307/com.waylau.hmos.netmanagerhandleurl I 00001/
MainAbilitySlice: onConnectionPropertiesChanged
04-27 10:29:38.210 10071-12306/com.waylau.hmos.netmanagerhandleurl I 00001/
MainAbilitySlice: ResponseCode: 200
04-27 10:29:38.372 10071-12306/com.waylau.hmos.netmanagerhandleurl I 00001/
MainAbilitySlice: connection disconnect
04-27 10:29:38.372 10071-10071/com.waylau.hmos.netmanagerhandleurl I 00001/
MainAbilitySlice: end open
```

界面效果如图15-2所示。

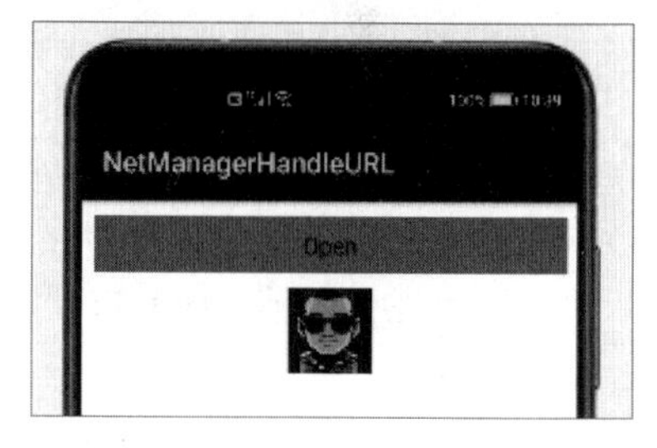

图15-2　界面效果

15.3 实战：使用当前网络进行Socket数据传输

本节演示如何实现Socket数据传输。

15.3.1 接口说明

应用使用当前网络进行Socket数据传输，主要涉及NetManager和NetHandle两个类。

1. NetManager

NetManager的主要接口如下。

- getByName(String host)：解析主机名，获取其IP地址。
- bindSocket(Socket socket)：绑定Socket到该数据网络。

2. NetHandle

NetHandle的主要接口如下。

- bindSocket(DatagramSocket socket)：绑定DatagramSocket到该数据网络。

15.3.2 创建应用

为了演示进行Socket数据传输的功能，创建一个名为“NetManagerHandleSocket”的应用。在应用的界面上，通过点击按钮，来触发进行Socket数据传输的操作。

15.3.3 声明权限

修改配置文件，声明使用网络的权限如下：

```
// 声明权限
"reqPermissions": [
    {
    "name": "ohos.permission.GET_NETWORK_INFO"
    },
    {
    "name": "ohos.permission.SET_NETWORK_INFO"
    },
    {
    "name": "ohos.permission.INTERNET"
    }
]
```

由于上述权限不是敏感权限，因此，不需要在代码中进行显式声明。

15.3.4 修改ability_main.xml

修改ability_main.xml内容如下：

```
<?xml version="1.0" encoding="utf-8"?>
<DirectionalLayout
    xmlns:ohos="http://schemas.huawei.com/res/ohos"
    ohos:height="match_parent"
    ohos:width="match_parent"
    ohos:orientation="vertical">

    <Button
```

```
        ohos:id="$+id:button_start"
        ohos:height="match_content"
        ohos:width="match_parent"
        ohos:background_element="#F76543"
        ohos:layout_alignment="horizontal_center"
        ohos:margin="10vp"
        ohos:padding="10vp"
        ohos:text="Start Server"
        ohos:text_size="30fp"
        />

    <Button
        ohos:id="$+id:button_open"
        ohos:height="match_content"
        ohos:width="match_parent"
        ohos:background_element="#F76543"
        ohos:layout_alignment="horizontal_center"
        ohos:margin="10vp"
        ohos:padding="10vp"
        ohos:text="Open"
        ohos:text_size="30fp"
        />

</DirectionalLayout>
```

界面预览效果如图15-3所示。

上述代码，设置了“Start Server”“Open”按钮，以备设置点击事件，以触发Socket数据传输的相关操作。

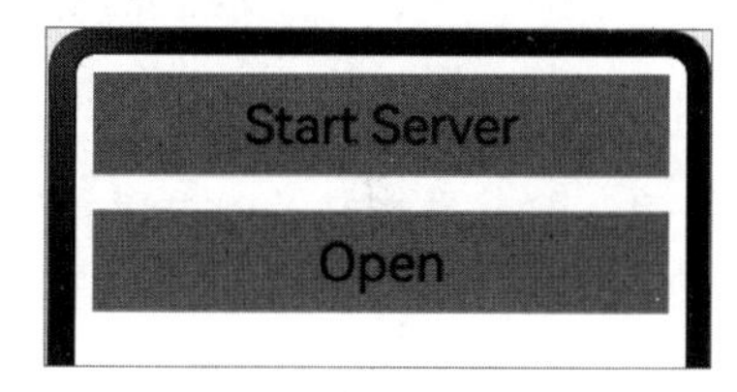

图15-3　界面预览效果

15.3.5　修改MainAbilitySlice

修改MainAbilitySlice内容如下：

```
package com.waylau.hmos.netmanagerhandlesocket.slice;

import com.waylau.hmos.netmanagerhandlesocket.ResourceTable;
import ohos.aafwk.ability.AbilitySlice;
import ohos.aafwk.content.Intent;
import ohos.agp.components.Button;
import ohos.agp.components.Text;
import ohos.app.dispatcher.TaskDispatcher;
import ohos.app.dispatcher.task.TaskPriority;
import ohos.hiviewdfx.HiLog;
import ohos.hiviewdfx.HiLogLabel;
import ohos.net.*;
```

```
import java.net.*;

public class MainAbilitySlice extends AbilitySlice {
    private static final String TAG = MainAbilitySlice.class.getSimpleName();
    private static final HiLogLabel LABEL_LOG =
            new HiLogLabel(HiLog.LOG_APP, 0x00001, TAG);

    private final static String HOST = "127.0.0.1";
    private final static int PORT = 8551;

    private TaskDispatcher dispatcher;

    @Override
    public void onStart(Intent intent) {
        super.onStart(intent);
        super.setUIContent(ResourceTable.Layout_ability_main);

        // 为按钮设置点击事件回调
        Button buttonStart =
                (Button) findComponentById(ResourceTable.Id_button_start);
        buttonStart.setClickedListener(listener -> initServer());

        Button buttonOpen =
                (Button) findComponentById(ResourceTable.Id_button_open);
        buttonOpen.setClickedListener(listener -> open());

        dispatcher = getGlobalTaskDispatcher(TaskPriority.DEFAULT);
    }

    private void open() {
        HiLog.info(LABEL_LOG, "before open");

        // 启动线程任务
        dispatcher.syncDispatch(() -> {
            NetManager netManager = NetManager.getInstance(null);

            if (!netManager.hasDefaultNet()) {
                HiLog.error(LABEL_LOG, "netManager.hasDefaultNet() failed");
                return;
            }

            NetHandle netHandle = netManager.getDefaultNet();

            // 通过 Socket 绑定来进行数据传输
            DatagramSocket socket = null;
```

```
            try {
                // 绑定到 Socket
                InetAddress address = netHandle.getByName(HOST);
                socket = new DatagramSocket();
                netHandle.bindSocket(socket);

                // 发送数据
                String data = "Welcome to waylau.com";
                DatagramPacket request = new DatagramPacket(data.
                    getBytes("utf-8"), data.length(), address, PORT);
                socket.send(request);

                // 显示到界面
                HiLog.info(LABEL_LOG, "send data: " + data);
            } catch (Exception e) {
                HiLog.error(LABEL_LOG, "send IOException: " + e.toString());
            } finally {
                if (null != socket) {
                    socket.close();
                }
            }
        });

        HiLog.info(LABEL_LOG, "end open");
    }

    private void initServer() {
        HiLog.info(LABEL_LOG, "before initServer");

        // 启动线程任务
        dispatcher.asyncDispatch(() -> {
            NetManager netManager = NetManager.getInstance(null);

            if (!netManager.hasDefaultNet()) {
                HiLog.error(LABEL_LOG, "netManager.hasDefaultNet() failed");
                return;
            }

            NetHandle netHandle = netManager.getDefaultNet();

            // 通过 Socket 绑定来进行数据传输
            DatagramSocket socket = null;
            try {
                // 绑定到 Socket
                InetAddress address = netHandle.getByName(HOST);
                socket = new DatagramSocket(PORT, address);
```

```
            netHandle.bindSocket(socket);
            HiLog.info(LABEL_LOG, "wait for receive data");

            // 接收数据
            byte[] buffer = new byte[1015];
            DatagramPacket response = new DatagramPacket(buffer,
                                                         buffer.length);
            socket.receive(response);
            int len = response.getLength();
            String data = new String(buffer, "utf-8").substring(0, len);

            // 显示到界面
            HiLog.info(LABEL_LOG, "receive data: " + data);
        } catch (Exception e) {
            HiLog.error(LABEL_LOG, "send IOException: " + e.toString());
        } finally {
            if (null != socket) {
                socket.close();
            }
        }
    });

    HiLog.info(LABEL_LOG, "end initServer");
}

private NetStatusCallback callback = new NetStatusCallback() {
    public void onAvailable(NetHandle handle) {
        HiLog.info(LABEL_LOG, "onAvailable");
    }

    public void onBlockedStatusChanged(NetHandle handle, boolean blocked) {
        HiLog.info(LABEL_LOG, "onBlockedStatusChanged");
    }

    public void onLosing(NetHandle handle, long maxMsToLive) {
        HiLog.info(LABEL_LOG, "onLosing");
    }

    public void onLost(NetHandle handle) {
        HiLog.info(LABEL_LOG, "onLosing");
    }

    public void onUnavailable() {
        HiLog.info(LABEL_LOG, "onUnavailable");
    }
```

```
        public void onCapabilitiesChanged(NetHandle handle, NetCapabilities
            networkCapabilities) {
            HiLog.info(LABEL_LOG, "onCapabilitiesChanged");
        }

        public void onConnectionPropertiesChanged(NetHandle handle,
            ConnectionProperties connectionProperties) {
            HiLog.info(LABEL_LOG, "onConnectionPropertiesChanged");
        }
    };

    @Override
    public void onActive() {
        super.onActive();
    }

    @Override
    public void onForeground(Intent intent) {
        super.onForeground(intent);
    }
}
```

上述代码解释如下。

- onStart方法中初始化了任务分发器TaskDispatcher。
- 在Button上设置了点击事件。
- initServer方法用于启动Socket服务端。
- open方法用于发送Socket数据。
- 网络操作比较耗时，为了避免阻塞主线程，需要将Socket的操作都放置到独立的线程任务里面去执行。

15.3.6 运行

运行应用后，点击界面按钮“Start Server”以触发启动Socket服务端的操作执行。此时，控制台输出内容如下：

```
04-27 12:52:32.488 8310-8310/com.waylau.hmos.netmanagerhandlesocket I 00001/
MainAbilitySlice: before initServer
04-27 12:52:32.489 8310-8310/com.waylau.hmos.netmanagerhandlesocket I 00001/
MainAbilitySlice: end initServer
04-27 12:52:32.495 8310-11157/com.waylau.hmos.netmanagerhandlesocket I 00001/
MainAbilitySlice: wait for receive data
```

点击界面按钮“Open”以触发发送Socket数据的操作执行。此时，控制台输出内容如下：

```
04-27 12:52:37.827 8310-8310/com.waylau.hmos.netmanagerhandlesocket I 00001/
MainAbilitySlice: before open
04-27 12:52:37.833 8310-11677/com.waylau.hmos.netmanagerhandlesocket I 00001/
MainAbilitySlice: send data: Welcome to waylau.com
04-27 12:52:37.837 8310-8310/com.waylau.hmos.netmanagerhandlesocket I 00001/
MainAbilitySlice: end open
04-27 12:52:37.837 8310-11157/com.waylau.hmos.netmanagerhandlesocket I 00001/
MainAbilitySlice: receive data: Welcome to waylau.com
```

15.4 实战：流量统计

应用通过调用API接口，可以获取蜂窝网络、所有网卡、指定应用或指定网卡的数据流量统计值。本节演示如何实现数据流量统计。

15.4.1 接口说明

应用进行流量统计，所使用的接口主要由DataFlowStatistics提供。DataFlowStatistics的主要接口有以下几种。

- getCellularRxBytes()：获取蜂窝数据网络的下行流量。
- getCellularTxBytes()：获取蜂窝数据网络的上行流量。
- getAllRxBytes()：获取所有网卡的下行流量。
- getAllTxBytes()：获取所有网卡的上行流量。
- getUidRxBytes(int uid)：获取指定UID的下行流量。
- getUidTxBytes(int uid)：获取指定UID的上行流量。
- getIfaceRxBytes(String nic)：获取指定网卡的下行流量。
- getIfaceTxBytes(String nic)：获取指定网卡的上行流量。

15.4.2 创建应用

为了演示数据流量统计的功能，创建一个名为“DataFlowStatistics”的应用。

在应用的界面上，通过点击按钮，来触发数据流量统计的操作。

15.4.3 声明权限

修改配置文件，声明使用网络的权限如下：

```
// 声明权限
"reqPermissions": [
    {
    "name": "ohos.permission.GET_NETWORK_INFO"
    },
    {
    "name": "ohos.permission.SET_NETWORK_INFO"
    },
    {
    "name": "ohos.permission.INTERNET"
    }
]
```

由于上述权限不是敏感权限，因此，不需要在代码中进行显式声明。

15.4.4 修改ability_main.xml

修改ability_main.xml内容如下：

```
<?xml version="1.0" encoding="utf-8"?>
<DirectionalLayout
    xmlns:ohos="http://schemas.huawei.com/res/ohos"
    ohos:height="match_parent"
    ohos:width="match_parent"
    ohos:alignment="center"
    ohos:orientation="vertical">

    <Button
        ohos:id="$+id:button_open"
        ohos:height="match_content"
        ohos:width="match_parent"
        ohos:background_element="#F76543"
        ohos:layout_alignment="horizontal_center"
        ohos:margin="10vp"
        ohos:padding="10vp"
        ohos:text="Open"
        ohos:text_size="40vp"
        />

</DirectionalLayout>
```

界面预览效果如图15-4所示。

Open

图15-4 界面预览效果

上述代码解释如下。

- 设置了“Open”按钮，以备设置点击事件，以触发打开链接的相关操作。

- Image组件，用于展示读取图片信息。

15.4.5 修改MainAbilitySlice

修改MainAbilitySlice内容如下：

```
package com.waylau.hmos.dataflowstatistics.slice;


import com.waylau.hmos.dataflowstatistics.ResourceTable;
import ohos.aafwk.ability.AbilitySlice;
import ohos.aafwk.content.Intent;
import ohos.agp.components.Button;
import ohos.agp.components.Image;
import ohos.app.dispatcher.TaskDispatcher;
import ohos.app.dispatcher.task.TaskPriority;
import ohos.hiviewdfx.HiLog;
import ohos.hiviewdfx.HiLogLabel;
import ohos.media.image.ImageSource;
import ohos.media.image.PixelMap;
import ohos.media.image.common.PixelFormat;
import ohos.net.*;

import java.io.InputStream;
import java.net.*;
import java.util.concurrent.atomic.AtomicLong;

public class MainAbilitySlice extends AbilitySlice {
    private static final String TAG = MainAbilitySlice.class.getSimpleName();
    private static final HiLogLabel LABEL_LOG =
            new HiLogLabel(HiLog.LOG_APP, 0x00001, TAG);

    private TaskDispatcher dispatcher;
    private Image image;

    @Override
    public void onStart(Intent intent) {
        super.onStart(intent);
        super.setUIContent(ResourceTable.Layout_ability_main);

        // 为按钮设置点击事件回调
        Button buttonOpen =
                (Button) findComponentById(ResourceTable.Id_button_open);
        buttonOpen.setClickedListener(listener -> open());
```

```
        dispatcher = getGlobalTaskDispatcher(TaskPriority.DEFAULT);
    }

    private void open() {
        HiLog.info(LABEL_LOG, "before open");

        // 获取所有网卡的下行流量
        AtomicLong rx = new AtomicLong(DataFlowStatistics.getAllRxBytes());

        // 获取所有网卡的上行流量
        AtomicLong tx = new AtomicLong(DataFlowStatistics.getAllTxBytes());

        // 启动线程任务
        dispatcher.syncDispatch(() -> {
            NetManager netManager = NetManager.getInstance(getContext());

            if (!netManager.hasDefaultNet()) {
                return;
            }
            NetHandle netHandle = netManager.getDefaultNet();

            // 可以获取网络状态的变化
            netManager.addDefaultNetStatusCallback(callback);

            // 通过 openConnection 来获取 URLConnection
            HttpURLConnection connection = null;
            try {
                String urlString = "https://waylau.com/images/waylau_181_181.jpg";
                URL url = new URL(urlString);

                URLConnection urlConnection = netHandle.openConnection(url,
                        java.net.Proxy.NO_PROXY);
                if (urlConnection instanceof HttpURLConnection) {
                    connection = (HttpURLConnection) urlConnection;
                }
                connection.setRequestMethod("GET");
                connection.setReadTimeout(10000);
                connection.setConnectTimeout(10000);
                connection.connect();

                // 之后可进行 url 的其他操作
                int code = connection.getResponseCode();
                HiLog.info(LABEL_LOG, "ResponseCode: %{public}s", code);
            } catch (Exception e) {
                e.printStackTrace();
            } finally {
```

```
                connection.disconnect();

                // 获取所有网卡的下行流量
                rx.set(DataFlowStatistics.getAllRxBytes() - rx.get());

                // 获取所有网卡的上行流量
                tx.set(DataFlowStatistics.getAllTxBytes() - tx.get());

                HiLog.info(LABEL_LOG, "connection disconnect, rx: %{public}s,
                           tx: %{public}s",
                        rx.get(), tx.get());
            }
        });

        HiLog.info(LABEL_LOG, "end open");
    }

    private NetStatusCallback callback = new NetStatusCallback() {
        public void onAvailable(NetHandle handle) {
            HiLog.info(LABEL_LOG, "onAvailable");
        }

        public void onBlockedStatusChanged(NetHandle handle, boolean blocked) {
            HiLog.info(LABEL_LOG, "onBlockedStatusChanged");
        }

        public void onLosing(NetHandle handle, long maxMsToLive) {
            HiLog.info(LABEL_LOG, "onLosing");
        }

        public void onLost(NetHandle handle) {
            HiLog.info(LABEL_LOG, "onLosing");
        }

        public void onUnavailable() {
            HiLog.info(LABEL_LOG, "onUnavailable");
        }

        public void onCapabilitiesChanged(NetHandle handle, NetCapabilities
                                          networkCapabilities) {
            HiLog.info(LABEL_LOG, "onCapabilitiesChanged");
        }

        public void onConnectionPropertiesChanged(NetHandle handle,
            ConnectionProperties connectionProperties) {
            HiLog.info(LABEL_LOG, "onConnectionPropertiesChanged");
```

```
        }
    };

    @Override
    public void onActive() {
        super.onActive();
    }

    @Override
    public void onForeground(Intent intent) {
        super.onForeground(intent);
    }
}
```

上述代码，基本与“NetManagerHandleURL”应用的代码一致，只是多了获取所有网卡的上行和下行流量的逻辑。

15.4.6 运行

运行应用后，点击界面按钮“Open”以触发操作的执行。此时，控制台输出内容如下：

```
04-27 15:47:34.545 9634-9634/com.waylau.hmos.dataflowstatistics I 00001/
MainAbilitySlice: before open
04-27 15:47:34.600 9634-10327/com.waylau.hmos.dataflowstatistics I 00001/
MainAbilitySlice: onAvailable
04-27 15:47:34.600 9634-10327/com.waylau.hmos.dataflowstatistics I 00001/
MainAbilitySlice: onCapabilitiesChanged
04-27 15:47:34.600 9634-10327/com.waylau.hmos.dataflowstatistics I 00001/
MainAbilitySlice: onConnectionPropertiesChanged
04-27 15:47:36.404 9634-10325/com.waylau.hmos.dataflowstatistics I 00001/
MainAbilitySlice: ResponseCode: 200
04-27 15:47:36.418 9634-10325/com.waylau.hmos.dataflowstatistics I 00001/
MainAbilitySlice: connection disconnect, rx: 1144951, tx: 1123654
04-27 15:47:36.419 9634-9634/com.waylau.hmos.dataflowstatistics I 00001/
MainAbilitySlice: end open
```

可以看到，已经输出流量的信息了。

15.5 实战：在Stage模型中通过HTTP请求数据

本节演示如何在Stage模型中通过HTTP来向Web服务请求数据。为了演示该功能，创建一个名为“ArkTSHttp”的应用。在应用的界面上，通过点击按钮，来触发HTTP请求数据的操作。

15.5.1 准备一个HTTP服务接口

HTTP服务接口地址为：https://waylau.com/data/people.json。通过调用该接口，可以返回如下JSON格式的数据：

```
[{"name": "Michael"},
{"name": "Andy Huang","age": 25,"homePage": "https://waylau.com/books"},
{"name": "Justin","age": 19},
{"name": "Way Lau","age": 35,"homePage": "https://waylau.com"}]
```

15.5.2 添加使用Button组件来触发点击

在初始化的Text组件的下方增加一个Button组件，以实现“请求”按钮。代码如下：

```
build() {
Row() {
    Column() {
    Text(this.message)
        .fontSize(38)
        .fontWeight(FontWeight.Bold)

    // 请求
    Button(('请求'), { type: ButtonType.Capsule })
        .width(140)
        .fontSize(40)
        .fontWeight(FontWeight.Medium)
        .margin({ top: 20, bottom: 20 })
        .onClick(() => {
            this.httpReq()
        })
    }
    .width('100%')
}
.height('100%')
}
```

当触发onClick事件时，会执行httpReq方法。

15.5.3 发起HTTP请求

httpReq方法实现如下：

```
// 导入http模块。
import http from '@ohos.net.http';
```

```
// 为节约篇幅，省略部分代码

private httpReq() {
    // 创建 httpRequest 对象
    let httpRequest = http.createHttp();

    let url = "https://waylau.com/data/people.json";

    // 发起 HTTP 请求
    let promise = httpRequest.request(
        // 请求 url 地址
        url,
        {
        // 请求方式
        method: http.RequestMethod.GET,
        // 可选，默认为 60s
        connectTimeout: 60000,
        // 可选，默认为 60s
        readTimeout: 60000,
        // 开发者根据自身业务需要添加 header 字段
        header: {
            'Content-Type': 'application/json'
        }
    });

    // 处理响应结果
    promise.then((data) => {
        if (data.responseCode === http.ResponseCode.OK) {
            console.info('Result:' + data.result);
            console.info('code:' + data.responseCode);
            this.message = JSON.stringify(data.result);
        }
    }).catch((err) => {
        console.info('error:' + JSON.stringify(err));
    });
}
```

上述代码演示了发起HTTP请求的基本流程。

- 导入http模块。
- 创建httpRequest对象。需要注意的是，每一个httpRequest对象对应一个http请求任务，不可复用。
- 通过httpRequest对象发起HTTP请求。

- 处理HTTP请求返回的结果，赋值给message变量。
- 界面重新渲染显示了最新的message变量值。

15.5.4 声明权限

需要在module.json5文件中声明使用网络的权限“ohos.permission.INTERNET”，示例如下：

```
{
  "module": {

    // 为节约篇幅，省略部分代码
    "requestPermissions": [
      {
        "name": "ohos.permission.INTERNET"
      }
    ]
}
```

15.5.5 运行

运行应用，显示的界面效果如图15-5所示。

点击“请求”按钮后发起HTTP请求，返回的结果显示在了界面上，效果如图15-6所示。

图 15-5　运行应用显示的界面效果

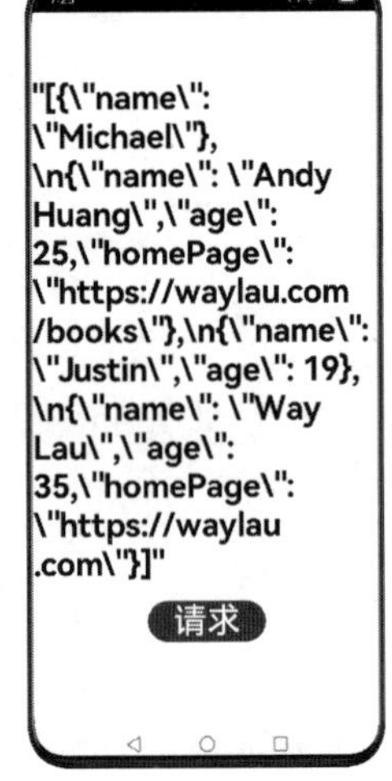

图 15-6　发起HTTP请求效果

15.6 Web组件概述

ArkUI为我们提供了Web组件来加载网页，借助它我们就相当于在自己的应用程序里嵌入一个浏览器，从而非常轻松地展示各种各样的网页。

15.6.1 加载本地网页

使用Web组件来加载本地网页非常简单，只需要创建一个Web组件，并传入两个参数就可以了。其中src指定引用的网页路径，controller为组件的控制器，通过controller绑定Web组件，用于实现对Web组件的控制。

比如，在main/resources/rawfile目录下有一个HTML文件index.html，那么可以通过$rawfile引用本地网页资源。示例代码如下：

```
// 导入 webview
import  webview from '@ohos.web.webview'

@Entry
@Component
struct SecondPage {
  // 创建 WebviewController
  controller: webview.WebviewController = new webview.WebviewController()

  build() {
    Column() {
      Web({ src: $rawfile('index.html'), controller: this.controller })
    }
  }
}
```

15.6.2 加载在线网页

下面的示例是使用Web组件来加载在线网页的代码：

```
// 导入 webview
import  webview from '@ohos.web.webview'

@Entry
@Component
struct WebComponent {
  // 创建 WebviewController
  controller: webview.WebviewController = new webview.WebviewController()
  build() {
    Column() {
      Web({ src: 'https://waylau.com/', controller: this.controller })
    }
  }
}
```

访问在线网页时还需要在module.json5文件中声明网络访问权限：ohos.permission.INTERNET。

15.6.3 加载沙箱路径下的本地资源文件

通过globalthis获取沙箱路径，代码如下：

```
// 导入 webview
import web_webview from '@ohos.web.webview'
let url = 'file://' + globalThis.filesDir + '/xxx.html'
```

```
@Entry
@Component
struct WebComponent {
  // 创建 WebviewController
  controller: web_webview.WebviewController = new web_webview.
WebviewController()
  build() {
    Column() {
      // 加载沙箱路径文件
      Web({ src: url, controller: this.controller })
    }
  }
}
```

15.6.4 网页缩放

有的网页可能不能很好地适配手机屏幕，需要对其缩放才能有更好的效果，开发者可以根据需要给Web组件设置zoomAccess属性，zoomAccess用于设置是否支持手势进行缩放，默认允许执行缩放。Web组件默认支持手势进行缩放。代码如下：

```
Web({ src:'https://waylau.com/', controller:this.controller })
    .zoomAccess(true)
```

还可以使用zoom(factor: number)方法设置网站的缩放比例。其中factor表示缩放倍数，下面示例，当点击一次按钮时，页面放大为原来的1.5倍。代码如下：

```
// 导入 webview
import  webview from '@ohos.web.webview'

@Entry
@Component
struct WebComponent {
  // 创建 WebviewController
  controller: webview.WebviewController = new webview.WebviewController()
  factor: number = 1.5;

  build() {
    Column() {
      Button('zoom')
        .onClick(() => {
          this.controller.zoom(this.factor);
        })
      Web({ src: 'https://waylau.com/', controller: this.controller })
```

```
        }
    }
}
```

15.6.5 文本缩放

如果需要对文本进行缩放，可以使用textZoomAtio(textZoomAtio: number)方法。其中，textZoomAtio用于设置页面的文本缩放百分比，默认值为100，表示100%。以下示例代码将文本放大为原来的1.5倍：

```
Web({ src:'https://waylau.com/', controller:this.controller })
    .textZoomAtio(150)
```

15.6.6 Web组件事件

Web组件还提供了处理Javascript的对话框、网页加载进度及各种通知与请求事件的方法。例如，onProgressChange可以监听网页的加载进度，onPageEnd在网页加载完成时触发该回调，且只在主frame触发，onConfirm则在网页触发confirm警告弹窗时触发回调。

15.6.7 Web和JavaScript交互

在开发专为适配Web组件的网页时，可以实现Web组件和JavaScript代码之间的交互。Web组件可以调用JavaScript方法，JavaScript也可以调用Web组件里面的方法。

1. Web组件调用JavaScript方法

如果希望加载的网页在Web组件中运行JavaScript，则必须为Web组件启用JavaScript功能，默认情况下是允许JavaScript执行的。

可以在Web组件onPageEnd事件中添加runJavaScript方法。事件是网页加载完成时的回调，runJavaScript方法可以执行HTML中的JavaScript脚本。

示例如下：

```
// 导入webview
import  webview from '@ohos.web.webview'

@Entry
@Component
struct WebComponent {
  // 创建WebviewController
  controller: webview.WebviewController = new webview.WebviewController()
```

```
  @State webResult: string = ''
  build() {
    Column() {
      Text(this.webResult).fontSize(20)
      Web({ src: $rawfile('index.html'), controller: this.controller })
      .javaScriptAccess(true)
      .onPageEnd(e => {
        this.controller.runJavaScript({
          script: 'test()',
          callback: (result: string)=> {
            this.webResult = result
          }});
      })
    }
  }
}

<!-- index.html -->
<!DOCTYPE html>
<html>
  <meta charset="utf-8">
  <body>
  </body>
  <script type="text/javascript">
  function test() {
      return "This value is from index.html"
  }
  </script>
</html>
```

当页面加载完成时，触发onPageEnd事件，调用HTML文件中的test方法并将结果返回给Web组件。

2. JavaScript 调用 Web 组件方法

可以使用registerJavaScriptProxy将Web组件中的JavaScript对象注入window对象中，这样网页中的JavaScript就可以直接调用该对象了。需要注意的是，要想让registerJavaScriptProxy方法生效，须调用refresh方法。

下面的示例将ets文件中的对象testObj注入了window对象中：

```
// 导入 webview
import  webview from '@ohos.web.webview'

@Entry
@Component
struct WebComponent{
```

```
  @State dataFromHtml: string = ''
  // 创建 WebviewController
  controller: webview.WebviewController = new webview.WebviewController();

  testObj = {
    test: (data) => {
      this.dataFromHtml = data
      return "ArkUI Web Component";
    },
    toString: () => {
      console.log('Web Component toString');
    }
  }

  build() {
    Column() {
      Text(this.dataFromHtml).fontSize(20)
      Row() {
        Button('Register JavaScript To Window').onClick(() => {
          this.controller.registerJavaScriptProxy({
            object: this.testObj,
            name: "objName",
            methodList: ["test", "toString"],
          });
          this.controller.refresh();
        })
      }

      Web({ src: $rawfile('index.html'), controller: this.controller })
        .javaScriptAccess(true)
    }
  }
}
```

其中object表示参与注册的对象，name表示注册对象的名称为objName，与window中调用的对象名一致；methodList表示参与注册的应用侧JavaScript对象的方法，包含test、toString两个方法。在HTML中使用的时候，直接使用objName调用methodList里面对应的方法即可，示例如下：

```
// index.html
<!DOCTYPE html>
<html>
<meta charset="utf-8">
<body>
<button onclick="htmlTest()"> 调用 Web 组件里面的方法 </button>
</body>
```

```
<script type="text/javascript">
    function htmlTest() {
        str = objName.test("param from Html");
    }
</script>
</html>
```

还可以使用deleteJavaScriptRegister删除通过registerJavaScriptProxy注册到window上的指定name的应用侧JavaScript对象。

15.6.8 处理页面导航

使用浏览器浏览网页时，可以执行返回、前进、刷新等操作，Web组件同样支持这些操作。可以使用backward()返回上一个页面，使用forward()前进一个页面，也可以使用refresh()刷新页面，使用clearHistory()清除历史记录。

示例如下：

```
Button("前进").onClick(() => {
    this.controller.forward()
})
Button("后退").onClick(() => {
    this.controller.backward()
})
Button("刷新").onClick(() => {
    this.controller.refresh()
})
Button("停止").onClick(() => {
    this.controller.stop()
})
Button("清除历史").onClick(() => {
    this.controller.clearHistory()
})
```

15.7 实战：在Stage模型中通过Web组件加载在线网页

本节演示如何通过Web组件来加载在线网页。为了演示该功能，创建一个名为“ArkTSWebComponent”的应用。在应用的界面上，通过点击按钮，来触发加载网页的操作。

15.7.1 准备一个在线网页地址

在线网页地址为：https://waylau.com/。通过浏览器访问该地址，可以看到图 15-7 所示的网页。

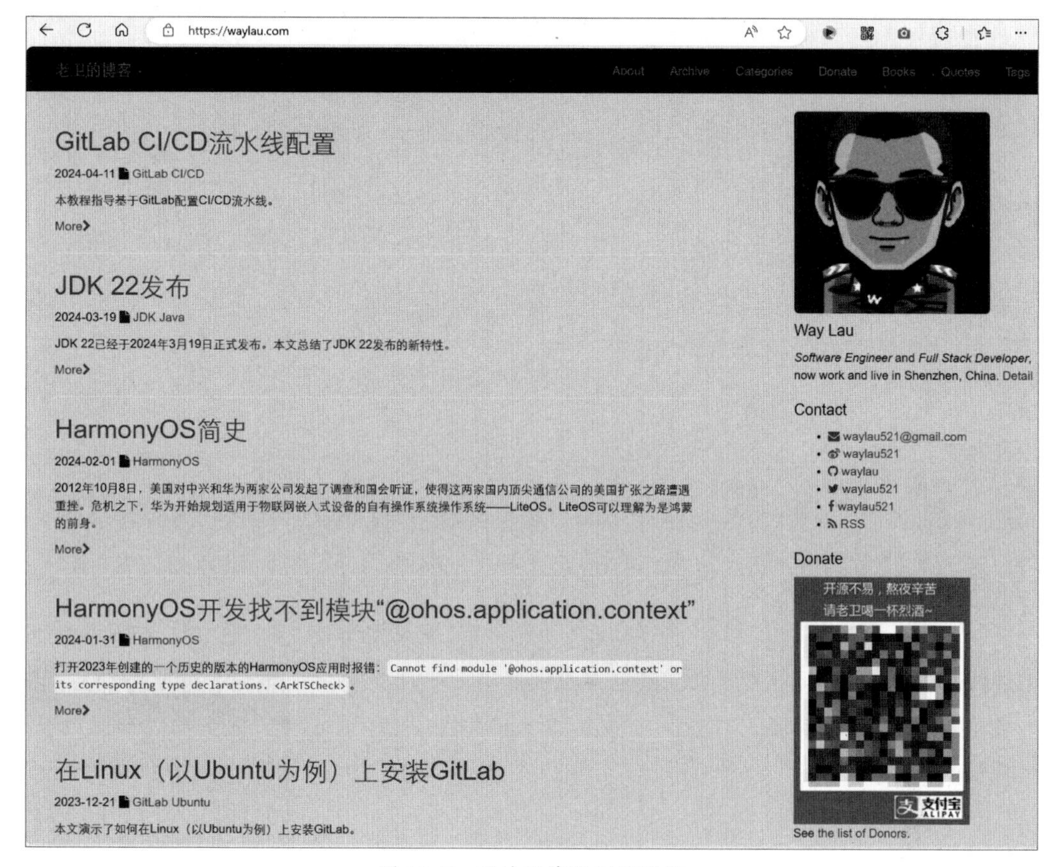

图 15-7 通过浏览器访问网页

15.7.2 声明网络访问权限

需要在 module.json5 文件中声明使用网络的权限 ohos.permission.INTERNET，示例如下：

```
{
  "module": {

    // 为节约篇幅，省略部分代码
    "requestPermissions": [
      {
        "name": "ohos.permission.INTERNET"
      }
    ]
}
```

15.7.3 发起HTTP请求

httpReq方法实现如下：

```
// 导入 webview
import  webview from '@ohos.web.webview'


@Entry
@Component
struct Index {

  // 创建 WebviewController
  controller: webview.WebviewController = new webview.WebviewController();

  build() {
    Column() {
      // 添加 Web 组件
      Web({ src: 'https://waylau.com/', controller: this.controller })
    }
  }
}
```

上述代码，演示了发起HTTP请求的基本流程：

- 创建WebviewController；
- 添加Web组件。

15.7.4 运行

运行应用，显示的界面效果如图15-8所示。Web组件将在线网页加载到应用里面了。

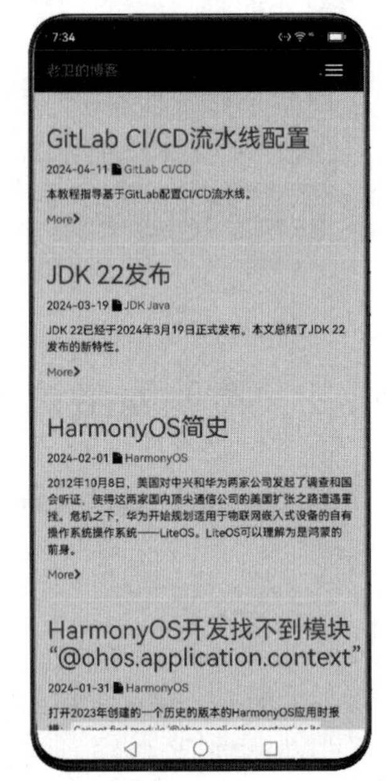

图15-8　运行应用显示的界面效果

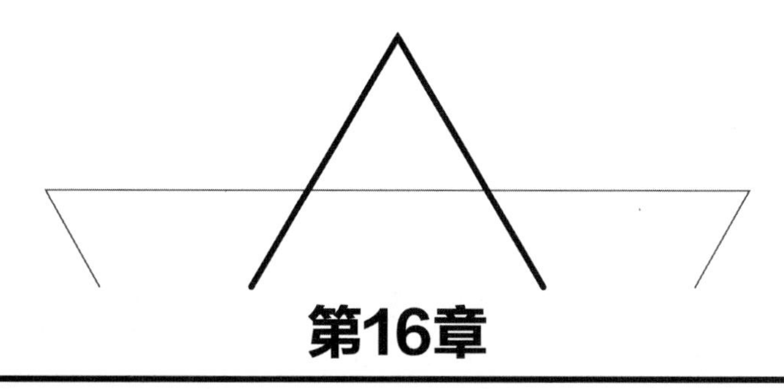

第16章

综合案例1：JS实现智能穿戴应用

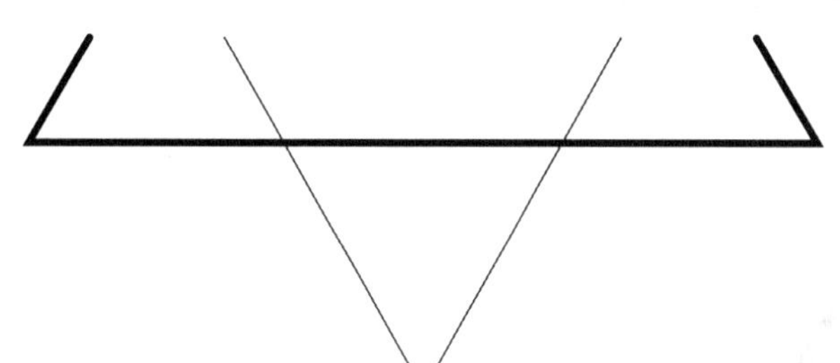

本章将演示一个完整的智能穿戴应用——数字华容道游戏。

16.1 案例概述

本章所演示的是一个完整的智能穿戴应用——数字华容道游戏。魔方、法国的钻石棋、中国的华容道，被誉为“智力游戏界的三个不可思议”。

16.1.1 传统华容道游戏

华容道游戏来源于三国故事。曹操兵败赤壁后中了诸葛亮设下的计谋，被迫落入华容道，最后被关羽放走。图 16-1 展示的是传统华容道游戏的实物图。

传统华容道游戏的游戏规则，是按照“曹瞒兵败走华容，正与关公狭路逢。只为当初恩义重，放开金锁走蛟龙”这一故事情节定下的：

- 通过移动各个棋子，帮助曹操从初始位置移到棋盘最下方中部，从出口逃走；
- 不允许跨越棋子，还要设法用最少的步数把曹操移到出口。

图 16-1　传统华容道游戏的实物图

16.1.2 数字华容道游戏

数字华容道玩法与传统华容道游戏类似，只是棋子改为了数字。玩家只需将数字按顺序排列即可，比较考验玩家的大脑和手速。图 16-2 展示的是数字华容道游戏的实物图。

数字华容道游戏的目的是用较少的步数，在最短时间内将棋盘上的数字棋子，按照从左到右，从上到下的顺序重新排列整齐。图 16-3 展示的是数字华容道游戏完成时的画面。

图 16-2　数字华容道游戏的实物图

图 16-3　数字华容道游戏完成时的画面

本章所要演示开发的是数字华容道游戏，将实现以下效果：

- 数字华容道游戏界面比较简单，看起来很清爽；
- 操作简单；
- 可以设置不同的等级；
- 能够在手表上运行该游戏。

16.2 代码实现

本节演示如何在Lite Wearable设备上实现数字华容道游戏。

为了演示数字华容道游戏的功能，创建一个名为“KlotskiJs”的Lite Wearable设备类型的应用。

16.2.1 技术重点

要实现数字华容道游戏，涉及以下技术重点。

- 数字华容道游戏布局；
- 画布的使用；
- stack的使用；
- 定时器的使用。

16.2.2 整体布局

数字华容道游戏的布局代码定义在index.hml中。代码如下：

```
<div class="container">
    <stack class="stack">
        <!--canvas 为游戏主界面 -->
        <canvas class="canvas" ref="canvas" onswipe="swipeGrids"></canvas>
        <!--tip-container 为提示界面 -->
        <div class="tip-container" show="{{isShowTip}}">
            <text class="success-text">
                Success!
            </text>
            <text class="cost-time">
                Cost: {{timeSeconds}}
            </text>
        </div>
    </stack>
    <input type="button" value="Restart" class="button" onclick="restart"/>
</div>
```

在上述代码中，主要分为两个界面：游戏主界面和提示界面。这两个界面放置到了stack布局里面。默认情况下，只显示游戏主界面，而提示界面是隐藏的。

提示界面又分为两部分，上面是“Success!”，下面用于展示耗时情况。

16.2.3 整体样式

数字华容道游戏的样式代码定义在index.css中。代码如下：

```
.container {
    flex-direction: column;
    justify-content: center;
    align-items: center;
    width: 450px;
    height: 450px;
}

.stack {
    width: 305px;
    height: 305px;
    margin-top: 40px;
}

.canvas {
    width: 305px;
    height: 305px;
    background-color: #FFFFFF;
}

.tip-container {
    width: 305px;
    height: 305px;
    flex-direction: column;
    justify-content: center;
    align-items: center;
    background-color: #008000;
}

.success-text {
    font-size: 38px;
    color: black;
}

.cost-time {
    font-size: 30px;
```

```
    text-align: center;
    width: 300px;
    height: 40px;
    margin-top: 10px;
}

.button {
    width: 150px;
    height: 40px;
    background-color: blue;
    font-size: 30px;
    margin-top: 10px;
}
```

提示界面效果如图16-4所示。

图16-4 提示界面效果

16.2.4 游戏核心逻辑

游戏核心逻辑定义在index.js中。

1. 变量和常量

变量和常量的定义代码如下：

```
var grids; // 网格
var context; // canvas上下文
var timer; // 定时器
const SIDELEN = 70; // 网格边长
const MARGIN = 5; // 网格间距
const ORIGINAL_GRIDS = [[1, 2, 3, 4],
                        [5, 6, 7, 8],
                        [9, 10, 11, 12],
                        [13, 14, 15, 0]];
const DIRECTIONS = ["left", "up", "right", "down"]; // 方向
const ORIGINAL_TIME = '0.0'; // 初始时间
```

各定义如注释所述，这里不再赘述。其中，ORIGINAL_GRIDS是一个二维数组，用来定义游戏的网格内的数字。

2. 生命周期

Lite Wearable应用生命周期主要有两个：应用创建时调用的onCreate和应用销毁时触发的onDestroy。这两个生命周期方法定义在app.js中，一般不需要更改。app.js代码如下：

```
export default {
    onCreate() {
        console.info("Application onCreate");
```

```
    },
    onDestroy() {
        console.info("Application onDestroy");
    }
};
```

一个应用中可能会有多个页面，一个页面一般包括onInit、onReady、onShow和onDestroy等在页面创建、显示和销毁时会触发调用的事件。

- onInit：表示页面的数据已经准备好，可以使用js文件中的data数据。
- onReady：表示页面已经编译完成，可以将界面显示给用户。
- onShow：JS UI只支持应用同时运行并展示一个页面，当打开一个页面时，上一个页面就销毁了。当一个页面显示的时候，会调用onShow。
- onHide：页面消失时被调用。
- onDestroy：页面销毁时被调用。

在本示例中，我们主要用到如下生命周期。

- onInit方法初始化了grids、timer的数据。
- onReady获取到了canvas的上下文。
- onShow方法用于初始化网格、绘制网格、启动定时器。

代码如下：

```
export default {
    data: {
        timeSeconds: ORIGINAL_TIME,
        isShowTip: false
    },
    onInit() {
        grids =  JSON.parse(JSON.stringify(ORIGINAL_GRIDS)); // 深拷贝
        timer = null;
    },
    onReady() {
        context = this.$refs.canvas.getContext('2d');
    },
    onShow() {
        // 初始化网格
        this.initGrids();

        // 绘制网格
        this.drawGrids();

        // 启动定时器
        timer = setInterval(this.costTime, 100);
    },
```

```
    ...
};
```

需要注意的是，grids的数据来源于ORIGINAL_GRIDS，但不能直接通过“=”来赋值，如下：

```
grids =  ORIGINAL_GRIDS; // 浅拷贝
```

上述方式虽然在初始化时能够将grids的值赋上，但由于这种方式本质上是一种浅拷贝，因此，后续程序修改grids时，会同时修改ORIGINAL_GRIDS的值，这显然是不对的。因此，本案例采用的是一种深拷贝的方式。

3. 初始化网格

初始化网格initGrids()方法如下：

```
initGrids() {
    // 搞乱格子
    for (let i = 0; i < 100; i++) {
        let randomIndex = Math.floor(Math.random() * 4);
        let direction = DIRECTIONS[randomIndex];
        this.updateGrids(direction);
    }
},
updateGrids(direction) {
    let x;
    let y;
    for (let row = 0; row < 4; row++) {
        for (let column = 0; column < 4; column++) {
            if (grids[row][column] == 0) {
                x = row;
                y = column;
                break;
            }
        }
    }
    let temp;
    if (this.isShowTip == false) {
        if (direction == 'left' && (y + 1) < 4) {
            temp = grids[x][y + 1];
            grids[x][y + 1] = grids[x][y];
            grids[x][y] = temp;
        } else if (direction == 'right' && (y - 1) > -1) {
            temp = grids[x][y - 1];
            grids[x][y - 1] = grids[x][y];
            grids[x][y] = temp;
        } else if (direction == 'up' && (x + 1) < 4) {
            temp = grids[x + 1][y];
```

```
            grids[x + 1][y] = grids[x][y];
            grids[x][y] = temp;
        } else if (direction == 'down' && (x - 1) > -1) {
            temp = grids[x - 1][y];
            grids[x - 1][y] = grids[x][y];
            grids[x][y] = temp;
        }
    }
}
```

在上述代码中，grids在ORIGINAL_GRIDS数据的基础上做了重新编排，打乱了原来的数据位置。每次编码都会随机从DIRECTIONS数组中取一个方向值，而后按照该方向值移动格子的位置。

4. 绘制网格

绘制网格drawGrids()方法如下：

```
drawGrids() {
    for (let row = 0; row < 4; row++) {
        for (let column = 0; column < 4; column++) {
            let gridStr = grids[row][column].toString();

            context.fillStyle = "#BBB509";

            let leftTopX = column * (MARGIN + SIDELEN) + MARGIN;
            let leftTopY = row * (MARGIN + SIDELEN) + MARGIN;

            context.fillRect(leftTopX, leftTopY, SIDELEN, SIDELEN);

            // 非0网格的特殊处理
            if (gridStr != "0") {
                context.fillStyle = "#0000FF";
                context.font = "30px";

                let offsetX = (4 - gridStr.length) * (SIDELEN / 8);
                let offsetY = (SIDELEN - 30) / 2;

                context.fillText(gridStr, leftTopX + offsetX, leftTopY + offsetY);
            }
        }
    }
}
```

上述代码按照4行4列的方式通过canvas上下文绘制网格图像。每个网格的边长和间距都已经定义在了MARGIN和SIDELEN常量中。

针对非0网格需要做特殊处理。非0网格和0网格的区别在于，非0网格还需要显示网格数字。

5. 启动定时器

通过setInterval方法启动定时器，并定时执行costTime()方法。代码如下：

```
// 启动定时器
timer = setInterval(this.costTime, 100);
...

costTime() {
    this.timeSeconds = (Math.floor(parseFloat(this.timeSeconds) * 10 + 1) / 10);
    if (this.timeSeconds % 1 == 0) {
        this.timeSeconds = this.timeSeconds + ".0";
    }
}
```

在上述代码中，每隔0.1秒（100毫秒）执行一次costTime()方法。costTime()方法的本质是实现了0.1秒的递增。

6. 网格滑动事件处理

滑动网格会触发onswipe事件。事件处理定义在swipeGrids方法中。代码如下：

```
swipeGrids(event) {
    // 按滑动的方向更新网格
    this.updateGrids(event.direction);

    // 绘制网格

    if (this.isSuccess()) {
        clearInterval(timer);
        this.isShowTip = true;
    }
},
isSuccess() {
    ...
}
```

在上述代码中，按滑动的方向更新网格并重新绘制网格。同时，会根据isSuccess()方法判断游戏是否成功。如果是，则清理掉定时器，并将isShowTip置为true，以显示提示界面。

7. 游戏完成判断

根据isSuccess()方法判断游戏是否成功。isSuccess()方法实现如下：

```
isSuccess() {
    // 判断 grids 与 ORIGINAL_GRIDS 是否相等，相等则游戏完成
    for (let row = 0; row < 4; row++) {
        for (let column = 0; column < 4; column++) {
            if (grids[row][column] != ORIGINAL_GRIDS[row][column]) {
```

```
                    return false;
                }
            }
        }
        return true;
}
```

上述代码判断grids与ORIGINAL_GRIDS是否相等，相等则游戏完成。

8. 重新开始游戏

“Restart”按钮会触发重新开始游戏的restart()方法。restart()方法实现如下：

```
restart() {
    this.timeSeconds = ORIGINAL_TIME;
    this.isShowTip = false;

    this.onShow();
}
```

上述代码将timeSeconds和isShowTip置为初始值，而后再执行onShow()方法。

16.3 应用运行

初次启动应用，将会出现图16-5所示的游戏界面。

滑动网格以进行游戏。

当游戏完成，则会出现图16-6所示的提示界面。

图16-5 初次进入游戏的界面

图16-6 提示界面

此时，游戏完成并会统计出游戏耗时情况。可以选择点击“Restart”按钮重新开始。

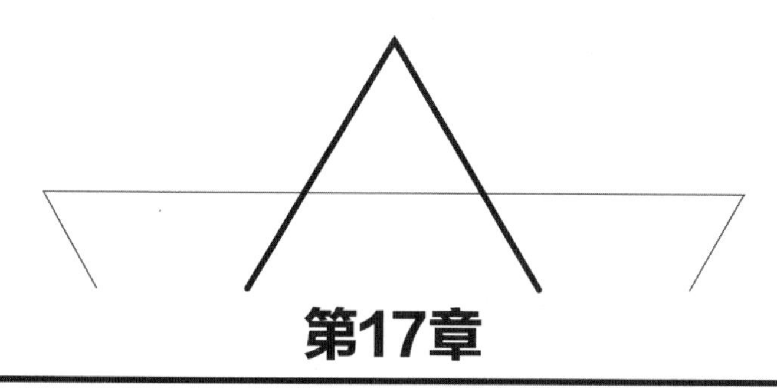

第17章

综合案例2：Java实现智慧屏应用

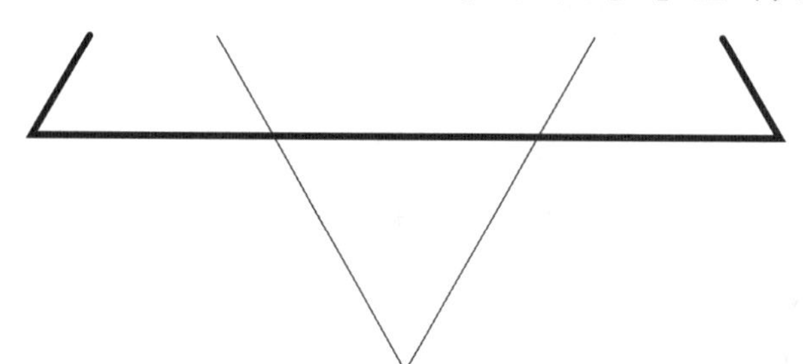

本章将演示一个完整的智慧屏应用——视频播放器。

17.1 案例概述

在“第13章 视频”一章，我们已经初步了解了视频处理的相关API。本章将会演示如何实现一款具有完整视频播放功能的智慧屏应用——视频播放器。

本章所要演示开发的视频播放器，将实现以下功能：

- 获取并展示视频列表；
- 从视频列表中选择视频进行播放；
- 展示播放进度；
- 全屏播放；
- 视频可以暂停；
- 支持播放上一个或下一个视频。

17.2 代码实现

本节演示如何在TV设备上实现视频播放器功能。

17.2.1 技术重点

要实现视频播放器功能，涉及以下技术重点。

- 视频播放器布局；
- 读取本地文件资源；
- Player类的使用；
- Slider组件的使用；
- 全屏显示；
- 任务分发器。

17.2.2 创建应用

为了演示视频播放器的功能，创建一个名为“VideoPlayer”的TV设备类型的应用。

为演示本应用，事先在rawfile目录下放置待播放的视频资源，如图17-1所示。

图 17-1　视频资源放置的位置

17.2.3 设置布局

修改ability_main.xml内容如下：

```
<?xml version="1.0" encoding="utf-8"?>
<DirectionalLayout
    xmlns:ohos="http://schemas.huawei.com/res/ohos"
    ohos:height="match_parent"
    ohos:width="match_parent"
    ohos:background_element="#000">

    <DirectionalLayout
        ohos:id="$+id:layout_player"
        ohos:height="400vp"
        ohos:width="match_parent"
        ohos:layout_alignment="top">
    </DirectionalLayout>

    <DirectionalLayout
        ohos:id="$+id:layout_menu"
        ohos:height="140vp"
        ohos:width="match_parent"
        ohos:layout_alignment="bottom"
        ohos:orientation="vertical">

        <Text
            ohos:id="$+id:text_current_name"
            ohos:height="24vp"
            ohos:width="match_parent"
            ohos:text="Big BuckBunny"
            ohos:text_color="#FFFFFF"
            ohos:text_size="24vp"/>

        <Slider
            ohos:id="$+id:player_progress"
            ohos:height="50vp"
            ohos:width="match_parent"
            ohos:max="100"
            ohos:min="0"
            ohos:padding="10vp"
            ohos:progress_element="#f00"/>

        <DirectionalLayout
            ohos:height="60vp"
            ohos:width="match_parent"
```

```
            ohos:orientation="horizontal"
            >

            <Button
                ohos:id="$+id:button_previous"
                ohos:height="40vp"
                ohos:width="0vp"
                ohos:background_element="#F76543"
                ohos:layout_alignment="horizontal_center"
                ohos:margin="10vp"
                ohos:padding="10vp"
                ohos:text="Previous"
                ohos:text_size="44"
                ohos:weight="1"
                />

            <Button
                ohos:id="$+id:button_play_pause"
                ohos:height="40vp"
                ohos:width="0vp"
                ohos:background_element="#F76543"
                ohos:layout_alignment="horizontal_center"
                ohos:margin="10vp"
                ohos:padding="10vp"
                ohos:text="Play"
                ohos:text_size="44"
                ohos:weight="1"
                />

            <Button
                ohos:id="$+id:button_next"
                ohos:height="40vp"
                ohos:width="0vp"
                ohos:background_element="#F76543"
                ohos:layout_alignment="horizontal_center"
                ohos:margin="10vp"
                ohos:padding="10vp"
                ohos:text="Next"
                ohos:text_size="44"
                ohos:weight="1"
                />
        </DirectionalLayout>
    </DirectionalLayout>
</DirectionalLayout>
```

界面预览效果如图 17-2 所示。

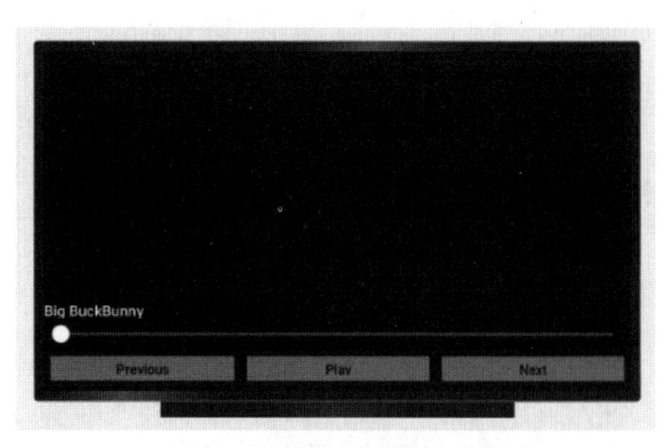

图17-2　界面预览效果

上述布局解释如下。

- 上方为播放区。
- 下方为菜单区。

菜单区又分为三部分：

- 最上面显示视频的名称；
- 中间是视频的播放进度条；
- 最下面是操作按钮。

17.2.4　设置全屏

运行应用，查看界面的实际效果，界面效果如图17-3所示。

图17-3　界面效果

可以发现，界面上方"VideoPlayer"字样的应用名称一栏，占据了非常大的空间，极其影响美观。因此，需要去掉这个应用名称一栏。

修改配置文件，在module属性中，增加metaData属性，修改如下：

```
"module": {
    "package": "com.waylau.hmos.videoplayer",
    "name": ".MyApplication",
    "deviceType": [
        "tv"
    ],
    "distro": {
        "deliveryWithInstall": true,
        "moduleName": "entry",
        "moduleType": "entry"
    },
    "abilities": [
        {
        "skills": [
            {
            "entities": [
                "entity.system.home"
            ],
            "actions": [
                "action.system.home"
            ]
            }
```

```
        ],
        "orientation": "landscape",
        "name": "com.waylau.hmos.videoplayer.MainAbility",
        "icon": "$media:icon",
        "description": "$string:mainability_description",
        "label": "VideoPlayer",
        "type": "page",
        "launchType": "standard"
        }
    ],
    // 增加 metaData 属性
    "metaData":{
        "customizeData":[
        {
            "name": "hwc-theme",
            "value": "androidhwext:style/Theme.Emui.NoTitleBar",
            "extra":""
        }
        ]
    }
}
```

再次运行应用，界面全屏效果如图 17-4 所示。

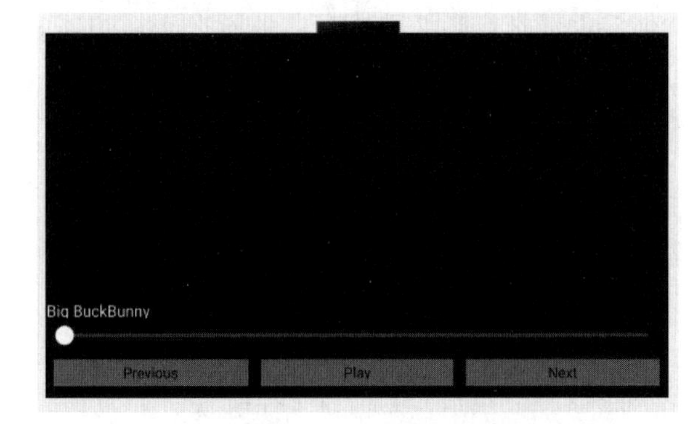

图 17-4　界面全屏效果

17.2.5　应用的主体逻辑

修改 MainAbilitySlice 内容，以实现视频播放器的主体逻辑。

代码如下：

```
package com.waylau.hmos.videoplayer.slice;

import com.waylau.hmos.videoplayer.Video;
import com.waylau.hmos.videoplayer.PlayerStateEnum;
import com.waylau.hmos.videoplayer.ResourceTable;
import ohos.aafwk.ability.AbilitySlice;
import ohos.aafwk.content.Intent;
import ohos.agp.components.*;
import ohos.agp.components.surfaceprovider.SurfaceProvider;
import ohos.agp.graphics.Surface;
import ohos.agp.graphics.SurfaceOps;
import ohos.app.dispatcher.TaskDispatcher;
```

```
import ohos.global.resource.RawFileDescriptor;

import ohos.hiviewdfx.HiLog;
import ohos.hiviewdfx.HiLogLabel;
import ohos.media.common.Source;
import ohos.media.player.Player;

import java.io.IOException;
import java.util.ArrayList;
import java.util.Arrays;
import java.util.List;

public class MainAbilitySlice extends AbilitySlice {
    private static final String TAG = MainAbilitySlice.class.getSimpleName();
    private static final HiLogLabel LABEL_LOG =
            new HiLogLabel(HiLog.LOG_APP, 0x00001, TAG);

    private Text currentName;       // 当前视频名称
    private int currentIndex = 0;   // 当前视频索引
    private Slider playerProgress;
    private List<Video> videoList; // 视频列表
    private SurfaceProvider surfaceProvider;
    private Player player;
    private DirectionalLayout playerLayout;
    private TaskDispatcher dispatcher;
    private Surface surface;
    private Button buttonQuery;
    private Button buttonPlayPause;
    private Button buttonNext;
    private PlayerStateEnum playerState = PlayerStateEnum.STOP;

    @Override
    public void onStart(Intent intent) {
        super.onStart(intent);
        super.setUIContent(ResourceTable.Layout_ability_main);

        // 添加点击事件来触发访问数据
        buttonQuery = (Button) findComponentById(ResourceTable.Id_button_previous);
        buttonQuery.setClickedListener(listener -> {
            try {
                this.doPrevious();
            } catch (IOException e) {
                e.printStackTrace();
            }
        });
```

```
    buttonPlayPause = (Button) findComponentById(ResourceTable.Id_button_
                                                    play_pause);
    buttonPlayPause.setClickedListener(listener -> {
        try {
            this.doPlayPause();
        } catch (IOException e) {
            e.printStackTrace();
        }
    });

    buttonNext = (Button) findComponentById(ResourceTable.Id_button_next);
    buttonNext.setClickedListener(listener -> {
        try {
            this.doNext();
        } catch (IOException e) {
            e.printStackTrace();
        }
    });

    currentName = (Text) findComponentById(ResourceTable.Id_text_current_name);

    // 跑马灯效果
    currentName.setTruncationMode(Text.TruncationMode.AUTO_SCROLLING);
    currentName.setAutoScrollingCount(Text.AUTO_SCROLLING_FOREVER);
    currentName.startAutoScrolling();

    playerProgress = (Slider) findComponentById(ResourceTable.Id_player_
                                                   progress);

    playerLayout
            = (DirectionalLayout) findComponentById(ResourceTable.Id_
                                                        layout_player);

    player = new Player(this);

    dispatcher = getUITaskDispatcher();

    // 初始化 SurfaceProvider
    initSurfaceProvider();

    // 初始化视频列表
    initData();
}

@Override
public void onStop() {
```

```
        super.onStop();

        // 关闭、释放资源
        release();
    }

    private void release() {
        HiLog.info(LABEL_LOG, "release");

        if (player != null) {
            player.stop();
            player.release();
            player = null;
        }
    }

    private void initSurfaceProvider() {
        HiLog.info(LABEL_LOG, "before initSurfaceProvider");

        surfaceProvider = new SurfaceProvider(this);
        surfaceProvider.getSurfaceOps().get().addCallback(
            new VideoSurfaceCallback());
        surfaceProvider.pinToZTop(true);
        surfaceProvider.setWidth(ComponentContainer.LayoutConfig.
            MATCH_CONTENT);
        surfaceProvider.setHeight(ComponentContainer.LayoutConfig.MATCH_PARENT);
        surfaceProvider.setTop(0);

        playerLayout.addComponent(surfaceProvider);

        HiLog.info(LABEL_LOG, "end initSurfaceProvider");
    }

    ...
```

上述代码解释如下。

- onStart方法初始化了Button、Player、SurfaceProvider、DirectionalLayout、Surface等组件和布局。
- initData方法用于初始化视频数据列表。
- onStop执行release方法用于关闭、释放资源。
- currentName用于显示视频的名称，并设置了跑马灯效果。

PlayerStateEnum是枚举类，用于表示播放器的状态。代码如下：

```
package com.waylau.hmos.videoplayer;
```

```
public enum PlayerStateEnum {
    PLAY,
    PAUSE,
    STOP
}
```

VideoSurfaceCallback是个回调类。代码如下：

```
private class VideoSurfaceCallback implements SurfaceOps.Callback {
    @Override
    public void surfaceCreated(SurfaceOps surfaceOps) {
        if (surfaceProvider.getSurfaceOps().isPresent()) {
            surface = surfaceProvider.getSurfaceOps().get().getSurface();
        }

        HiLog.info(LABEL_LOG, "surfaceCreated");
    }

    @Override
    public void surfaceChanged(SurfaceOps surfaceOps, int i, int i1, int i2) {
        HiLog.info(LABEL_LOG, "surfaceChanged, %{public}s, %{public}s,
        %{public}s",
                i, i1, i2);
    }

    @Override
    public void surfaceDestroyed(SurfaceOps surfaceOps) {
        HiLog.info(LABEL_LOG, "surfaceDestroyed");
    }
}
```

17.2.6 初始化视频数据

初始化视频数据的方法initData代码如下：

```
private void initData() {
    HiLog.info(LABEL_LOG, "before initData");

    Video video1 = new Video("Trailer", "trailer.mp4");
    Video video2 = new Video("Captain Marvel", "captain_marvel.mp4");
    Video video3 = new Video("Big BuckBunny", "big_buck_bunny.mp4");

    videoList = new ArrayList<>(
            Arrays.asList(
```

```
                    video1,
                    video2,
                    video3));

    int size = videoList.size();

    HiLog.info(LABEL_LOG, "end initData, size: %{public}s", size);
}
```

上述列表存储了待播放的视频的名称和文件的名称。

其中，Video类代码如下：

```
package com.waylau.hmos.videoplayer;

public class Video {
    private String name;
    private String filePath;
    public Video(String name, String filePath) {
        this.name = name;
        this.filePath = filePath;
    }
    public String getName() {
        return name;
    }
    public void setName(String name) {
        this.name = name;
    }
    public String getFilePath() {
        return filePath;
    }
    public void setFilePath(String filePath) {
        this.filePath = filePath;
    }
}
```

17.2.7 播放、暂停视频

播放、暂停视频的方法doPlayPause代码如下：

```
private void doPlayPause() throws IOException {
    Video video = videoList.get(currentIndex);
    currentName.setText(video.getName());
    if (playerState == PlayerStateEnum.PLAY) {
        HiLog.info(LABEL_LOG, "before pause");
        player.pause();
```

```
            playerState = PlayerStateEnum.PAUSE;
            buttonPlayPause.setText("Play");
            HiLog.info(LABEL_LOG, "end pause");
        } else if (playerState == PlayerStateEnum.PAUSE) {
            player.play();
            playerState = PlayerStateEnum.PLAY;
            buttonPlayPause.setText("Pause");
        } else {
            start(video.getFilePath());
            playerState = PlayerStateEnum.PLAY;
            buttonPlayPause.setText("Pause");
        }
    }
```

上述代码解释如下。

- 主体是通过playerState来判断当前的播放状态的。
- 如果当前是播放中的状态（PlayerStateEnum.PLAY），则执行player的pause()方法。
- 如果当前是暂停的状态（PlayerStateEnum.PAUSE），则执行player的play()方法。
- 其他状态则是通过start方法重新加载视频文件进行播放。

start方法如下：

```
private void start(String fileName) throws IOException {
    HiLog.info(LABEL_LOG, "before start: %{public}s", fileName);

    RawFileDescriptor filDescriptor = getResourceManager()
        .getRawFileEntry("resources/rawfile/" + fileName).openRawFileDescriptor();

    Source source = new Source(filDescriptor.getFileDescriptor(),
         filDescriptor.getStartPosition(), filDescriptor.getFileSize());

    player.setSource(source);
    player.setVideoSurface(surface);
    player.setPlayerCallback(new PlayerCallback());
    player.prepare();
    player.play();

    HiLog.info(LABEL_LOG, "end start");
}
```

上述start方法，根据指定视频名称来将视频源设置到player上。

其中，player.setPlayerCallback方法用来设置回调PlayerCallback。PlayerCallback代码如下：

```
private class PlayerCallback implements Player.IPlayerCallback {
    @Override
    public void onPrepared() {
```

```
        HiLog.info(LABEL_LOG, "onPrepared");
        // 延迟 1 秒，再刷新进度
        Runnable runnable = new Runnable() {
            @Override
            public void run() {
                HiLog.info(LABEL_LOG, "before playerProgress");
                int progressValue = player.getCurrentTime() * 100 / player.
                    getDuration();
                playerProgress.setProgressValue(progressValue);
                HiLog.info(LABEL_LOG, "playerProgress: %{public}s", progressValue);
                dispatcher.delayDispatch(this, 200);
            }
        };
        dispatcher.asyncDispatch(runnable);
    }
    @Override
    public void onMessage(int i, int i1) {
        HiLog.info(LABEL_LOG, "onMessage");
    }
    @Override
    public void onError(int i, int i1) {
        HiLog.info(LABEL_LOG, "onError, i: %{public}s, i1: %{public}s", i, i1);
    }
    @Override
    public void onResolutionChanged(int i, int i1) {
        HiLog.info(LABEL_LOG, "onResolutionChanged");
    }
    @Override
    public void onPlayBackComplete() {
        HiLog.info(LABEL_LOG, "onPlayBackComplete");
    }
    @Override
    public void onRewindToComplete() {
        HiLog.info(LABEL_LOG, "onRewindToComplete");
    }
    @Override
    public void onBufferingChange(int i) {
        HiLog.info(LABEL_LOG, "onBufferingChange");
    }
    @Override
    public void onNewTimedMetaData(Player.MediaTimedMetaData mediaTimedMetaData) {
        HiLog.info(LABEL_LOG, "onNewTimedMetaData");
    }
    @Override
    public void onMediaTimeIncontinuity(Player.MediaTimeInfo mediaTimeInfo) {
        HiLog.info(LABEL_LOG, "onMediaTimeIncontinuity");
```

```
        }
}
```

17.3 应用运行

初次启动应用，点击“Play”，将会出现如图 17-5 所示的播放器的界面。

点击“Pause”，将会暂停播放器的播放，出现如图 17-6 所示的播放器的界面。

图 17-5　播放器的播放界面

图 17-6　播放器的暂停界面

点击“Next”，将会播放下一个视频，出现如图 17-7 所示的播放器的界面。

点击“Previous”，将会播放上一个视频，出现如图 17-8 所示的播放器的界面。

图 17-7　播放下一个视频

图 17-8　播放上一个视频

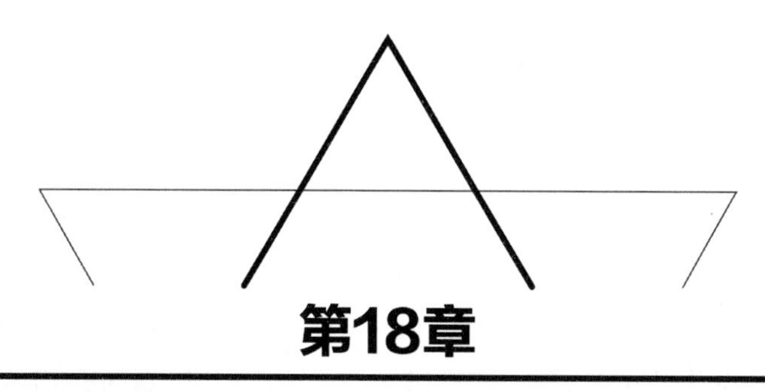

第18章

综合案例3：Java实现手机应用

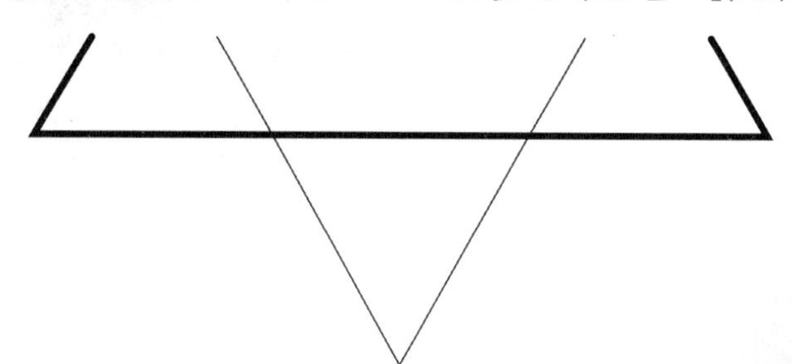

本章将演示如何创建一个完整的手机应用——俄罗斯方块游戏。

18.1 案例概述

本章所演示的是一个完整的手机应用——俄罗斯方块游戏。

18.1.1 俄罗斯方块游戏概述

俄罗斯方块（俄语为Тетрис，英语为Tetris）是一款家喻户晓的游戏，无论是在掌上游戏机，还是在计算机或是移动终端都能见到这款游戏的身影。

俄罗斯方块最初由阿列克谢·帕基特诺夫在苏联设计和编程，是一款益智类视频游戏。游戏发布于1984年6月6日，那时他在莫斯科科学计算机中心工作。游戏的名字源于希腊数字4的前缀tetra-（所有游戏中的米诺牌都被称为tetromino，包含4个块），以及网球tennis（帕基特诺夫喜爱的运动）。

俄罗斯方块游戏随处可见，几乎所有的视频游戏机和计算机操作系统，以及图形计算器、手机便携式媒体播放器、掌上电脑、网络音乐播放器，都可用于玩这个游戏，甚至还可以在非媒体产品，如示波器、建筑物上玩。图18-1所示的是某大楼所实现的俄罗斯方块游戏。

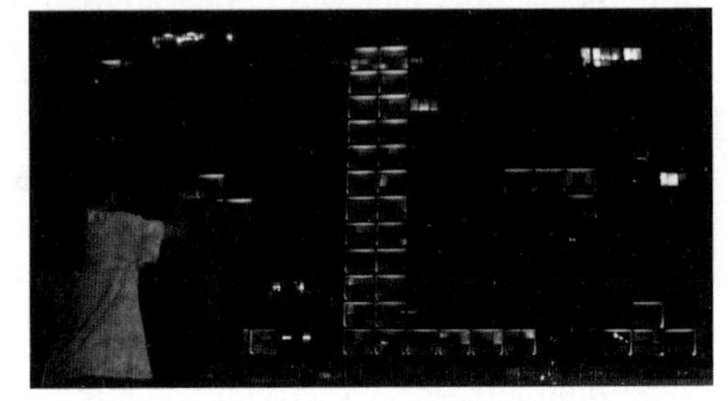

图18-1 在大楼上所实现的俄罗斯方块游戏

18.1.2 俄罗斯方块游戏规则

当俄罗斯方块游戏开启后，由小方格组成的不同形状的方块陆续从屏幕上方落下来，玩家通过调整方块的位置和方向，使它们在屏幕底部拼出完整的一条或几条。这些完整的横条会随即消失，给新落下来的方块腾出空间，与此同时，玩家得到分数奖励。没有被消除的方块不断堆积起来，一旦堆到屏幕顶端，玩家便告输，游戏结束。

本章所实现的俄罗斯方块游戏，玩家可以做如下操作：

- 向左移动方块；
- 向右移动方块；
- 以90度为单位旋转方块；
- 重启游戏。

18.2 代码实现

本节演示如何在手机上实现俄罗斯方块游戏功能。

为了演示俄罗斯方块游戏的功能，创建一个名为“Tetris”的Phone设备类型的应用。

18.2.1 技术重点

要实现俄罗斯方块游戏，涉及以下技术重点：

- 俄罗斯方块游戏布局；
- Canvas类的使用；
- 全屏显示；
- 任务分发器。

18.2.2 设置布局

修改ability_main.xml内容如下：

```
<?xml version="1.0" encoding="utf-8"?>
<DirectionalLayout
    xmlns:ohos="http://schemas.huawei.com/res/ohos"
    ohos:height="match_parent"
    ohos:width="match_parent"
    ohos:background_element="#000"
    ohos:orientation="vertical">

    <DirectionalLayout
        ohos:id="$+id:layout_tip"
        ohos:height="60vp"
        ohos:width="match_parent"
        ohos:layout_alignment="top"
        ohos:orientation="horizontal"
        ohos:margin="10vp"
        ohos:padding="10vp"
        >

        <Text
            ohos:id="$+id:text_score_label"
            ohos:height="match_parent"
            ohos:width="80vp"
            ohos:text_color="#FFFFFF"
            ohos:text_size="24vp"
            ohos:text="Score:"/>

        <Text
            ohos:id="$+id:text_score_value"
            ohos:height="match_parent"
            ohos:width="100vp"
            ohos:text_color="#FFFFFF"
```

```
            ohos:text_size="24vp"
            ohos:text="0"/>
        <Text
            ohos:id="$+id:text_gameover"
            ohos:height="match_parent"
            ohos:width="match_content"
            ohos:text_color="#FFFFFF"
            ohos:text_size="24vp"
            ohos:text="Start!"/>
    </DirectionalLayout>

    <DirectionalLayout
        ohos:id="$+id:layout_game"
        ohos:height="534vp"
        ohos:width="match_parent"
        ohos:orientation="vertical"
        ohos:background_element="#696969"
        >

    </DirectionalLayout>

    <DirectionalLayout
        ohos:id="$+id:layout_menu"
        ohos:height="60vp"
        ohos:width="match_parent"
        ohos:layout_alignment="bottom"
        ohos:orientation="horizontal"
        >

        <Button
            ohos:id="$+id:button_left"
            ohos:height="40vp"
            ohos:width="0vp"
            ohos:background_element="#87CEFA"
            ohos:layout_alignment="horizontal_center"
            ohos:margin="10vp"
            ohos:padding="10vp"
            ohos:text="←"
            ohos:text_size="44"
            ohos:weight="1"
            />

        <Button
            ohos:id="$+id:button_right"
            ohos:height="40vp"
            ohos:width="0vp"
```

```
            ohos:background_element="#87CEFA"
            ohos:layout_alignment="horizontal_center"
            ohos:margin="10vp"
            ohos:padding="10vp"
            ohos:text=" → "
            ohos:text_size="44"
            ohos:weight="1"
            />

        <Button
            ohos:id="$+id:button_shift"
            ohos:height="40vp"
            ohos:width="0vp"
            ohos:background_element="#87CEFA"
            ohos:layout_alignment="horizontal_center"
            ohos:margin="10vp"
            ohos:padding="10vp"
            ohos:text=" ∽ "
            ohos:text_size="44"
            ohos:weight="1"
            />
        <Button
            ohos:id="$+id:button_restart"
            ohos:height="40vp"
            ohos:width="0vp"
            ohos:background_element="#87CEFA"
            ohos:layout_alignment="horizontal_center"
            ohos:margin="10vp"
            ohos:padding="10vp"
            ohos:text="R"
            ohos:text_size="44"
            ohos:weight="1"
            />
    </DirectionalLayout>

</DirectionalLayout>
```

界面预览效果如图18-2所示。

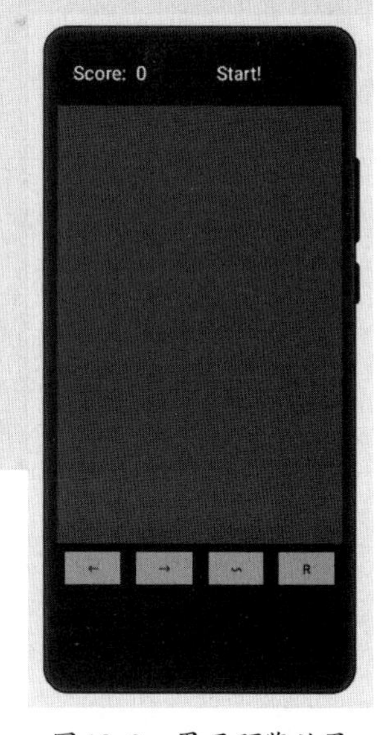

图18-2　界面预览效果

上述布局大体分为三部分：

- 上方为提示区，提示当前得分和游戏状态；
- 中部为游戏画面区；
- 下方为菜单区。从左至右分别代表“左移”“右移”“转换”“重置”四个功能按钮。

18.2.3 设置全屏

运行应用，查看界面的实际效果。界面效果如图 18-3 所示。

图 18-3 界面的实际效果

可以发现，界面上方“Tetris”字样的应用名称一栏，占据了非常大的空间，极其影响美观。而且，导致下方的菜单区显示不全。因此，需要去掉这个应用名称一栏。

修改配置文件，在“module”属性中，增加“metaData”属性，修改如下：

```
"module": {
    "package": "com.waylau.hmos.tetris",
    "name": ".MyApplication",
    "deviceType": [
      "phone"
    ],
    "distro": {
      "deliveryWithInstall": true,
      "moduleName": "entry",
      "moduleType": "entry"
    },
    "abilities": [
      {
        "skills": [
          {
            "entities": [
              "entity.system.home"
            ],
            "actions": [
              "action.system.home"
            ]
          }
        ],
        "orientation": "unspecified",
        "name": "com.waylau.hmos.tetris.MainAbility",
        "icon": "$media:icon",
        "description": "$string:mainability_description",
        "label": "Tetris",
        "type": "page",
        "launchType": "standard"
      }
    ],
    // 增加 metaData 属性
    "metaData":{
      "customizeData":[
```

```
        {
          "name": "hwc-theme",
          "value": "androidhwext:style/Theme.Emui.NoTitleBar",
          "extra":""
        }
      ]
    }
  }
```

再次运行应用，界面全屏效果如图 18-4 所示。

18.2.4 应用的主体逻辑

修改 MainAbilitySlice 内容，以实现俄罗斯方块游戏的主体逻辑。

代码如下：

图 18-4 界面全屏效果

```
package com.waylau.hmos.tetris.slice;

import com.waylau.hmos.tetris.Grid;
import com.waylau.hmos.tetris.ResourceTable;
import ohos.aafwk.ability.AbilitySlice;
import ohos.aafwk.content.Intent;
import ohos.agp.components.*;
import ohos.agp.render.Canvas;
import ohos.agp.render.Paint;
import ohos.agp.utils.Color;
import ohos.agp.utils.RectFloat;
import ohos.app.dispatcher.TaskDispatcher;
import ohos.hiviewdfx.HiLog;
import ohos.hiviewdfx.HiLogLabel;

public class MainAbilitySlice extends AbilitySlice {
    private static final String TAG = MainAbilitySlice.class.getSimpleName();
    private static final HiLogLabel LABEL_LOG =
            new HiLogLabel(HiLog.LOG_APP, 0x00001, TAG);

    private static final int LENGTH = 100;// 方格的边长
    private static final int MARGIN = 2;// 方格的间距
    private static final int GRID_NUMBER = 4;// 方块所占方格的数量，固定为 4

    private DirectionalLayout layoutGame;// 布局
    private Text textScoreValue;
    private Text textGameover;

    private int[][] grids;// 描绘方格颜色的二维数组
```

```
private int currentRow;// 向下移动的行数
private int currentColumn;// 向左右移动的列数，减 1 表示左移，加 1 表示右移
private String tipValue = "Start!";
private TaskDispatcher taskDispatcher;
private Grid currentGrid;
private boolean isRunning;
private int scoreValue = 0;

@Override
public void onStart(Intent intent) {
    super.onStart(intent);
    super.setUIContent(ResourceTable.Layout_ability_main);

    // 初始化布局
    layoutGame =
            (DirectionalLayout) findComponentById(ResourceTable.Id_
                                                    layout_game);

    // 初始化组件
    Button buttonLeft =
            (Button) findComponentById(ResourceTable.Id_button_left);
    buttonLeft.setClickedListener(listener -> goLeft(grids));

    Button buttonRight =
            (Button) findComponentById(ResourceTable.Id_button_right);
    buttonRight.setClickedListener(listener -> goRight(grids));

    Button buttonShift =
            (Button) findComponentById(ResourceTable.Id_button_shift);
    buttonShift.setClickedListener(listener -> shiftGrids(grids));

    Button buttonRestart =
            (Button) findComponentById(ResourceTable.Id_button_restart);
    buttonRestart.setClickedListener(listener -> {
        initialize();
        startGame();
    });

    textScoreValue =
            (Text) findComponentById(ResourceTable.Id_text_score_value);
    textGameover =
            (Text) findComponentById(ResourceTable.Id_text_gameover);

    // 初始化游戏
    initialize();
```

```
        // 启动游戏
        startGame();
    }

    ...
```

上述代码解释如下。

- onStart方法初始化了DirectionalLayout、Button、Text等组件和布局。
- initialize方法用于初始化游戏的状态和数据。
- startGame方法用于启动游戏。

Grid是方格类，用于表示游戏中的方格，代码如下：

```
package com.waylau.hmos.tetris;

public class Grid {

    private int[][] currentGrids;// 当前方块的形态
    private int rowNumber;       // 方块的总行数
    private int columnNumber;    // 方块的总列数
    private int currentGridColor;// 当前方格的颜色
    private int columnStart;     // 方块的第一个方格所在二维数组的列数

    private Grid(int[][] currentGrids, int rowNumber, int columnNumber,
                 int currentGridColor, int columnStart) {
        this.currentGrids = currentGrids;
        this.rowNumber = rowNumber;
        this.columnNumber = columnNumber;
        this.currentGridColor = currentGridColor;
        this.columnStart = columnStart;
    }

    ...
```

Grid各字段含义详见注释。

18.2.5 初始化游戏

初始化游戏的方法initialize代码如下：

```
private void initialize() {
    // 初始化变量
    scoreValue = 0;
    tipValue = "Start!";
    isRunning = false;
```

```
    taskDispatcher = null;

    // 显示提示
    showTip();

    // 初始化网格,
    // 15×10 的二维数组
    // 数组元素都是 0
    grids = new int[15][10];
    for (int row = 0; row < 15; row++) {
        for (int column = 0; column < 10; column++) {
            grids[row][column] = 0;
        }
    }

    // 创建网格数据
    createGrids(grids);

    // 画出网格
    drawGrids(grids);
}
```

上述代码解释如下。

- 初始化了变量，显示提示。
- 初始化网格。该网格本质上是一个15×10的二维数组，数组元素值都是0。
- 同时调用createGrids和drawGrids来在界面上画出网格。

showTip方法代码如下：

```
private void showTip() {
    textScoreValue.setText(scoreValue + "");
    textGameover.setText(tipValue);
}
```

18.2.6 创建网格数据

创建网格数据的方法createGrids代码如下：

```
private void createGrids(int[][] grids) {
    currentColumn = 0;
    currentRow = 0;

    // 当有任一行全部填满颜色方块时，消去该行
    eliminateGrids(grids);
```

```
    // 判断游戏是否完成：
    // 完成时就停止定时器，并提示结束文本；
    // 未完成时就生成新的颜色方块
    if (isGameOver(grids)) {
        tipValue = "Game Over!";
        isRunning = false;
        taskDispatcher = null;
        showTip();
    } else {
        //随机生成一个颜色方块
        currentGrid = Grid.generateRandomGrid();
        int[][] currentGrids = currentGrid.getCurrentGrids();
        int currentGridColor = currentGrid.getCurrentGridColor();

        //将颜色方块对应的Grids添加到二维数组中
        for (int row = 0; row < GRID_NUMBER; row++) {
            grids[currentGrids[row][0] + currentRow][currentGrids[row][1] +
                currentColumn] = currentGridColor;
        }
    }
}
```

上述代码解释如下。

- 判断是否有任一行全部填满颜色方块，如果有则执行eliminateGrids方法消除该行。
- 调用isGameOver方法进行游戏是否完成的判断：完成时就停止定时器，并提示结束文本；未完成时就生成新的颜色方块初始化网格。
- Grid.generateRandomGrid()用来生成随机的方块。

isGameOver方法代码如下：

```
private boolean isGameOver(int[][] grids) {
    //新生成的颜色方块覆盖原有的颜色方块则游戏结束
    if (currentGrid != null) {
        int[][] currentGrids = currentGrid.getCurrentGrids();

        for (int row = 0; row < GRID_NUMBER; row++) {
            if (grids[currentGrids[row][0] + currentRow][currentGrids[row][1] +
                currentColumn] != 0) {
                return true;
            }
        }
    }

    return false;
}
```

eliminateGrids方法代码如下：

```
private void eliminateGrids(int[][] grids) {
    boolean k;
    for (int row = 14; row >= 0; row--) {
        k = true;

        //判断是否有任一行全部填满颜色方块
        for (int column = 0; column < 10; column++) {
            if (grids[row][column] == 0) {
                k = false;
            }
        }

        //消去全部填满颜色方块的行
        if (k) {
            // 加分
            this.scoreValue++;
            this.showTip();

            // 且所有方格向下移动一格
            for (int i = row - 1; i >= 0; i--) {
                for (int j = 0; j < 10; j++) {
                    grids[i + 1][j] = grids[i][j];
                }
            }
            for (int n = 0; n < 10; n++) {
                grids[0][n] = 0;
            }
        }
    }

    drawGrids(grids);
}
```

eliminateGrids方法内部先判断是否有任一行全部填满颜色方块，有则消除全部填满颜色方块的行，同时得分scoreValue加1，并且所有的方格向下移动一格。

Grid.generateRandomGrid()方法代码如下：

```
package com.waylau.hmos.tetris;

public class Grid {
    private static final int[][] RedGrids1 =
            {{0, 3}, {0, 4}, {1, 4}, {1, 5}};//红色方块形态1
    private static final int[][] RedGrids2 =
            {{0, 5}, {1, 5}, {1, 4}, {2, 4}};//红色方块形态2
```

```
private static final int[][] GreenGrids1 =
        {{0, 5}, {0, 4}, {1, 4}, {1, 3}};// 绿色方块形态 1
private static final int[][] GreenGrids2 =
        {{0, 4}, {1, 4}, {1, 5}, {2, 5}};// 绿色方块形态 2
private static final int[][] CyanGrids1 =
        {{0, 4}, {1, 4}, {2, 4}, {3, 4}};// 蓝绿色方块形态 1
private static final int[][] CyanGrids2 =
        {{0, 3}, {0, 4}, {0, 5}, {0, 6}};// 蓝绿色方块形态 2
private static final int[][] MagentaGrids1 =
        {{0, 4}, {1, 3}, {1, 4}, {1, 5}};// 品红色方块形态 1
private static final int[][] MagentaGrids2 =
        {{0, 4}, {1, 4}, {1, 5}, {2, 4}};// 品红色方块形态 2
private static final int[][] MagentaGrids3 =
        {{0, 3}, {0, 4}, {0, 5}, {1, 4}};// 品红色方块形态 3
private static final int[][] MagentaGrids4 =
        {{0, 5}, {1, 5}, {1, 4}, {2, 5}};// 品红色方块形态 4
private static final int[][] BlueGrids1 =
        {{0, 3}, {1, 3}, {1, 4}, {1, 5}};// 蓝色方块形态 1
private static final int[][] BlueGrids2 =
        {{0, 5}, {0, 4}, {1, 4}, {2, 4}};// 蓝色方块形态 2
private static final int[][] BlueGrids3 =
        {{0, 3}, {0, 4}, {0, 5}, {1, 5}};// 蓝色方块形态 3
private static final int[][] BlueGrids4 =
        {{0, 5}, {1, 5}, {2, 5}, {2, 4}};// 蓝色方块形态 4
private static final int[][] WhiteGrids1 =
        {{0, 5}, {1, 5}, {1, 4}, {1, 3}};// 白色方块形态 1
private static final int[][] WhiteGrids2 =
        {{0, 4}, {1, 4}, {2, 4}, {2, 5}};// 白色方块形态 2
private static final int[][] WhiteGrids3 =
        {{0, 5}, {0, 4}, {0, 3}, {1, 3}};// 白色方块形态 3
private static final int[][] WhiteGrids4 =
        {{0, 4}, {0, 5}, {1, 5}, {2, 5}};// 白色方块形态 4
private static final int[][] YellowGrids =
        {{0, 4}, {0, 5}, {1, 5}, {1, 4}};// 黄色方块形态 1

public static Grid generateRandomGrid() {
    // 随机生成一个颜色方块
    double random = Math.random();
    if (random >= 0 && random < 0.2) {
        if (random >= 0 && random < 0.1)
            return createRedGrids1();
        else
            return createRedGrids2();
    } else if (random >= 0.2 && random < 0.4) {
        if (random >= 0.2 && random < 0.3)
            return createGreenGrids1();
```

```
        else
            return createGreenGrids2();
    } else if (random >= 0.4 && random < 0.45) {
        if (random >= 0.4 && random < 0.43)
            return createCyanGrids1();
        else
            return createCyanGrids2();
    } else if (random >= 0.45 && random < 0.6) {
        if (random >= 0.45 && random < 0.48)
            return createMagentaGrids1();
        else if (random >= 0.48 && random < 0.52)
            return createMagentaGrids2();
        else if (random >= 0.52 && random < 0.56)
            return createMagentaGrids3();
        else
            return createMagentaGrids4();
    } else if (random >= 0.6 && random < 0.75) {
        if (random >= 0.6 && random < 0.63)
            return createBlueGrids1();
        else if (random >= 0.63 && random < 0.67)
            return createBlueGrids2();
        else if (random >= 0.67 && random < 0.71)
            return createBlueGrids3();
        else
            return createBlueGrids4();
    } else if (random >= 0.75 && random < 0.9) {
        if (random >= 0.75 && random < 0.78)
            return createWhiteGrids1();
        else if (random >= 0.78 && random < 0.82)
            return createWhiteGrids2();
        else if (random >= 0.82 && random < 0.86)
            return createWhiteGrids3();
        else
            return createWhiteGrids4();
    } else {
        return createYellowGrids();
    }
}

// 以下为各种颜色各种形状的方块
private static Grid createRedGrids1() {
    return new Grid(RedGrids1, 2, 3, 1, 3);
}

private static Grid createRedGrids2() {
    return new Grid(RedGrids2, 3, 2, 1, 4);
```

```
}

private static Grid createGreenGrids1() {
    return new Grid(GreenGrids1, 2, 3, 2, 3);
}

private static Grid createGreenGrids2() {
    return new Grid(GreenGrids2, 3, 2, 2, 4);
}

private static Grid createCyanGrids1() {
    return new Grid(CyanGrids1, 4, 1, 3, 4);
}

private static Grid createCyanGrids2() {
    return new Grid(CyanGrids2, 1, 4, 3, 3);
}

private static Grid createMagentaGrids1() {
    return new Grid(MagentaGrids1, 2, 3, 4, 3);
}

private static Grid createMagentaGrids2() {
    return new Grid(MagentaGrids2, 3, 2, 4, 4);
}

private static Grid createMagentaGrids3() {
    return new Grid(MagentaGrids3, 2, 3, 4, 3);
}

private static Grid createMagentaGrids4() {
    return new Grid(MagentaGrids4, 3, 2, 4, 4);
}

private static Grid createBlueGrids1() {
    return new Grid(BlueGrids1, 2, 3, 5, 3);
}

private static Grid createBlueGrids2() {
    return new Grid(BlueGrids2, 3, 2, 5, 4);
}

private static Grid createBlueGrids3() {
    return new Grid(BlueGrids3, 2, 3, 5, 3);
}
```

```
    private static Grid createBlueGrids4() {
        return new Grid(BlueGrids4, 3, 2, 5, 4);
    }

    private static Grid createWhiteGrids1() {
        return new Grid(WhiteGrids1, 2, 3, 6, 3);
    }

    private static Grid createWhiteGrids2() {
        return new Grid(WhiteGrids2, 3, 2, 6, 4);
    }

    private static Grid createWhiteGrids3() {
        return new Grid(WhiteGrids3, 2, 3, 6, 3);
    }

    private static Grid createWhiteGrids4() {
        return new Grid(WhiteGrids4, 3, 2, 6, 4);
    }

    private static Grid createYellowGrids() {
        return new Grid(YellowGrids, 2, 2, 7, 4);
    }

    ...
}
```

18.2.7 绘制网格

绘制网格的方法drawGrids代码如下：

```
private void drawGrids(int[][] grids) {
    Component.DrawTask task = new Component.DrawTask() {
        @Override
        public void onDraw(Component component, Canvas canvas) {
            Paint paint = new Paint();

            for (int row = 0; row < 15; row++) {
                for (int column = 0; column < 10; column++) {
                    // grids 的值表示不同的颜色
                    int grideValue = grids[row][column];
                    Color color = null;

                    switch (grideValue) {
                        case 0:
```

```
                        color = Color.GRAY;
                        break;
                    case 1:
                        color = Color.RED;
                        break;
                    case 2:
                        color = Color.GREEN;
                        break;
                    case 3:
                        color = Color.CYAN;
                        break;
                    case 4:
                        color = Color.MAGENTA;
                        break;
                    case 5:
                        color = Color.BLUE;
                        break;
                    case 6:
                        color = Color.WHITE;
                        break;
                    case 7:
                        color = Color.YELLOW;
                        break;
                    default:
                        break;
                }

                paint.setColor(color);

                RectFloat rectFloat =
                    new RectFloat(30 + column * (LENGTH + MARGIN),
                                  40 + row * (LENGTH + MARGIN),
                                  30 + LENGTH + column * (LENGTH + MARGIN),
                                  40 + LENGTH + row * (LENGTH + MARGIN));

                //绘制方格
                canvas.drawRect(rectFloat, paint);
            }
        }
    }
};

layoutGame.addDrawTask(task);
}
```

上述代码解释如下。

- 通过实现 Component.DrawTask 来自定义游戏界面组件。
- Canvas 类用于绘制画面。
- 游戏界面组件最终添加到 layoutGame 布局上。

18.2.8 启动游戏

startGame() 方法用于启动游戏，代码如下：

```
private void startGame() {
    // 设置游戏状态
    isRunning = true;

    // 派发任务
    taskDispatcher = getUITaskDispatcher();
    taskDispatcher.asyncDispatch(new GameTask());
}
```

上述代码解释如下。

- 设置游戏状态 isRunning 为 true。
- 通过 getUITaskDispatcher 获取任务分发器。
- 通过任务分发器来执行 GameTask。

GameTask 类代码如下：

```
private class GameTask implements Runnable {

    @Override
    public void run() {
        HiLog.info(LABEL_LOG, "before GameTask run");

        int[][] currentGrids = currentGrid.getCurrentGrids();
        int currentGridColor = currentGrid.getCurrentGridColor();
        int rowNumber = currentGrid.getRowNumber();

        // 如果方块能下移则下移，否则重新随机生成新的方块
        if (couldDown(currentGrids, rowNumber)) {
            // 将原来的颜色方块清除
            for (int row = 0; row < GRID_NUMBER; row++) {
                grids[currentGrids[row][0] +
                        currentRow][currentGrids[row][1] + currentColumn] = 0;
            }

            currentRow++;

            // 重新绘制颜色方块
```

```
            for (int row = 0; row < GRID_NUMBER; row++) {
                grids[currentGrids[row][0] + currentRow][currentGrids[row][1]
                    + currentColumn] = currentGridColor;
            }
        } else {
            createGrids(grids);
        }

        drawGrids(grids);

        if (isRunning) {
            taskDispatcher.delayDispatch(this, 750);
        }

        HiLog.info(LABEL_LOG, "end GameTask run");
    }
}
```

上述代码解释如下。

- 先判断方块能否下移，如果能则执行下移，否则重新随机生成新的方块。
- 通过isRunning判断游戏是否还在执行，如果是则执行分发新任务，从而实现任务的定时执行。

couldDown方法如下：

```
private boolean couldDown(int[][] currentGrids, int rowNumber) {
    boolean k;

    // 如果方块向下移动到下边界，则返回 false
    if (currentRow + rowNumber == 15) {
        return false;
    }

    // 当下边缘方块再下一格为空时则可以下移
    for (int row = 0; row < GRID_NUMBER; row++) {
        k = true;
        for (int i = 0; i < GRID_NUMBER; i++) {
            if (currentGrids[row][0] + 1 == currentGrids[i][0]
                    && currentGrids[row][1] == currentGrids[i][1]) {
                    // 找出非下边缘方块
                k = false;
            }
        }

        // 当任一下边缘方块再下一格不为空时返回 false
        if (k) {
            if (grids[currentGrids[row][0] +
```

```
                currentRow + 1][currentGrids[row][1] + currentColumn] != 0)
                return false;
        }
    }

    return true;
}
```

18.2.9 左移操作

goLeft方法用于响应“左移”按钮操作，代码如下：

```
private void goLeft(int[][] grids) {
    int currentGridColor = currentGrid.getCurrentGridColor();
    int[][] currentGrids = currentGrid.getCurrentGrids();
    int columnStart = currentGrid.getColumnStart();

    // 当方块能向左移动时则左移
    if (couldLeft(currentGrids, columnStart)) {
        // 将原来的颜色方块清除
        for (int row = 0; row < GRID_NUMBER; row++) {
            grids[currentGrids[row][0] + currentRow][currentGrids[row][1] +
                    currentColumn] = 0;
        }

        currentColumn--;

        // 重新绘制颜色方块
        for (int row = 0; row < GRID_NUMBER; row++) {
            grids[currentGrids[row][0] + currentRow][currentGrids[row][1] +
                    currentColumn] = currentGridColor;
        }
    }

    drawGrids(grids);
}
```

上述代码解释如下。

- 先判断当方块能否执行向左移动，如果能则执行左移。
- 左移实际逻辑是，将原来的颜色方块清除，而后重新绘制颜色方块。

couldLeft方法代码如下：

```
private boolean couldLeft(int[][] currentGrids, int columnStart) {
    boolean k;
```

```
    //如果方块向左移动到左边界，则返回 false
    if (currentColumn + columnStart == 0) {
        return false;
    }

    //当左边缘方块再左一格为空时则可以左移
    for (int column = 0; column < GRID_NUMBER; column++) {
        k = true;
        for (int j = 0; j < GRID_NUMBER; j++) {
            //找出非左边缘方块
            if (currentGrids[column][0] == currentGrids[j][0]
                    && currentGrids[column][1] - 1 == currentGrids[j][1]) {
                k = false;
            }
        }

        //当任一左边缘方块再左一格不为空时返回 false
        if (k) {
            if (grids[currentGrids[column][0] + currentRow]
                [currentGrids[column][1] + currentColumn - 1] != 0)
                return false;
        }
    }

    return true;
}
```

18.2.10 右移操作

goRight方法用于响应“右移”按钮操作，代码如下：

```
private void goRight(int[][] grids) {
    int currentGridColor = currentGrid.getCurrentGridColor();
    int[][] currentGrids = currentGrid.getCurrentGrids();
    int columnNumber = currentGrid.getColumnNumber();
    int columnStart = currentGrid.getColumnStart();

    //当方块能向右移动时则右移
    if (couldRight(currentGrids, columnNumber, columnStart)) {
        //将原来的颜色方块清除
        for (int row = 0; row < GRID_NUMBER; row++) {
            grids[currentGrids[row][0] + currentRow][currentGrids[row][1] +
currentColumn] = 0;
        }
```

```
            currentColumn++;

            // 重新绘制颜色方块
            for (int row = 0; row < GRID_NUMBER; row++) {
                grids[currentGrids[row][0] + currentRow][currentGrids[row][1] +
                        currentColumn] = currentGridColor;
            }
        }

        drawGrids(grids);
    }
```

上述代码解释如下。

- 先判断当方块能否执行向右移动，如果能则执行右移。
- 右移实际逻辑是，将原来的颜色方块清除，而后重新绘制颜色方块。

couldRight方法代码如下：

```
private boolean couldRight(int[][] currentGrids, int columnNumber, int
columnStart) {
    boolean k;

    // 如果方块向右移动到右边界，则返回 false
    if (currentColumn + columnNumber + columnStart == 10) {
        return false;
    }

    // 当右边缘方块再右一格为空时则可以右移
    for (int column = 0; column < GRID_NUMBER; column++) {
        k = true;
        for (int j = 0; j < GRID_NUMBER; j++) {
            // 找出非右边缘方块
            if (currentGrids[column][0] == currentGrids[j][0]
                    && currentGrids[column][1] + 1 == currentGrids[j][1]) {
                k = false;
            }
        }

        // 当任一右边缘方块再右一格不为空时返回 false
        if (k) {
            if (grids[currentGrids[column][0] + currentRow]
                [currentGrids[column][1] + currentColumn + 1] != 0)
                return false;
        }
    }
```

```
    return true;
}
```

18.2.11 转换操作

shiftGrids 方法用于响应“转换”按钮操作，代码如下：

```
private void shiftGrids(int[][] grids) {
    int[][] currentGrids = currentGrid.getCurrentGrids();
    int columnNumber = currentGrid.getColumnNumber();
    int columnStart = currentGrid.getColumnStart();

    //将原来的颜色方块清除
    for (int row = 0; row < GRID_NUMBER; row++) {
        grids[currentGrids[row][0] + currentRow][currentGrids[row][1] +
                currentColumn] = 0;
    }

    if (columnNumber == 2 && currentColumn + columnStart == 0) {
        currentColumn++;
    }

    //根据 Grids 的颜色值调用改变方块形状的 "chang+Color+Grids" 函数
    currentGrid = Grid.shiftGrid(currentGrid);
    int currentGridColor = currentGrid.getCurrentGridColor();
    currentGrids = currentGrid.getCurrentGrids();

    //重新绘制颜色方块
    for (int row = 0; row < GRID_NUMBER; row++) {
        grids[currentGrids[row][0] + currentRow][currentGrids[row][1] +
                currentColumn] = currentGridColor;
    }

    drawGrids(grids);
}
```

上述代码解释如下。

- 转换实际逻辑是将原来的颜色方块清除，而后重新绘制颜色方块。
- 新的颜色方块通过 Grid.shiftGrid 方法获得。

Grid.shiftGrid 方法代码如下：

```
public static Grid shiftGrid(Grid grid) {
    int currentGridColor = grid.getCurrentGridColor();
```

```
    int[][] currentGrids = grid.getCurrentGrids();

    // 根据 Grids 的颜色值调用改变方块形状的 "chang+Color+Grids" 函数
    if (currentGridColor == 1) {
        return changRedGrids(currentGrids);
    } else if (currentGridColor == 2) {
        return changeGreenGrids(currentGrids);
    } else if (currentGridColor == 3) {
        return changeCyanGrids(currentGrids);
    } else if (currentGridColor == 4) {
        return changeMagentaGrids(currentGrids);
    } else if (currentGridColor == 5) {
        return changeBlueGrids(currentGrids);
    } else if (currentGridColor == 6) {
        return changeWhiteGrids(currentGrids);
    } else {
        return null;
    }
}

private static Grid changRedGrids(int[][] currentGrids) {
    if (currentGrids == RedGrids1) {
        return createRedGrids2();
    } else if (currentGrids == RedGrids2) {
        return createRedGrids1();
    } else {
        return null;
    }
}

private static Grid changeGreenGrids(int[][] currentGrids) {
    if (currentGrids == GreenGrids1) {
        return createGreenGrids2();
    } else if (currentGrids == GreenGrids2) {
        return createGreenGrids1();
    } else {
        return null;
    }
}

private static Grid changeCyanGrids(int[][] currentGrids) {
    if (currentGrids == CyanGrids1) {
        return createCyanGrids2();
    } else if (currentGrids == CyanGrids2) {
        return createCyanGrids1();
    } else {
```

```
            return null;
        }
    }

    private static Grid changeMagentaGrids(int[][] currentGrids) {
        if (currentGrids == MagentaGrids1) {
            return createMagentaGrids2();
        } else if (currentGrids == MagentaGrids2) {
            return createMagentaGrids3();
        } else if (currentGrids == MagentaGrids3) {
            return createMagentaGrids4();
        } else if (currentGrids == MagentaGrids4) {
            return createMagentaGrids1();
        } else {
            return null;
        }
    }

    private static Grid changeBlueGrids(int[][] currentGrids) {
        if (currentGrids == BlueGrids1) {
            return createBlueGrids2();
        } else if (currentGrids == BlueGrids2) {
            return createBlueGrids3();
        } else if (currentGrids == BlueGrids3) {
            return createBlueGrids4();
        } else if (currentGrids == BlueGrids4) {
            return createBlueGrids1();
        } else {
            return null;
        }
    }

    private static Grid changeWhiteGrids(int[][] currentGrids) {
        if (currentGrids == WhiteGrids1) {
            return createWhiteGrids2();
        } else if (currentGrids == WhiteGrids2) {
            return createWhiteGrids3();
        } else if (currentGrids == WhiteGrids3) {
            return createWhiteGrids4();
        } else if (currentGrids == WhiteGrids4) {
            return createWhiteGrids1();
        } else {
            return null;
        }
    }
```

18.2.12　重置操作

“重置”按钮操作所执行的代码如下：

```
Button buttonRestart =
            (Button) findComponentById(ResourceTable.Id_button_restart);
buttonRestart.setClickedListener(listener -> {
    initialize();
    startGame();
});
```

上述操作的本质是再次执行一次 initialize 和 startGame 方法。

18.3 应用运行

初次启动应用，则可以直接进行游戏。游戏界面如图 18-5 所示。

当游戏结束时，则会提示“Game Over!”字样，并统计得分，界面如图 18-6 所示。

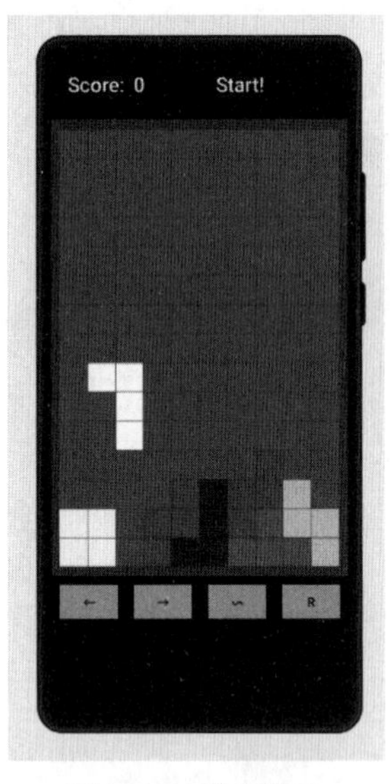

图 18-5　游戏界面

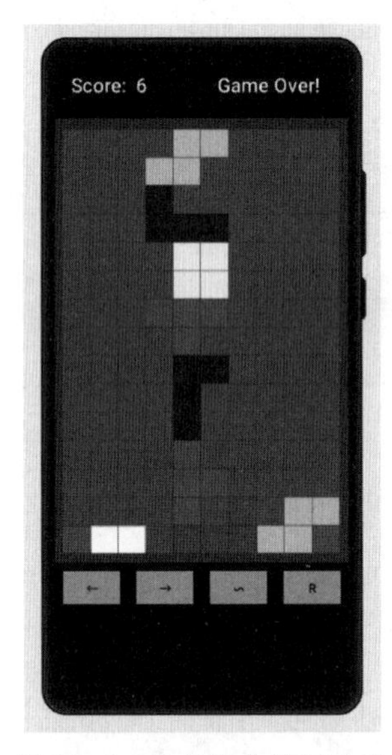

图 18-6　游戏结束界面效果

点击“重置”按钮可以重新开始游戏。

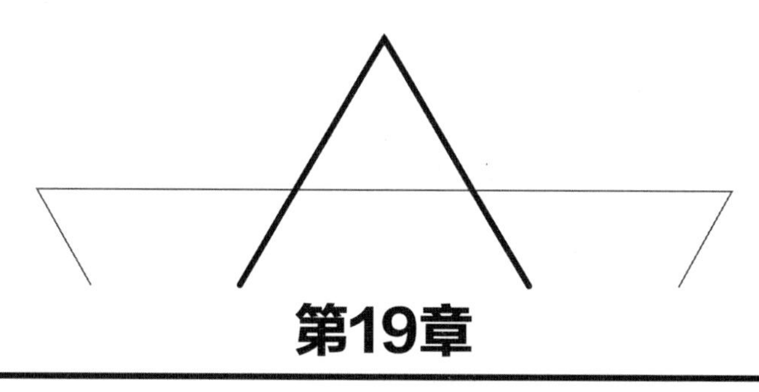

第19章

综合案例4：ArkTS实现手机应用

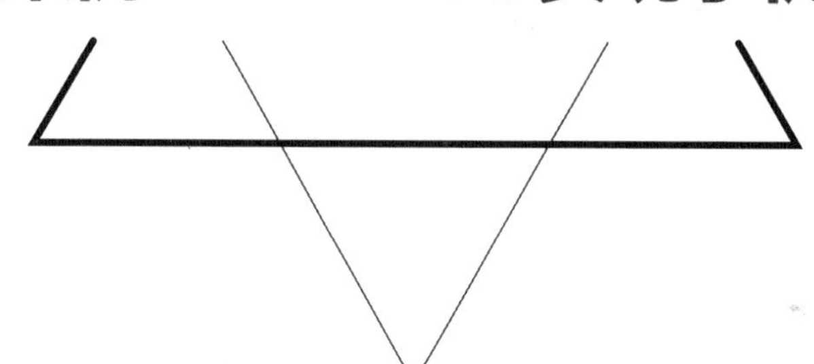

本章是一个实战章节，结合前面所介绍的知识点来实现一个类似于微信的 App。

19.1 仿微信应用概述

本节将基于HarmonyOS提供的组件来实现类似于微信界面效果的应用“ArkUIWeChat”。

微信界面主要包含四部分，即微信、联系人、发现、我。本章所演示的例子也要实现这四个部分。

19.1.1 “微信”页面

“微信”页面是微信应用的首页，主要用于展示联系人之间的沟通信息。

图 19-1是“微信”页面的效果图。

19.1.2 “联系人”页面

“联系人”页面展示用户所关联的联系人。

图 19-2是“联系人”页面的效果图。

19.1.3 “发现”页面

“发现”页面是微信进入其他子程序的入口。

图 19-3是“发现”页面的效果图。

图 19-1 “微信”页面的效果图

图 19-2 “联系人”页面的效果图

图 19-3 “发现”页面的效果图

19.1.4 “我”页面

“我”页面是展示用户个人信息的页面。

图19-4是“我”页面的效果图。

图19-4 “我”页面的效果图

19.2 实战：“微信”页面

本节演示如何实现“微信”页面。

“微信”页面主要展示联系人的沟通记录列表。列表的每个项，都包含联系人头像、联系人名称、联系人聊天记录及时间。

19.2.1 创建“微信”页面ChatPage

在pages创建ChatPage.ets作为“微信”页面。“微信”页面主要分为标题栏及沟通记录列表，因此，核心的代码也分为这两部分。代码如下：

```
import {ChatItemStyle, WeChatTitle} from '../model/CommonStyle'
import {getContactInfo} from '../model/WeChatData'
import {Person} from '../model/Person'

@Component
export struct ChatPage {
  private contactList: Person[] = getContactInfo()

  build() {
    Column() {
      // 标题
      WeChatTitle({ text: "微信" })

      // 列表
      List() {
        ForEach(this.contactList, item => {
          ListItem() {
            ChatItemStyle({
              WeChatImage: item.WeChatImage,
              WeChatName: item.WeChatName,
              ChatInfo: item.ChatInfo,
              time: item.time
            })
          }
```

```
          }, item => item.id.toString())
        }
        .height('100%')
        .width('100%')
      }
    }
  }
```

在上述代码中，通过ForEach来遍历getContactInfo()所返回的联系人数据，并生成ChatItemStyle数据项。

19.2.2　定义联系人Person

联系人是用Person类作为表示。在ets目录下，创建model目录，并在该model目录创建Person.ets。代码如下：

```
let personId = 0;

export class Person {
  id: string;
  WeChatImage: string;
  WeChatName: string;
  ChatInfo: string;
  time: string;

  constructor(WeChatImage: string, WeChatName: string, ChatInfo: string,
              time: string) {
    this.id = `${personId++}`
    this.WeChatImage = WeChatImage;
    this.WeChatName = WeChatName;
    this.ChatInfo = ChatInfo;
    this.time = time;
  }
}
```

Person内包含头像、名称、聊天记录及时间等信息。

19.2.3　定义联系人数据

在model目录下创建WeChatData.ets作为联系人数据。代码如下：

```
import {Person} from './Person'

const ContactInfo: any[] = [
```

```
  {
    "WeChatImage": "person (1).jpg",
    "WeChatName": "枫",
    "ChatInfo": "缓缓飘落的枫叶像思念，我点燃烛光温暖岁末的秋天",
    "time": "18:30"
  },
  {
    "WeChatImage": "person (2).jpg",
    "WeChatName": "珊瑚海",
    "ChatInfo": "转身离开，分手说不出来",
    "time": "17:29"
  },
  {
    "WeChatImage": "person (3).jpg",
    "WeChatName": "听妈妈的话",
    "ChatInfo": "听妈妈的话别让她受伤，想快快长大才能保护她",
    "time": "17:28"
  },
  // 为了节约篇幅，此处省略部分数据
  {
    "WeChatImage": "person (15).jpg",
    "WeChatName": "一路向北",
    "ChatInfo": "我一路向北，离开有你的季节",
    "time": "10:16"
  }
]

export function getContactInfo(): Array<Person> {
  let contactList: Array<Person> = []
  ContactInfo.forEach(item => {
    contactList.push(new Person(item.WeChatImage, item.WeChatName, item.
                ChatInfo, item.time))
  })

  return contactList;
}

export const WeChatColor:string = "#ededed"
```

19.2.4 定义样式

在model目录下创建CommonStyle.ets作为样式类。在该类中定义标题的样式，代码如下：

```
@Component
export struct WeChatTitle {
```

```
  private text: string

  build() {
    Flex({ alignItems: ItemAlign.Center, justifyContent: FlexAlign.Center }) {
      Text(this.text).fontSize('18fp').padding('20px')
    }.height('120px').backgroundColor(WeChatColor)
  }
}
```

在CommonStyle中定义沟通记录的样式，代码如下：

```
@Component
export struct ChatItemStyle {
  WeChatImage: string;
  WeChatName: string;
  ChatInfo: string;
  time: string;

  build() {
    Column() {
      Flex({ alignItems: ItemAlign.Center, justifyContent: FlexAlign.Start }) {
        Image($rawfile(this.WeChatImage)).width('120px').height('120px').
              margin({ left: '50px', right: "50px" })

        Column() {
          Text(this.WeChatName).fontSize('16fp')
          Text(this.ChatInfo).fontSize('12fp').width('620px').
                fontColor("#c2bec2").maxLines(1)
        }.alignItems(HorizontalAlign.Start).flexGrow(1)

        Text(this.time).fontSize('12fp')
          .margin({ right: "50px" }).fontColor("#c2bec2")

      }
      .height('180px')
      .width('100%')

      Row() {
        Text().width('190px').height('3px')
        Divider()
          .vertical(false)
          .color(WeChatColor)
          .strokeWidth('3px')
      }
      .height('3px')
      .width('100%')
```

```
      }

    }
}
```

最终，沟通记录的样式效果如图19-5所示。

图19-5　沟通记录的样式效果图

19.3 实战："联系人"页面

本节演示如何实现"联系人"页面。

"联系人"页面主要展示联系人列表。列表的每个项，都包含联系人头像、联系人名称。因此，实现方式与"微信"页面类似。

19.3.1 创建"联系人"页面ContactPage

在pages创建ContactPage.ets作为"联系人"页面。"联系人"页面主要分为标题栏及联系人列表，因此，核心的代码也分为这两部分。代码如下：

```
import {ContactItemStyle, WeChatTitle} from '../model/CommonStyle'
import {Person} from '../model/Person'
import {getContactInfo, WeChatColor} from '../model/WeChatData'

@Component
export struct ContactPage {
  private contactList: Person[] = getContactInfo()

  build() {
    Column() {
      // 标题
      WeChatTitle({ text: "通讯录" })

      // 列表
      Scroll() {
        Column() {
          // 分类
          ContactItemStyle({ imageSrc: "new_friend.png", text: "新的朋友" })
```

```
        ContactItemStyle({ imageSrc: "group.png", text: "群聊" })
        ContactItemStyle({ imageSrc: "biaoqian.png", text: "标签" })
        ContactItemStyle({ imageSrc: "gonzh.png", text: "公众号" })

        // 企业联系人
        Text("        我的企业及企业联系人").fontSize('12fp').
            backgroundColor(WeChatColor).height('80px').width('100%')
        ContactItemStyle({ imageSrc: "qiye.png", text: "企业微信联系人" })

        // 微信好友
        Text("        我的微信好友").fontSize('12fp').
            backgroundColor(WeChatColor).height('80px').width('100%')
        List() {
          ForEach(this.contactList, item => {
            ListItem() {
              ContactItemStyle({ imageSrc: item.WeChatImage, text: item.
                                    WeChatName })
            }
          }, item => item.id.toString())
        }
      }
    }

    }.alignItems(HorizontalAlign.Start)
    .width('100%')
    .height('100%')
  }
}
```

联系人列表又细分为三个部分：分类、企业联系人、微信好友。

19.3.2 定义样式

在CommonStyle中定义联系人的样式，代码如下：

```
@Component
export struct ContactItemStyle {
  private imageSrc: string
  private text: string

  build() {
    Column() {
      Flex({ alignItems: ItemAlign.Center, justifyContent: FlexAlign.Center }) {
        Image($rawfile(this.imageSrc)).width('100px').height('100px').margin({
left: '50px' })
```

```
          Text(this.text).fontSize('15vp').margin({ left: '40px' }).flexGrow(1)
        }
        .height('150px')
        .width('100%')

        Row() {
          Text().width('190px').height('3px')
          Divider()
            .vertical(false)
            .color(WeChatColor)
            .strokeWidth('3px')
        }
        .height('3px')
        .width('100%')
      }
    }
  }
```

最终，联系人的样式效果如图19-6所示。

图19-6　联系人的样式效果

19.4 实战：“发现”页面

本节演示如何实现“发现”页面。

“发现”页面是微信进入其他子程序的入口。每个子程序本质上也是一个列表项。

19.4.1 创建“发现”页面DiscoveryPage

在pages创建DiscoveryPage.ets作为“发现”页面。“发现”页面主要分为标题栏及子程序列表，因此，核心的代码也分为这两部分。代码如下：

```
import {WeChatItemStyle, MyDivider, WeChatTitle} from '../model/CommonStyle'

@Component
export struct DiscoveryPage {
  build() {
    Column() {
      // 标题
```

```
      WeChatTitle({ text: "发现" })

      // 列表
      WeChatItemStyle({ imageSrc: "moments.png", text: "朋友圈" })
      MyDivider()

      WeChatItemStyle({ imageSrc: "shipinghao.png", text: "视频号" })
      MyDivider({ style: '1' })
      WeChatItemStyle({ imageSrc: "zb.png", text: "直播" })
      MyDivider()

      WeChatItemStyle({ imageSrc: "sys.png", text: "扫一扫" })
      MyDivider({ style: '1' })
      WeChatItemStyle({ imageSrc: "yyy.png", text: "摇一摇" })
      MyDivider()

      WeChatItemStyle({ imageSrc: "kyk.png", text: "看一看" })
      MyDivider({ style: '1' })
      WeChatItemStyle({ imageSrc: "souyisou.png", text: "搜一搜" })
      MyDivider()

      WeChatItemStyle({ imageSrc: "fujin.png", text: "附近" })
      MyDivider()

      WeChatItemStyle({ imageSrc: "gw.png", text: "购物" })
      MyDivider({ style: '1' })
      WeChatItemStyle({ imageSrc: "game.png", text: "游戏" })
      MyDivider()

      WeChatItemStyle({ imageSrc: "xcx.png", text: "小程序" })
      MyDivider()
    }.alignItems(HorizontalAlign.Start)
    .width('100%')
    .height('100%')
  }
}
```

子程序用WeChatItemStyle定义样式，并通过MyDivider来进行分割。

19.4.2 定义样式

在CommonStyle中定义子程序的样式，代码如下：

```
@Component
export struct WeChatItemStyle {
```

```
  private imageSrc: string
  private text: string
  private arrow: string = "arrow.png"

  build() {
    Column() {
      Flex({ alignItems: ItemAlign.Center, justifyContent: FlexAlign.Center })
{
        Image($rawfile(this.imageSrc)).width('75px').height('75px').margin({
              left: '50px' })
        Text(this.text).fontSize('15vp').margin({ left: '40px' }).flexGrow(1)
        Image($rawfile(this.arrow))
          .margin({ right: '40px' })
          .width('75px')
          .height('75px')
      }
      .height('150px')
      .width('100%')
    }.onClick(() => {
      if (this.text === "视频号") {
        router.push({ uri: 'pages/VideoPage' })
      }
    })
  }
}
```

子程序主要由三部分组成：图标、名称及箭头。

在CommonStyle中定义分割线的样式，代码如下：

```
@Component
export struct MyDivider {
  private style: string = ""

  build() {
    Row() {
      Divider()
        .vertical(false)
        .color(WeChatColor)
        .strokeWidth(this.style == "1" ? '3px' : '23px')
    }
    .height(this.style == "1" ? '3px' : '23px')
    .width('100%')
  }
}
```

最终，子程序的样式效果如图19-7所示。

朋友圈 >

图 19-7　子程序的样式效果图

19.5 实战：“我”页面

本节演示如何实现“我”页面。

“我”页面展示用户人的信息。

在pages创建MyPage.ets作为“我”页面。“我”页面主要分为用户信息部分及菜单列表，因此，核心的代码也分为这两部分。代码如下：

```
import {WeChatItemStyle, MyDivider} from '../model/CommonStyle'

@Component
export struct MyPage {
  private imageTitle: string = "title.png"

  build() {
    Column() {
      // 用户信息部分
      Image($rawfile(this.imageTitle)).height(144).width('100%')

      // 列表
      WeChatItemStyle({ imageSrc: "pay.png", text: "服务" })
      MyDivider()

      WeChatItemStyle({ imageSrc: "favorites.png", text: "收藏" })
      MyDivider({ style: '1' })
      WeChatItemStyle({ imageSrc: "moments2.png", text: "朋友圈" })
      MyDivider({ style: '1' })
      WeChatItemStyle({ imageSrc: "video.png", text: "视频号" })
      MyDivider({ style: '1' })
      WeChatItemStyle({ imageSrc: "card.png", text: "卡包" })
      MyDivider({ style: '1' })
      WeChatItemStyle({ imageSrc: "emoticon.png", text: "表情" })
      MyDivider()

      WeChatItemStyle({ imageSrc: "setting.png", text: "设置" })
      MyDivider()
    }.alignItems(HorizontalAlign.Start)
    .width('100%')
```

```
    .height('100%')
  }
}
```

与“发现”页面的子程序类似，“我”页面同样也是使用了 WeChatItemStyle、MyDivider。

19.6 实战：组装所有页面

在应用的 Index 页面，我们需要将“微信”“联系人”“发现”“我”四个页面组装在一起，并实现自由切换。此时，就可以使用 HarmonyOS 的 Tabs 组件作为导航栏。

19.6.1 将Tabs组件作为导航栏

将 Tabs 组件作为导航栏，代码实现如下：

```
import { ChatPage } from './ChatPage'
import { ContactPage } from './ContactPage'
import { DiscoveryPage } from './DiscoveryPage'
import { MyPage } from './MyPage'

@Entry
@Component
struct Index {
  @Provide currentPage: number = 0
  @State currentIndex: number = 0;

  build() {
    Column() {
      Tabs({
        index: this.currentIndex,
        barPosition: BarPosition.End
      }) {
        TabContent() {
          ChatPage()
        }
        .tabBar(this.TabBuilder(' 微信 ', 0, $r('app.media.wechat2'), $r('app.
                media.wechat1')))

        TabContent() {
          ContactPage()
        }
        .tabBar(this.TabBuilder(' 联系人 ', 1, $r('app.media.contacts2'),
```

```
$r('app.media.contacts1')))

        TabContent() {
          DiscoveryPage()
        }
        .tabBar(this.TabBuilder('发现', 2, $r('app.media.find2'), $r('app.
                media.find1')))

        TabContent() {
          MyPage()
        }
        .tabBar(
          this.TabBuilder('我', 3, $r('app.media.me2'), $r('app.media.me1'))
        )
      }
      .barMode(BarMode.Fixed)
      .onChange((index: number) => {
        this.currentIndex = index;
      })
    }
  }

  ...
```

对于底部导航栏，一般作为应用主页面功能区分，为了实现更好的用户体验，会组合文字及对应语义图标表示页签内容，在这种情况下，需要自定义导航页签的样式。代码如下：

```
@Builder TabBuilder(title: string, targetIndex: number, selectedImg: Resource,
normalImg: Resource) {
  Column() {
    Image(this.currentIndex === targetIndex ? selectedImg : normalImg)
      .size({ width: 25, height: 25 })
    Text(title)
      .fontColor(this.currentIndex === targetIndex ? '#1698CE' : '#6B6B6B')
  }
  .width('100%')
  .height(50)
  .justifyContent(FlexAlign.Center)
}
```

导航栏在选中时会呈现高亮的效果，如图 19-8 所示。

图 19-8　导航栏效果图

19.6.2 使用Swiper组件实现页面滑动

除了通过底部导航栏实现页面切换外，还可以使用Swiper组件来左右滑动页面，从而实现页面切换。代码如下：

```
@Component
struct HomeTopPage {
  @Consume currentPage: number

  build() {
    Swiper() {
      ChatPage()
      ContactPage()
      DiscoveryPage()
      MyPage()
    }
    .onChange((index: number) => {
      this.currentPage = index
    })
    .index(this.currentPage)
    .loop(false)
    .indicator(false)
    .width('100%')
    .height('100%')
  }
}
```

附录　本书第1版与第2版的差异对比

更新

更新全书代码示例和截图、上一版的勘误、部分章节的描述。

删除

1.4　获取开发支持

2.4　DevEco Studio功能介绍

2.5　DevEco Studio 常见问题小结

3.3　在真机中运行应用

4.5.2　app对象的内部结构

4.5.3　deviceConfig对象的内部结构

4.5.4　module对象的内部结构

5.1.6　Ability的配置

5.2　Ability的三层架构

5.4.4　设置PayAbilitySice样式布局

5.6.2　增加PayAbilitySice

5.6.3　新增PayAbilitySice样式布局

5.6.4　实现AbilitySice之间的路由和导航

第6章　Ability任务调度

第8章　剪切板

第10章　用JS开发UI

第11章　多模输入UI开发

第15章　相机

第16章　音频

第17章　媒体会话管理

第18章　媒体数据管理

19.6　生物特征识别认证概述

19.7　生物特征识别运作机制

19.8　生物特征识别约束与限制

19.9　生特征识别开发流程

第20章　二维码

第21章　通用文字识别

第22章　蓝牙

第23章　WLAN

第25章　电话服务

第28章　综合案例1：车机应用

新增

- 1.4　HarmonyOS 2新特性概述
- 1.5　HarmonyOS 3新特性概述
- 1.6　HarmonyOS 4新特性概述
- 1.7　Java与ArkTS如何抉择
- 3.4　使用ArkTS创建一个新应用
- 5.1.4　UIAbility
- 5.1.5　ExtensionAbility
- 5.2　Stage模型介绍
- 5.13　实战：Stage模型Ability内页面的跳转和数据传递
- 5.14　Want概述
- 5.15　实战：通过显式Want启动Ability
- 5.16　实战：通过隐式Want打开应用管理
- 6.6　Stage模型访问控制开发步骤
- 6.7　实战：访问控制授权
- 7.7　实战：Stage模型的订阅、发布、取消公共事件
- 第8章　用ArkUI开发UI
- 11.9　实战：Stage模型的关系型数据库开发
- 11.10　实战：Stage模型的首选项开发
- 13.6　Stage模型的视频开发

- 13.7　实战：实现Stage模型的视频播放器
- 14.5　实现Stage模型的图片开发
- 15.5　实战：在Stage模型中通过HTTP请求数据
- 15.6　Web组件概述
- 15.7　实战：在Stage模型中通过Web组件加载在线网页
- 第19章　综合案例4：ArkTS实现手机应用

参考文献

[1] HarmonyOS文档[EB/OL].https://developer.harmonyos.com/cn/docs/documentation/doc-guides/harmonyos-features-0000000000011907，2021-01-01/2021-04-24.

[2] 柳伟卫. Node.js企业级应用开发实战[M]. 北京：北京大学出版社，2020.

[3] 柳伟卫. 跟老卫学HarmonyOS开发[EB/OL].https://github.com/waylau/harmonyos-tutorial，2020-12-13/2021-04-24.

[4] 柳伟卫. 分布式系统常用技术及案例分析[M]. 北京：电子工业出版社，2017.

[5] 柳伟卫. Java核心编程[M]. 北京：清华大学出版社，2020.

[6] 柳伟卫. Netty原理解析与开发实战[M]. 北京：北京大学出版社，2020.

[7] 二维码[EB/OL].https://baike.baidu.com/item/二维码，2021-02-10/2021-02-10.

[8] 光学字符识别[EB/OL].https://baike.baidu.com/item/光学字符识别，2021-02-10/2021-02-10.

[9] 柳伟卫. Cloud Native分布式架构原理与实践[M]. 北京：北京大学出版社，2019.

[10] 无线局域网[EB/OL].https://baike.baidu.com/item/无线局域网，2021-02-10/2021-02-10.

[11] 鸿蒙生态应用开发白皮书V2.0[EB/OL].https://developer.huawei.com/consumer/cn/doc/harmonyos-bps，2023-12-10/2023-12-17.

[12] 柳伟卫. 鸿蒙HarmonyOS手机应用开发实战[M]. 北京：清华大学出版社，2022.

[13] 柳伟卫. HarmonyOS题库[EB/OL].https://github.com/waylau/harmonyos-exam，2023-12-10/2023-12-17.

[14] 柳伟卫. 鸿蒙系统实战短视频App从0到1掌握HarmonyOS[EB/OL].https://coding.imooc.com/class/674.html，2023-12-10/2023-12-17.

[15] 柳伟卫. 鸿蒙HarmonyOS应用开发入门[M]. 北京：清华大学出版社，2024.

[16] 柳伟卫. 2024鸿蒙零基础快速实战-仿抖音App开发（ArkTS版）[EB/OL].https://coding.imooc.com/class/674.html，2024-04-29/2024-05-05.